Strategic Management and Leadership for Systems Development in Virtual Spaces

Christian Graham
University of Maine, USA

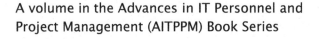

A volume in the Advances in IT Personnel and
Project Management (AITPPM) Book Series

Published in the United States of America by
Business Science Reference (an imprint of IGI Global)
701 E. Chocolate Avenue
Hershey PA, USA 17033
Tel: 717-533-8845
Fax: 717-533-8661
E-mail: cust@igi-global.com
Web site: http://www.igi-global.com

Library of Congress Cataloging-in-Publication Data

Names: Graham, Christian, 1972- editor.
Title: Strategic management and leadership for systems development in virtual
  spaces / Christian Graham, editor.
Description: Hershey : Business Science Reference, 2016. | Includes
  bibliographical references and index.
Identifiers: LCCN 2015037480| ISBN 9781466696884 (hardcover) | ISBN
  9781466696891 (ebook)
Subjects: LCSH: Strategic planning. | Leadership. | Telecommuting--Management.
Classification: LCC HD30.28 .S733427 2016 | DDC 658.4/092--dc23 LC record available at http://lccn.loc.gov/2015037480

This book is published in the IGI Global book series Advances in IT Personnel and Project Management (AITPPM) (ISSN: 2331-768X; eISSN: 2331-7698)

British Cataloguing in Publication Data
A Cataloguing in Publication record for this book is available from the British Library.

All work contributed to this book is new, previously-unpublished material. The views expressed in this book are those of the authors, but not necessarily of the publisher.

For electronic access to this publication, please contact: eresources@igi-global.com.

Advances in IT Personnel and Project Management (AITPPM) Book Series

Sanjay Misra
Covenant University, OTA, Nigeria
Ricardo Colomo-Palacios
Østfold University College, Norway

ISSN: 2331-768X
EISSN: 2331-7698

MISSION

Technology has become an integral part of organizations in every sector, contributing to the way in which large enterprises, small businesses, government agencies, and non-profit organizations operate. In the midst of this revolution, it is essential that these organizations have a thorough knowledge of how to implement and manage IT projects as well as an understanding of how to attract and supervise the employees associated with these projects.

The **Advances in IT Personnel and Project Management (AITPPM)** book series aims to provide current research on all facets of IT Project Management including factors to consider when managing and working with IT personnel. Books within the AITPPM book series will provide managers, IT professionals, business leaders, and upper-level students with the latest trends, applications, methodologies, and literature available in this field.

COVERAGE

- Cost-Effective Methods for Project Management
- Agile Project Management
- IT Strategy
- Requirements Management
- Project Sponsorship
- Project Planning
- Measuring Project Success
- IT Career Development
- Communication between Managers and IT Personnel
- IT Personnel Management

IGI Global is currently accepting manuscripts for publication within this series. To submit a proposal for a volume in this series, please contact our Acquisition Editors at Acquisitions@igi-global.com or visit: http://www.igi-global.com/publish/.

Titles in this Series

For a list of additional titles in this series, please visit: www.igi-global.com

Modern Techniques for Successful IT Project Management
Shang Gao (Zhongnan University of Economics and Law, China) and Lazar Rusu (Stockholm University, Sweden)
Business Science Reference • copyright 2015 • 374pp • H/C (ISBN: 9781466674738) • US $225.00 (our price)

Handbook of Research on Effective Project Management through the Integration of Knowledge and Innovation
George Leal Jamil (Informações em Rede, Brazil) Sérgio Maravilhas Lopes (CETAC.MEDIA - Porto and Aveiro Universities, Portugal) Armando Malheiro da Silva (Porto University, Portugal) and Fernanda Ribeiro (Porto University, Portugal)
Business Science Reference • copyright 2015 • 602pp • H/C (ISBN: 9781466675360) • US $285.00 (our price)

Handbook of Research on Technology Project Management, Planning, and Operations
Terry T. Kidd (Texas A&M University, USA)
Information Science Reference • copyright 2009 • 610pp • H/C (ISBN: 9781605664002) • US $265.00 (our price)

Managing Very Large IT Projects in Businesses and Organizations
Matthew Guah (Erasmus University Rotterdam, The Netherlands)
Information Science Reference • copyright 2009 • 358pp • H/C (ISBN: 9781599045467) • US $165.00 (our price)

www.igi-global.com

701 E. Chocolate Ave., Hershey, PA 17033
Order online at www.igi-global.com or call 717-533-8845 x100
To place a standing order for titles released in this series, contact: cust@igi-global.com
Mon-Fri 8:00 am - 5:00 pm (est) or fax 24 hours a day 717-533-8661

Table of Contents

Section 3
Innovative Solutions and Games for Crises, Virtual Leading, and Effective Remote Teams

Detailed Table of Contents

Section 1
Leadership, Virtual Teams, Knowledge

Chapter 1

    Kijpokin Kasemsap, Suan Sunandha Rajabhat University, Thailand

This chapter aims to examine the roles of virtual team and information technology (IT) in global business, thus describing the theoretical and practical overviews of virtual team and IT; the importance of virtual team in global business; and the importance of IT in global business. The applications of virtual team and IT are necessary for modern organizations that seek to serve suppliers and customers, increase business performance, strengthen competitiveness, and achieve continuous success in global business. Therefore, it is essential for modern organizations to examine their virtual team and IT applications, develop a strategic plan to regularly check their practical advancements, and immediately respond to virtual team and IT needs of customers in modern organizations. The chapter argues that applying virtual team and IT has the potential to increase organizational performance and reach strategic goals in global business.

Chapter 2

    Julia Eisenberg, Pace University, USA
    Jennifer Gibbs, Rutgers University, USA
    Niclas Erhardt, University of Maine, USA

This chapter reviews current trends in the literature related to the influence of vertical and shared leadership styles in the context of virtual teams, unpacking the influence of team structure and task structure to better understand the mechanisms influencing team effectiveness. The authors start by reviewing key features of virtual teams and different aspects of leadership and its influence in the virtual team environment. They argue that both vertical and shared leadership have strengths and limitations, and both styles may complement one another. The authors discuss the influence of leadership on virtual team processes and outcomes and examine contingency factors related to team and task structure in order to identify the boundary conditions for the effectiveness of vertical and shared leadership. The chapter offers a conceptual framework to guide future research in this domain.

With globalization of economy and increase of global competition to acquire rare resources, the organizations have moved towards geographical distribution to achieve competitive advantage. Users and project teams at various places within various countries with different national and local cultures throughout the world work on projects at the environment concerning geographical distribution. On the other hand, with increase of advancements in communication, the distributed project teams have been witnessed with more expansion, known with "virtual teams". When members of virtual project team from various organizations and time zones attend in the projects, they will be more likely affiliated to electronic media such as email. With regard to virtualization of IT projects, the present study aims to develop a model for virtual project management with an emphasis on information technology projects, including several elements in geographically distributed environments. The final model of virtual project management of information technology projects was represented.

This chapter offers a working definition of the concepts of virtual, management, leadership, and team, and proposes pragmatic tools and solutions to management and leadership challenges in virtual, distributed team situations. Practical experiences are surveyed, including scenarios of remote team, remote team member, distributed learning, and traveling manager. Descriptions of tools and techniques are offered, along with a set of guiding concepts and principles to apply to any virtual leadership situation.

Advances in Information and Communication Technologies (ICT) are creating new opportunities for organizations to build and manage virtual teams. Such teams are composed of employees with unique skills, located a distance from each other, who must collaborate to accomplish important organizational tasks. As such, it is very important for organizations to identify and develop skills that critical for virtual teams to succeed. Participation in and management of virtual teams comes with its own unique challenges and opportunities. This chapter explores virtual teams, their benefits and challenges to organizations, and provide ways to ensure that virtual team members and leaders in their organizations have the skills, competencies and tools needed to succeed. Specific recommendations to improve skills of virtual teams are also provided.

Chapter 6

Organizations are heavily investing in virtual teams to enhance their performance and competitiveness. These types of teams are made possible by advances in computer-mediated communication and software that allows people to work collaboratively on projects without being co-located or even working at the same time. Managing teams and collaborating online presents unique challenges. Maintaining a productive virtual team requires more than just the willingness of global participants, but even more so the tools to conduct and manage virtual projects. It is therefore important to incorporate online collaboration skills into the IT curriculum at the university level. This chapter provides a general overview of virtual teams; today's collaborative tools, and discuss expertise necessary for virtual teams to be successful.

Chapter 7

Tacit knowledge transfer and knowledge creation represents perhaps the best means of sustainable competitive advantage through continual innovation. As organizations become more distributed in their different offices, virtual teams become more common and valuable. The question of how these virtual teams can effectively transfer tacit knowledge and create new knowledge thus becomes of importance to organizations. This chapter focuses on this issue and presents supporting evidence related to tacit knowledge transfer and creation, virtual teams, and how businesses can effectively harness capacity of virtual teams to transfer valuable tacit knowledge and create new knowledge.

Chapter 8

Electronic collaboration was born with the new technologies, which establish a more harmonious balance of organizations in an increasingly global, open and competitive digital economy, called, nowadays, "the economy of the crowds". This economy has caused changes in the organizations of the century, as new administrative principles. In this context, organizations use new business models to achieve its objectives to a meager cost. Similarly, they have managed the integration of different levels and optimizing performance of the entire organization together through electronic media and online collaboration. This work shows the areas of the different levels and forms of organizational electronic collaboration.

Section 2
Information System: Design, Methodologies, and Strategic Thinking

Chapter 9

As more projects require the specialized technical skills of those who work in virtual environments due to dispersed geographic locations, project managers of these distributed virtual teams (DVT) must gain insight into achieving project success amongst team members who hold varying operational and world perspectives. When organizational managers decide to implement virtual teams (VT), can they develop strategies to overcome the lack of social interaction, cultural differences, and preconceived notions that can hinder the development of a collaborative and cohesive team? In addition, leading DVTs in a manner that encourages collaboration, diversity, competency building, open communication, and overcoming feelings of isolation must be met in this technology-based environment. This chapter addresses the dilemma of managers in which they must have a clear understanding of what communication and relationship-building techniques and management systems are best suited.

The business environment today is transforming towards a collaborative context compounded by multi-organizational cooperation and related information system infrastructures. This chapter aims to examine Enterprise Resource Planning (ERP) systems development and emerging practices in the management of multi-organizational enterprises and identify the circumstances under which the so-called 'ERPIII' systems fit into the Virtual Enterprise paradigm; and vice versa. An empirical inductive study was conducted using case studies from successful companies in the UK and China. Data were collected through 48 semi-structured interviews and analyzed using the Grounded-Theory based Methodology (GTM) to derive a set of 29 tentative propositions which were then validated via a questionnaire survey to further propose a novel conceptual framework referred to as the 'Dynamic Enterprise Reference Grid for ERP (DERG-ERP)'; which can be used for innovative decision-making about how ERP information systems and multi-organizational enterprises – particularly the Virtual Enterprise may be co-developed.

Humans have a rich awareness of locations and situations that directs how we interpret and interact with our surroundings. The principle aim of this paper is to create 'Information Spaces' where people will use their awareness to search, browse and learn. In the same way that they navigate in a physical environment, they will navigate through knowledge. An information space is a type of design in which representations of information objects are situated in a principled space. In this chapter we present an architecture based on the principles of electrostatistics, which presents a model for design of information spaces. Our model gives an easy conceptual framework to reason about how information can be represented as well as secure ways of extracting and storing information leading to a design which are easily scalable in virtual team environments.

Failure diagnosis in large and complex information systems (LCIS) is a critical task due to respect the safe development of these systems. A discrete event system (DES) approach to the problem of failure diagnosis of LCIS is presented in this chapter. A classic solution to solve DES's diagnosis is a stochastic Petri nets. Unfortunately, the solution of a stochastic Petri net is severely restricted by the size of its underlying Markov chain. On the other hand, it has been shown that foraging behavior of ant colonies can give rise to the shortest path, which will reduce the state explosion of stochastic Petri net. Therefore, a new model of stochastic Petri net, based on foraging behavior of real ant colonies is introduced in this paper. This model can contribute to the diagnosis, the performance analysis and design of information systems.

Chapter 13

Nitasha Hasteer, Amity University, India
Abhay Bansal, Amity University, India
B. K. Murthy, Centre for Development of Advanced Computing, India

Production of quality software requires selecting the right development strategy. The process and development strategies for creating software have evolved over the years to cope with the changing paradigms. Cloud computing models have made provisioning of the computing capabilities and access to configurable pooled resources as convenient as having access to the common utilities. With the recent advancements in the use of social media and advent of software development through crowdsourcing, the need to comprehend and analyze the traditional process models of software development, with regard to the changed paradigm have become ever more necessary. The changes in the way software are being created and the continuous evolution in the processes of development and deployment has created a need to understand the development process models. This chapter provides an insight on the transition from the conventional process models of software development to the software development methodology being used to develop software through crowdsourcing.

Chapter 14

Frank Stowell, University of Portsmouth, UK
Shavindrie Cooray, Curry College, USA

Recent research adds support to the view that the way that individuals act as part of a virtual group is different from behavior in face-to-face meetings. Specifically researchers have discovered that conflicts are more prevalent within virtual teams as opposed to face to face teams. This is because research has shown that participants are more likely to change their initial points of view (shaped by personal values, biases and experience) when discussions are held in a face to face environment rather than through virtual means. This insight raises doubts upon the effectiveness of CMCs as an instrument of organizational cohesion. In this paper we reflect upon this position and attempt to discover if these concerns can be overcome through the employment of Systems methods used in organizational inquiry. We do this through an evaluation of the results of a preliminary study between Curry College in Boston, Massachusetts, USA and Richmond University in London, UK.

Section 3
Innovative Solutions and Games for Crises, Virtual Leading, and Effective Remote Teams

Chapter 15

    Hemant Purohit, George Mason University, USA
    Mamta Dalal, InCrisisRelief.org, India
    Parminder Singh, Twitter, USA
    Bhavana Nissima, InCrisisRelief.org, India
    Vijaya Moorthy, InCrisisRelief.org, India
    Arun Vemuri, Google, USA
    Vidya Krishnan, InCrisisRelief.org, India
    Raheel Khursheed, Twitter, USA
    Surendran Balachandran, InCrisisRelief.org, India
    Harsh Kushwah, InCrisisRelief.org, India
    Aashish Rajgaria, InCrisisRelief.org, India

Crisis times are characterized by a dynamically changing and evolving need set that should be evaluated and acted upon with the least amount of latency. Though the established practice of response to rescue and relief operations is largely institutionalized in norms and localized; there is a vast sea of surging goodwill and voluntary involvement that is available globally to be tapped into and channelized for maximum benefit in the initial hours and days of the crisis. This is made possible with the availability of real-time, collaborative communication platforms such as those facilitated by Facebook, Google and Twitter. They enable building and harnessing real-time communities as an amorphous force multiplier to collate, structure, disseminate, follow-through, and close the loop between on-ground and off-ground coordination on information, which aids both rescue as well relief operations of ground response organizations. At times of emergencies, amorphous online communities of citizens come into existence on their own, sharing a variety of skill sets to assist response, and contribute immensely to relief efforts during earthquakes, epidemics, floods, snow-storms and typhoons. Since the Haiti earthquake in 2010 to the most recent Ebola epidemic, online citizen communities have participated enthusiastically in the relief and rehabilitation process. This chapter draws from real world experience, as authors joined forces to set up JKFloodRelief.org initiative, to help the government machinery during floods in the state of Jammu & Kashmir (JK) in India in September 2014. The authors discuss the structure and nature of shared leadership in virtual teams, and benefits of channelizing global goodwill into a purposeful, and sustained effort to tide over the initial hours when continued flow of reliable information will help in designing a better response to the crisis. The authors discuss the lessons learned into 5 actionable dimensions: first, setting up response-led citizen communities with distributed leadership structure, in coordination with the on-ground teams. Second, communicating clearly and consistently about sourcing, structuring, and disseminating information for both internal team challenges, solutions, and plans with shared goal-preserving policies, as well as external public awareness. Third, developing partner ecosystem, where identifying, opening communication lines, and involving key stakeholders in community ecosystem - corporates, nonprofits, and government provide a thrust for large-scale timely response. Fourth, complementing and catalyzing offline efforts by providing a public outlet for accountability of the efforts, which recognizes actions in both off-ground and on-ground environments for volunteers, key stakeholders and citizens. Lastly, the fifth dimension is about follow-up & closure, with regrouping for assessing role, next steps, and proper acknowledgement of various stakeholders for a sustainable partnership model, in addition to communicating outcome of the efforts transparently with every stakeholder including citizen donors to ensure accountability. With the extensive description of each of these dimensions via narrative of experiences from the JKFloodRelief.org initiative, the authors aim to provide a structure of lessons learned that can help replicate such collaborative initiatives of citizens and organizations during crises across the world.

Chapter 16

The concept of entrepreneurial leadership in any challenging industry involves fusing the concepts of a strategic approach of the management change, enhancing capabilities for continuously, creating and appropriating support, and development of value and competitive advantages in the company and technological growth. The challenge for shipping and transportation industry is to build a compatibility to continuously explore and reduce the threats of secure and safety and enhance new successful and competitive opportunities. Shipping Entrepreneurial Leadership style that scanned The twenty one attributes of outstanding leadership dimensions in the GLOBE project scales to identify nineteen attributes of their extraordinary performance, The chapter categorizes ten shipping behavior likely to be expected in loading on the two roles that impede the shipping entrepreneurial leaders' performance enactment scenario and Eight leadership behavior likely to be facilitated in their roles towered encouraging the team performance.

Chapter 17

In this chapter we examine how virtual team trust and effectiveness may be improved through the transformative power of serious games and creative process. To start we explore the pervasive lack of emotional intelligence within the workplace at an individual level and which we call 'the EQ Gap'. This is followed by an examination of challenges faced by both traditional and virtual teams. We then consider how the same EQ Gap also manifests in both traditional and virtual teams as well. Indeed, it's worse for the latter. This leads to a review of the kinds of EQ training needed for both team types. A discussion then follows as to how serious games, play, and creativity can help virtual teams in particular to become more emotionally intelligent, trusting, and ultimately more collaborative. A brief case study of a serious game called Prelude is shared to illustrate these findings in a practical context.

Preface

MANAGEMENT AND LEADERSHIP OF VIRTUAL TEAMS

In thinking about why I wanted to put together this book I realized I had an opportunity as a business and information systems educator to "marry" the two fields of study together and acknowledge the fact that with the advent of computing and telecommunication technologies the way we as humans work with one another has dramatically changed in the last 30 years. Today, rather than punching in on a time clock, many of us are working from our homes, our cars, even the beach now because of major advances computing and telecommunication technologies. This ability to work from anywhere at any time allowed a new type of worker to emerge: *the telecommuter*. According to Neufeld and fang (2005) telecommuters are those individuals that work outside of the conventional workplace. Telecommuters still had to work with other telecommuters to get work done and hence a new type of organizational team emerged next: *The virtual team*. Since the emergence of virtual teams (VTs), their use in organizations has become a very common practice. According to Zivick (2012) organizations are using VTs to accomplish a variety of business goals. Kayworth and Leidner (2000) reported that organizations derive many benefits from the use of VTs. Some of these benefits include: cost reductions, cycle-time reductions, integration of distant members, and improved decision-making and problem-solving skills. Virtual teams over the last several years has had many closely related definitions. According to Brandt, England, and Ward (2011) VTs are made up of individuals working together who may not have ever met and very often do not meet face-to-face during assigned projects. Green and Roberts (2010) add that VTs are individuals separated by time and space, have little face-to-face contact, and are completely dependent upon computers and telecommunication technologies to communicate with one another. Martins, Gilson, and Maynard (2004) defined VTs as "teams whose members use technology to varying degrees in working across locational, temporal, and relational boundaries to accomplish an interdependent task" (p. 808). It should be noted that as computing has become more portable, computer and telecommunication technologies are no longer limited to desktop computers with an Internet or network connection. According to Kock and Nosek (2005) VT members can and are using mobile devices such as smart phones and Internet enabled tablets with texting capability, email, and social networks to accomplish organizational tasks and assigned projects as well.

A very important contribution acknowledged in the research literature related to virtual teams is their adaptability. According to Brandt, et al (2011) organizations have reported that VTs must have the ability to be assembled quickly and be adaptable enough to meet each projects goals. Zivick (2012) reported that VTs are brought together to take full advantage of all the organizations human resources irrespective of their physical location. This means that organizations can quickly assemble staff possessing the necessary skills and knowledge to complete a specific task or serve an on-going project at any time from any geographic location.

Despite the increased use of VTs in organizations, VTs have not emerged as the definitive solution to all problems in accomplishing organizational tasks. Blackburn, Furst, and Rosen (2003) acknowledged that there is "no guarantee that VTs will reach their full potential and that as many VTs fail as they do succeed in assigned tasks and projects. VTs experience many of the same problems face-to-face teams must contend with. According to Marks, Sabella, Burke, and Zaccaro (2002) a few of these problems include: poor team member composition, incomplete knowledge of project goals, and poor coordination processes. Ongoing research in VTs has found that one problem unique to VTs is the lack of interpersonal skills and human relations. Gonzales, Nardi, and Mark (2009) validated this when they found that while collaborative technology had improved over the years with better human-computer-interaction (HCI) design, approaches on how to "work better" within collaborative environments had still not been sufficiently addressed.

Several other factors contribute to the success or failure of VTs. Brandt et al (2011) stated that VTs have many dynamics contributing to the success or failure of virtual team projects. Some of these are: *Trust, Cultural differences, Communication, Social skills, Mission and goal clarity, Rewards and recognition, Time.*

It's probably obvious that VT members should trust each other if they plan on working well with one another however, *how is trust established* when perhaps team members have never met one another? What role does *cultural differences* play in work strategies, work ethics, and even social etiquette when working in VTs? How do VTs *communicate* within the virtual space? Do all team members communicate at a designated synchronous time or do the team members communicate asynchronously, contributing to discussions at their own time? How are *social skills* conveyed or not conveyed via email, collaborative software, or video-conferencing? Can project managers convey *rewards, recognition, mission, and goal clarity* effectively and efficiently to subordinates in virtual spaces? How do VTs manage *time* when team member can literally be all over the world spread out across several different time zones?

SYSTEMS DEVELOPMENT

While virtual teams can be engaged in any number of types of business projects, the focus of this book was on systems development projects in virtual spaces. What is systems development? In 1990, information systems researchers Nunamaker and Chen described systems development as five stages:

1. **Concept Design:** Which is the adaption and union of technology and theoretic advances into possible practical applications.
2. **Constructing the Architecture of the System:** Meaning what hardware, software, and processes must be in place for the system to work.
3. **Prototyping:** A preliminary model of the system to be built.
4. The actual system being developed (this would be the product development stage). And finally,
5. The system built is transferred into operation within the business or organization that needed the information system (Nunmaker Jr., J.f. and Chen, M., 1990).

More recently, systems development has been defined in the context of information systems development. According to Doolin and McLeod (2012) described information systems development as a "complex organizational activity involving multiple stakeholders who interact with various artifacts in order to facilitate understanding and cooperation across diverse knowledge domains". Ismail and King (2014) defined information systems development again as a five stage process: *1) Definition of needs (information requirements), 2) Selection of hardware and software, 3) Implementation of major systems, 4) System maintenance and problem-solving, and 5) Planning of future IT deployment.* Regardless of the definition of systems development / information systems development many information systems researchers report that systems development is wrought with problems. Information systems development is still a difficult process that often times ends in failure. This is not to say that information systems projects don't get completed, but rather that they *"have failed"* in the sense that these projects often go over the time originally allotted to be completed, they often go over budget, and they often do not do everything that the original planned system requirements had intended the system to do. The many difficulties associated with systems development is now compounded by the use of VTs that often do not have the benefit of physical presence and or centralized leadership.

SYSTEMS DEVELOPMENT AND VIRTUAL SPACES: THE WORLD TODAY

To find evidence that systems development is a difficult process that often ends in failure, one only has to look at the initial roll out of the United States HealthCare.gov website. The website was the result of legislation that was put into place in 2010 called the Affordable Care Act which sought to provide affordable access to healthcare for citizens of the United States through a series of health care insurance exchanges. HealthCare.gov was the primary destination for businesses and citizens to go to review these newly created exchanges and sign up for affordable health care insurance. The HealthCare.gov website was definitely not ready for the amount of traffic that hit it in the first few days! According to Lahm (2013) users of the website experienced the site "freezing and crashing". What was initially described as a few minor glitches in the system quickly turned to being described as, "a website with severe problems and severe security flaws." Lahm went on to report that as of November 1st, 2013, while the website had improved, at that time it was only working at about 80% functionality. Even the Whitehouse administration reluctantly admitted in a report titled: "Health insurance exchanges: An update from the administration," (2013) that the rollout of the HealthCare.gov website was largely a debacle!

What went wrong with the rollout of the website? A lot! According to Venkatesh, Hoehle, and Aljafari (2014) the website had serious problems with its data storage capabilities, telecommunications, interoperability with other systems, and user interface issues. Additionally, according to the website BALLOTPEDIA, a private consulting firm, McKinsey & Co. reported two other important aspects of the websites failure: *1) Enough time for testing and revising the website was not provided, and 2) the project to build the website did not have one single leader and decision-maker.* Despite decades of telecommuting, virtual teams, and advancements in computing and telecommunication technology combined with the Whitehouse's and Silicon Valley's best and brightest, this system development project failed epically! (Healthcare.gov website rollout, 2013).

This brings us to the purpose of this book titled: *Strategic Management and Leadership for Systems Development within Virtual Spaces.* This book addresses from multiple perspectives virtual teams, virtual environments, virtual leadership, project management, and novel new methodologies for systems development in virtual spaces. The goal of this book for me was to bring together researchers, scholars, practitioners, and managers to confront and address problems, share best practices, and report novel new strategies and ongoing research that seeks to improve systems development within virtual spaces. My hope is that readers of this book will find useful tools and strategies to work better within virtual spaces, lead better within virtual spaces, and perhaps even build on the research found in this book to develop new theories on strategic management and leadership for systems development in virtual spaces.

BOOK CHAPTERS OVERVIEW

The book, "Strategic Management and Leadership for Systems Development in Virtual Spaces" is organized into three sections. The first section discusses the roles of virtual teams and information technology in global business, reviews virtual leadership and virtual team collaboration, globalized project management, and knowledge creation, sharing, and transfer. The second section of the book discusses the design of global enterprise systems, conceptual framework for innovative strategic thinking, the design of information spaces and retrieval of information using electrostatics in virtual spaces, failure diagnosis and a new formalism for safe development of information systems, and crowdsourced software development. The third section concludes the book by sharing applications of ideas, tools, and games used to improve systems development, leadership, and collaboration within virtual spaces.

Chapter 1, "Examining the Roles of Virtual Team and Information Technology in Global Business," examines the roles of virtual team and information technology (IT) in global business describing the theoretical and practical overview of the virtual team, the relationships among the virtual team, conflict management, emotion management, the applications of IT, the importance of virtual team in global business, and the importance of IT in global business.

Chapter 2, "The Role of Vertical and Shared Leadership in Virtual Team Collaboration," reviews current trends in the literature related to the influence of vertical and shared leadership styles in the context of virtual teams, unpacking the influence of team structure and task structure to better understand the mechanisms influencing virtual team effectiveness.

Chapter 3, "Providing a Model for Virtual Project Management with an Emphasis on IT Projects," presents a study that aimed to develop a model for virtual project management with an emphasis on information technology projects, including several elements in geographically distributed environments. The final model of virtual project management of information technology projects is presented.

Chapter 4, "The Virtual Leader: Developing Skills to Lead and Manage Distributed Teams," offers a working definition of the concepts of virtual, management, leadership, team, and proposes pragmatic tools and solutions to management and leadership challenges in virtual, distributed team situations.

Chapter 5, "Skill Building for Virtual Teams," explores virtual teams, their benefits and challenges to organizations, and provide ways to ensure that virtual team members and leaders in their organizations have the skills, competencies and tools needed to succeed. Specific recommendations to improve skills of virtual teams are also provided.

Chapter 6, "E-Tools for E-Team: The Importance of Social Ties and Knowledge Sharing," provides a general overview of virtual teams; today's collaborative tools, and discuss expertise necessary for virtual teams to be successful.

Chapter 7, "Knowledge Transfer and Knowledge Creation in Virtual Teams," focuses on this issue and presents supporting evidence related to tacit knowledge transfer and creation, virtual teams, and how businesses can effectively harness capacity of virtual teams to transfer valuable tacit knowledge and create new knowledge.

Chapter 8, "Electronic Collaboration in Organizations," provides a literature review of the term that has been coined "the economy of crowds" and new administrative principles and business models now needed to achieve organizational goals.

Chapter 9, "Developing Project Team Cohesiveness in a Virtual Environment," addresses the dilemma of managers who must have a clear understanding of what communication and relationship-building techniques and management systems are best suited for distributed virtual teams. The results of the research presented in this chapter aims to limit many of the social and human relations problems associated with virtual teams.

Chapter 10, "Designing and Managing ERP Systems for Virtual Enterprise Strategy: A Conceptual Framework for Innovative Strategic Thinking," aims to examine Enterprise Resource Planning (ERP) systems development and emerging practices in the management of multi-organizational enterprises and identify the circumstances under which the so-called 'ERPIII' systems fit into the Virtual Enterprise paradigm; and vice versa. An empirical inductive study was conducted using case studies from successful companies in the UK and China.

Chapter 11, "Design of Information Spaces and Retrieval of Information Using Electrostatics in Virtual Spaces," creates 'Information Spaces' where people will use their awareness to search, browse and learn. In this chapter the authors present an architecture based on the principles of electrostatistics, which presents a model for design of information spaces. Our model gives an easy conceptual framework to reason about how information can be represented as well as secure ways of extracting and storing information leading to a design which is easily scalable in virtual team environments.

Chapter 12, "A New Formalism for Diagnosis and Safe Development of Information Systems," reports that failure diagnosis in large and complex information systems (LCIS) is a critical task necessary to respect the safe development of these systems. A discrete event system (DES) approach to the problem of failure diagnosis of LCIS is presented in this chapter. The author takes a novel approach to improve failure diagnosis using a modified stochastic Petri net approach that analyzes the foraging behavior of ant colonies. With this new novel approach, the author presents a new model of stochastic Petri net that can contribute to the better diagnosis, performance analysis, and design of information systems.

Chapter 13, "Software Process Paradigms and Crowdsourced Software Development: An Overview," provides an insight on the transition from the conventional process models of software development to the software development methodology being used to develop software through crowdsourcing.

Chapter 14, "Using Soft Systems Ideas within Virtual Teams," reviews recent research that adds force to the view that the way that individuals act as part of a virtual group is different from behavior in face-to-face meetings. Specifically researchers have discovered that conflicts are more prevalent within virtual teams as opposed to face to face teams. In this paper the authors reflects upon this position and attempt to discover if these concerns can be overcome through the employment of Systems methods used in organizational inquiry.

Chapter 15, "Empowering Crisis Response-Led Citizen Communities: Lessons Learned from JKFloodRelief.org Initiative," shares the experiences of researchers and engineers who developed an information system that uses the goodwill and online volunteerism that is an available resource ready to be tapped when responding to crises. This chapter discusses the structure and nature of shared leadership in virtual teams and the benefits of channelizing global goodwill into a purposeful and sustained effort to tide over the initial hours when continued flow of reliable information will help in designing a better response to the crisis.

Chapter 16, "Virtual Shipping: Entrepreneurial Leadership Styles in Maritime and Shipping Industry," discusses the concept of entrepreneurial leadership in any challenging industry involves fusing the concepts of a strategic approach of the management change, enhancing capabilities for continuously, creating and appropriating support, and development of value and competitive advantages in the company with technological growth. The challenge for shipping and transportation industry is to build a compatibility to continuously explore and reduce the threats of secure and safety and enhance new successful and competitive opportunities. The chapter categorizes ten shipping behaviors likely to be expected in loading on the two roles that impede the shipping entrepreneurial leaders' performance enactment scenario and eight leadership behaviors likely to be facilitated in their roles towered encouraging the team performance.

Chapter 17, "Virtual Strangers No More: Serious Games and Creativity for Effective Remote Teams," examines how virtual team trust and effectiveness may be improved through the transformative power of serious games and creative process.

CONCLUSION

With the rapid proliferation of virtual teams stakeholders from every industry, from business people and project managers to scholars and academics, are confronting the difficult problems that come from managing the systems development process within virtual spaces. In this context, this book is intended to fill existing gaps in the literature from many different perspectives related to systems development with virtual spaces. The target audience for this book is composed of professionals, researchers, and scholars working in the fields of information systems, information systems development, information science, information management, virtual teams, team effectiveness, leadership, strategic management, sociology, e-collaboration, and information technology. Moreover, this book should provide insights and support executives concerned with the management of virtual teams, leadership and e-collaboration. My hope is that this book will prove to be a valuable resource and comprehensive guide on how to improve strategic management and leadership for systems development within virtual spaces.

REFERENCES

Blackburn, R., Furst, S., & Rosen, B. (2003). Building a winning virtual team. In *Virtual teams that work* (pp. 95–120). Academic Press.

Brandt, V., England, W., & Ward, S. (2011). Virtual teams. *Research Technology Management*, *54*(6), 62–63.

Doolin, B., & McLeod, L. (2012). Sociomateriality and boundary objects in information systems development. *European Journal of Information Systems*, *21*(5), 570–586.

Gonzales, V. M. G., Nardi, B., & Mark, G. (2009). Ensembles: Understanding the instantiation of activities. *Information Technology & People*, *22*(2), 109–131.

Green, D., & Roberts, G. (2010). Personnel implications of public sector virtual organizations. *Public Personnel Management*, *39*(1), 47–57.

Health Insurance Exchanges: An Update from the Administration. (2013, November 6). United States Senate Committee on Finance. Retrieved November 9, 2013, from http://www.finance.senate.gov/hearings/hearing/?id=3dd91089-5056-a032-5290-f158359b9247

Healthcare.gov Website Rollout. (2014, March 31). In *Ballotpedia*. Retrieved August 8, 2015, from http://ballotpedia.org/Healthcare.gov_website_rollout

Ismail, N. A., & King, M. (2014). Factors influencing the alignment of accounting information systems in small and medium sized Malaysian manufacturing firms. *Journal of Information Systems and Small Business*, *1*(1-2), 1–20.

Kayworth, T., & Leidner, D. (2000). The global virtual manager: A prescription for success. *European Management Journal*, *18*(2), 183–194.

Kock, N., & Nosek, J. (2005). Expanding the boundaries of e-collaboration. *IEEE Transactions on Professional Communication*, *48*(1), 1–9.

Lahm, R. J. Jr. (2013). Obamacare and small business: Delays and "glitches" exacerbate uncertainty and economic consequences. *Journal of Management and Marketing Research*, *16*, 1–16.

Marks, M. A., Sabella, M. J., Burke, C. S., & Zaccaro, S. J. (2002). The impact of cross-training on team effectiveness. *The Journal of Applied Psychology*, *87*(1), 3.

Martins, L. L., Gilson, L. L., & Maynard, M. T. (2004). Virtual teams: What do we know and where do we go from here? *Journal of Management*, *30*(6), 805–835.

Neufeld, D. J., & Fang, Y. (2005). Individual, social and situational determinants of telecommuter productivity. *Information & Management*, *42*(7), 1037–1049.

Nunamaker, J. F., Jr., & Chen, M. (1990, January). Systems development in information systems research. In System Sciences, 1990. In *Proceedings of the Twenty-Third Annual Hawaii International Conference* (Vol. 3, pp. 631-640). IEEE.

Venkatesh, V., Hoehle, H., & Aljafari, R. (2014). A usability evaluation of the Obamacare website. *Government Information Quarterly*, *31*(4), 669–680.

Zivick, J. (2012). Mapping global virtual team leadership actions to organizational roles. *Business Review (Federal Reserve Bank of Philadelphia)*, *19*(2), 18–25.

Acknowledgment

I'd like to express my sincere gratitude to the contributors of this book. Your research and findings are what made this book possible. I'd like to also thank the Editorial Advisory Board and Reviewers for their support and insightful comments on the chapters presented in this book. Finally, thank you to everyone at IGI Global who provided me with the opportunity to write this book.

Christian Graham
University of Maine, USA

Section 1
Leadership, Virtual Teams, Knowledge

Chapter 1
Examining the Roles of Virtual Team and Information Technology in Global Business

Kijpokin Kasemsap
Suan Sunandha Rajabhat University, Thailand

ABSTRACT

This chapter aims to examine the roles of virtual team and information technology (IT) in global business, thus describing the theoretical and practical overviews of virtual team and IT; the importance of virtual team in global business; and the importance of IT in global business. The applications of virtual team and IT are necessary for modern organizations that seek to serve suppliers and customers, increase business performance, strengthen competitiveness, and achieve continuous success in global business. Therefore, it is essential for modern organizations to examine their virtual team and IT applications, develop a strategic plan to regularly check their practical advancements, and immediately respond to virtual team and IT needs of customers in modern organizations. The chapter argues that applying virtual team and IT has the potential to increase organizational performance and reach strategic goals in global business.

INTRODUCTION

Virtual teams are becoming more and more important and continuing to grow in popularity (Olariu & Aldea, 2014). Understanding the influence of leaders in virtual team settings can help organizations in allocating resources and sorting teams' priorities (Saafein & Shaykhian, 2014). Multinational companies are beginning to adopt a virtual team strategy, due to the benefits of cost reduction and performance improvement (Ferreira, de Lima, & da Costa, 2012). Virtual teams are an organizational form in which an overlay of information and communication technologies (ICT) enables departures from traditional, face-to-face, organizational forms (Haines, 2014). Virtual team members have developed a new technique which reduces the ambiguity of virtual communications in order to influence their team members (Wadsworth & Blanchard, 2015).

DOI: 10.4018/978-1-4666-9688-4.ch001

Economic growth of most of the industries depends largely on the development of operating technologies within the industries (Prabhakaran, Lathabai, & Changat, 2015). The increasing business competitiveness leads industries to differentiate themselves in terms of technology (Reis & Freitas, 2014). ICT has been a major technological innovation for developed countries and is increasingly spreading to developing countries (Kossaï & Piget, 2014). Allameh et al. (2011) stated that ICT is one of the strategic factors toward the improvement of business and human resource productivity. Organizations are relying on Internet services as well as information systems to enhance business operations, facilitate management decision-making, and deploy business strategies (Kankanhalli, Teo, Tan, & Wei, 2003).

The strength of this chapter is on the thorough literature consolidation of virtual team and IT. The extant literature of virtual team and IT provides a contribution to practitioners and researchers by describing a comprehensive view of the functional applications of virtual team and IT to appeal to different segments of virtual team and IT in order to maximize the business impact of virtual team and IT in global business.

BACKGROUND

Virtual work settings cause organizational challenges such as maintaining remote leadership, managing cultural differences, and developing trust relationships among the teams (Saafein & Shaykhian, 2014). Virtual teams need to deal with such as communication difficulties, decreased cohesion, and high level of conflicts among teams (Staples & Zhao, 2006). Virtual team members are physically separated from one another and they rely on technological devices for communication and information exchange (Ghaffari, Sheikhahmadi, & Safakish, 2014). Virtual teams are thought to be differently experienced and to have poor outcomes because there is little or no face-to-face interaction and a tendency for virtual team members to use different communication techniques for forming relationships (Haines, 2014).

The diffusion of new technologies can have a significant impact on economic growth and development (Kossaï & Piget, 2014). Allameh et al. (2011) indicated that the application of IT has a great stance among basic industries since it plays an important role in different industries such as productivity, social services and job opportunity improvement. The continued diffusion of new ICT is an example of the dynamics of technological change and economic development (Freeman & Soete, 1997). ICT is used as a production technology to improve labor productivity and coordination within the firm (Raymond & St-Pierre, 2005). ICT is considered to be the main engine of growth in the knowledge economy (Kossaï & Piget, 2014). Firms invest in ICT in order to become more successful in highly competitive markets (Koivunen, Hätonen, & Välimäki, 2008).

VIRTUAL TEAM AND INFORMATION TECHNOLOGY IN GLOBAL BUSINESS

This section describes the theoretical and practical overviews of virtual team and IT; the importance of virtual team in global business; and the importance of IT in global business.

Overview of Virtual Team

Organizational teams mainly bring numerous benefits for both management and employees, including a method of pooling ideas, facilitating communication, and improving workflow (Lin, 2011). Transformational usage of computer-mediated interactions and online collaboration technology platforms among team members collaborating in virtual teams practically contextualizes multiple settings for the terminology of virtual teams in organizations (Mansor, Mirahsani, & Saidi, 2012). Virtual teams are teams whose members are mediated by time, distance, or technology and whose members are interdependent, working together on a common task (Driskell, Radtke, & Salas, 2003). Luse et al. (2014) stated that virtual teams are characterized as a group of people with unique skills who interdependently work but are geographically separated and they need technological mediums to communicate.

The technologies of virtual teams vary as to how much they incorporate the media dimensions of co-presence, visibility, audibility, contemporality, simultaneity, sequentiality, reviewability, and revisability (Clark & Brennan, 1991). Members of virtual teams communicate through various ICT including telephone, video and audio conferencing, chat rooms and instant messaging, file and application sharing, and other virtual reality options (Hertel, Geister, & Konradt, 2005; Olson & Olson, 2000). Virtual teams are a group of geographically, organizationally and time dispersed workers brought together by IT to accomplish one or more organization tasks (Powell, Piccoli, & Ives, 2004).

Members of virtual teams often collaborate within and across institutional boundaries (Schiller, Mennecke, Nah, & Luse, 2014). Virtual teams are comprised of members who are located in more than one physical location (El-Kassrawy, 2014). Virtual teams represent interdependent groups of individuals who work across space, time, and geographical boundaries with communication links that are heavily dependent upon advanced IT (Hambley, O'Neill, & Kline, 2007). Geographically distributed team members can easily withhold information from one another (Rosen, Furst, & Blackburn, 2007).Virtual teams have become basic units in business organizations, and their activities are ubiquitous and have received considerable attention from social and organizational psychologists (Bosch-Sijtsema, 2007).

Advances in ICT have allowed organizations to facilitate virtual team member coordination in a cost effective manner (Pendharkar, 2013). Team performance and knowledge sharing remain important issues for interpersonal relationships within virtual teams (Levy, Loebbecke, & Powell, 2003). ICT, the dispersion of members around the world, and the relationships between team members of different status may affect the use and effectiveness of influence tactics in virtual teams (Wadsworth & Blanchard, 2015). Influence tactics are how people enact power over others (Lines, 2007; Yukl & Tracey, 1992).

Martins et al. (2004) defined virtual team as team whose members use technology to varying degrees in working across locational, temporal, and relational boundaries to accomplish an interdependent task. Nemiro (2001) indicated the five traits of virtual teams: geographical separation (members being separated from one another), alternative methods of communication (members using new information technologies in order to collaborate), reduced information richness (the majority of methods used to communicate having low levels of information richness), loose boundaries, and dynamic membership (membership in virtual teams being adaptable and temporary).

The interpersonal relationships within virtual teams are comprised of two elements: cooperation and competition relationship (Baruch & Lin, 2012). Regarding coopetition relationships, all teaming activities should aim for the establishment of a beneficial partnership with one another in the team, including the coworkers who may be considered as a competitor (Zineldin, 2004). Coopetition is important among intra-organizational partners and inter-team parties (or inter-organizational parties), and these interactions are key for a firm's long-term viability (Luo, Slotegraaf, & Pan, 2006). If both the elements co-exist, then the relationship between the members is considered coopetition (Bengtsson & Kock, 2000).

Concerning advances in IT facilitating information sharing, more companies are moving toward using virtual teams (Baruch & Lin, 2012). The use of technology can allow a virtual team to recruit additional members (Lipnack & Stamps, 1997), which leads to increased communication requirements. Virtual teams work collaboratively in geographically dispersed locations but still share the same interests, goals, needs and practices that define face-to-face teams (Chiu, Hsu, & Wang, 2006). The handling of technological issues such as adaptation and regular use of communication tools as another challenge that faces virtual teams (Indiramma & Anandakumar, 2009).

Because members of virtual teams have fewer tools available for developing relationships than face-to-face teams, they must rely on categorization processes (McKnight, Cummings, & Chervany, 1998) and their experience from other settings (Jarvenpaa & Leidner, 1999). Virtual teams need to deal with such as communication difficulties, decreased cohesion, and high level of conflicts among virtual teams (Staples & Zhao, 2006). Individuals working in anonymous groups believe that their group's decision are less effective, and are often less satisfied as compared to members of teams' face-to-face meetings (Reinig & Mejias, 2004; Valacich, Dennis, & Nunamaker, 1992). In order to prevent problems arising regarding anonymous communication, individuals supervising virtual teams would prefer team members to be identifiable (Midha & Nandedkar, 2012). Knowledge of the identity of a specific source enhances the source credibility and accountability for virtual teams (Rains, 2007).

Virtual team leaders stimulate followers to be high performers through effective use of motivational and communication skills on individual and group levels (Bolman & Deal, 2003). Virtual team leaders may need different leadership characteristics from leaders of traditional teams when setting goals, offering support and guidance, and inspiring team members (Saafein & Shaykhian, 2014). Challenges to good communication between virtual team members are obtaining timely communication feedback (Jarvenpaa & Leidner, 1999), building trust between team members (Fan, Suo, Feng, & Liu, 2011), building a common sense of purpose between team members (Blackburn, Furst, & Rosen, 2003) and team members' access to rich communication media (Kayworth & Leidner, 2000).

Overview of Information Technology

Information Technology (IT) is an integral part to support, sustain, and grow a business (Mohamed & Singh, 2012). With the rapid development of IT, the innovative IT strategy has become an important topic of research in the era of electronic business (Yeh, Lee, & Pai, 2012). IT as a new human technology is rapidly affecting business and productivity (Zafiropoulos, Vrana, & Paschaloudis, 2006). Most companies heavily invest on IT for getting better business feedbacks (Williams & Williams, 2007). ICT encompasses computer and communication technologies, shareable technical platforms and databases, networking technologies, broadcast media, and audio and video processing and transmission (Chung & Hossain, 2010; Schaper & Pervan, 2007).

The mixed empirical results concerning financial performance achieve from IT integration (Chapman & Kihn, 2009; Hunton, Lippincott, & Reck, 2003; Poston & Grabski, 2001). Successful IT integration can deliver IT resources in support of the new roles and functions of workers as a result of redesigned and tightened business processes (Rockart, Ear, & Ross, 1996). Integrated IT not only enables process automation, but also can provide the ability to disseminate timely and accurate information, resulting in improved managerial and employee decision-making (Hitt, Wu, & Zhou, 2002). Manufacturing plants will reap the greatest financial performance benefits from investments in activity-based cost control systems when combined with IT integration (Maiga, Nilsson, & Jacobs, 2014).

Technology can improve business productivity (Yang, Lee, & Lee, 2007). IT has a considerable effect on globalization and is a revolution in information, knowledge, and organizational changes (Pavic, Koh,

Simpson, & Padmore, 2007). IT applications can be used in formalizing knowledge and distributing it in modern business (Okumus, 2013). IT should be managed to assist core business in achieving organizational mission and vision (Sharma & Baoku, 2013). IT alignment has an indirect effect on operational performance through information sharing (Ye & Wang, 2013).

Brady et al. (2008) stated that obtaining high efficiency and effectiveness in organizations requires investment in IT components such as the Internet (Deeter-Schmelz & Kennedy, 2004), office automation (Geiger & Turley, 2005), and management information system (Li, 1995). The Internet is a global network of computers which independently work and offers a variety of useful tools such as e-mail and website (Obra, Camara, & Melendez, 2002). Most companies effectively use the Internet to boost national economy (Martin & Matlay, 2001) with a much more profitability than telephone and fax (Grandon & Pearson, 2004).

The use of ICT has been found to increase the productivity of research and development (R&D) in innovative industries (Hollenstein, 2004). Productivity improvement, service quality improvement, cost reduction, individuals' satisfaction, and long-term profitability are among the expectations of practitioners and researchers dealing with ICT (Law & Jogaratnam, 2005). IT helps companies to find new ways of improving their market by providing their customers with their immediate needs (Phuong, 2008).

Importance of Virtual Team in Global Business

Virtual teams have to be viewed from dimension of human resource strategic management and approached from the perspective of human resource practitioners in order to develop theories and frameworks to measure team performance of human resources or intellectual assets of an organization who are not co-located during their team collaboration, considering performance measurement tools and theories available in face-to-face teams context such as economic value added (EVA) determinants and measures (Mansor et al., 2012).

Virtual teams create challenges to leaders such as leading remotely, building trust relationship among team members without personal face-to-face interaction, and using technologies that facilitate virtual collaboration (Neufeld, Wan, & Fang 2008). Aldea et al. (2012) stated that the tendencies of business globalization have force organizations to develop ICT facilities for the development of virtual teams. Leaders of virtual support teams face various challenges which differ from those facing leaders in traditional face-to- face environment. Leaders' inability to meet their teams face-to-face may limit their influence on team members (Piccoli, Powell, & Ives, 2004).

Due to the challenges that arise when learning and working in electronic settings, the satisfaction of team members tends to be lower compared to face-to-face teams (Furumo & Pearson, 2007). Thus, the evidences of team members' low satisfaction are acknowledged in virtual teams (Martins et al., 2004). When virtually collaborating, members' satisfaction may be moderated by the team's level of virtuality (Gurtner, Kolbe, & Boos, 2007). The effects of employed communication media is often influenced by the nature of task and the composition of team (Cicei, 2012).

A team leader is the person who defines the project objectives, selects participants to assure a broad-based team composition, conduct training regarding communications technologies, evaluate the team outputs, coordinate the team processes, clarifies confusions, solves operational problems, and eliminates conflicts (Chang, 2011). Challenges in virtual environment require leadership styles that have different personal skill sets, leadership strategies, and communication techniques (Neufeld et al., 2008). Leaders in virtual teams tend to be informal, convincing, and possess effective communication skills (Beranek,

Broder, Reinig, Romano, & Sump, 2005). Leaders may face challenges to setup, design, manage, and operate virtual teams, which may result in less productivity and low performance (Gibson & Gibbs, 2006).

Global virtual team leaders have to lead from a distance (Zander, Zettinig, & Makela, 2013). Management, control, coordination, jobs, and operation s are distributed among different human agent teams that work virtual combination of virtual and face-to-face (Aart, Wielinga, & Schreiber, 2004; Bélanger & Watson-Manheim, 2006). The macro economic developments have led organizations to increase inter-firm cooperation (Gluckler & Schrott, 2007). The inter-firm cooperation often requires teamwork from employees in different firms, and physically collocating team members, while effective, is often expensive and impractical (Poltrock & Engelbeck, 1999).

Virtual teams offer several advantages over their traditional face-to-face counterparts (Midha & Nandedkar, 2012). In order to enhance the depth of information access and task performance effectiveness, virtual teams overcome space and time constraints which limit their collocated face-to-face counterparts (McGrath & Hollingshead, 1994). Virtual teams make managerial communication more efficient; reduce the expense and difficulty concerning distributed work (Galegher & Kraut, 1994). Virtual teams enable organizational flexibility, improve resource utilization, reduce cost, and expand organizational knowledge generation by bringing together diverse range of knowledge, skills, and competencies (McLean, 2007). Virtual teams pose challenges due to lack of trust between team members, language barriers, temporal constraints, individual goals, institutional constraints (legal and political), and choice of technology (Aart et al., 2004; Bélanger & Watson-Manheim, 2006; Fan et al., 2011).

Many organizations use teaming arrangements to push team members to both compete and cooperate with each other, leading to a major challenge for organizations that seek to manage their team workflows and performance (Tsai, 2002). An effective virtual team coordination strategy will involve supervision, standardization of work, standardization of output, standardization of skills and mutual adjustment (Aart et al., 2004). Good communication between team members lead to successful operation of a team (Anderson, McEwan, Bal, & Carletta, 2007) and poor communication leads to poor performance (Thompson & Couvert, 2003). The improved communication tends to bind virtual teams, which leads to low dissolution rates; and long-term teams are generally more productive than short-term teams (Ortiz de Guinea, Webster, & Staples, 2012).

Virtual worlds have been shown to be a powerful environment for collaboration and communication (Schultze & Orlikowski, 2010). Virtual teams can effectively function both within and across organizational boundaries (Levina & Vaast, 2008; Romano, Pick, & Roztocki, 2010). Within-boundary virtual teams are preferred over cross-boundary virtual teams if satisfaction with the collaboration process is of the highest priority (Schiller et al., 2014). The increased communication requirements need special handling to ensure improved productivity and performance of virtual team (Anderson et al., 2007). Zibetti et al. (2012) stated that IT-based virtual team meeting scheduling may increase team member motivation and coordination.

Virtual teams-related boundaries such as corporate culture (Krishna, Sahay, & Walsham, 2004), formal organizational boundaries (Espinosa, Cummings, Wilson, & Pearce, 2003), spatial and temporal boundaries (O'Leary & Cummings, 2007; Van den Bulte & Moenaert, 1998), and functional boundaries (Birnholtz & Finholt, 2007; Espinosa et al., 2003) account for a team's success as well as pose challenges to effective team outcomes. Working in virtual teams can assure a series of advantages consisting in the enhancement in creativity and innovation of products, the facilitation of learning, and the development of positive attitudes facing the tasks that need to be achieved (Levenson & Cohen, 2003). Virtual teams can support knowledge sharing between members and sustain the application of obtained information and skills (Johnson et al., 2002).

Importance of Information Technology in Global Business

Yılmaz et al. (2015) stated that since the advent of computer and Internet technologies, the continuous and rapid developments occurred in technology-based applications have become an important part of the reality of the life for many people in the world. ICT has substantial impact on economic growth and productivity (Choi & Hoon Yi, 2009; Litan & Rivlin, 2001; Roller & Waverman, 2001). Access to ICT through cell phones, the Internet, and electronic media has exponentially increased around the world (Graham & Nikolova, 2013).

There has been considerable research about the impact of ICT on organizational performance and importance of ICT as an essential enabler of business growth and economic development has been widely confirmed using both a growth accounting and at the firm-level approach (Jorgenson & Vu, 2007; Oliner & Sichel, 2003). IT is fundamental to a firm's growth and represents a primary tool for enhancing business-to-business (B2B) sales force performance (Limbu, Jayachandran, & Babin, 2014). ICT enhances the sales force activities, thus employing market intelligence, managing their customer contacts, submitting sales call reports, and developing sales forecasts (Gohmann, Guan, Barker, & Faulds, 2005). IT plays an important role in modern business regarding the accounting function (Efendi, Mulig, & Smith, 2006). IT has transformed the nature of business and accounting practice (Hunton, 2002).

The link between IT and organizational performance is important for the information systems researchers and practitioners (Stoel & Muhanna, 2009). IT is universal because modern IT crosses organizational activities, and has become aligned with business activities (Ko & Fink, 2010). Xue et al. (2012) identified the positive effect of IT on organizational innovation. The use of technology has caused critical dependency on IT, thus involving a complex mix of organizational and technical perspectives (Sethibe, Campbell, & McDonald, 2007).

IT has a mixed record regarding organizational performance and user satisfaction, thus playing an important role in modern organizations (Peslak, 2012). Kudyba and Vitaliano (2003) stated that the enhanced flow of information brought about by IT integration and cost control systems throughout the company practically empowers decision makers to consolidate operations by reducing unnecessary waste and increasing profitability. There is a positive impact of ICT, human resource practices, and organizational reorganization on productivity (Black & Lynch, 2001). Business strategy and IT strategy remain as critical determinants of an organizational success in the foreseeable future (Brown, 2006).

Investing in ICT has an impact on productivity and contributes to reducing prices and improving services (Brynjolfsson & Hitt, 2004). Kleis et al. (2012) found the positive relationship between IT and intangible output, and stated that the use of IT in both organizational innovation and knowledge creation processes is the most critical factor in an organization's long-term success. IT can provide organizational members with quick and effective access to the right amounts of information (Hope & Hope, 1997). If connectivity is enhanced through IT, organizational members can more easily share individual interpretations of the cost information, thus making consensus development more efficient (Maiga et al., 2014).

FUTURE RESEARCH DIRECTIONS

The strength of this chapter is on the thorough literature consolidation of virtual team and IT. The extant literature of virtual team and IT provides a contribution to practitioners and researchers by describing a comprehensive view of the functional applications of virtual team and IT to appeal to different segments of virtual team and IT in order to maximize the business impact of virtual team and IT in global business. The classification of the extant literature in the domains of virtual team and IT will provide the potential opportunities for future research. Future research direction should broaden the perspectives in the implementation of virtual team and IT to be utilized in the knowledge-based organizations.

Practitioners and researchers should accept the usefulness of a more multidisciplinary approach toward research activities in implementing virtual team and IT in terms of knowledge management-related variables (i.e., knowledge-sharing behavior, knowledge creation, organizational learning, learning orientation, and motivation to learn). It will be useful to bring additional disciplines together (i.e., strategic management, marketing, finance, and human resources) to support a more holistic examination of virtual team and IT in order to combine or transfer existing theories and approaches to inquiry in this area.

CONCLUSION

This chapter aimed to examine the roles of virtual team and IT in global business, thus describing the theoretical and practical overviews of virtual team and IT; the importance of virtual team in global business; and the importance of IT in global business. The applications of virtual team and IT are necessary for modern organizations that seek to serve suppliers and customers, increase business performance, strengthen competitiveness, and achieve continuous success in global business. Leadership has an influence on the performance of virtual teams.

One of the main concerns of virtual team is the reliability of communication, perceived as the most important factor concerning performance of virtual support teams. Leaders of virtual teams carry the responsibilities to satisfy their bosses, subordinates, and external customers in a complex distributed environment that is highly dependent on IT perspectives. Leaders of virtual support teams may need to assume a coordinating role to ensure effective collaboration and communication among virtual team members. It is essential for modern organizations to examine their virtual team and IT applications, develop a strategic plan to regularly check their practical advancements, and immediately respond to virtual team and IT needs of customers in modern organizations. The chapter argues that applying virtual team and IT has the potential to improve organizational performance and gain sustainable competitive advantage in global business.

REFERENCES

Aart, C. J., Wielinga, B., & Schreiber, G. (2004). Organizational building blocks for design of distributed intelligent system. *International Journal of Human-Computer Studies*, *61*(5), 567–599. doi:10.1016/j.ijhcs.2004.03.001

Aldea, C. C., Popescu, A. D., Draghici, A., & Draghici, G. (2012). ICT tools functionalities analysis for the decision making process of their implementation in virtual engineering teams. *Procedia Technology*, *5*, 649–658. doi:10.1016/j.protcy.2012.09.072

Allameh, S. M., Momeni, Z. M., Esfahani, Z. S., & Bardeh, M. K. (2011). An assessment of the effect of information communication technology on human resource productivity of Mobarekeh Steel Complex in Isfahan (IRAN). *Procedia Computer Science*, *3*, 1321–1326. doi:10.1016/j.procs.2011.01.010

Anderson, A. H., McEwan, R., Bal, J., & Carletta, J. (2007). Virtual team meetings: An analysis of communication and context. *Computers in Human Behavior*, *23*(5), 2558–2580. doi:10.1016/j.chb.2007.01.001

Baruch, Y., & Lin, C. P. (2012). All for one, one for all: Coopetition and virtual team performance. *Technological Forecasting & Social Change*, *79*(6), 1155–1168. doi: 10.1016/j.techfore.2012.01.008

Bélanger, F., & Watson-Manheim, M. B. (2006). Virtual teams and multiple media: Structuring media use to attain strategic goals. *Group Decision and Negotiation*, *15*(4), 299–321. doi:10.1007/s10726-006-9044-8

Bengtsson, M., & Kock, S. (2000). "Coopetition" in business networks – to cooperate and compete simultaneously. *Industrial Marketing Management*, *29*(5), 411–426. doi:10.1016/S0019-8501(99)00067-X

Beranek, P. M., Broder, J., Reinig, B. A., Romano, N. C., & Sump, S. (2005). Management of virtual project teams: Guidelines for team leaders. *Communications of the AIS*, *16*(10), 247–259.

Bilgihan, A., Okumus, F., Nusair, K., & Kwun, D. (2011). Information technology applications and competitive advantage in hotel companies. *Journal of Hospitality and Tourism Technology*, *2*(2), 139–153. doi:10.1108/17579881111154245

Birnholtz, J., & Finholt, T. (2007). Cultural challenges to leadership in cyberinfrastructure development. In S. Weisband (Ed.), *Leadership at a distance: Research in technology-supported work* (pp. 195–207). New York, NY: Lawrence Erlbaum Associates.

Blackburn, R. S., Furst, S. A., & Rosen, B. (2003). Building a winning virtual team. In C. Gibson & S. Cohen (Eds.), *Virtual teams that work: Creating conditions for effective virtual teams* (pp. 95–120). San Francisco, CA: Jossey–Bass.

Bolman, L. E., & Deal, T. E. (2003). *Reframing organizations: Artistry, choice, and leadership*. San Francisco, CA: Jossey-Bass.

Bosch-Sijtsema, P. (2007). The impact of individual expectations and expectation conflicts on virtual teams. *Group & Organization Management*, *32*(3), 358–388. doi:10.1177/1059601106286881

Brady, M., Fellenz, M., & Brookes, R. (2008). Researching the role of information communication technology (ICT) in contemporary marketing practices. *Journal of Business and Industrial Marketing*, *23*(2), 108–114. doi:10.1108/08858620810850227

Brown, W. C. (2006). IT governance, architectural competency, and the Vasa. *Information Management & Computer Security*, *14*(2), 140–154. doi:10.1108/09685220610655889

Brynjolfsson, E., & Hitt, L. (2004). Computing productivity: Firm-level evidence. *The Review of Economics and Statistics, 85*(4), 793–808. doi:10.1162/003465303772815736

Campbell, B., Kay, R., & Avison, D. (2005). Strategic alignment: A practitioner's perspective. *Journal of Enterprise Information Management, 18*(6), 653–664. doi:10.1108/17410390510628364

Chang, C. M. (2011). New organizational designs for promoting creativity: A case study of virtual teams with anonymity and structured interactions. *Journal of Engineering and Technology Management, 28*(4), 268–282. doi:10.1016/j.jengtecman.2011.06.004

Chapman, C. S., & Kihn, L. A. (2009). Information systems integration, enabling control and performance. *Accounting, Organizations and Society, 34*(2), 151–169. doi:10.1016/j.aos.2008.07.003

Chiu, C. M., Hsu, M. H., & Wang, E. T. G. (2006). Understanding knowledge sharing in virtual communities: An integration of social capital and social cognitive theories. *Decision Support Systems, 42*(3), 1872–1888. doi:10.1016/j.dss.2006.04.001

Choi, C., & Hoon Yi, M. (2009). The effect of the Internet on economic growth: Evidence cross-country panel data. *Economics Letters, 105*(1), 39–41. doi:10.1016/j.econlet.2009.03.028

Chung, K. S. K., & Hossain, L. (2010). Towards a social network model for understanding information and communication technology use for general practitioners in rural Australia. *Computers in Human Behavior, 26*(4), 562–571. doi:10.1016/j.chb.2009.12.008

Cicei, C. C. (2012). Assessing members' satisfaction in virtual and face-to-face learning teams. *Procedia: Social and Behavioral Sciences, 46*, 4466–4470. doi:10.1016/j.sbspro.2012.06.278

Clark, H. H., & Brennan, S. E. (1991). Grounding in communication. In L. B. Resnick, J. M. Levine, & S. D. Teasley (Eds.), *Perspectives on socially shared cognition* (pp. 127–149). Washington, DC: American Psychological Association. doi:10.1037/10096-006

Deeter-Schmelz, D., & Kennedy, K. (2004). Buyer-seller relationships and information sources in an e-commerce world. *Journal of Business and Industrial Marketing, 19*(3), 188–196. doi:10.1108/08858620410531324

Driskell, J. E., Radtke, P. H., & Salas, E. (2003). Virtual teams: Effects of technological mediation on team performance. *Group Dynamics, 7*(4), 297–323. doi:10.1037/1089-2699.7.4.297

Efendi, J., Mulig, E., & Smith, L. (2006). Information technology and systems research published in major accounting academic and professional journals. *Journal of Emerging Technologies in Accounting, 3*(1), 117–128. doi:10.2308/jeta.2006.3.1.117

El-Kassrawy, Y. A. (2014). The impact of trust on virtual team effectiveness. *International Journal of Online Marketing, 4*(1), 11–18. doi:10.4018/ijom.2014010102

Espinosa, J., Cummings, J., Wilson, J., & Pearce, B. (2003). Team boundary issues across multiple global firms. *Journal of Management Information Systems, 19*(4), 157–190.

Fan, Z. P., Suo, W. L., Feng, B., & Liu, Y. (2011). Trust estimation in a virtual team: A decision support method. *Expert Systems with Applications, 38*(8), 10240–10251. doi:10.1016/j.eswa.2011.02.060

Ferreira, P. G. S., de Lima, E. P., & da Costa, S. E. G. (2012). Perception of virtual team's performance: A multinational exercise. *International Journal of Production Economics*, *140*(1), 416–430. doi:10.1016/j.ijpe.2012.06.025

Freeman, C., & Soete, L. (1997). *The economics of industrial innovation*. Cambridge, MA: MIT Press.

Furumo, K., & Pearson, M. (2007). Gender based communication styles, trust and satisfaction in virtual teams. *Journal of Information, Information Technology, and Organizations*, *2*, 47–60.

Galegher, J., & Kraut, R. E. (1994). Computer-mediated communication for intellectual teamwork: An experiment in group writing. *Information Systems Research*, *5*(2), 110–138. doi:10.1287/isre.5.2.110

Geiger, S., & Turley, D. (2005). Personal selling as knowledge-based activity: Communities of practice in the sales force. *Irish Journal of Management*, *26*(1), 61–71.

Ghaffari, M., Sheikhahmadi, F., & Safakish, G. (2014). Modeling and risk analysis of virtual project team through project life cycle with fuzzy approach. *Computers & Industrial Engineering*, *72*, 98–105. doi:10.1016/j.cie.2014.02.011

Gibson, C. B., & Gibbs, J. L. (2006). Unpacking the concept of virtuality: The effects of geographic dispersion, electronic dependence, dynamic structure, and national diversity on team innovation. *Administrative Science Quarterly*, *51*(3), 451–495.

Gluckler, J., & Schrott, G. (2007). Leadership and performance in virtual teams: Exploring brokerage in electronic communication. *International Journal of e-Collaboration*, *3*(3), 31–52. doi:10.4018/jec.2007070103

Gohmann, S. F., Guan, J., Barker, R. M., & Faulds, D. J. (2005). Perceptions of sales force automation: Differences between sales force and management. *Industrial Marketing Management*, *34*(4), 337–343. doi:10.1016/j.indmarman.2004.09.014

Graham, C., & Nikolova, M. (2013). Does access to information technology make people happier? Insights from well-being surveys from around the world. *Journal of Socio-Economics*, *44*, 126–139. doi:10.1016/j.socec.2013.02.025

Grandon, E., & Pearson, J. M. (2004). Electronic commerce adoption: An empirical study of small and medium US business. *Information & Management*, *42*(1), 197–216. doi:10.1016/j.im.2003.12.010

Gurtner, A., Kolbe, M., & Boos, M. (2007). Satisfaction in virtual teams and in organizations. The *Electronic Journal for Virtual Organizations and Networks*. [Special Issue]. *The Limits of Virtual Work*, *9*, 9–29.

Haines, R. (2014). Group development in virtual teams: An experimental reexamination. *Computers in Human Behavior*, *39*, 213–222. doi:10.1016/j.chb.2014.07.019

Hambley, L. A., O'Neill, T. A., & Kline, T. J. B. (2007). Virtual team leadership: The effects of leadership style and communication medium on team interaction styles and outcomes. *Organizational Behavior and Human Decision Processes*, *103*(1), 1–20. doi:10.1016/j.obhdp.2006.09.004

Hertel, G., Geister, S., & Konradt, U. (2005). Managing virtual teams: A review of current empirical research. *Human Resource Management Review, 15*(1), 69–95. doi:10.1016/j.hrmr.2005.01.002

Hitt, L., Wu, D., & Zhou, X. (2002). Investment in enterprise resource planning: Business impact and productivity measures. *Journal of Management Information Systems, 19*(1), 71–98.

Hollenstein, H. (2004). Determinants of the adoption of information and communication technologies. *Structural Change and Economic Dynamics, 15*(3), 315–342. doi:10.1016/j.strueco.2004.01.003

Hope, J., & Hope, T. (1997). *Competing in the third wave: The ten management issues of the information age*. Boston, MA: Harvard Business School Press.

Hunton, J. (2002). Blending information and communication technology with accounting research. *Accounting Horizons, 16*(1), 55–67. doi:10.2308/acch.2002.16.1.55

Hunton, J. E., Lippincott, B., & Reck, J. L. (2003). Enterprise resource planning (ERP) systems: Comparing firm performance of adopters and non-adopters. *International Journal of Accounting Information Systems, 4*(3), 165–184. doi:10.1016/S1467-0895(03)00008-3

Indiramma, M. M., & Anandakumar, K. R. (2009). Behavioral analysis of team members in virtual organization based on trust dimension and learning. *Proceedings of World Academy of Science: Engineering & Technology, 39*(3), 269–274.

Jarvenpaa, S. L., & Leidner, D. E. (1999). Is anybody out there? Antecedents of trust in global virtual teams. *Journal of Management Information Systems, 14*(4), 29–64.

Johnson, S. D., Suriya, C., Won Yoon, S., Berrett, J. V., & La Fleur, J. (2002). Team development and group processes of virtual learning teams. *Computers & Education, 39*(4), 379–393. doi:10.1016/S0360-1315(02)00074-X

Jorgenson, D. W., & Vu, K. (2007). Information technology and the world growth resurgence. *German Economic Review, 8*(2), 125–145. doi:10.1111/j.1468-0475.2007.00401.x

Kankanhalli, A., Teo, H. H., Tan, B. C. Y., & Wei, K. K. (2003). An integrative study of information systems security effectiveness. *International Journal of Information Management, 23*(2), 139–154. doi:10.1016/S0268-4012(02)00105-6

Kleis, L., Chwelos, P., Ramirez, R. V., & Cockburn, I. (2012). Information technology and intangible output: The impact of IT investment on innovation productivity. *Information Systems Research, 23*(1), 42–59. doi:10.1287/isre.1100.0338

Ko, D., & Fink, D. (2010). Information technology governance: An evaluation of the theory-practice gap. *Corporate Governance, 10*(5), 662–674. doi:10.1108/14720701011085616

Koivunen, M., Hätonen, H., & Välimäki, M. (2008). Barriers to facilitators influencing the implementation of an interactive internet-portal application for patient education in psychiatric hospital. *Patient Education and Counseling, 70*(3), 412–419. doi:10.1016/j.pec.2007.11.002 PMID:18079085

Kossaï, M., & Piget, P. (2014). Adoption of information and communication technology and firm profitability: Empirical evidence from Tunisian SMEs. *The Journal of High Technology Management Research, 25*(1), 9–20. doi:10.1016/j.hitech.2013.12.003

Krishna, S., Sahay, S., & Walsham, G. (2004). Cross-cultural issues in global software outsourcing. *Communications of the ACM, 47*(4), 62–66. doi:10.1145/975817.975818

Kudyba, S., & Vitaliano, D. (2003). Information technology and corporate profitability: A focus on operating efficiency. *Information Resources Management Journal, 16*(1), 1–13. doi:10.4018/irmj.2003010101

Law, R., & Jogaratnam, G. (2005). A study of hotel information technology applications. *International Journal of Contemporary Hospitality Management, 17*(2–3), 170–180. doi:10.1108/09596110510582369

Levenson, A., & Cohen, S. (2003). Meeting the performance challenge: Calculating return of investment for virtual teams. In C. Gibson & S. Cohen (Eds.), *Virtual teams that work: Creating conditions for virtual team effectiveness* (pp. 145–174). San Francisco, CA: Jossey–Bass.

Levina, N., & Vaast, E. (2008). Innovating or doing as told? Status differences and overlapping boundaries in offshore collaboration. *Management Information Systems Quarterly, 32*(2), 307–332.

Levy, M., Loebbecke, C., & Powell, P. (2003). SMEs, co-opetition and knowledge sharing: The role of information systems. *European Journal of Information Systems, 12*(1), 3–17. doi:10.1057/palgrave.ejis.3000439

Li, E. (1995). Marketing information systems in US companies: A longitudinal analysis. *Information & Management, 28*(1), 13–31. doi:10.1016/0378-7206(94)00030-M

Limbu, Y. B., Jayachandran, C., & Babin, B. J. (2014). Does information and communication technology improve job satisfaction? The moderating role of sales technology orientation. *Industrial Marketing Management, 43*(7), 1236–1245. doi:10.1016/j.indmarman.2014.06.013

Lin, C. P. (2011). Modeling job effectiveness and its antecedents from a social capital perspective: A survey of virtual teams within business organizations. *Computers in Human Behavior, 27*(2), 915–923. doi:10.1016/j.chb.2010.11.017

Lines, R. (2007). Using power to install strategy: The relationships between expert power, position power, influence tactics and implementation success. *Journal of Change Management, 7*(2), 143–170. doi:10.1080/14697010701531657

Lipnack, J., & Stamps, J. (1997). *Virtual teams: Reaching across space, time and organizations wit technology.* New York, NY: John Wiley & Sons.

Litan, R. E., & Rivlin, A. M. (2001). Projecting the economic impact of the internet. *The American Economic Review, 91*(2), 313–317. doi:10.1257/aer.91.2.313

Luo, X., Slotegraaf, R. J., & Pan, X. (2006). Cross-functional "coopetition": The simultaneous role of cooperation and competition within firms. *Journal of Marketing, 70*(2), 67–80. doi:10.1509/jmkg.70.2.67

Luse, A., McElroy, J. C., Townsend, A. M., & DeMarie, S. (2013). Personality and cognitive style as predictors of preference for working in virtual teams. *Computers in Human Behavior, 29*(4), 1825–1832. doi:10.1016/j.chb.2013.02.007

Maiga, A. S., Nilsson, A., & Jacobs, F. A. (2014). Assessing the interaction effect of cost control systems and information technology integration on manufacturing plant financial performance. *The British Accounting Review, 46*(1), 77–90. doi:10.1016/j.bar.2013.10.001

Mansor, N. N. B., Mirahsani, S., & Saidi, M. I. (2012). Investigating possible contributors towards "organizational trust" in effective "virtual team" collaboration context. *Procedia: Social and Behavioral Sciences, 57*, 283–289. doi:10.1016/j.sbspro.2012.09.1187

Martin, L., & Matlay, H. (2001). Blanket' approaches to promoting ICT in small firms: Some lessons from the DTI ladder adoption model in the UK. *Internet Research: Electronic Networking Applications and Policy, 11*(5), 399–410. doi:10.1108/EUM0000000006118

Martins, L. L., Gilson, L. L., & Maynard, M. T. (2004). Virtual teams: What do we know and where do we go from here? *Journal of Management, 30*(6), 805–835. doi:10.1016/j.jm.2004.05.002

McGrath, J. E., & Hollingshead, A. B. (1994). *Groups interacting with technology*. Thousand Oaks, CA: Sage Publications.

McKnight, D. H., Cummings, L. L., & Chervany, N. L. (1998). Initial trust formation in new organizational relationships. *Academy of Management Review, 23*(3), 473–490.

McLean, J. (2007). Managing global virtual teams. *British Journal of Administrative Management, 59*(2), 16–17.

Midha, V., & Nandedkar, A. (2012). Impact of similarity between avatar and their users on their perceived identifiability: Evidence from virtual teams in Second Life platform. *Computers in Human Behavior, 28*(3), 929–932. doi:10.1016/j.chb.2011.12.013

Mohamed, N., & Singh, J. K. G. (2012). A conceptual framework for information technology governance effectiveness in private organizations. *Information Management & Computer Security, 20*(2), 88–106. doi:10.1108/09685221211235616

Nemiro, J. E. (2001). Assessing the climate for creativity in virtual teams. In M. Beyerlein, D. Johnson, & S. Beyerlein (Eds.), *Virtual teams: Advances in interdisciplinary studies of work teams* (pp. 59–84). Bingley, UK: Emerald Group Publishing. doi:10.1016/S1572-0977(01)08019-0

Neufeld, D., Wan, Z., & Fang, Y. (2008). Remote leadership, communication effectiveness and leader performance. *Journal of Group Decision and Negotiation, 19*(3), 227–246. doi:10.1007/s10726-008-9142-x

O'Leary, M., & Cummings, J. (2007). The spatial, temporal, and configurational characteristics of geographic dispersion in teams. *Management Information Systems Quarterly, 31*(3), 433–452.

Obra, A., Camara, S., & Melendez, A. (2002). Internet usage and competitive advantage: The impact of the Internet on an old economy industry in Spain. *Benchmarking: An International Journal, 12*(5), 391–401.

Okumus, F. (2013). Facilitating knowledge management through information technology in hospitality organizations. *Journal of Hospitality and Tourism Technology, 4*(1), 64–80. doi:10.1108/17579881311302356

Olariu, C., & Aldea, C. C. (2014). Managing processes for virtual teams: A BPM approach. *Procedia: Social and Behavioral Sciences, 109*, 380–384. doi:10.1016/j.sbspro.2013.12.476

Oliner, S. D., & Sichel, D. E. (2003). Information technology and productivity: Where are we now and where are we going? *Journal of Policy Modeling, 25*(5), 477–503. doi:10.1016/S0161-8938(03)00042-5

Olson, G. M., & Olson, J. S. (2000). Distance matters. *Human-Computer Interaction, 15*(2–3), 139–179. doi:10.1207/S15327051HCI1523_4

Ortiz de Guinea, A., Webster, J., & Staples, D. S. (2012). A meta-analysis of the consequences of virtualness on team functioning. *Information & Management, 49*(6), 301–308. doi:10.1016/j.im.2012.08.003

Pavic, S., Koh, S. C. L., Simpson, M., & Padmore, J. (2007). Could e-business create a competitive advantage in UK SMEs? *Benchmarking: An International Journal, 14*(3), 320–351. doi:10.1108/14635770710753112

Pendharkar, P. C. (2013). Genetic learning of virtual team member preferences. *Computers in Human Behavior, 29*(4), 1787–1798. doi:10.1016/j.chb.2013.02.015

Peslak, A. R. (2012). An analysis of critical information technology issues facing organizations. *Industrial Management & Data Systems, 112*(5), 808–827. doi:10.1108/02635571211232389

Phuong, T. (2008). Internet use, customer relationships and loyalty in the Vietnamese travel industry. *Asia Pacific Journal of Marketing and Logistics, 20*(2), 190–210. doi:10.1108/13555850810864551

Piccoli, G., Powell, A., & Ives, B. (2004). Virtual teams: Team control structure, work processes, and team effectiveness. *Information Technology & People, 15*(4), 389–406.

Poltrock, S. E., & Engelbeck, G. (1999). Requirements for a virtual collocation environment. *Information and Software Technology, 41*(6), 331–339. doi:10.1016/S0950-5849(98)00066-4

Poston, R., & Grabski, S. (2001). Financial impact of enterprise resource planning implementations. *International Journal of Accounting Information Systems, 2*(4), 271–294. doi:10.1016/S1467-0895(01)00024-0

Powell, A., Piccoli, G., & Ives, B. (2004). Virtual teams: A review of current literature and directions for future research. *The Data Base for Advances in Information Systems, 35*(1), 6–36. doi:10.1145/968464.968467

Prabhakaran, T., Lathabai, H. H., & Changat, M. (2015). Detection of paradigm shifts and emerging fields using scientific network: A case study of Information Technology for Engineering. *Technological Forecasting & Social Change: An International Journal, 91*, 124–145. doi:10.1016/j.techfore.2014.02.003

Rains, S. A. (2007). The impact of anonymity on perceptions of source credibility and influence in computer-mediated group communication: A test of two competing hypotheses. *Communication Research, 34*(1), 100–125. doi:10.1177/0093650206296084

Raymond, L., & St-Pierre, J. (2005). Antecedents and performance outcomes of advanced manufacturing systems sophistication in SMEs. *International Journal of Operations & Production Management, 25*(6), 514–533. doi:10.1108/01443570510599692

Reinig, B. A., & Mejias, R. J. (2004). The effects of national culture and anonymity on flaming and criticalness in GSS supported discussions. *Small Group Research, 35*(6), 698–723. doi:10.1177/1046496404266773

Reis, R. A. D., & Freitas, M. D. C. D. (2014). Critical factors on information technology acceptance and use: An analysis on small and medium Brazilian clothing industries. *Procedia Computer Science, 31*, 105–114. doi:10.1016/j.procs.2014.05.250

Rockart, J. F., Ear, M. J., & Ross, J. W. (1996). Eight imperatives for the new IT organization. *MIT Sloan Management Review, 38*(1), 43–56.

Rodriguez, O. F., Fernandez, F., & Torres, R. S. (2011). Impact of information technology certifications in Puerto Rico. *Management Research: The Journal of the Iberoamerican Academy of Management, 9*(2), 137–153.

Roller, L. H., & Waverman, L. (2001). Telecommunications infrastructure and economic development: A simultaneous approach. *The American Economic Review, 91*(4), 909–923. doi:10.1257/aer.91.4.909

Romano, N. C. Jr, Pick, J. B., & Roztocki, N. (2010). A motivational model for technology-supported cross-organizational and cross-border collaboration. *European Journal of Information Systems, 19*(2), 117–133. doi:10.1057/ejis.2010.17

Rosen, B., Furst, S., & Blackburn, R. (2007). Overcoming barriers to knowledge sharing in virtual teams. *Organizational Dynamics, 36*(3), 259–273. doi:10.1016/j.orgdyn.2007.04.007

Saafein, O., & Shaykhian, G. A. (2014). Factors affecting virtual team performance in telecommunication support environment. *Telematics and Informatics, 31*(3), 459–462. doi:10.1016/j.tele.2013.10.004

Schaper, L. K., & Pervan, G. P. (2007). ICTs & OTs: A model of information and communications technology acceptance and utilisation by occupational therapists. *International Journal of Medical Informatics, 76*(1), S212–S221. doi:10.1016/j.ijmedinf.2006.05.028 PMID:16828335

Schiller, S. Z., Mennecke, B. E., Nah, F. F. H., & Luse, A. (2014). Institutional boundaries and trust of virtual teams in collaborative design: An experimental study in a virtual world environment. *Computers in Human Behavior, 35*, 565–577. doi:10.1016/j.chb.2014.02.051

Schultze, U., & Orlikowski, W. J. (2010). Research commentary – Virtual worlds: A performative perspective on globally distributed, immersive work. *Information Systems Research, 21*(4), 810–821. doi:10.1287/isre.1100.0321

Sharma, G., & Baoku, L. (2013). Customer satisfaction in Web 2.0 and information technology development. *Information Technology & People, 26*(4), 347–367. doi:10.1108/ITP-12-2012-0157

Staples, D. D., & Zhao, L. (2006). The effects of cultural diversity in virtual teams versus face-to-face teams. *Group Decision and Negotiation, 15*(4), 389–406. doi:10.1007/s10726-006-9042-x

Stoel, M. D., & Muhanna, W. A. (2009). IT capabilities and firm performance: A contingency analysis of the role of industry and IT capability type. *Information & Management, 46*(4), 181–189. doi:10.1016/j.im.2008.10.002

Thompson, L., & Couvert, M. (2003). Teamwork online: The effects of computer conferencing on perceived confusion, satisfaction and post discussion accuracy. *Group Dynamics*, *7*(2), 135–151. doi:10.1037/1089-2699.7.2.135

Tsai, W. (2002). Social structure of "coopetition" within a multiunit organization: Coordination, competition, and intraorganizational knowledge sharing. *Organization Science*, *13*(2), 179–190. doi:10.1287/orsc.13.2.179.536

Valacich, J., Dennis, A. R., & Nunamaker, A. F. Jr. (1992). Group size and anonymity effects on computer-mediated idea generation. *Small Group Research*, *23*(1), 49–73. doi:10.1177/1046496492231004

Van den Bulte, C., & Moenaert, R. (1998). The effects of R&D team co-location on communication patterns among R&D, marketing, and manufacturing. *Management Science*, *44*(11), 1–18. doi:10.1287/mnsc.44.11.S1

Wadsworth, M. B., & Blanchard, A. L. (2015). Influence tactics in virtual teams. *Computers in Human Behavior*, *44*, 386–393. doi:10.1016/j.chb.2014.11.026

Williams, M. D., & Williams, J. (2007). A change management approach to evaluating ICT investment initiatives. *Journal of Enterprise Information Management*, *20*(1), 32–50. doi:10.1108/17410390710717129

Xue, L., Ray, G., & Sambamurthy, V. (2012). Efficiency or innovation: How do industry environments moderate the effects of firms' IT assets portfolios. *Management Information Systems Quarterly*, *36*(2), 509–528.

Yang, K. H., Lee, S. M., & Lee, S. (2007). Adoption of information and communication technology. *Industrial Management & Data Systems*, *107*(9), 1257–1275. doi:10.1108/02635570710833956

Ye, F., & Wang, Z. (2013). Effects of information technology alignment and information sharing on supply chain operational performance. *Computers & Industrial Engineering*, *65*(3), 370–377. doi:10.1016/j.cie.2013.03.012

Yeh, C. H., Lee, G. G., & Pai, J. C. (2012). How information system capability affects e-business information technology strategy implementation: An empirical study in Taiwan. *Business Process Management Journal*, *18*(2), 197–218. doi:10.1108/14637151211225171

Yılmaz, F. G. K., Yılmaz, R., Ozturk, H. T., Sezer, B., & Karademir, T. (2015). Cyberloafing as a barrier to the successful integration of information and communication technologies into teaching and learning environments. *Computers in Human Behavior*, *45*, 290–298. doi:10.1016/j.chb.2014.12.023

Yukl, G., & Tracey, J. B. (1992). Consequences of influence tactics used with subordinates, peers, and the boss. *The Journal of Applied Psychology*, *77*(4), 525–535. doi:10.1037/0021-9010.77.4.525

Zafiropoulos, C., Vrana, V., & Paschaloudis, D. (2006). Research in brief the Internet practices analysis from Greece. *International Journal of Contemporary Hospitality Management*, *18*(2), 156–163. doi:10.1108/09596110610646709

Zander, L., Zettinig, P., & Makela, K. (2013). Leading global virtual teams to success. *Organizational Dynamics*, *42*(3), 228–237. doi:10.1016/j.orgdyn.2013.06.008

Zibetti, E., Chevalier, A., & Eyraud, R. (2012). What type of information displayed on digital scheduling software facilitates reflective planning tasks for students? Contributions to the design of a school task management tool. *Computers in Human Behavior, 28*(2), 591–607. doi:10.1016/j.chb.2011.11.005

Zineldin, M. (2004). Co-opetition: The organization of the future. *Marketing Intelligence & Planning, 22*(7), 780–789. doi:10.1108/02634500410568600

ADDITIONAL READING

Akoumianakis, D. (2014). Ambient affiliates in virtual cross-organizational tourism alliances: A case study of collaborative new product development. *Computers in Human Behavior, 30*, 773–786. doi:10.1016/j.chb.2013.03.012

Asakiewicz, C. (2011). Business investments in IT: Managing integration risks. *IT Professional Magazine, 13*(4), 41–45. doi:10.1109/MITP.2010.138

Boulesnane, S., & Bouzidi, L. (2013). The mediating role of information technology in the decision-making context. *Journal of Enterprise Information Management, 26*(4), 387–399. doi:10.1108/JEIM-01-2012-0001

Chang, H. H., Chuang, S. S., & Chao, S. H. (2011). Determinants of cultural adaptation, communication quality, and trust in virtual teams' performance. *Total Quality Management, 22*(3), 305–329. doi:10.1080/14783363.2010.532319

Chang, Y. B., & Gurbaxani, V. (2012). The impact of IT-related spillovers on long-run productivity: An empirical analysis. *Information Systems Research, 23*(3), 868–886. doi:10.1287/isre.1110.0381

Cheshin, A., Rafaeli, A., & Bos, N. (2011). Anger and happiness in virtual teams: Emotional influences of text and behavior on others' affect in the absence of non-verbal cues. *Organizational Behavior and Human Decision Processes, 116*(1), 2–16. doi:10.1016/j.obhdp.2011.06.002

Costello, T. (2011). 2011 IT tech and strategy trends. *IT Professional Magazine, 13*(1), 61–65.

Dimitrios, N. K., Sakas, D. P., & Vlachos, D. S. (2013). Analysis of strategic leadership models in information technology. *Procedia: Social and Behavioral Sciences, 73*, 268–275. doi:10.1016/j.sbspro.2013.02.052

Dong, J. Q. (2011). User acceptance of information technology innovations in the remote areas of China. *Journal of Knowledge-based Innovation in China, 3*(1), 44–53. doi:10.1108/17561411111120864

Faiola, A., Newlon, C., Pfaff, M., & Smyslova, O. (2013). Correlating the effects of flow and telepresence in virtual worlds: Enhancing our understanding of user behavior in game-based learning. *Computers in Human Behavior, 29*(3), 1113–1121. doi:10.1016/j.chb.2012.10.003

Fransen, J., Kirschner, P. A., & Erkens, G. (2011). Mediating team effectiveness in the context of collaborative learning: The importance of team and task awareness. *Computers in Human Behavior, 27*(3), 1103–1113. doi:10.1016/j.chb.2010.05.017

Goh, S., & Wasko, M. (2012). The effects of leader–member exchange on member performance in virtual world teams. *Journal of the Association for Information Systems, 13*(10), 861–885.

Haines, R., & Mann, J. E. C. (2011). A new perspective on de-individuation via computer-mediated communication. *European Journal of Information Systems, 20*(2), 156–167. doi:10.1057/ejis.2010.70

Hazen, B. T., & Byrd, T. A. (2012). Toward creating competitive advantage with logistics information technology. *International Journal of Physical Distribution & Logistics Management, 42*(1), 8–35. doi:10.1108/09600031211202454

Hosseini, M. R., & Chileshe, N. (2013). Global virtual engineering teams (GVETs): A fertile ground for research in Australian construction projects context. *International Journal of Project Management, 31*(8), 1101–1117. doi:10.1016/j.ijproman.2013.01.001

Iveroth, E., & Bengtsson, F. (2014). Changing behavior towards sustainable practices using Information Technology. *Journal of Environmental Management, 139*, 59–68. doi:10.1016/j.jenvman.2013.11.054 PMID:24681365

Jacks, T., Palvia, P., Schilhavy, R., & Wang, L. (2011). A framework for the impact of IT on organizational performance. *Business Process Management Journal, 17*(5), 846–870. doi:10.1108/14637151111166213

Joe, S. W., Tsai, Y. H., Lin, C. P., & Liu, W. P. (2014). Modeling team performance and its determinants in high-tech industries: Future trends of virtual teaming. *Technological Forecasting & Social Change: An International Journal, 88*(1), 16–25. doi:10.1016/j.techfore.2014.06.012

Ker, J. I., Wang, Y., Hajli, M. N., Song, J., & Ker, C. W. (2014). Deploying lean in healthcare: Evaluating information technology effectiveness in U.S. hospital pharmacies. *International Journal of Information Management, 34*(4), 556–560. doi:10.1016/j.ijinfomgt.2014.03.003

Kim, C., Lee, S. G., & Kang, M. (2012). I became an attractive person in the virtual world: Users' identification with virtual communities and avatars. *Computers in Human Behavior, 28*(5), 1663–1669. doi:10.1016/j.chb.2012.04.004

Kohli, R., Devaraj, S., & Ow, T. T. (2012). Does information technology investment influence a firm's market value? The case of non-publicly traded healthcare firms. *Management Information Systems Quarterly, 36*(4), 1145–1163.

Laudon, K. C., & Laudon, J. P. (2012). *Management information systems*. Upper Saddle River, NJ: Pearson Education.

Lusher, D., Kremer, P., & Robins, G. (2014). Cooperative and competitive structures of trust relations in teams. *Small Group Research, 45*(1), 3–36. doi:10.1177/1046496413510362

Martinez-Nunez, M., & Perez-Aguiar, W. S. (2014). Efficiency analysis of information technology and online social networks management: An integrated DEA-model assessment. *Information & Management, 51*(5), 712–725. doi:10.1016/j.im.2014.05.009

McGinnes, S., & Kapros, E. (2015). Conceptual independence: A design principle for the construction of adaptive information systems. *Information Systems, 47*, 33–50. doi:10.1016/j.is.2014.06.001

Mennecke, B. E., Triplett, J. L., Hassall, L. M., Conde, Z. J., & Heer, R. (2011). An examination of a theory of embodied social presence in virtual worlds. *Decision Sciences*, *42*(2), 413–450. doi:10.1111/j.1540-5915.2011.00317.x

Nah, F. F. H., Eschenbrenner, B., & DeWester, D. (2011). Enhancing brand equity through flow and telepresence: A comparison of 2D and 3D virtual worlds. *Management Information Systems Quarterly*, *35*(3), 731–747.

Olson, D. L., & Wu, D. D. (2011). Multiple criteria analysis for evaluation of information system risk. *Asia-Pacific Journal of Operational Research*, *28*(1), 25–39. doi:10.1142/S021759591100303X

Pai, F. Y., & Huang, K. I. (2011). Applying the Technology Acceptance Model to the introduction of healthcare information systems. *Technological Forecasting & Social Change: An International Journal*, *78*(4), 650–660. doi:10.1016/j.techfore.2010.11.007

Saeed, K. A. (2012). Evaluating the value of collaboration systems in collocated teams: A longitudinal analysis. *Computers in Human Behavior*, *28*(2), 552–560. doi:10.1016/j.chb.2011.10.027

Sanchez, M. A., Macada, A. C. G., & Sagardoy, M. D. V. (2014). A strategy-based method of assessing information technology investments. *International Journal of Managing Projects in Business*, *7*(1), 43–60. doi:10.1108/IJMPB-12-2012-0073

Seadle, M. (2012). Thirty years of information technology. *Library Hi Tech*, *30*(4), 557–564. doi:10.1108/07378831211285040

Steizel, S., & Rimbau-Gilabert, E. (2013). Upward influence tactics through technology-mediated communication tools. *Computers in Human Behavior*, *29*(2), 462–472. doi:10.1016/j.chb.2012.04.024

Tong, Y., Yang, X., & Teo, H. H. (2013). Spontaneous virtual teams: Improving organizational performance through information and communication technology. *Business Horizons*, *56*(3), 361–375. doi:10.1016/j.bushor.2013.01.003

Turan, A. H., & Palvia, P. C. (2014). Critical information technology issues in Turkish healthcare. *Information & Management*, *51*(1), 57–68. doi:10.1016/j.im.2013.09.007

Uflacker, M., & Zeier, A. (2011). A semantic network approach to analyzing virtual team interactions in the early stages of conceptual design. *Future Generation Computer Systems*, *27*(1), 88–99. doi:10.1016/j.future.2010.05.006

Yan, Y., Davison, R. M., & Mo, C. (2013). Employee creativity formation: The roles of knowledge seeking, knowledge contributing and flow experience in Web 2.0 virtual communities. *Computers in Human Behavior*, *29*(5), 1923–1932. doi:10.1016/j.chb.2013.03.007

Yonazi, E., Kelly, T., Halewood, N., & Blackman, C. (2012). *The transformational use of information and communication technologies in Africa*. Washington, DC: The World Bank.

KEY TERMS AND DEFINITIONS

Business: An organization or economic system where goods and services are exchanged for one another or for money.

Communication: The imparting or interchange of thoughts, opinions, or information by speech, writing, or signs.

Competitive Advantage: A superiority gained by an organization when it can provide the same value as its competitors but at a lower price, or can charge higher prices by providing greater value through differentiation.

Information and Communication Technology: The study of the technology used to handle information and aid communication.

Information Technology: The development, implementation, and maintenance of computer hardware and software systems to organize and communicate information electronically.

Innovation: The newly introduced something such as a new method and device.

Internet: The single worldwide computer network that interconnects other computer networks.

Technology: The branch of knowledge that deals with the creation and use of technical means and their interrelation with life, society, and the environment.

Virtual Team: A group of individuals spread across different time zones, cultures, languages or, ethnicities which are united by a common goal.

Chapter 2
The Role of Vertical and Shared Leadership in Virtual Team Collaboration

Julia Eisenberg
Pace University, USA

Jennifer Gibbs
Rutgers University, USA

Niclas Erhardt
University of Maine, USA

ABSTRACT

This chapter reviews current trends in the literature related to the influence of vertical and shared leadership styles in the context of virtual teams, unpacking the influence of team structure and task structure to better understand the mechanisms influencing team effectiveness. The authors start by reviewing key features of virtual teams and different aspects of leadership and its influence in the virtual team environment. They argue that both vertical and shared leadership have strengths and limitations, and both styles may complement one another. The authors discuss the influence of leadership on virtual team processes and outcomes and examine contingency factors related to team and task structure in order to identify the boundary conditions for the effectiveness of vertical and shared leadership. The chapter offers a conceptual framework to guide future research in this domain.

INTRODUCTION

Technological advancements have enabled unprecedented growth of virtual teams within organizations. About 66% of multinational companies rely on virtual teams (Society for Human Resource Management, 2012) and their popularity is expected to keep increasing as information and communication technologies (ICTs) evolve (Gilson, Maynard, Jones Young, Vartiainen, & Hakonen, 2015). Virtual teams offer a number of benefits ranging from providing access to remote expertise to reducing costs. However, they

DOI: 10.4018/978-1-4666-9688-4.ch002

are also associated with a number of challenges (Erhardt & Gibbs, 2014). Leadership has been advocated for decades to address issues in traditional collocated teams (Avolio, Walumbwa, & Weber, 2009; Day, Fleenor, Atwater, Sturm, & McKee, 2014; Hiller, DeChurch, Murase, & Doty, 2011); thus it is not surprising that leaders are also called upon to resolve virtual team issues. A key source of contention in the literature, however, has been whether virtual teams require unique leadership competencies, and if so, what style of leadership works best.

Some research advocates strong vertical leadership to provide virtual teams with structure (e.g., Hambley, O'Neill, & Kline, 2007; Kayworth & Leidner, 2002). Other research emphasizes more decentralized forms of shared leadership, which is associated with distributed authority in which team members share responsibilities and take on leadership roles to lead each other toward mutually established goals (e.g., Hoch & Kozlowski, 2014; Pearce & Conger, 2003). Research on leadership, specifically in virtual teams, seems to be moving away from traditional leadership styles that are rooted in hierarchy to embrace the notion of shared leadership, defined as "an emergent and dynamic team phenomenon whereby leadership roles and influence are distributed among team members" (D'Innocenzo, Mathieu, & Kukenberger, 2014, p. 5). While shared leadership may work quite well in a number of different situations, its presence does not necessarily replace vertical leadership styles (Hill, 2005). In fact, a paradox exists in that shared leadership may depend on a formal leader to empower team members to take on shared leadership roles.

Research has often taken an "either-or" approach, and leadership studies tend to contrast vertical and shared leadership (e.g. Hoch & Kozlowski, 2014; Pearce, Yoo, & Alavi, 2004). The authors argue here that this is an overly simplistic view of leadership in virtual teams, in that both shared and vertical leadership are necessary in virtual teams. Rather than regarding vertical leadership and shared leadership as mutually exclusive, the authors recognize that both styles can co-exist and that both may be needed to some extent. The aim of this chapter is to examine the boundaries of both vertical and shared leadership effectiveness and highlight contingency factors that may limit its influence across the complexities of virtual collaboration.

Contingency theory (Galbraith, 1972) is used to frame this chapter. By drawing on contingency theory, the authors attempt to flesh out key factors that may drive and undermine the effectiveness of different types of leadership. Two overarching categories of contingency factors are considered: team structure and task structure. *Team structure* is conceptualized here using dimensions of virtuality that have been shown in earlier research to influence collaborative processes among virtual team members (Gibson & Gibbs, 2006); vertical leadership is proposed to be more effective for teams that are higher in virtuality. *Task structure* is conceptualized here as the extent to which a team has task interdependence and complexity; shared leadership is proposed to be more effective for teams with more interdependent and complex tasks. The authors start out by reviewing the literature associated with leadership in the context of virtual collaboration and then develop a contingency framework to integrate the effects of leadership across team and task structures. The authors conclude the chapter by examining the implications and future research directions for leadership in virtual teams.

BACKGROUND

Research on teamwork has long noted the importance of teams as knowledge integrative mechanisms (Erhardt, 2011; Grant, 1996). Over the last few years, teamwork has become increasingly virtual with the use of ICTs in the workplace. However, while enabling access to expertise and facilitating collaboration, virtual work is characterized by a number of discontinuities (Watson-Manheim, Chudoba, & Crowston, 2002). For example, virtual team structure is associated with a number of management challenges, some of which are especially relevant to teams working on knowledge-intensive work including new organizational designs (Erhardt, Martin-Rios, & Way, S. A. (2009).) and the development of information systems, due to the heightened reliance and importance placed on communication and knowledge sharing (i.e., tacit knowledge) for successful team outcomes.

Collaborative relationships are especially challenging in the context of virtual teams due to their reliance on ICTs, and they are exacerbated by geographical dispersion (Hill, 2005). The nature of the team's structure is fundamentally important to projects related to information systems, which rely on integrating team member expertise and benefit from leadership that encourages individual performance and teamwork, promotes coordination and facilitates goal setting (Faraj & Sambamurthy, 2006). However, because teams focused on the development of information systems are often composed of highly skilled team members who must actively share their knowledge with one other, supplemental leadership styles may be needed to promote member interaction (Hoch & Dulebohn, 2013). Some suggest that "one way to think about virtual team environments is as contexts (i.e., modus operandi) in which anyone can participate and have an equal influence in the team" (Zigurs, 2003, p. 339), recommending a more shared approach to leadership.

As teams working on knowledge-intensive tasks are increasingly geographically dispersed and include members from various locations, it is important to explore the boundaries of effectiveness associated with both vertical and shared leadership styles. Specifically, teams that are higher in virtuality may be harder to lead using just one leadership style. The greater complexity of such teams may require a more multifaceted combination of vertical and shared leadership styles.

Further, virtual teams vary greatly in the extent to which they are virtual, such that talking about virtual teams as a monolithic category for which a single leadership style is best suited may not make sense. The basis for effective teamwork is some form of leadership that helps the team identify a plan, coordinate actions, resolve disagreements, stay on track, communicate, and hold team members accountable for individual responsibilities. However, the type of leadership required may depend on the degree of virtuality of the team and its task type. Therefore, leaders need to understand the specifics of the given context and the extent of virtuality across a multitude of dimensions and adapt their styles accordingly (Zigurs, 2003).

To examine the effects of vertical and shared leadership styles in virtual teams, the authors adopt a multidimensional definition of virtuality as composed of the features of geographical dispersion, electronic dependence, national diversity, and dynamic structure (Gibson & Gibbs, 2006), which are often associated with greater complexity. Gluesing and Gibson (2002) argue that global virtual teams involve more complexity because they are embedded in multiple geographical, cultural, and temporal contexts. Connaughton and Shuffler (2007) also note that multinational multicultural teams are more complex on the dimensions of culture and distribution or distance. Finally, Chudoba, Wynn, Lu, and Watson-Mannheim (2005) regard virtuality as involving discontinuities in geographical location, time zone, culture, and organizational affiliation, which also increase the level of complexity in virtual teams. Lean communication and limited social presence and context cues through ICTs, cultural misunderstandings, and logistical issues such as coordination across time zones add additional layers of complexity for leaders in virtual teams (Kayworth & Leidner, 2002).

LEADING VIRTUAL TEAMS

The influence of leadership has been of interest for quite some time and has received a lot of attention in the literature (for a recent review see Day, Fleenor, Atwater, Sturm, & McKee, 2014). Leadership has been found to play a central role in overcoming virtual team challenges and has been identified as an important research opportunity (for a review see Gilson et al., 2015). Not surprisingly, some of the styles recognized for their effectiveness in traditional collocated settings do not have the same influence in virtual settings, and leadership theories are not easily transferred to the more complex virtual context (Carte, Chidambaram, & Becker, 2006).

Vertical and shared leadership styles have been contrasted in earlier studies to examine their influence in virtual teams (Hoch & Kozlowski, 2014; Pearce et al., 2004). Hoch and Kozlowski (2014) found that shared leadership influenced virtual teams more consistently across virtual team contexts than vertical leadership. Similarly, Muethel and colleagues demonstrated shared leadership's effectiveness in influencing virtual team performance (Muethel & Hoegl, 2010). Furthermore, a number of other studies that have compared vertical and shared leadership have concluded that shared leadership is a better predictor of team effectiveness than vertical leadership (Pearce et al., 2004; Pearce & Sims, 2002). However, some scholars have highlighted a need to go beyond a narrow focus on vertical or shared leadership as that is an oversimplification, instead recommending the examination of both (Pearce et al., 2004; Pearce, 2004). Giuri and colleagues claim that vertical leaders have an important task of managing the virtual team's "distributed authority," among members engaged in shared leadership, arguing that lack of centralized coordination may be "a major source of failure" for many projects (Giuri, Rullani, & Torrisi, 2008, p. 306). Such contrasting findings may in part be due to the great variability of the virtual team context, which makes a leadership style that may be effective in one context lose its efficacy in another context.

While leadership style is important in influencing team dynamics and outcomes, its influence is likely to be contingent on a number of factors. An approach to leadership that may be effective in one type of team with a unique set of team and task structural components may be less influential in another type of team with a different set of structural components. While many suggest that leadership of virtual teams is more complex than that of traditional face-to-face teams (e.g., Bell & Kozlowski, 2002; Gibson & Cohen, 2003), in some cases, team members have been found to be more satisfied and leaders better at decoding messages from team members when leaders are geographically distant (Henderson, 2008) and the effect of leadership on team performance has been found to be more pronounced among virtual teams than in face-to-face teams (Purvanova & Bono, 2009).

Vertical Leadership and Virtual Teams

Traditional vertical leadership (sometimes referred to as hierarchical) has been recognized by a number of studies for its effectiveness in collocated settings (Wang, Oh, Courtright, & Colbert, 2011) and it has been suggested to be a crucial mechanism for increasing virtual team effectiveness (Bell & Kozlowski, 2002; Malhotra, Majchrzak, & Rosen, 2007). Vertical types of leadership (e.g., transformational, transactional), which explicitly or implicitly assume a leader-follower relationship, offer a range of advantages such as providing an overall vision, role clarity, individual performance assessments, and overall structure, which may help address negative team processes such as conflict and social loafing and help ensure team success. Coordination of effort among various contributors, including those outside the immediate team, has been lauded as one of the important benefits of having a centralized vertical leader in virtual teams (Giuri et al., 2008). Further, in a study of global virtual teams, effective vertical team leaders were able to address issues by taking on a variety of simultaneous leadership duties from mentoring to asserting authority, effectively communicating with followers, and fostering relationships among virtual team members (Kayworth & Leidner, 2002, p. 7). In global teams, successful vertical leaders focus on "managing external linkages, setting direction, managing team operations, and acquiring adequate resources and organizational support" (Joshi & Lazarova, 2005, p. 285), thus helping the team focus and centralize its operations. A review of virtual team leadership suggests effective vertical leadership is characterized by establishing norms early, maintaining control, and encouraging emergent leadership (Connaughton & Daly, 2005). The vertical leader, or centralized authority figure, may provide stability and structure, help team members address the lack of role clarity associated with the geographically dispersed context of virtual teams, and diminish social loafing by increasing individual accountability.

While a number of different vertical leadership styles have been examined in the context of virtual teams (Avolio & Kahai, 2003; Hambley et al., 2007; Sosik, Avolio, Kahai, & Jung, 1998), transformational leadership has been suggested to be particularly effective in building committed and high performing virtual teams (Hambley et al., 2007; Purvanova & Bono, 2009). Transformational leadership is defined as a process of leading team members towards mutual goals in which the leader motivates employees to work towards achieving higher level self-actualization objectives (Burns, 1978; Pawar & Eastman, 1997). Inspirational (a form of transformational) leadership was found to be particularly effective in dispersed team settings as it helped foster a collective identity that strengthened trust and team commitment (Joshi, Lazarova, & Liao, 2009). However, some research questions the extent of a vertical leader's influence across all contexts, suggesting the need to examine the influence of shared leadership (Hoch & Kozlowski, 2014).

Shared Leadership and Virtual Teams

While vertical leadership has been around much longer, shared leadership is gaining momentum. The literature is increasingly advocating the benefits of shared leadership, particularly in virtual teams (Hill, 2005; Hoch & Kozlowski, 2014; Muethel & Hoegl, 2010; Pearce et al., 2004; Robert & You, 2013; Shuffler, Wiese, Salas, & Burke, 2010). In a recent review of vertical taxonomies of leadership, Yukl called for more research on shared leadership (Yukl, 2012). The recent scholarly attention may be attributed to a broader organizational trend to flatten hierarchy and shift more power into the hands of lower echelons of workers (Erhardt et al., 2009). A recent review of 25 years' worth of studies related to leadership and its influence on outcomes suggested that shared leadership is among leadership styles with "the potential to make significant advances in our understanding of leadership phenomena and outcomes and should continue to be studied" (Hiller, DeChurch, Murase, & Doty, 2011, p. 1170).

A number of different terms and corresponding definitions are associated with leadership that is "distributed," "participative," or "shared." Shared leadership has been defined as "mutual influence embedded in the interactions among team members" (Carson, Tesluk, & Marrone, 2007, p. 1218) as well as "a dynamic, interactive influence process among individuals in groups for which the objective is to lead one another to the achievement of group or organizational goals or both" (Pearce & Conger, 2003, p. 1). Empirical evidence of the value of shared leadership has been associated with collaborative decision making, leading one another towards common goals and knowledge sharing (Day, Gronn, & Salas, 2004; Pearce & Conger, 2003) and enhancing both team and organizational outcomes (Hoch & Dulebohn, 2013).

While some studies have suggested that virtual teams require greater distribution of leadership responsibilities among team members (Shuffler et al., 2010), highlighting the particular importance of shared leadership (Hill, 2005), a gap exists in the literature in terms of the implications of shared leadership for various types of virtual teams (Carte et al., 2006; Robert & You, 2013), including those with different team and task structures. Shared leadership may be particularly effective in addressing some of the challenges associated with virtuality by facilitating team member interaction (Hoch & Kozlowski, 2014; Pearce & Conger, 2003). First, shared leadership is associated with more involved team members, and therefore may have a positive influence on the quality and quantity of team communication. Moreover, self-management has been suggested to improve the performance of virtual teams (Kirkman, Rosen, Tesluk, & Gibson, 2004). Second, since virtual work is often associated with higher levels of cognitive tasks (Hoch & Kozlowski, 2014), shared leadership may be particularly relevant due to the higher complexity in such teams (Pearce & Conger, 2003). Third, since shared leadership has been associated with stronger interpersonal relationships, trust, and collaboration (Pearce & Conger, 2003), it may be instrumental in helping facilitate other processes and in turn team performance. Finally, shared leadership has been suggested to be particularly influential in interdependent, complex, and creative contexts (Ensley, Hmieleski, & Pearce, 2006; Pearce & Ensley, 2004) and more effective than vertical leadership for teams characterized by greater task complexity (Cox, Pearce, & Perry, 2003; D'Innocenzo et al., 2014).

Research suggests that shared leadership in virtual teams may be especially useful in providing essential supplemental mechanisms to vertical leadership (Bell & Kozlowski, 2002; Hoch & Kozlowski, 2014; Pearce et al., 2004), rather than replacing it altogether. Vertical leadership styles have been suggested to facilitate shared leadership (Hoch & Dulebohn, 2013). In this way, these two styles may be compatible rather than mutually exclusive. Both vertical and shared leadership styles have important implications for virtual team performance but there is still lack of clarity about the effectiveness of each of these styles across different types of teams. The aim of this chapter is to explore and contrast the structural mechanisms for both styles, demonstrating that the sphere of influence for vertical and shared leadership will be constrained by contingency factors associated with team and task structure.

CONTINGENCY MODEL

Our review of the virtual teams literature suggests that findings are mixed on the relative merits of traditional vertical or shared leadership styles in virtual teams. For example, Hoch and Kozlowski (2012) found that shared leadership was associated with better performance in global virtual teams. Research has also found that shared leadership is harder to foster in virtual teams but that it has benefits when achieved by promoting crucial affective components related to team effectiveness such as facilitating cohesion and shared vision (Shuffler et al., 2010). Other studies have found evidence for the benefits of transformational leadership in virtual teams (e.g. Connaughton & Daly, 2004; Joshi & Lazarova, 2005; Kayworth & Leidner, 2002).

Rather than viewing vertical leadership as outdated and shared leadership as a superior new form, the greater complexity of today's virtual teams is likely to require a combination of both vertical and shared leadership to maximize the value of leadership processes (Day et al., 2004; Ensley et al., 2006; Pearce & Sims, 2002). Vertical leaders play an important range of roles that cannot be dismissed or replaced by shared leadership. A review of the virtual teams literature suggests that an effective virtual team leader is characterized as someone who is more flexible in letting team members take the lead when appropriate (Powell, Piccoli, & Ives, 2004). However, allowing team members to take the lead is quite different than relinquishing all forms of control and reinforces the role of the vertical leader in facilitating computer-mediated communication (CMC). Furthermore, conflict is more likely to occur in virtual teams (Furumo, 2009), making conflict resolution an essential component of leaders of such teams (Wakefield, Leidner, & Garrison, 2008). The need to take charge and resolve conflict reinforces the need for a clearly designated vertical leader to take on the role of an arbitrator.

In a study of virtual teams, the effects of leadership styles have been suggested to be contingent on factors such as the communication environment (Huang, Kahai, & Jestice, 2010). The authors believe there are other contingency factors that warrant examination. Furthermore, to reconcile conflicting findings related to leadership in virtual teams, it is important to examine contingency factors missing from this discourse that may impact team performance. Team and task structure have been suggested to be quite influential in virtual teams (Hinds & Kiesler, 1995; Hoch & Dulebohn, 2013; Hoch & Kozlowski, 2014) and have been examined for their interaction with leadership-related dynamics (Cordery & Soo, 2008; D'Innocenzo et al., 2014). Therefore, the authors posit that the effectiveness of a particular leadership style is contingent on the team and task structure of a virtual team.

We draw on contingency theory (Galbraith, 1972; Miller, 1988) to consider how team and task structure may offer additional insights to understand the relationship between leadership and team performance and help resolve inconsistencies found in research contrasting vertical and shared leadership. In studying the effectiveness of leadership in virtual teams, a contingency approach is useful because rather than assuming a "one-size-fits-all" leadership style, it acknowledges a number of influential factors that will determine the scope of leader's effectiveness in responding to a wide range of complex organizational contexts (Kayworth & Leidner, 2002). Figure 1 outlines the conceptual model and will guide the subsequent discussion.

Contingency Factor I: Team Structure

Virtual team structures are often characterized by a combination of the four dimensions suggested by Gibson and Gibbs (2006): *geographical dispersion, electronic dependence, dynamic structure,* and *na-*

Figure 1. Theoretical model

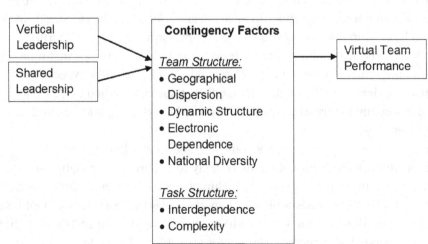

tional diversity. The first dimension, *geographical dispersion,* signifies the extent to which team members are collocated or spread out across multiple locations. Virtual teams are often utilized to enable access to geographically dispersed expertise, and thus are more likely to be working on projects in which vertical leadership has a critical role in influencing virtual team effectiveness (Bell & Kozlowski, 2002). Virtual team leaders are relied on to provide a clear set of goals and direction (Hambley & Kline, 2007; Kayworth & Leidner, 2002; Malhotra et al., 2007) but they must also facilitate greater engagement by members to foster team performance (Bell & Kozlowski, 2002; Zaccaro & Burke, 1998).

Interestingly, vertical leaders may provide increased responsibility to team members to facilitate a number of team processes virtually. Collective responsibility that is distributed among team members is associated with shared leadership behaviors. However, geographical dispersion may make it harder for team members to foster the types of bonds and relationships that would facilitate shared leadership. Without having sufficient information about fellow teammates, whether about their expertise or personalities, it may be difficult for team members to effectively work together, necessitating the presence of a vertical leader who can provide and reinforce direction for the team as well as foster increased communication and collaboration across multiple office locations. For example, in a study of software development project teams operating virtually, leaders had to take on critical tasks such as providing a vision, attracting new contributors, facilitating division of responsibilities, and helping participants avoid conflict (Giuri et al., 2008; Lerner & Tirole, 2002).

Electronic dependence, the second dimension of virtuality examined, is based on the extent to which a team communicates using technology and depends on it for collaborative interactions (Gibson & Cohen, 2003; Gibson & Gibbs, 2006; Griffith, Sawyer, & Neale, 2003). Most modern teams, regardless if they are geographically dispersed or co-located, engage in teamwork through some level of technology use. Electronic dependence is likely to limit important success factors for team performance such as interpreting feedback and other information by diminishing nonverbal cues (Desanctis & Monge, 1999; Gibson & Gibbs, 2006). Additionally, virtual teams' reliance on CMC has been suggested to diminish the effectiveness of communication processes as well as other key team dynamics (Cramton & Orvis, 2003; Hill, 2005).

CMC is associated with decreased individualization and may lead to feelings of isolation (Connaughton & Daly, 2004). Effective vertical leaders, particularly transformational leaders, have been suggested to have a stronger effect on team performance in virtual teams, in part by helping teams overcome communication constraints associated with the virtual environment (Purvanova & Bono, 2009), thus helping team members overcome some of the challenges associated with CMC. Furthermore, newer technologies such as enterprise social media have been found to help reduce vertical barriers and enable direct communication among leaders and lower-level employees, which may reduce miscommunication and increase the speed of communication across hierarchical boundaries (Gibbs, Eisenberg, Rozaidi, & Gryaznova, 2015), facilitating communication, in turn helping address some of the challenges associated with electronic dependence. When well-managed, CMC may help team members overcome time, space, and organizational boundaries (Millward & Kyriakidou, 2004).

Some of the communication-related issues may be at the root of the constant changes within a team associated with composition, roles, and relationships that are characteristic of the modern work environment, which is increasingly *structurally dynamic* (Brown & Eisenhardt, 1995; Gibson & Gibbs, 2006). Issues related to dynamic structure may pose particular challenges for virtual team leaders (Malhotra et al., 2007) and can greatly influence leadership effectiveness (Hoch & Kozlowski, 2014). Vertical team leaders may help structurally dynamic teams adjust to the constant changes and facilitate a focus on goals, while helping team members cooperate with each other (Faraj & Sambamurthy, 2006).

The final virtual dimension outlined by Gibson & Gibbs (2006) involves diversity, which raises additional challenges for virtual teamwork and leadership within it. Conflicts might be particularly problematic in virtual teams where employees not only collaborate across geographical boundaries but also must overcome differences associated with their national backgrounds. The added value of diversity hypothesis suggests that national background shapes how individuals perceive, process and operate in their work (Erhardt, Werbel & Shrader, 2003; Jackson, Joshi, & Erhardt, 2003), which may make it harder for them to collaborate. *National diversity* is a common feature in many virtual teams and particularly in global virtual teams, potentially exacerbating the challenges to collaborative processes that are crucial to successful teamwork. The more diverse the composition of the team, the more likely members are to have trouble collaborating, at least initially. National diversity has also been connected to team communication constraints (Earley & Gibson, 2002), potentially in part because members of nationally diverse teams may have different communication norms and preferences (Gibson & Gibbs, 2006). Diversity challenges are likely to require vertical leadership.

Some have suggested that in global team environments, shared leadership may be quite influential (Gibbs & Boyraz, 2015). However, shared leadership may be particularly hard to foster in environments where culturally, employees do not feel comfortable being more engaged by contributing to their team in ways outside their delineated job descriptions. To some, engaging in shared leadership may be viewed as questioning their vertical leader's authority and thus counterproductive. As the team environment increases in complexity, this is likely to bring about greater uncertainty about team processes and relationships. This is likely to make the need for a vertical leader more salient to help facilitate team norms and provide clear direction and shared vision. Research suggests that in teams that are highly virtual in structure, effective leadership requires a vertical approach through clarifying structure, responsibilities, and direction and thus facilitating greater shared consensus and vision (Heckman, Crowston, & Misiolek, 2007; Kayworth & Leidner, 2002; Powell et al., 2004). Furthermore, effective virtual team leaders are able to balance multiple complexities and contradictions, assert their authority, foster regular and prompt communication, as well as help clarify team member responsibilities (Kayworth & Leidner, 2002), which highlights the important role of the vertical leadership style. This leads to the first proposition:

Proposition 1: The influence of vertical leadership on team performance will be contingent upon the degree of virtuality, such that it is more effective in highly virtual teams in terms of the dimensions of geographical dispersion, dynamic structure, electronic dependence, and national diversity.

Contingency Factor II: Task Structure

In addition to examining team structure, literature has highlighted the importance of task structure including interdependence and task complexity (Hinds & Kiesler, 1995; Hoch & Dulebohn, 2013; Hoch & Kozlowski, 2014). *Task interdependence* has been found to influence the effects of leadership in virtual teams (Hoch & Kozlowski, 2012). Generally, interdependence among members of a unit has been classified into one of the following hierarchy of types: pooled/ independent, sequential, reciprocal, and intensive (Van De Ven, Delbecq, & Koenig, 1976; see Erhardt, 2011 for a similar framing of team interdependence). The higher up on the hierarchy and further away from being independent and closer to intensive, the more complex and costly the workflow becomes and the more it relies on collaboration and coordination (Van de Ven et al., 1976). Task interdependence is differentiated based on the following types of interconnectedness: "task (the flow of work between actors), role (the position of actors engaged in concerted action), social (mutual needs or goals of actors) and knowledge (the differentiated expertise of actors)" (Van de Ven et al., 1976, p. 324).

In order to discuss teamwork in a meaningful way, acknowledging task interdependence among team members is essential, and perhaps even more so regarding the functioning of virtual teams. Interdependence is a key moderator of virtual team processes, suggesting that it can positively influence team performance (Gilson et al., 2015; Maynard & Gilson, 2013). Levels of interdependence among virtual team members may influence their reliance on each other, making them feel closer and thus increasing their sense of perceived proximity, which may be a more accurate measure of the effects of virtuality on team member perceptions (Wilson, Boyer O'Leary, Metiu, & Jett, 2008).

There is a natural connection between shared leadership and interdependence in teams. Pearce and Conger (2003) suggest that shared leadership "helps meet the need for coordinated effort under conditions of interdependence" particularly because peer-to- peer communication – associated with shared leadership – should help increase awareness of team member interdependence, reinforcing the need for quality collaboration (Pearce & Conger, 2003, p. 65). Further, research on self-managed teams suggests that virtual teams, which often include experts focused on interdependent work, may work better when they have more control in the hands of the team members rather than a vertical leader (Manz, 1986). In interdependent work, the coordinating role of a vertical leader may lack sufficient flexibility, which is why shared leadership is likely to be more appropriate. Team members may know more about each other's dependencies and how to best manage them, reinforcing the need for a shared leadership approach to coordination and leadership. This is also likely the case for teams focused on complex tasks as there is an increased need for team members to provide boundary spanning and interaction than in less interdependent, more modular teams (Erhardt, 2011). There is also more need and opportunity for team members to interact and develop shared leadership behaviors in more interdependent as opposed to more modular virtual teams.

Task complexity is a second important dimension of task structure to evaluate in studying team dynamics. As task complexity increases, shared leadership is more likely to be effective in facilitating task completion than vertical leadership (Cox et al., 2003; D'Innocenzo et al., 2014). Evaluating task complexity may be particularly relevant for virtual teams as they are often utilized to address complex tasks and include cross-functional team members, presenting unique leadership challenges (Malhotra et al., 2007). On the one hand, the virtual team environment may diminish challenges associated with complex tasks (Kock & Lynn, 2012) by helping "digest" ideas of others at a pace that may allow for time to identify and diminish disagreements (Malhotra et al., 2007). On the other hand, there are fewer opportunities to form relationships with team members, interact informally, engage in knowledge sharing, and brainstorm ideas to foster creativity.

A review of the virtual teams literature suggests that lateral communication is facilitated by reducing vertical structure and decentralization (Ebrahim, Ahmed, & Taha, 2009). In contrast to less complex tasks that can be performed through weaker and asynchronous linkages among team members with decreased reliance on collaborative relationships and knowledge sharing, more complex tasks are more dynamic and require greater reliance on connections among team members (Bell & Kozlowski, 2002). This highlights the potential importance of shared leadership, which is associated with greater formal and informal interactions among team members. Furthermore, Kirkman et al. (2004) found that highly virtual teams (those relying heavily on ICTs) performed better in non-routine learning-oriented tasks when they had high levels of team empowerment (i.e., shared leadership). Their study outlines four key dimensions for driving empowerment:

1. Potency (the belief that the team can be effective);
2. Meaningfulness (team members feeling an intrinsic caring for their task;
3. Autonomy (the belief that team members have freedom to make decisions); and
4. Impact (the perception that the task makes a significant contribution).

Kirkman et al.'s study suggests that empowering team members becomes more essential for non-routine tasks the more virtual the team is. Task complexity requires more collaboration and thus greater shared leadership. For virtual teams working on tasks that require high levels of collaboration, such as complex non-routine tasks, it is important to have high levels of interaction and proactive behavior (Hill, 2005). Hill suggests that such behavior is associated with shared leadership and may in turn increase task based trust, positively influencing the quality of virtual team interactions and knowledge sharing (Hill, 2005; Jarvenpaa, Knoll, & Leidner, 1998).

Teams focused on complex tasks often employ team members from a variety of functional backgrounds. Exacerbating problems related to virtuality, cross-functional teams may have to overcome issues related to different thought worlds, which are associated with differences in approaches. training and perspectives in team members with different functional backgrounds (Dougherty, 1992). This may make it harder for members of differing functional backgrounds to understand and trust each other. Yet, trust has been suggested to be associated with a higher level of knowledge sharing (McEvily, Perrone, & Zaheer, 2003; Szulanski, 2000), making diminished levels of trust particularly problematic for teams working on knowledge-intensive and complex tasks. Furthermore, as trust among virtual team members may help facilitate open communication and thus enable discussion of differences, potentially helping to avoid conflict (Gibson & Manuel, 2003; Hill, 2005), lack of trust may be quite problematic for teams that may already be dealing with higher levels of conflict due to differences in their functional backgrounds.

Trust has been found to be especially challenging in virtual settings given the absence of nonverbal cues (Jarvenpaa & Leidner, 1999) and as vertical leadership is based on a trustworthy role model (Bass, 1985; Burns, 1978), this style of leadership may be particularly affected by the virtual environment. However, shared leadership may help facilitate greater trust and cohesion among team members because of the stronger bonds it creates among team members (Hoch & Kozlowski, 2014; Pearce & Conger, 2003). Research has also shown that informal communication among teams focused on knowledge-intensive work may facilitate shared understanding and in turn knowledge sharing (Majchrzak, Rice, King, Malhotra, & Sulin, 2000). The increased informal communication associated with shared leadership may help foster increased knowledge sharing, thus facilitating teamwork on complex tasks. This highlights the important role of having multiple team members take on leadership roles to enhance collaboration among team members and in turn improve overall team performance. This leads to the second proposition:

Proposition 2: The influence of shared leadership on team performance will be contingent upon the task structure, such that it is more effective in teams characterized by high task interdependence and task complexity.

DISCUSSION

This chapter reviews the state of the art related to vertical and shared leadership of virtual teams. Rather than regarding the two styles as an "either-or" approach or arguing for a "one-size-fits-all" model of leadership in virtual teams as prior literature often does, the authors regard vertical and shared leadership as complementing one another and argue that virtual teams require a combination of the two. Drawing on contingency theory, the authors propose a model where team structure and task structure shape the effectiveness of vertical and shared leadership and constitute boundary conditions in influencing virtual team performance.

Specifically, the authors propose that vertical leadership may be more effective in virtual teams characterized by greater levels of virtuality and thus complexity, such as those with greater geographical dispersion, electronic dependence, dynamic structure, and national diversity. Formally designated vertical leaders may be more effective in providing the needed direction, dealing with conflict, unifying members and facilitating relationships in teams operating in environments with greater structural complexity. Further, the authors propose that shared leadership will be more effective in teams with greater task interdependence and task complexity. In teams where team members are highly interdependent or working on complex tasks requiring greater levels of collaboration and knowledge sharing, shared leadership is likely to be more effective.

A theoretical contribution of this chapter is to review various dimensions of virtual teams that affect the effectiveness of leadership in influencing teams. The authors develop a contingency model that articulates conditions under which vertical and shared leadership are particularly effective in virtual teams. This helps to reconcile prior literature on leadership in virtual teams, which is fraught with disagreement on whether strong, vertical leadership is needed to provide structure in virtual teams or whether as decentralized, network forms, shared leadership is more effective in such teams. The authors also highlight the paradox that while shared leadership may be a great supplement to vertical leadership, especially in more interdependent and complex task environments, it may necessitate vertical leader's involvement to help shared leadership emerge and be sustainable over time. In this way, vertical and shared leadership are not opposing styles are they are often conceptualized in the literature, but they are compatible and complementary as each brings differing benefits.

Today's fast paced trends related to outsourcing the development of products and systems across multiple locations highlight the importance of virtual teams and as such the need to study this context further. In developing information systems, a number of different issues related to the team structure and task structure are relevant. Teams are assembled by bringing in members for their expertise to work on interdependent and complex tasks, highlighting the importance of shared leadership. However, because these types of teams are operating across virtual contexts, characterized by the complex collaborative environment that is associated with multiple virtual team dimensions, presence of a vertical leader may be necessary to help establish norms, provide direction, and address conflict and issues among team members.

Given the complexity of virtual teams, it is unlikely that a certain style of leadership is always most influential across various contexts. For example, local cultural context may have important implications in terms of the extent of team member participation. The more local experts engage in shared leadership behavior, the more likely the team is to benefit from their knowledge of the local environment. Therefore, in teams characterized by interdependence and complexity of tasks, reliant on each other's input and knowledge sharing, shared leadership is more likely to effectively address issues related to the intricacies of the local context than vertical leadership. However, the greater the virtuality of the team, the harder it may be to arrive at a consensus and avoid conflict, highlighting the need for a vertical leader to provide direction towards a unified set of goals for the team. In sum, managers and practitioners should look to vertical leaders to help facilitate shared leadership in order to maximize the efficiency of complex virtual teams by having a formal authority figure guide the team in conjunction with multiple members who share leadership responsibilities. This in turn may help to bridge some of the gaps associated with complex work and improve overall virtual team efficiency.

FUTURE RESEARCH DIRECTIONS

Our focus in this chapter was to review factors that influence collaborative processes in virtual teams and contrast the influence of vertical and shared leadership team effectiveness. Modern organizations are characterized by constant changes that influence the work environment, including interactions among virtual team members, highlighting a number of possible future research directions. As technology becomes ever more advanced, globalization trends further increase national diversity, and dynamic structure becomes even more prevalent, it is likely theories of teamwork will continue to evolve. Perhaps some of the complexities associated with virtual teams will diminish, reducing the need for vertical leadership. However, it is also likely that new complexities will emerge. For example, one factor the authors did not account for was generational diversity; as more technology savvy millennials enter the workforce, complexities emerging from different comfort levels with and attitudes toward CMC may emerge. Will younger workers feel more comfortable in computer-mediated environments since they grew up using electronic devices to communicate? How will they react to leadership enacted or communicated through CMC?

The multitude of changes anticipated to follow organizational development of the future may also influence how virtuality will influence team processes and outcomes. Widespread usage of emerging communication technologies such as enterprise social media (ESM) may help bridge some of the differences among colleagues and perhaps help connect team members across various dimensions of virtuality (Gibbs et al., 2015) but they can also enable employees, particularly those who want to minimize interactions, to hide behind technology and disengage (Gibbs, Rozaidi, & Eisenberg, 2013). Future research should examine the ways in which the use of different technologies shapes the effectiveness of leadership styles across virtual team settings. Further, more empirical research is needed to better understand the influence of vertical and shared leadership and how these styles work in tandem to maximize virtual team effectiveness. Finally, the authors examined complexity in virtual teams, however, they did not consider size as a key factor for leadership. Many growth-oriented organizations grapple with the goal of increasing efficiency while maintaining innovation capabilities and flexibility, and some organizations have succeeded such as Apple. Zappos, currently experimenting with the notion of *holacracy*, is replacing top-down leadership and taking shared leadership to the extreme. Jobs are defined around work and decisions are made locally where they occur, with a fluid and constantly updated organizational structure with self-organizing teams. The authors encourage future research on this topic to explore the interface of vertical and shared leadership in maintaining the "startup feel".

CONCLUSION

There is considerable evidence that addressing challenges associated with virtual collaboration requires a special approach. Theories that were based on studies of traditional face-to-face teams may no longer be valid. Applying traditional theories to understand virtual teams is likely to lead to misconceptions and erroneous conclusions. As more work is becoming virtual, it is important to examine factors that facilitate leadership effectiveness in complex teams collaborating across distance and time and using CMC. While shared leadership has emerged as a positive new force influencing teams (D'Innocenzo et al., 2014) and particularly virtual teams with high levels of team member interdependence (Hill, 2005), it may not always be the ideal leadership approach. In virtual teams characterized by higher levels of structural complexity, vertical leadership may be necessary to provide direction and steer the team towards achievement of goals, resolve conflict, and facilitate communication. Shared leadership is likely to complement vertical leadership but not be sufficient as the sole approach. Given the continuous technology advancement anticipated to continue changing the rules of the game at an ever growing pace, vertical leadership may become even more important in helping teams quickly and efficiently address challenges, while enabling team members to take on more shared leadership roles to help address escalating complexity of tasks.

REFERENCES

Avolio, B., Walumbwa, F. O., & Weber, T. J. (2009). Leadership: Current theories, research, and future directions. *Annual Review of Psychology*, *60*(1), 421–449. doi:10.1146/annurev.psych.60.110707.163621 PMID:18651820

Avolio, B. J., & Kahai, S. (2003). Adding the "e" to e-leadership: How it may impact your leadership. *Organizational Dynamics*, *31*(4).

Bass, B. M. (1985). *Leadership and performance beyond expectations*. New York: Free Press.

Bell, B. S., & Kozlowski, S. W. J. (2002). A typology of virtual teams: Implications for effective leadership. *Group & Organization Management*, *27*(1), 14–49. doi:10.1177/1059601102027001003

Brown, S. L., & Eisenhardt, K. M. (1995). Product development: Past research, present findings, and future directions. *Academy of Management Review*, *20*(2), 343.

Burns, J. M. (1978). *Leadership*. New York: Harper & Row.

Carson, J. B., Tesluk, P. E., & Marrone, J. A. (2007). Shared leadership in teams: An investigation of antecedent conditions and performance. *Academy of Management Journal*, *50*(5), 1217–1234. doi:10.2307/20159921

Carte, T. A., Chidambaram, L., & Becker, A. (2006). Emergent leadership in self-managed virtual teams. *Group Decision and Negotiation*, *15*(4), 323–343. doi:10.1007/s10726-006-9045-7

Connaughton, S. L., & Daly, J. (2004). Leading from afar: Strategies for effectively leading virtual teams. In S. H. Godar & S. P. Ferris (Eds.), *Virtual and Collaborative Teams: Process, Technologies and Practice* (pp. 49–75). Hershey, PA: Idea Group Publishing. doi:10.4018/978-1-59140-204-6.ch004

Connaughton, S. L., & Daly, J. (2005). Leadership in the new millennium: Communicating beyond temporal, spacial, and geographical boundaries. In P. Kalbfleisch (Ed.), *Communication Yearbook 29* (pp. 187–213). Mahwah, NJ: Psychology Press.

Cordery, J. L., & Soo, C. (2008). Overcoming impediments to virtual team effectiveness. *Human Factors, 18*(5), 487–500. doi:10.1002/hfm.20119

Cox, J., Pearce, C. L., & Perry, M. (2003). Toward a model of shared leadership and distributed influence in the innovation process. In C. L. Pearce & J. A. Conger (Eds.), *Shared leadership: Reframing the hows and whys of leadership* (p. 48). Thousand Oaks, CA: Sage Publications. doi:10.4135/9781452229539.n3

Cramton, C. D., & Orvis, K. L. (2003). Overcoming barriers to information sharing in virtual teams. In C. B. Gibson & S. G. Cohen (Eds.), *Virtual teams that work: Creating conditions for virtual team effectiveness* (pp. 21–36). San Francisco, CA: Jossey-Bass.

D'Innocenzo, L., Mathieu, J. E., & Kukenberger, M. R. (2014). A meta-analysis of different forms of shared leadership-team performance relations. *Journal of Management*. doi:10.1177/0149206314525205

Day, D. V., Fleenor, J. W., Atwater, L. E., Sturm, R. E., & McKee, R. (2014). Advances in leader and leadership development: A review of 25 years of research and theory. *The Leadership Quarterly, 25*(1), 63–82. doi:10.1016/j.leaqua.2013.11.004

Day, D. V., Gronn, P., & Salas, E. (2004). Leadership capacity in teams. *The Leadership Quarterly, 15*(6), 857–880. doi:10.1016/j.leaqua.2004.09.001

Desanctis, G., & Monge, P. (1999). Communication processes for virtual organizations. *Organization Science, 10*(6), 693–703. doi:10.1287/orsc.10.6.693

Dougherty, D. (1992). Interpretive barriers to successful product innovation in large firms. *Organization Science, 3*(2), 179–202. doi:10.1287/orsc.3.2.179

Earley, C. P., & Gibson, C. (2002). *Multinational work teams - a new perspective*. Mahwah, NJ: Lawrence Erlbaum Associates, Inc.

Ebrahim, A., Ahmed, S., & Taha, Z. (2009). Virtual teams: A literature review. *Australian Journal of Basic and Applied Sciences, 3*(3), 1–9.

Ensley, M. D., Hmieleski, K. M., & Pearce, C. L. (2006). The importance of vertical and shared leadership within new venture top management teams: Implications for the performance of startups. *The Leadership Quarterly, 17*(3), 217–231. doi:10.1016/j.leaqua.2006.02.002

Erhardt, N. (2011). Is it all about teamwork? Understanding processes in team-based knowledge work. *Management Learning, 42*(1), 87–112. doi:10.1177/1350507610382490

Erhardt, N., & Gibbs, J. L. (2014). The Dialectical Nature of Impression Management in Knowledge Work: Unpacking Tensions in Media Use Between Managers and Subordinates. *Management Communication Quarterly, 28*(May issue), 155–186. doi:10.1177/0893318913520508

Erhardt, N. L., Martin-Rios, C., & Way, S. A. (2009). From bureaucratic forms towards team-based knowledge work systems: Implications for human resource management. *International Journal of Collaborative Enterprise*, *1*(2), 160–179. doi:10.1504/IJCENT.2009.029287

Erhardt, N. L., Werbel, J. D., & Shrader, C. B. (2003). Board of director diversity and firm financial performance. *Corporate Governance: An International Review*, *11*(2), 102–111. doi:10.1111/1467-8683.00011

Faraj, S., & Sambamurthy, V. (2006). Leadership of information systems development projects. *IEEE Transactions on Engineering Management*, *53*(2), 238–249. doi:10.1109/TEM.2006.872245

Furumo, K. (2009). The impact of conflict and conflict management style on deadbeats and deserters in virtual teams. *Journal of Computer Information Systems*, *49*, 66–73.

Galbraith, J. (1972). Organization design: An information processing view. In J. Lorsch & P. Lawrence (Eds.), *Organization planning: Cases and concepts* (pp. 49–74). Homewood, IL: Richard D. Irwin, Inc.

Gibbs, J., & Boyraz, M. (2015). International HRM's role in managing global teams. In D. G. Collings, G. Wood, & P. Caligiuri (Eds.), *The Routledge companion to international human resource management* (pp. 532–551). New York, NY: Routledge.

Gibbs, J., Eisenberg, J., Rozaidi, N. A., & Gryaznova, A. (2015). The "megapozitiv" role of enterprise social media in enabling cross-boundary communication in a distributed russian organization. *The American Behavioral Scientist*, *59*(1), 75–102. doi:10.1177/0002764214540511

Gibbs, J., Rozaidi, N. A., & Eisenberg, J. (2013). Overcoming the "ideology of openness": Probing the affordances of social media for organizational knowledge sharing. *Journal of Computer-Mediated Communication*, *19*(1), 102–120. doi:10.1111/jcc4.12034

Gibson, C., & Cohen, S. (2003). *Virtual teams that work: Creating conditions for virtual team effectiveness*. San Francisco, CA: Jossey Bass.

Gibson, C., & Gibbs, J. (2006). Unpacking the concept of virtuality: The effects of geographic dispersion, electronic dependence, dynamic structure, and national diversity on team innovation. *Administrative Science Quarterly*, *51*, 451–495.

Gibson, C., & Manuel, J. (2003). Building trust. In C. B. Gibson & S. Cohen (Eds.), *Virtual teams that work: Creating conditions for virtual team effectiveness* (pp. 59–86). San Francisco, CA: John Wiley & Sons Inc.

Gilson, L., Maynard, M. T., Jones Young, N. C., Vartiainen, M., & Hakonen, M. (2015). Virtual teams research: 10 years, 10 themes, and 10 opportunities. *Journal of Management*, *41*(5), 1313–1337. doi:10.1177/0149206314559946

Giuri, P., Rullani, F., & Torrisi, S. (2008). Explaining leadership in virtual teams: The case of open source software. *Information Economics and Policy*, *20*(4), 305–315. doi:10.1016/j.infoecopol.2008.06.002

Grant, R. M. (1996). Toward a knowledge-based theory of the firm. *Strategic Management Journal*, *17*(S2), 109–122. doi:10.1002/smj.4250171110

Griffith, T. L., Sawyer, J. E., & Neale, M. A. (2003). Virtualness and knowledge in teams: Managing the love triangle of organizations, individuals, and information technology. *Management Information Systems Quarterly, 27*(2), 265–287.

Hambley, L., & Kline, T. (2007). Virtual team leadership: Perspectives from the field. *International Journal of e-Collaboration, 3*(1), 40–64. doi:10.4018/jec.2007010103

Hambley, L., O'Neill, T., & Kline, T. (2007). Virtual team leadership: The effects of leadership style and communication medium on team interaction styles and outcomes. *Organizational Behavior and Human Decision Processes, 103*(1), 1–20. doi:10.1016/j.obhdp.2006.09.004

Heckman, R., Crowston, K., & Misiolek, N. (2007). A structurational perspective on leadership in virtual teams. In K. Crowston, & S. Seiber (Eds.), *Proceedings of the IFIP Working Group 8.2/9.5 Working Conference on Virtuality and Virtualization* (pp. 151–168). Portland, OR: Springer. doi:10.1007/978-0-387-73025-7_12

Henderson, L. (2008). The impact of project managers' communication competencies. *Project Management Journal, 39*(June), 48–59. doi:10.1002/pmj.20044

Hill, N. S. (2005). Leading together, working together: The role of team shared leadership in building collaborative capital in virtual teams. *Advances in Interdisciplinary Studies of Work Teams, 11*(05), 183–209. doi:10.1016/S1572-0977(05)11007-3

Hiller, N. J., DeChurch, L., Murase, T., & Doty, D. (2011). Searching for outcomes of leadership: A 25-year review. *Journal of Management, 37*(4), 1137–1177. doi:10.1177/0149206310393520

Hinds, P., & Kiesler, S. (1995). Communication across boundaries: Work, structure, and use of communication technologies in a large organization. *Organization Science, 6*(4), 373–393. doi:10.1287/orsc.6.4.373

Hoch, J. E., & Dulebohn, J. H. (2013). Shared leadership in enterprise resource planning and human resource management system implementation. *Human Resource Management Review, 23*(1), 114–125. doi:10.1016/j.hrmr.2012.06.007

Hoch, J. E., & Kozlowski, S. W. J. (2014). Leading virtual teams: Hierarchical leadership, structural supports, and shared team leadership. *The Journal of Applied Psychology, 99*(3), 390–403. doi:10.1037/a0030264 PMID:23205494

Huang, R., Kahai, S., & Jestice, R. (2010). The contingent effects of leadership on team collaboration in virtual teams. *Computers in Human Behavior, 26*(5), 1098–1110. doi:10.1016/j.chb.2010.03.014

Jackson, S., Joshi, A., & Erhardt, N. (2003). Recent research on team and organizational diversity: Swot analysis and implications. *Journal of Management, 29*(6), 801–830. doi:10.1016/S0149-2063(03)00080-1

Jarvenpaa, S., Knoll, K., & Leidner, D. E. (1998). Is anybody out there? Antecedents of trust in global virtual teams. *Journal of Management Information Systems, 14*(4), 29–64.

Jarvenpaa, S., & Leidner, D. E. (1999). Communication and trust in global virtual teams. *Organization Science, 10*(6), 791–815. doi:10.1287/orsc.10.6.791

Joshi, A., & Lazarova, M. (2005). Do "global" teams need "global" leaders? Identifying leadership competencies in multinational teams. In *Managing Multinational Teams. Global Perspectives.*

Joshi, A., Lazarova, M. B., & Liao, H. (2009). Getting everyone on board: The role of inspirational leadership in geographically dispersed teams. *Organization Science, 20*(1), 240–252. doi:10.1287/orsc.1080.0383

Kayworth, T. R., & Leidner, D. E. (2002). Leadership effectiveness in global virtual teams. *Journal of Management Information Systems, 18*(3), 7–40.

Kirkman, B. L., Rosen, B., Tesluk, P., & Gibson, C. (2004). The impact of team empowerment on virtual team performance: The moderating role of face-to-face interaction. *Academy of Management Journal, 47*(2), 175–192. doi:10.2307/20159571

Kock, N., & Lynn, G. S. (2012). Electronic media variety and virtual team performance: The mediating role of task complexity coping mechanisms. *IEEE Transactions on Professional Communication, 55*(4), 325–344. doi:10.1109/TPC.2012.2208393

Lerner, J., & Tirole, J. (2002). Some simple economics of open source. *The Journal of Industrial Economics, L*(2), 197–234.

Majchrzak, A., Rice, R. E., King, N., Malhotra, A., & Sulin, B. (2000). Computer-mediated inter-organizational knowledge-sharing: Insights from a virtual team innovating using a collaborative tool. *Information Resources Management Journal, 13*(2), 44–53. doi:10.4018/irmj.2000010104

Malhotra, A., Majchrzak, A., & Rosen, B. (2007). Leading virtual teams. *The Academy of Management Perspectives, 21*(February), 60–71. doi:10.5465/AMP.2007.24286164

Manz, C. C. (1986). Self-leadership: Toward an expanded theory of self-influence processes in organizations. *Academy of Management Review, 11*(3), 585–600.

Maynard, M. T., & Gilson, L. (2013). The role of shared mental model development in understanding virtual team effectiveness. *Group & Organization Management, 39*(1), 3–32. doi:10.1177/1059601113475361

McEvily, B., Perrone, V., & Zaheer, A. (2003). Trust as an organizing principle. *Organization Science, 14*(1), 91–103. doi:10.1287/orsc.14.1.91.12814

Miller, D. (1988). Relating porter's business strategies to environment and structure: Analysis and performance implications. *Academy of Management Journal, 31*(2), 280–308. doi:10.2307/256549

Millward, L., & Kyriakidou, O. (2004). Effective virtual teamwork. In S. H. Godar & S. P. Ferris (Eds.), *Leading from Afar: Strategies for Effectively Leading Virtual Teams* (pp. 20–34). Hershey, PA: Idea Group Publishing.

Muethel, M., & Hoegl, M. (2010). Cultural and societal influences on shared leadership in globally dispersed teams. *Journal of International Management, 16*(3), 234–246. doi:10.1016/j.intman.2010.06.003

Pawar, B., & Eastman, K. K. (1997). The nature and implications contextual influences on transformational leadership: A conceptual examination. *Academy of Management Review, 22*(1), 80–109.

Pearce, C. L. (2004). The future of leadership: Combining vertical and shared leadership to transform knowledge work. *The Academy of Management Executive, 18*(1), 47–57. doi:10.5465/AME.2004.12690298

Pearce, C. L., & Conger, J. a. (2003). *Shared leadership: Reframing the hows and whys of leadership.* Thousand Oaks, CA: Sage Publications.

Pearce, C. L., & Ensley, M. D. (2004). *A reciprocal and longitudinal investigation of the innovation process: The central role of shared vision in product and process innovation teams (ppits).* Academic Press.

Pearce, C. L., & Sims, H. P. (2002). Vertical versus shared leadership as predictors of the effectiveness of change management teams: An examination of aversive, directive, transactional, transformational, and empowering leader behaviors. *Group Dynamics, 6*(2), 172–197. doi:10.1037/1089-2699.6.2.172

Pearce, C. L., Yoo, Y., & Alavi, M. (2004). Leadership, social work, and virtual teams: The relative influence of vertical vs. shared leadership in the nonprofit section. In R. Reggio & S. Smith Orr (Eds.), *Improving leadership in nonprofit organizations.* San Francisco: Jossey-Bass.

Powell, A., Piccoli, G., & Ives, B. (2004). Virtual teams: A review of current literature and directions for future research. *ACM SIGMIS Database, 35*(1), 6–36. doi:10.1145/968464.968467

Purvanova, R., & Bono, J. (2009). Transformational leadership in context: Face-to-face and virtual teams. *The Leadership Quarterly, 20*(3), 343–357. doi:10.1016/j.leaqua.2009.03.004

Robert, L., & You, S. (2013). Are you satisfied yet? shared leadership, trust and individual satisfaction in virtual teams. In iConference 2013 (pp. 461–466).

Shuffler, M., Wiese, C., Salas, E., & Burke, S. (2010). Leading one another across time and space: Exploring shared leadership functions in virtual teams. *Revista de Psicología del Trabajo y de las Organizaciones, 26*(1), 3–17. doi:10.5093/tr2010v26n1a1

Society for Human Resource Management. (2012). *Virtual teams.* Retrieved from http://www.shrm.org/research/surveyfindings/articles/pages/virtualteams.aspx

Sosik, J., Avolio, B., Kahai, S., & Jung, D. I. (1998). Computer-supported work group potency and effectiveness: The role of transformational leadership, anonymity, and task interdependence. *Computers in Human Behavior, 14*(3), 491–511. doi:10.1016/S0747-5632(98)00019-3

Szulanski, G. (2000). The process of knowledge transfer: A diachronic analysis of stickiness. *Organizational Behavior and Human Decision Processes, 82*(1), 9–27. doi:10.1006/obhd.2000.2884

Van De Ven, A. H., Delbecq, A., & Koenig, R. (1976). Determinants of coordination modes within organizations. *American Sociological Review, 41*(2), 322–338. doi:10.2307/2094477

Wakefield, R. L., Leidner, D. E., & Garrison, G. (2008). A model of conflict, leadership, and performance in virtual teams. *Information Systems Research, 19*(4), 434–455. doi:10.1287/isre.1070.0149

Wang, G., Oh, I. S., Courtright, S. H., & Colbert, E. (2011). Transformational leadership and performance across criteria and levels: A meta-analytic review of 25 years of research. *Group & Organization Management, 36*(2), 223–270. doi:10.1177/1059601111401017

Watson-Manheim, M. B., Chudoba, K. M., & Crowston, K. (2002). Discontinuities and continuities: A new way to understand virtual work. *Information Technology & People*, *15*(3), 191–209. doi:10.1108/09593840210444746

Wilson, J. M., Boyer O'Leary, M., Metiu, A., & Jett, Q. R. (2008). Perceived proximity in virtual work: Explaining the paradox of far-but-close. *Organization Studies*, *29*(7), 979–1002. doi:10.1177/0170840607083105

Yukl, G. (2012). Effective leadership behavior: What we know and what questions need more attention. *The Academy of Management Perspectives*, *26*(November), 66–85. doi:10.5465/amp.2012.0088

Zaccaro, S. J., & Burke, C. (1998). *Team versus crew leadership: Differences and similarities*. Academic Press.

Zigurs, I. (2003). Leadership in virtual teams: Oxymoron or opportunity? *Organizational Dynamics*, *31*(4), 339–351. doi:10.1016/S0090-2616(02)00132-8

ADDITIONAL READING

Bell, B. S., & Kozlowski, S. W. J. (2002). A typology of virtual teams: Implications for effective leadership. *Group & Organization Management*, *27*(1), 14–49. doi:10.1177/1059601102027001003

D'Innocenzo, L., Mathieu, J. E., & Kukenberger, M. R. (2014). A meta-analysis of different forms of shared leadership-team performance relations. *Journal of Management*. doi:10.1177/0149206314525205

Gibson, C., & Gibbs, J. (2006). Unpacking the concept of virtuality: The effects of geographic dispersion, electronic dependence, dynamic structure, and national diversity on team innovation. *Administrative Science Quarterly*, *51*, 451–495.

Hambley, L., & Kline, T. (2007). Virtual team leadership: Perspectives from the field. *International Journal of e-Collaboration*, *3*(1), 40–64. doi:10.4018/jec.2007010103

Hill, N. S. (2005). Leading together, working together: The role of team shared leadership in building collaborative capital in virtual teams. *Advances in Interdisciplinary Studies of Work Teams*, *11*(5), 183–209. doi:10.1016/S1572-0977(05)11007-3

Hoch, J. E., & Kozlowski, S. W. J. (2014). Leading virtual teams: Vertical leadership, structural supports, and shared team leadership. *The Journal of Applied Psychology*, *99*(1), 1–14. PMID:24079670

Kayworth, T. R., & Leidner, D. E. (2002). Leadership effectiveness in global virtual teams. *Journal of Management Information Systems*, *18*(3), 7–40.

KEY TERMS AND DEFINITIONS

Computer-Mediated Communication (CMC): Communication that occurs electronically.

Contingency Factors: Assumption that there is no one best way to organize and instead effectiveness is contingent on a variety of factors associated with the specific context.

Information and Communication Technologies (ICTs): Technologies that are used to communicate information.

Interdependence: Depending on other team member(s) for completing work related assignments.

Shared Leadership: Team members influence each other and share leadership responsibilities in assuring performance of their team.

Vertical Leadership: Formally nominated leader is in charge of leading the team.

Virtual Teams: Geographically dispersed team members collaborate using various ICTs.

Chapter 3
Providing a Model for Virtual Project Management with an Emphasis on IT Projects

Hamed Nozari
Islamic Azad University, Iran

Meisam Jafari-Eskandari
Payam Noor University, Iran

Seyed Esmaeil Najafi
Islamic Azad University, Iran

Alireza Aliahmadi
Iran University of Science and Technology, Iran

ABSTRACT

With globalization of economy and increase of global competition to acquire rare resources, the organizations have moved towards geographical distribution to achieve competitive advantage. Users and project teams at various places within various countries with different national and local cultures throughout the world work on projects at the environment concerning geographical distribution. On the other hand, with increase of advancements in communication, the distributed project teams have been witnessed with more expansion, known with "virtual teams". When members of virtual project team from various organizations and time zones attend in the projects, they will be more likely affiliated to electronic media such as email. With regard to virtualization of IT projects, the present study aims to develop a model for virtual project management with an emphasis on information technology projects, including several elements in geographically distributed environments. The final model of virtual project management of information technology projects was represented.

INTRODUCTION

Advancements in information technology together with rapid growth of internet have led to development of globalization. These advancements have caused failure of time and geographical barriers to the organizations which seek business and economic growth (Guillen, 2001). Progressive realization of world economy has led to increase of interaction and cohesion of economic systems which generally do not enable to coexist due to time and place limitations. As the result of increasing interactions in these

DOI: 10.4018/978-1-4666-9688-4.ch003

systems, most of organizations have developed flexible and dynamic structures which can be immediately adjusted with customers' needs (Jarvenpaa & Leidner, 1994). Information technology plays a major role in this transfer, which emerges as an enabler in new organizational form. Currently, information technology which has been considered as a substantial element in organizational processes has infiltrated most organizations, emerging in the global work teams (Armin. B. Cremers et al, 2005). Emergence and use of work teams have caused a substantial change at work environment. Global work teams which rely on information technology for their interactions are recognized as virtual teams. Through virtual teams, organizations enable to share experts' and personnel's skills and knowledge beyond time and place barriers. Project managers might be at a position enabling to select the best resources for their project at any place (Goiuld, 2004). Virtual teams might be distributed at building, region and/or countries. In point of view of Gold (2005), virtual teams imply the teams from individuals who firstly interact with each other through electronic devices, but they might need face-to-face sessions later. Anyhow, the team members might never have a face-to-face interaction among the geographically distributed organizations. In general, it can define the virtual teams as a group of individuals who work in the same projects but in more than one workplace, and severly adhere to information and communication technologies (Guido Hertel, 2005). Enviornmental changes have taught the organizations to search for the best resources around the world for the purpose of acquiring the competitive advantage (Canie L. et al, 2004). Mergers and acquisitions, emerging markets in different geographic locations, needing to reduction of cost, needing to reduction of time to enter into the market, and production cycle time are the reasons which justify the need to virtual teams. Today, virtual project teams have become an integral part of management forms, and the organizations are required ensuring achievement of project goals concerning the new environment. Virtual project teams require organizing their team to the full extent of organizing in traditional teams. These teams have raised a new challenge for the project managers for delivery of high-quality products (Bruce J., Arolio G., 2002). Kenneth David Strong (2004) in a study at the area of leadership in the virtual environment has more likely strived to identify the leadership roles and behaviors. For this purpose, he has referred to the leadership components including common vision and electronic monitoring. In his study, he has examined various types of behavioral leadership models concerning virtual teams.

Hertdeller (2005) has proposed a model for life cycle of virtual teams, and also has pointed out elements of their life cycle by dividing the stages of virtual teams' life cycle to various phases.

Yuhyunge Shin (2004) in a model well known with "person-environment fit model" within virtual organizations has examined virtual environment and behaviors of the members in the virtual teams. However, the components such as confidence, independence, diversity, virtual communication, and knowledge and job satisfaction have been pointed out in this model, this research has been firstly fulfilled at the area of virtual organization and the concepts defined in his model have been secondly classified at the area of characteristics of members, where the major elements such ascyberculture, virtual leadership and other elements have not been pointed out in this model. Louise (2004) in a model well known with "I-P-O" has classified operation of virtual teams in terms of inputs, outputs and processes.

Terri Williard (2001) has examined the elements such as virtual communication and cyberculture, and emphasized on the relationship between these two elements in this research. Ultimately, despite the previous studies, this study has aimed to propose a conceptual model for the virtual project management at the area of information technology. With regard to the studies represented to date concerning the investigations into the elements and components of virtual project management, this study has aimed to represent the elements and components of virtual project management in form of a model. Further, this study specifically has focused on the areas of information technology projects, where this can be a different point compared to previous studies.

THEORETICAL BACKGROUND OF RESEARCH

Working in a distributed team in different places with different time zones is not considered as a new phenomenon. There are a large body of evidences which show how people have cooperated with each other in long distances far from each other in the long past. (Bruce J. et al, 2003; Cathrine Darnell Crampton, Sheila S., 2005; B. Parkinson, P. Hudson, 2001). Anyhow, with rapid development of information and communication technology in the long lost past, Working in a distributed team can be fulfilled rapider and simpler. Virtualization is the main feature of working in a distributed team, which has been grounded on the basis for information and communication technologies. Various forms of virtualization for working in a distributed team can be separated from each other by means of the number of individuals and the extent to which they have interaction with each other. In this regard, firstly telecommuting is taken into account that a part or full extent of it is fulfilled out of organization by means of information and communication services (Frank, 2002). In following, virtual group is taken into account in which several different types of telecommuting are combined with each other from remote area and each member presents report to a manager. In contrast, there is a virtual team in which the members of virtual group interact with each other to achieve a common goal. This differentiation exists between virtual team and virtual group in parallel with each other in differentiation between traditional group and traditional team. Ultimately, the virtual communicates are a larger entity for working in a distributed team in which the members cooperate with each other via internet by means of common goals, roles and norms. In the virtual teams, virtual communities are not handled by a virtual structure, yet they are developed by means of their members(Cristina Anna et al, 2004).

Separate from the aforementioned differences, certain definitions have not been proposed for virtual teams on whom researchers agree. Based on an investigation into the literature review of virtual teams and the definitions proposed by researchers, the definition of virtual teams include the characteristics below: 1-virtual teams consists of two or more individuals, 2-they cooperate with each other to achieve general goals, 3-the relationship and coordination among them is fulfilled based on information and communication technologies (Hy Ung Jan Ann, Hong Joo Lee, 2005). On the other hand, the projects have been considered as one of the reasons for formation of virtual teams.

Krill and Jól (2006) have defined virtual projects as an effort and collaboration to achieve particular aims or fulfill duties based on telecommuting. Formation of virtual teams has made the possibility of fulfillment of virtual projects close to the reality (O'Sallivan, 2003). In point of view of Aniston and Miller (2002), virtual teams are similar to traditional teams, with this difference that the members of team in the beginning of project interact with each other via electronic devices. There are several reasons which have led to increasing use of virtual teams. In point of view of the researchers such as Garcı́a-Parıs, & Wake, 2007; Gould, 2002; McNamara & Swenson, & Walsh, 2005, the most important reason for increasing use of virtual teams can be reduction of costs for fulfillment of virtual projects. Virtual teams use the electronic techniques which cause reducing the travel by the memerbers of the project, reducing overhead costs, reducing project scheduling, improving decision making (Evie Tastoglou, Evangelou Milious, 2005).

Outsourcing and the advantage of use of cheap labor market are accounted as the policies relating to the reduction of cost. Cheap labor market has caused the intra-organizational business processes which are fulfilled via traditional techniques within organizations, are fulfilled by the contractors in most of organizations. Globalization is another reason that induces the organizations to use virtual teams throughout the world (Collines, 2002).

Most of organizations have used the advancements in the information technologies as much as possible. Emergence of such an environment requires the inter-organizational cooperation for competition with the other networks developed by the competitors. In the global market, by increasing competition, the organizations require hiring the best labor forces that might not be necessarily in a geographic location (Edward F. Mcdonough, Kenneth B, 2001). In a slow movement of organization towards use of human resources distributed by means of virtual teams, the advantages such as overcoming time barriers and obtaining the best specialists are regarded (Evie Tastoglou, Evangelou Milious, 2005).

On the other hand, the same as the policies used in use of human resources in different geographic locations, the outcomes from hiring virtual teams can be analyzed at three levels: personal, organizational and social levels (Could, 2005). At the personal level, the potential advantages such as high flexibility and time control with top accountability, motivation at work, and strengthening and empowering team members will be followed. At this level, a challenge which is felt is witnessed in isolation and reducing interpersonal interactions, increasing misunderstanding, emergence of conflict, role ambiguity and conflict in goals. At organizational level, the virtual teams specifically enjoy strategic advantages. For instance, teams can hire the individuals based on their skills rather than their availability. Team can have a high flexibility in responsiveness to the market demand and set closer relationships with customers and suppliers. The potential challenge at this level includes the problems in supervision on the activities of members in virtual project, prevention from the effective development at time zone, additional costs for suitable technologies and further education programs. Ultimately, at social level, employing virtual teams can result in development of regions with low sub-structures and increase of their employment rate, cohesion of individuals with low movement capability, and reduction of environmental limitations by means of reducing air pollution and traffic (Durate, 2001). Stephen (2004) has represented a model describing the processes of creating successful virtual teams for virtual project managers. In his opinion, the managers who are in virtual teams know that creating successful virtual teams requires the leaders who not just create a view but also create a sub-structure to support working processes and communications. He has emphasized on the concepts such as creating a shared vision, establishing a sub-structure, assessment of members and selection appropriate reward system in his model (Joyce A. Thompson, 2006). According to Stephen's model, two other models have been also developed which are suggested to create effective distributed teams. These models include maturity and coordination models. The coordination model will help the managers to evaluate and select the members of team for performing their tasks in the best sub-structure and current organization. Maturity model displays a framework for transfer to the more advanced sub-structures which allows the virtual teams to acquire the highest performance level. Access to the highest performance level requires the teams concerning their aims, processes and skills (Bradford S. et al, 2002). In another research, the lifecycle model has been represented for the virtual project teams, that this model is associated to the virtual project teams concerning various areas.

Lifecycle indicates that the disadvantages due to new communication technologies can have a different dependence on phases of working teams; in this model, it is argued about human resource management problems which can affect high level of virtualization. In contrast to input-process-output (IPO) model, the lifecycle model specifies five phases with certain management tasks which are described in virtual work teams. The represented phases include preparation, start, performance management, training and development of team, closure and reintegration. This model is a dominant framework which has been represented to study the virtual project teams and cohesion of literature review on them. The input for representation of a group's conditions such as human or materials resources and processes indicates the interactions between members of project teams to fulfill their tasks. Outcomes represent the functional or non-functional consequences from the operations of a group (Hackman, 2002). McMahon in "virtual project management" has proposed an 8-step program for implementation of virtual project management. This framework includes the steps including creating top level of the project, architecture, breaking work and duties, planning, project rules, detailed planning, testing the concept of operations and implementation. McMahon (2001) in an article has represented a four-level framework for the cyberculture in the virtual project management, which is considered as the basis for this research, due to attention to the information technology projects. In this framework, the organization displays the first level. This level represents the first level of 8-step processes described previously. The second level represents the major layer of cyberculture in which the major elements below is are provided: breaking work, creating and regulating filtered project memory, definitions of product planning matrix, product deployment matrix, component deployment matrix. This level associated to steps, 2 and 3 has been defined in the process. The third layer represents the intermediate cyberculture layer which has been provided in the models of task, sub-structure and project rules. Finally, the fourth layer represents operational models and Conflict Management (Jeremy S. Laurey, Manesh S. Raisinghani, 2001).

THEORETICAL FRAMEWORK OF RESEARCH

With regard to overview of literature review and the related works at the area of information technology virtual teams and virtual projects, the elements and components below have been identified.

1. **Virtual Leadership:** leadership is the main challenge in virtual teams because managers of these teams are not in a place together with members. Hence, understanding various types of direct control seems difficult. The consequence of delegation management principles is considered as a part of classical management functions for the members of teams. Members of virtual teams just accept such management functions, due to the fact that access to the goals is much more difficult in virtual teams (Bruce J., Arolio G., 2002). The results of studies on virtual teams are characterized in three leadership attitudes that are different in the extent of independence of members of team: 1- electronic monitoring as an attempt to direct leadership beyond the physical distance, 2- Management by objective as an example for delegating leadership principles, 3- self- managing teams as an example of autonomous work teams (Zigaras, 2003).The components associated to virtual leadership include as follows:
    a. **Team Solidarity:** Team solidarity at virtual environment implies recruitment of members to team. In a study by Bruce et al.(2003), it can say that a higher level of team solidarity is seen in traditional teams rather than virtual teams. In virtual teams, team solidarity comes to realize by means of higher satisfaction level, more cohesive groups and the ability for knowledge sharing. On the other hand, team solidarity positively associates to the effectiveness of team (Bruce J. et al, 2003).

b. **Receive Continuous Feedback:** The performance associated to continuous feedback is one of major elements within project teams which is prioritized under distributed work environments and virtual work conditions, and this is due to the fact that receiving information on achievement of goals of other members of virtual team is much more difficult (D.kissler, Gray, 2000). Indeed, introducing continuous feedback especially graphical feedback on actual performance results in high performance in electronic brainstorming groups(Bruce J., Arolio G., 2002).

c. **Building Trust:** Trust is important for teams to achieve their goals, but importance of trust is much more among virtual teams due to their less face-to-face communications (Cathrine Darnell Crampton, Sheila S., 2005). The processes of building and maintaining trust is similar in all teams. In virtual teams, trust comes to realize by means of members' expectations and then delivery of consistent results that meet expectations. Nature of virtual environment is in a way that organizing accurate expectations from traditional teams will be much more difficult, and naturally there will be less control over revision of expectations of project stakeholders, whereby this will be the biggest challenge for virtual leaders (Durate, 2001).

d. **Development of Cooperation:** Cooperation in virtual environment associates to the tasks which have been structured, whereby division of tasks for the members of virtual team will be easier. In point of view of Strijbos(2004), development of cooperation and coordination in virtual team implies involving the members of virtual team in tasks to achieve a common goal, acquiring a decision and/or solution which can be repeated. Development of cooperation and coordination requires companionship, understanding, and trust (Cathrine Darnell Crampton, Sheila S., 2005).

e. **Delegation of Authority:** Although direct leadership strategies are possible in traditional teams, delegation of management functions to the members is more suitable in virtual teams. This attitude causes changing the role of team manager from traditional control to mediation and coaching functions. Management by objective is an example for the concept of delegative leadership which emphasizes on setting goals, cooperation and feedback (Bruce J., Arolio G., 2002).

f. **A Clear Definition of Objectives:** A clear definition of objectives is of planning processes which focuses on team effort. Research indicate that there is a positive relationship between setting clear goals and team unity, cooperation, the quality of decisions and the number of generated alternatives. In the virtual environment, due to time and place distance between members of team despite use of communication technologies, development of clear goals is much more difficult (Cathrine Darnell Crampton, Sheila S., 2005).

g. **Motivation of Team Members:** Lack of physical contact in virtual teams can result in various challenges in members' motivation, which can be due to difficulty in understanding common goals, feeling of lack of independence and low social control, low feedback, and difficulty of the trust. Processes of motivation of virtual team members can be measured based on model (concept of value*expectation)(Guido Hertel, 2005).

h. **Existential Philosophy:** Expressing existential philosophy of formation of virtual teams is one of factors which can be effective in the beginning of virtual project and formation of virtual teams (Edward F. Mcdonough, Kenneth B., 2001). This causes formation of team solidarity and clear understanding of the objectives and the reason for formation of virtual team. The importance of this issue derives from this fact that expressing existential philosophy of virtual team can reduce ambiguity in team's objectives and tasks, due to geographical distribution of members (Collines, 2002).

i.   **Knowledge Sharing:** Most of virtual projects especially information technology projects require different specialties. To reduce risk at virtual projects, individuals' experiences and skills at different geographical locations by means of the concepts such as virtual culture and formation of knowledge bank must be shared between members in project (David, 2004).

j.   **Electronic Performance Monitoring (EPM):** Using network technologies, EPM systems allow the managers to control the employees and examine the accuracy degree of log-in and log-out times. EPM, instead of Taylor principle, emphasizes on standardization, separation, and ease of work processes. Furthermore, EPM increases speed at performance of workers with high skills, yet it decreases speed at performance of workers with low skill (Bruce J., et al, 2003).

k.   **Rewarding Systems:** Development of motivation and rewarding systems is one of the important issues in the beinging of vritual projects. The same as traditional teams, the team-based incentives are important for an emphasis on the importance of cooperation in virtual teams(O'Sallivan, 2003). team-based rewards can result in strong motivation of virtual team members. In point of view of Lawler (2002), rewarding system must be in consistent with certain aspects of the virtual project, including objectives, tasks, interdependence, autonomy, diversity and degree of virtualization. The main purpose of this process is granting reward to the behaviors which are in line with strategy of virtual organization (Zigaras, 2003).

2. **Virtual Communication:** Virtual communication is one of the most fundamental activities which is fulfilled by virtual team. Virtual communications will be followed by challenges for the virtual project managers in implementation of virtual work environment (Bent, 2004). Project managers require having access to efficient instruments so as to maximize their relationship with members of virtual team. As members of virtual team are not in a certain location, use of new communication technologies seems essential. Previous studies indicate that virtual teams have desire to effective communications the same as traditional teams. If virtual teams dedicate sufficient time to set relationship and adapt themseleves with virtual communication media, their proccesses will be more effective(D. Kissler Gray, 2000). Virtual communications include the components as follows:

a.   **Geographical Distribution of Members:** Geographical distribution of members are classified into two groups: 1-teams with members who fulfill their tasks in a certain location or office, 2- teams with members who have been distributed at geographic locations and remote organizations in which at least 30% of members fulfill most of their tasks in more than one location or more locations (Ann Hong Joo Lee, 2005).

b.   **Time Difference between the Teams:** Time difference between the teams is considered as one dimension of virtual environment, which is attributed to a degree that members of virtual teach work non concurrently together and do their tasks(Cathrine Darnell Crampton, Sheila S., 2005). Due to time distribution of virtual team members, time management and its skills must be trained to the virtual project managers, in order that they must be able to use nonconcurrent communication technologies.

c.   **Communication Patterns:** Communication patterns imply key design elements which guarantee support systems from virtual teams properly. They can be defined as a set of communication workflows and normative definitions that are favorable and acceptable for virtual team. These patterns also include the meta-norms including the do's and don'ts actions (Fang Chon, et al, 2003).

d. **Synchronous Communication Technology:** Diversity in group decision support systems which has been developed in recent years supports cooperation between members of virtual teams, and being used as a tool for communication, sharing and learning (David, 2004). Synchronous communication technologies pave the way to set relationship between members of virtual teams at different geographical locations. These technologies include the tools such as group tools, video conferencing, electronic meeting systems, electronic brainstorming.

e. **Asynchronous Communication Technology:** With regard to distributed nature of virtual teams, the possibility to use synchronous communications will not be always possible. Asynchronous communication technologies pave the way to set low-cost and easy relationship as well as knowledge sharing between members of virtual team. Asynchronous communication technologies include the tools such as email, group calendar, and asynchronous database (Francis Lau, S. Sarker, 2004).

f. **Face-to-Face Meetings:** Due to distributed nature of virtual teams, the possibility to develop face-to-face meetings among members of virtual project goes beyond. Yet, the researchers believe that a face-to-face meeting must be developed in the beginning of the project in order to let the members get familiar with each other (Edward F. Mcdonough, Kenneth B., 2001).

3. **Cyberculture:** Cyberculture is a simple concept which has been drawn into attention by schoars at the age of communication. Cyberculture view is called to a framework which supports effective communications across sites on a single but distributed project (Cathrine Darnell Crampton, Sheila S., 2005). Cyberculture despite traditional culture is product-oriented, which is not supposed to be replaced with traditional culture. Cyberculture refers to the culture for special sites which provides the required information for setting relationship between information on tasks of sites. Cyberculture can be applied through website or joint directory system. It is worthwhile that the key difference between a Cyberculture and traditional culture lies on officialism. Experience indicates that an effective Cyberculture cannot be informal, i.e. it must be in written form (Durate, 2001) (Could, 2005).

a. **Diversity:** Virtual teams consist of the members with different cultural background. Hence, diversity is regarded as an important issue. Diversity associates to synergy with the hope that various skills and attitudes increase the effectiveness of team (D. Kissler, Gray, 2000).

b. **Language:** Virtual teams might be in different countries and locations with different languages. Any location, with respect to its language, uses its own symbols, which can result in unwanted conflict and improper understanding from each other (Cristina Anna, et al, 2004).

c. **Filtered Project Memory:** Filtered project memory refers to a method which employs a part of Cyberculture, aiming at providing special and important information of project to set coordination. The information include necessary information on how to do tasks, updated information on approaches of virtual organization, key words used in communications, identifying key problems at virtual project and strategies associated to it (Collines, 2002).

d. **Rules of Project:** Any project has its own rules, which includes the rules of communication and leadership and develops the basis for Cyberculture. The rules of project can be considered as a part of filtered project memory (M. Macmahon, 2004).

e.   **Architecture and Breaking Work:** Here, architecture implies components of a system and the rules which define the relationship between these components based on Cyberculture. When the term "breaking work" is used, it implies allocation of responsibilities across separate physical sites and organizations. Breaking work can be fulfilled through devolving responsibility across the site, where technical architecture can provide one of the best communications and coordination techniques. The team acquires a required attitude through an informal architecture so as to achieve the work expectations without special expectations. Yet, concerning effective architecture as the work communications, definition of work gap across separate sites must follow the definition of architecture (M. Macmahon, 2004; Bruce J., Arolio G., 2002).

f.   **Component-Product Development Matrix:** Component-product development matrix is conveyed as a strong communication tool to support improvement of the definition for work appointments, which is determined in the Cyber culture. These tools include a table in which

Figure 1. Conceptual model of research
Source: Researchers.

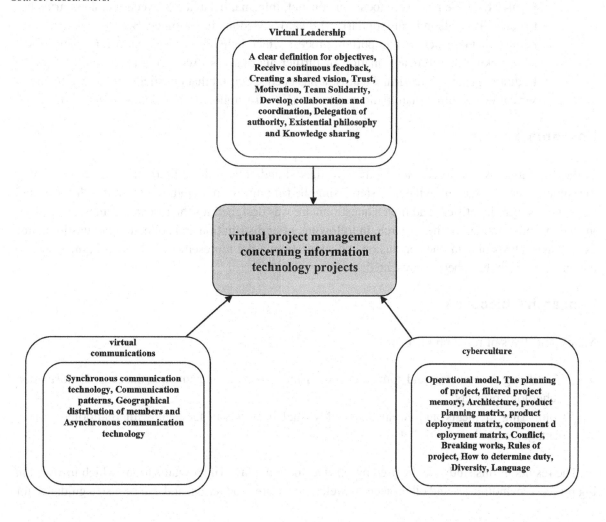

a list of special consequences of project has been represented. This technique is specifically used in the information technology projects. The information of this matrix is not generally found in the virtual project processes (Larsline, 2006).

g. **Operational Model:** This refers to the sub-cyberculture which uses a strong mechanism in relation to the work expectations within organization (Collines, 2002).

h. **How to Determine Duty:** Virtual project managers must pay a particular attention to how to break duties and their ownership across virtual team boundaries. The important point lies on the risk that exists in improper understanding and wrong interpretation of duties. Use of electronic written communication at this area is the idea proposed at this area. Nature of certain duties such as cohesion of deliveries to another virtual team requires cooperation between members of virtual teams (Joe Nandhakumar, R. Baskervil, 2001).

i. **The Planning of the Project:** The points such as definition of duties and appointment of them to the members of special teams must be considered in planning for a virtual project. These duties require a high coordination to use the resources which have been geographically distributed. The planning of the project is determined based on cyberculture (B. Parkinson, P. Hudson, 2001).

j. **Conflict:** These processes focus on conflict, informal behavior between members, trust between individuals and team solidarity (Edward F. Mcdonough, Kenneth B., 2001). Researchers believe that conflict is an important process which helps the virtual teams for a better decision making due to determination of further alternatives. According to the comparison of traditional teams with virtual teams, it has been specified that conflict more likely occurs in virtual teams rather than traditional teams (Evie Tastoglou, Evangelou Milious, 2005).

Research Method

In the present study, the researcher firstly investigated and collected the literature review concerning the subject area of research. In the next step, using the findings from literature review, the elements and components were identified, and then a questionnaire was designed for the research sample by designing a conceptual model for the research. In following, after distribution and collection of questionnaires as well as analysis of data, the conclusion and suggestions were represented. In Figure, 2, the stages to conduct this study have been represented:

Research Objectives

Major objectives of research include:

- Identifying the elements and components of virtual project management at the area of information technology projects.
- Comparing elements and components of virtual project management with traditional project management.

The results of this study can be used by service and manufacturing organizations which involved in big projects including gas and oil projects as well as software production, and can be also a guidance for

Table 1. An overview on literature review of research

Dimensions	Component	Joe Nandhkumar, Richard Basckerville, 2001		Susan Vonsild, 2000		Phillip D. Elis, 1999		Yuhyung Shin, 2004		Louis Martin, Lay L. Gilson, M. Travis Mayrar, 2005		G. Susanne Geister, Udo Konradt, 2005		Francis Lou, Suprateek Sarker, Sundeep Sanay, 2000		Jarvenpea & Leidner, 1998		William M. Hugnes, 2003		Louce A. Thompson, 2005		Parsert Kanawattanachai, Yaun, Jin Yoo, 2004	
		Dimension	Component	Primary	Secondary	Primary	Secondary	Primary	Secondary	Primary	Secondary	Primary	Secondary	Primary	Secondary	Primary	Secondary	Primary	Secondary	Primary	Secondary	Primary	Secondary
Virtual Leadership	Ensuring		*		*		*		*		*		*		*		*		*		*		*
	Knowledge sharing		*				*		*				*		*								*
	Team cohesion		*		*														*				
	Setting Goals						*				*						*						*
	Continuous feedback		*		*								*								*		
	Motivation	*	*	*	*	*		*		*		*	*	*	*			*		*	*	*	
	Reward Systems						*		*		*		*		*				*		*		*
	Shared vision		*				*		*		*		*						*				
	Existential philosophy		*												*								
	Delegate				*										*				*	*			
	Electronic monitoring function								*														
virtual Communications	Terms of Contact		*										*		*				*		*		*
	Communication patterns		*										*				*						*
	Communication technologies		*								*		*		*				*				
	Time difference		*		*		*		*		*		*		*		*		*		*		*
	Geographical distribution				*				*												*		
	Face to face meetings				*		*						*		*				*				

continued on following page

Table 1. Continued

Dimensions	Component	Joe Nandhkumar, Richard Baseckerville, 2001 — Dimension	Joe — Component	Susan Vonsild, 2000 — Primary	Susan — Secondary	Phillip D. Elis, 1999 — Primary	Phillip — Secondary	Yuhyung Shin, 2004 — Primary	Yuhyung — Secondary	Louis Martin, Lay L. Gilson, M. Travis Mayrar, 2005 — Primary	Louis — Secondary	G. Susanne Geister, Udo Konradt, 2005 — Primary	G. Susanne — Secondary	Francis Lou, Suprateek Sarker, Sundeep Samay, 2000 — Primary	Francis — Secondary	Jarvenpea & Leidner, 1998 — Primary	Jarvenpea — Secondary	William M. Hugnes, 2003 — Primary	William — Secondary	Louce A. Thompson, 2005 — Primary	Louce — Secondary	Parsert Kanawattana-chai, Yaum, Jin Yoo, 2004 — Primary	Parsert — Secondary
Cyberculture	Conflict		*						*						*				*		*		*
	Architecture		*		*								*		*				*				*
	The time		*						*		*		*				*		*		*		*
	Diversity		*		*				*		*		*		*				*		*		*
	Language				*				*				*		*				*		*		
	Geographical distribution	*		*		*		*		*													*
	Project rules				*		*		*		*												*
	The planning of the project						*		*		*		*		*								*
	Conflict		*		*								*		*				*		*		*
	Breaking the work						*																
	Operational Model														*				*				*

Source: Researchers.

Figure 2. The stages to conduct this study
Source: Researchers.

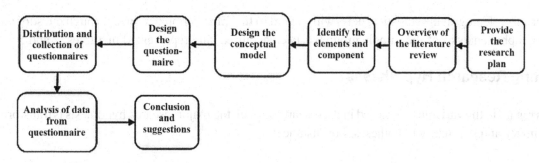

further studies and research institutes to get familiar with various sections in uses of information technology in project management (virtual project) and continue their investigations at the area of building a virtual environment for such projects.

Research Questions

With regard to the advantages of virtual project management which can cause reducing costs, increasing communication and coordination between members, the need to use virtual systems is felt in large and distributed projects around the world, especially at the area of information technology, as the result the present study aims to propose a suitable model for virtual project management at the area of information technology. The questions concerning the research subject are proposed as follows:

The main research question is as follows:

- What is the suitable model for virtual project management at the area of information technology?

The secondary research questions are as follows:

- What are the components of virtual project management at the area of information technology?
- Is there a difference between components of virtual project management at the area of information technology and common project management?

RESEARCH HYPOTHESES

With regard to the proposed model, the research hypotheses are defined as follows:

- Virtual project management model at area of information technology has been developed from various components.
- There is a significant difference between some components of virtual project management at area of information technology and common project management.

Data Analysis

Descriptive demographic statistics: here, with regard to the information obtained through questionnaire, demographic analysis of the sample group has been acquired, and represented in Tables, 2 and 3.

Testing Research Hypotheses

With regard to the variables identified in the research model, the major research hypothesis is transformed to primary and secondary hypotheses. For instance:

First Hypothesis: Virtual leadership is one of the major elements of virtual project management at the area of information technology. With regard to the five-point Likret scale and mean of score(3), the research hypotheses concerning the research data using the statistical hypotheses can be represented as follows:

$$\begin{cases} H\circ : \mu \leq 3 \\ H\circ : \mu > 3 \end{cases}$$

According to the statistical characteristics of this study, binomial test is used to test statistical hypotheses of research.

As shown in Table 4, as the significance level is less than error level(0.05), thus H0 is rejected and H1 is confirmed, and the first, second and third hypotheses of research have been accepted. Hence, it can deduce that virtual leadership, virtual communications and cyberculture are of the components of virtual project management in the field of information technology.

Table 2. Overview of frequency of sample group based on their education status

Cumulative Frequency Percent	Percent	Frequency	
8/10	8/10	11	Bachelor
4/80	6/69	71	Master degree
0/100	6/19	20	PhD
	0/100	102	Sum

Table 3. Overview of frequency of sample group based on their experience

Cumulative Frequency Percent	Percent	Frequency	
9/4	9/4	5	Less than one year
2/38	3/33	34	1-3 year
5/77	2/39	40	3-5 year
0/100	5/22	23	Over 5 years
	0/100	102	sum

Table 4. Results from analysis of binomial test

Result	Type of Test	Research Hypotheses
Rejecting H_0	Binomial test	Virtual leadership
Rejecting H_0	Binomial test	Virtual communications
Rejecting H_0	Binomial test	Cyberculture

Significance level (0.01).

Second Hypothesis: With regard to the components identified for each of dimensions, similar sub-hypotheses below are defined that binomial test has been used to test them. For instance, the hypotheses of this section are described as follows:

Trust is one of the leading components of virtual leadership.

With regard to the response spectrum to six-option questions and mean of score (2.5), it can test the research hypotheses by generalizing sample group to the population. The statistical hypotheses relating to the hypotheses above can be defined as follows:

$$\begin{cases} Ho : \mu \leq 2.5 \\ Ho : \mu > 2.5 \end{cases}$$

Results of binomial test for sub-hypotheses indicate that function electronic monitoring among the components of virtual leadership is rejected as the significance level for this component is greater than 0.05, and the rest of components are confirmed. Thereby, components of time difference, communication rules, face-to-face meetings in virtual communications and component of language in cyberculture cannot be considered as the components of research.

Third Hypothesis: With regard to the identified elements, the hypothesis "there is a significant difference between some components of virtual project management at the area of information technology and common project management". Thereby, there will be similar hypotheses as follows:

There is a significant difference between virtual project management and traditional project management. Here, the statistical hypothesis below has been represented below for which Mann-Whitney-Wilcoxon test has been used to test it.

H0: There is not a significant difference between virtual project management and traditional project management.
H1: There is a significant difference between virtual project management and traditional project management.

With regard to the obtained results, there is a significant difference on the identified elements (Virtual leadership, Virtual communications and cyberculture) at virtual and traditional environment. In following, using Friedman test, priority of elements of virtual leadership in virtual project management and traditional project management is represented as follows:

Table 5. Result of binomial test for sub-hypotheses

Element	Hypothesis	Significance Level	Result
Virtual leadership	Creating a shared vision is one component of virtual leadership	0.000	Rejecting
		0.122	Confirming
	function electronic monitoring	0.000	Rejecting
	Team Solidarity	0.000	Rejecting
	Receive continuous feedback	0.000	Rejecting
	Trust	0.000	Rejecting
	Develop collaboration and coordination	0.000	Rejecting
	Delegation of authority	0.000	Rejecting
	A clear definition for objectives	0.000	Rejecting
	Motivation of team members	0.000	Rejecting
	Existential philosophy	0.000	Rejecting
Virtual communications	Knowledge sharing	0.000	Rejecting
	Geographical distribution of members	0.141	Confirming
	Time difference	0.000	Rejecting
	Communication patterns	0.000	Rejecting
	Synchronous communication technology	0.000	Rejecting
		0.142	Confirming
	Asynchronous communication technology	0.071	Rejecting
Cyberculture		0.000	Rejecting
	Communication rules	0.101	Confirming
	Face-to-face meetings	0.000	Rejecting
	Diversity	0.000	Rejecting
	Language	0.000	Rejecting
	filtered project memory	0.000	Rejecting
	Architecture	0.000	Rejecting
	product planning matrix, product deployment matrix, component deployment matrix	0.000	Rejecting
	How to determine duty	0.000	Rejecting
	Rules of project	0.000	Rejecting
	The planning of project	0.000	Rejecting

$$\begin{cases} H\circ : \mu 1 = \mu 2 = ...\mu 1 \\ H1 : \end{cases}$$

There is a significant difference between at least each mean and other means.

The results of Friedman test indicate that there is a difference between calculated x^2 between each pair. As the result, hypothesis $H\circ$ is rejected. In other words, cyberculture and virtual communications enjoy a high importance at virtual and traditional environment.

Table 6. Result of Mann-Whitney-Wilcoxon test

Result of Test	Significance Level	Value of Statistics	Type of Statistics	Type of Test	
Rejecting H_0	0.032	453/8-	Z	Wilcoxon	Virtual leadership
Rejecting H_0	0.043	278/8-	Z	Wilcoxon	Virtual communication
Rejecting H_0	0.043	641/8-	Z	Wilcoxon	Virtual culture

Table 7. Results of Friedman test

Chi-Squared Test	Freedom Degree	Error Level	Result of Test	Hypothesis
11/64	10	05/0	Rejecting H_0	Leadership in virtual project management
801/146	10	05/0	Rejecting H_0	Leadership in traditional project management
07/44	6	05/0	Rejecting H_0	Communications in virtual project management
585/189	6	05/0	Rejecting H_0	Communications in traditional project management
741/108	10	05/0	Rejecting H_0	cyber Culture in virtual project management
270/173	10	05/0	Rejecting H_0	Cyberculture in traditional project management

Significance level: 0.00.

Furthermore, results of Friedman test indicates difference between calculated x^2 in components of traditional and virtual environment. These results indicate that the priority is given to the existential philosophy, creating a shared vision, and knowledge sharing in virtual leadership, yet the priority is given to the components of motivation of team members, existential philosophy and creating a shared vision in traditional environment. The first priority is given to the component of synchronous communication technology in virtual communications, yet the third priority is given to this component in traditional environment. Thereby, differences are seen in the priority of components in cyberculture.

Results of Research

Results of this study have been formulated based on the goals as follows:

- Identify effective elements and components in virtual project management at the area of information technology.
- Represent a model based on identified elements and components.
- Compare the elements and components in virtual project management and traditional project management.

Table 8. Results of Friedman test

Leadership Component				
Components	Average Ranking		Priority	
	Virtual	Traditional	Virtual	Traditional
Creating a shared vision	94/6	41/7	Second	Third
Team Correlation	23/5	76/5	Tenth	Eighth
Continuous feedback	57/6	75/5	Fourth	Ninth
Ensure	20/6	21/5	Seventh	Tenth
Delegation of authority	62/5	31/6	Ninth	Sixth
entrusting	53/6	98/5	Fifth	Seventh
Clearly definition of objectives	06/6	93/6	Eighth	Fourth
Motivation of team members	32/6	63/7	Sixth	First
Philosophy of existence	96/6	43/7	First	Second
Knowledge sharing	67/6	56/6	Third	Fifth
Components of the Virtual Community				
Geographical Distribution	16/4	99/1	Third	Fourth
Communication patterns	40/4	98/5	Second	First
Synchronous communication technology	74/4	96/4	First	Third
Asynchronous communication technology	07/4	48/5	Fourth	Second
Components of Cyberculture				
Diversity	06/6	48/6	Sixth	Second
Filtered Memory Project	50/6	41/5	Third	Fourth
Architecture and breaking of work	09/6	73/3	Fifth	Third
Development component Matrix -product	61/6	88/3	Second	Seventh
assigning	39/6	30/5	Fourth	Fifth
Conflict	09/7	35/7	First	First
Operational Model	98/5	97/3	Seventh	Sixth

In this regard, an overview on research generalities and literature review was represented so as to specify the importance of the issues such as virtual teams and virtual projects and identify the elements and components of virtual project management at the area of information technology projects and determine the early scenario of model which has been the major aim of researcher. The final model of research consists of three major elements including virtual leadership, virtual communications and cyberculture.

According to the results obtained from research, virtual leadership has been recognized as one the leading elements of virtual project management at the area of information technology which includes the components such as: A clear definition for objectives, Receive continuous feedback, Creating a shared vision, Trust, Motivation, Team Solidarity, Develop collaboration and coordination, Delegation of authority, Existential philosophy and Knowledge sharing. Further, it is deduced that there is no significant difference on leadership in traditional project management and virtual project management, i.e. leadership is considered as a leading element in both traditional project management and virtual project management.

Findings of this research indicate the priority of components of virtual leadership in both traditional project management and virtual project management at the area of information technology.

Results of research indicate that there are significant differences on components identified in traditional and virtual project management, so that higher priority is given to knowledge sharing and receive continuous feedback in virtual project management rather than traditional project management, and this derives from this fact that due to distributed nature of virtual project management and remote communication between members of virtual team, leadership requires receiving continuous feedback from members of team for the purpose of controlling effectiveness and reaching to the determined goals.

Further, as individuals with high skills have little interaction with each other, higher priority is given to knowledge sharing in virtual project management. On the other hand, with regard to the results of research, virtual communications has been recognized as one of leading elements in virtual project management, including the components such as Geographical distribution of members, Communication patterns, Synchronous communication technology, Asynchronous communication technology. Findings of research indicate that communication in both traditional project management and virtual project management at the area of information technology is considered as one of leading elements of project management at the area of information technology.

The results of this research lie on the priority of components obtained in traditional project management and virtual project management at the area of information technology, so that the priority is given to the Synchronous communication technology, Communication patterns, Geographical distribution of members and Asynchronous communication technology in virtual project management, yet the highest priority is given to face-to-face meetings in traditional project management.

Hence, it can perceive that the communication technologies play a major role in reducing low interactions between members in virtual team, yet a high priority is given to communication patterns in traditional project management. Cyberculture is the last element which has been confirmed in this study. The importance of Cyberculture lies on its documentary nature in virtual project management, which causes creating a knowledge database during the project. This element includes the components such as Operational model, The planning of project, filtered project memory, Architecture, product planning matrix, product deployment matrix, component deployment matrix, Conflict, Breaking works, Rules of project, How to determine duty, Diversity, Language.

As seen, most of components obtained in Cyberculture associate to documentation of projects and the rules on how to fulfill the activities of project. In this regard, diversity is recognized as a factor which is more likely witnessed in trans boundary and international projects that experts from various regions with cultural, language, time and place diversities help for achievement of goals of project.

In this project, it was specified that a different priority is given to culture in both virtual project management and traditional project management. Yet, there is a great difference on understanding culture in both virtual project management and traditional project management. In this regard, conflict has been recognized as a component with the same priority in both virtual project management and traditional project management. Yet, the next priority is given to the product planning matrix, product deployment matrix, component deployment matrix and filtered project memory in virtual project management, yet the least priority is given to these component in traditional project management.

REFERENCES

Laurey & Raisinghani. (2001). An impirical study of best practices in virtual team. *Information & Management*, 523–544.

Ann & Lee. (2005). Utilizing knowledge context in virtual collaborative works. *DSS Journal*, 35.

Anna et al. (2004). Building a project ontology with extreme collaboration and virtual design &construction. *Advanced Engineering Informatics*, 71-83.

Bell, B. S., & Kozlowski, S. (2002). A Typology of virtual teams:omplications for effective leadership. *Group & Organization Management*, 27(1), 14–49. doi:10.1177/1059601102027001003

Bruce, J., & Arolio, G. (2002). E-Leadership:implications for theory, research and practices. *Leadership Quarterly*, *11*(4).

Bruce, J. et al. (2003). Adding the "E" to leadership: How it may impact your leadership. *Organizational Dynamics*, *31*(4).

Canie, L. et al. (2004). Would the real project management language please stand up? *International Journal of Project Management*, 22-43.

Chaur, Paul, & Chillarege. (1996). Virtual project management for software. *NSF Workshop on Workflow &Process Automation.*

Chon, F. (2003). A collaborative project management architecture. In *Proceeding of the 36th Hawaii International Conference on System Science.*

Collines. (2002). Virtual & networked organizations. *Express Exec.*

Could, D. (2005). *Virtual team.* Retrieved from www.seanet.com/~daveg/vrteams.html

Crampton & Sheila. (2005). Relationships among geoghraphical dispersion,team processes and effectiveness in software development work team. *Journal of Business Research.*

Cremers. A., Kahler. H., & Rittenbruch. M. (2005). Supporting cooperation in a virtual organization. *Proceeding pf ICIS* (pp. 3-38).

Damm & Schindler. (2002). Security issue of a knowledge medium for distributed project work. *International Journal of Project Management*, 37-48.

David, K. (2004). Examining effective technology project leadership traits &behavior. *Computers in Human Behavior.*

Durate. (2001). *Mastering virtual team: Strategies, tools, and techniques the succeed.* Academic Press.

Evaristo, R., & Muntvold, E. (2002). Collaborative infrastructure formation in virtual projects. *Journal of Global Information Technology Management*, 5(2), 29–47. doi:10.1080/1097198X.2002.10856324

Garden & Bent. (2004). *Working together, apart the web as project infrastructure.* Retrieved from www. intranetjournal.com/feathures/idmo398.html

Hertel, G., Geister, S., & Konradt, U. (2005). Managing virtual team:a review of current empirical research. *Human Resource Management Review, 15*(1), 69–95. doi:10.1016/j.hrmr.2005.01.002

Jacobs, J. (2005). Exploring defect causes in product developed by virtual team. *Information and Software Technology*, 47–60.

Jones. (1997). *Virtual culture: Identity &communication in cyber society.* Sag Press.

Kissler & Gray. (2000). E-leadership. *Organizational Dynamics, 30*(2).

Larsline. (2006). *Virtual engineering teams: Strategy and implementation.* Retrieved from www.itcon,org/1007/3/paper.html

Lau & Sarker. (2004). On managing virtual teams. *HCSM, 2nd quarter.*

Levin, G., & Rad, P. (2007). *Key people skills for virtual project managers.* Retrieved from www.aapm.com

MacMahon. (2004). *Virtual Project Management for software.* McGraw-Hill.

Martins. (2004). Virtual team: What do we know and where do we go from here. *Journal of Management*, 30-41.

McDonough & Kenneth. (2001). An investigation of the use global virtual and collocated new product development teams. *Journal of Product Innovation Management.*

Nandhakumar & Baskervil. (2001). Trusting online:nurturing trust in virtual teams. *The 9th European Conference on Information Systems.* Slovenia.

Noll & Scachi. (1999). Supporting software development in virtual enterprise. *Journal of Digital Information, 1*(4).

O'sallivan. (2003). Dispersed collaboration in a multi-firm, multi-team, product-development projects. *Journal Technology Management*, 93-116.

Parkinson & Hudson. (2001). Extending the learning exprence using the web &a knowledge –based virtual environment. *Computer &Education*, 95-101.

Reich. (2000). Knowledge traps in IT projects. *CACM, 36*, 63-77.

Sulonen & Alho. (2000). Supporting virtual software project on the web. *Computer in Industry.*

Tastoglou & Milious. (2005). Virtual culture: Work and play on the internet. *Social Computer Review.*

Thompson. (2006). *Leading virtual team.* Retrieved from www.qualitydiyst.com/septoo/teams.html

Ulrich. (2002). *Managing virtual web organization in the, 20th century: Issues & challenges.* Ipea Great Press.

Zigaras, I. (2003). leadership in virtual team:oxymoron or opportunity. *Organizational Dynamics, 31*(4).

Chapter 4
The Virtual Leader:
Developing Skills to Lead and Manage Distributed Teams

Andrew Seely
University of Tampa, USA

ABSTRACT

This chapter offers a working definition of the concepts of virtual, management, leadership, and team, and proposes pragmatic tools and solutions to management and leadership challenges in virtual, distributed team situations. Practical experiences are surveyed, including scenarios of remote team, remote team member, distributed learning, and traveling manager. Descriptions of tools and techniques are offered, along with a set of guiding concepts and principles to apply to any virtual leadership situation.

INTRODUCTION

Creating a functioning team from a group of individuals includes the challenge of navigating personalities while holding the team's business purpose in sharp focus. Keeping individuals focused on tasks while providing for professional growth and worker satisfaction is only the first barrier; the mark of an experienced and dedicated manager is in developing the sense of teamwork and shared purpose that results in a team producing more than the sum of its parts. The manager loses many social cues and becomes reliant on "face value" for assessing meaning when communicating with individuals when distributed team members are not in the same location. Spread a group too far apart across space, time, and culture and the idea of "team" will become almost too abstract to matter, resulting in management, leadership, and, ultimately, productivity problems.

The concept of "leadership" has as many definitions as there are people to define it, as does the word "management." While the terms are frequently used interchangeably, as if they were synonyms, this chapter takes the approach that they are separate skills that are symbiotic. One may not be an effective manager without demonstrating some amount of leadership, and one will not be a valued leader if unable to perform general management functions. This chapter's working definition of leadership is to inspire

DOI: 10.4018/978-1-4666-9688-4.ch004

others to do the thing that needs to be done, while the definition of management is to provide structure, framework, tasking guidance, and administrative functions that give members of a team the tools they need to do the tasks that need to be done to meet the team's goals.

The concept of "virtual" is well-defined elsewhere in this book and has been used in the context of the human experience with computing technology as far back as the early 1990s (Heim, 1993). This chapter will generally consider the term "virtual" to mean "effectively present while physically absent." The definition of "team" is not considered contentious. This chapter will focus on the concept of team from the perspective of a "sense of team," where an individual contributor feels like a part of a greater whole with a sense of mutual accountability, and from the external perception of individual contributors making up a coherent unit that functions as a whole. It is proposed that these feelings and perceptions define the concept of team more than a team's tasking and output does. A team can be considered "who we are" as much as "what we produce," and that concept is the underlying assumption for this chapter.

This chapter will discuss practical case studies and present specific techniques for cultivating individual contributors and building a sense of common bond in virtual teams, drawing on personal experiences leading a remote technical team, experiences leading physical teams with one or more remote team members, and experiences teaching distance education courses to a world-wide audience for both U.S. and European universities. The baseline assumption is while there are common lessons that may be applied across all scenarios, each situation presents its own challenges and each individual team member will require a slightly different approach from a manager to get the best work, and the best team-work, from virtual, distributed teams.

BACKGROUND AND LITERATURE REVIEW

The literature regarding virtual technical teams is vast, spanning disciplines of Information Systems, Human Resources, Psychology, Project Management, Business, and Management. This brief literature review includes recent works that specifically impact and support the development of tools for developing leadership in a virtual team environment.

O'Leary, Wilson, & Metiu (2014) explore a concept of "perceived proximity," where through their research they have exposed differences in how individuals identify with others in relationships that are collocated or distant. An interesting finding of this research is the central aspect of emotions and feelings creating a baseline of closeness that is not dependent on geographic proximity. In fact, a counterintuitive finding in their results is the condition where colleagues would identify a closer relationship with a more distant person than with a physically proximate individual, a result of the richness of communication combined with better focus in the remote relationship compared to the bleed over of negative personal habits and hygiene in a shared office setting being seen as a detraction from a close relationship. The discussion and literature review of the usage of technology tools for communication highlights the ways that technology allows the development and reinforcement of shared identity through "dependability, reliability, accessibility, likability, and informality." This shared identity performs as a sort of mitigation of distance, allowing a perceived proximity that is notably closer than the actual distance between individuals.

Berry (2011) explores the impact of the asynchronous nature of electronic communications on the nature of teams, changing work patterns and relationships between individuals, and discusses why the skills required in a collocated setting are not entirely sufficient in geographically distributed environments.

He points out that virtual teams are not defined by the technology they use, but that all virtual teams must use some degree of technology and the "more virtual" a team is, the more technology-dependent it becomes. A very useful aspect of Berry's work is the definition of six characteristics of virtual teams and four categories of virtual interactions, which help to identify and codify different virtual team situations. Similar to O'Leary et al, Berry highlights the setting of norms by the virtual leader as a key aspect for developing a sense of teamwork. A successful virtual team focuses on collaboration, rather than the technology that enables the collaboration, but a standardized and efficient data storage and retrieval is a key enabling technology for any virtual collaboration and must be established early and with support from all stakeholders. Berry concludes that a manager's skill set must be more complex than a collocated manager's, largely to overcome miscommunication between team members that can degrade team performance.

Hoch and Kozlowski (2014) explore the relative pro and con of traditional, hierarchical management techniques against more federated and shared leadership approaches, with detailed discussion on how top-down leadership, shared leadership, and employee performance incentives can combine to influence team performance. The "structural support" of performance-based incentives to create improved focus in team members was proven in the research presented in this paper, though this approach was found to be less impactful in collocated team situations where personal leadership and proximate team members are more powerful motivators. A notable finding in their research in addition to the incentive approach was that shared leadership approaches resulted in improved team performance, regardless of the degree of virtualization in the team.

Sarker, et al (2011) note that individual performance in a virtual team context is a more relevant and important area of focus, as opposed to the more traditional focus on overall team productivity. The focus of Sarker's work is to understand the impact of trust in a virtual team, and to explore how trust is conveyed between actors in such a way as to influence individual and team productivity outputs. Their discussion on structural makeup of a team is especially interesting in the context of developing leadership core competencies in a distributed environment, as the location, linkages, and responsibilities of an individual contributor in a virtual team can have a measurable impact on the output of both the individual and the team. The salient conclusion of this study is that "communication leads to performance through trust." Key points of trust between individuals and between groups, and organizational structure's influence on communication support the virtual manager's understanding of the roots of team performance.

Benetytė and Jatuliavičienė (2014) continue Sarker's exploration into the influence of trust on the effectiveness of virtual teams. A very interesting overview of the components of trust as defined in the literature is presented, with the key aspects of vulnerability, reciprocity, fairness, and emotional content being discussed as foundational. Of special note to the topic of developing virtual leadership traits, Benetytė and Jatuliavičienė discuss the relationship of an individual's identification with the core competencies and core values of their group; the stronger this identification, the stronger the trust becomes between the individual and others in the group. It is noted that developing a common core value set is increasingly difficult as a team becomes increasingly diverse in cultural and national terms, but establishing this core team identity in terms of shared values is a foundation of team trust.

Purvanova (2014) notes the wide discrepancy between experimental and practical perceptions of the viability of virtual teams, and demonstrates that many of the negative conclusions on the effectiveness of the virtual approach are skewed by inaccurate base assumptions and conditions. In both corporate and academic environments, there is a strong perception that virtual is "good enough," but collocated is always better, a feeling that Puryanova demonstrates is clearly not supported in the scientific literature.

Puryanova's work also highlights a key point relevant to the management of virtual teams: The required use of and continued changing of the technologies key to the communication in the team presents a challenge of having to re-learn how to perform basic communication every time there is a technology change. Understanding and overcoming this reluctance early can prevent reluctance of team members to communicate frequently and effectively.

CASE STUDIES AND DISCUSSION

The following case studies are personal experiences of the author and represent practical experience reports in the roles of a remote manager, managing remote employees, being a remote professor with students remote but collocated, and teaching remote and distributed students

The Remote Manager

I was hired into a gaming services start-up company in California. The intent was for me to remain based in Florida and be part of the tier-two technical team as a remote employee. The time zones worked in everyone's favor: The team was made up of late-starters and I was an early-riser, allowing me to fill a coverage gap in the very-early morning hours of the operation. From the start, this opportunity was a new challenge as I had never been an individual contributor on a remote team, and I had never worked at a start-up company. I was concerned about being consistently left "out of the joke" from all the team members who were working in the same office, had common interests, and had known each other for years. The challenge grew exponentially when my role immediately expanded to include becoming the manager for the tier-one support team. I went from being a virtual team member to being a virtual manager.

I assessed my challenges from both roles. As an individual contributor, I needed to establish myself as a valuable producer in a technical and social environment in which I had no direct experience. In a job market where reputation carries immense weight, all I had from the start was a good resume and a strong interview. I had to identify, and quickly, what the key problems were, who was depending on me … and who was watching me. An early and critical task was to determine how and with whom to communicate, given the separation of 2,000 miles and four time zones.

As a new manager in the group, my challenges were significantly greater. I assessed my problem space in this way:

1. Create a positive impression of my own credibility,
2. Establish authority over and responsibility for my team,
3. Gain understanding of organization at large, my team's role in it,
4. Track down a nagging sense of pre-existing unhappiness and discord in the team.

I set short and long-term goals for this activity. For short-term goals, simply "not failing" was at the top of the list. The team was nominally managed by the head of operations, but in reality the group was mostly self-governed by their individual senses of duty and the pleasures they took in being part of a fun company. My short-term goal was to not get in the way while the team took care of the forest-fires that seemed to constantly burn through our operation.

Long-term goals started to gel after I was able to gain trust and articulate real problems. While it was far from obvious at first, the team's greatest struggle was with a lack of predictability. I strongly felt that by providing an organizational framework that would allow people to take time off, or even reliably sleep through the night, the team would have fewer emotional highs and lows and would produce a more consistent output.

My second long-term goal was to establish a sense of identity and pride for the team. I found that the time spent being "self-governed" resulted in a team whose members identified with themselves or with specific senior members on other teams in an unofficial mentor-protégé way, but they didn't have a shared sense of teamwork or identity for their own group. Understanding the nature of the trust each individual felt for others and for the organization, and predicting and building the trust each held for me would be a critical step for the team's overall success (Sarker, Ahuja, Sarker, & Kirkeby 2011)

The key points to understand were no different than with a physical team:

1. *Who are the players*? Getting to know them personally and establish rapport, which leads to trust.
2. *What are the tasks*? Understanding the fires and the long-term efforts is critical for prioritizing work.
3. *How do people communicate*? This is critical, given the distance between me, the remote manager, from them, the team who all work in the same location.

I spent considerable time getting to know the members of the team as individuals. Much like applying a social graph technique to find a common person between two strangers, I was able to find a common interest between myself and each person, and to understand the common interests that they had between each other. There is remarkable value, poorly documented in the literature, in being able to carry on a conversation with highly technical people using specific and obscure pop culture references: In-depth knowledge of cinematic touchstones like Star Wars, The Princess Bride, and The Simpsons can add considerable lubrication to a conversation. Cracking the ice in this way exposed a stable of quiet thoroughbreds. I discovered that we had a lot of motivated and opinionated hot-heads on the team, just waiting for someone to listen to them.

While there were significant geographic challenges, I made an early goal to remove the sense of distance, to give a sense of perceived proximity that would make me seem accessible and easy to communicate with (O'Leary, Wilson, & Metiu 2014). Attending every meeting, required or not, and working to establish relationships with the actors and players across every team was very time-consuming, but this approach helped to paint in a bigger-picture for the company. As a start-up company with all the inherent creativity and energy that the term implies, no one had taken time to plan for organizational growth. My personal ad-hoc organizational chart ultimately became a hotly-traded commodity across many teams struggling to understand their place in the bigger picture.

Telephone. Email. Internet Relay Chat. Instant Message. Mobile text messaging. Wikis. Virtual-world avatar meetings. In-office meetings. There were a lot of communication channels, and many of them were ephemeral and semi-private. It was critical to understand the best way to communicate for each actor, and how to do so in an enduring fashion that built trust and predictability. (Huettner, Brown, & James-Tanny, 2006)

Understanding the course heading of the whole ship helped me to guide the small team in strategic ways that gained the team credibility. This credibility fueled pride and confidence, and ultimately even more credibility. Two specific, and simple things we did:

1. Establishment of a dedicated, logged, private communication channel for the team created a sense of common voice that didn't exist before. Their voices had been lost in the din of the company and they never had a place to be themselves first.
2. The most novel act was to create a simple shift schedule. For a 24/7 team that had to adjust schedules on a regular basis, it was shocking that we didn't already have a schedule. Creating this schedule had a long-term stabilizing effect that resulted in increased confidence in each other and a stronger sense of team (Seely, 2014).

After a year of this effort, the result was a resilient and cohesive unit. The approach of tackling the individuals as human beings, understanding the effective public and private communications, and clearly identifying and prioritizing the short-fuse and long-ball needs of the company made the remote aspect of the remote management a negligible liability. Focusing on their sense of teamwork and pride overcame the "new manager" problem (Watkins, 2003) and resulted in more effective output from the team.

The Remote Teacher

I was offered an assignment to teach a semester at Oslo and Akershus University College of Applied Sciences (HiOA) in Oslo, Norway. The agreement was that I could start the first week in-residence, and then finish out the semester via electronic means from Florida. I was assigned a teaching assistant (TA) who could help with any sort of communications setup I needed as well as more standard TA activities like grading papers or helping students work through problem sets. The class was made up of 30 students from all over the world. While the course language was English, none of the students were native English-speakers.

As this was my first engagement with HiOA, I felt significant pressure to perform well, to "make an impact." This was also an early exercise in a remotely-taught class for the university. They were taking a double-gamble, so the visibility of my effort was high. I had to coordinate the timing across six time-zones, while navigating pre-existing working responsibilities and family impacts around the scheduled class meetings. At a bare minimum, I needed to deliver 16 high-quality, well-balanced lectures to a group of English-as-a-second-language students, with consistency and predictability. To meet my own standards, I needed to make the class feel like I was "there," and to develop real rapport with the students – despite being separated by time, distance, and culture.

Any telecommuter can attest to the challenges of working from home. Even though this was a weekly engagement and not a full telecommute activity, there were telecommuter challenges related to the environment. My young children, neighborhood dogs, postman at the door, and similar distractions were commonplace.

I decided early that consistency was going to be the key. There were variables I could not easily control, like students actually showing up to the classroom or the quality of the audio and video projection. In order to give students a real sense that they were part of a real class and not some misguided experiment, I had to be strictly on time with a predictable lecture. This simple rule was my baseline: consistent and boring would be successful, while brilliant but chaotic would be characterized as a failure. And, of course, consistent and brilliant would be the true goal.

As a teacher, I am always focused on my students' level of interest. It is not a requirement that a student love a topic or love a teacher, but it helps the learning if they at least find the delivery interesting. I believe that a student will engage with a topic if the teacher is engaged with the student on an individual level. In order to develop a sense of engagement, I needed multiple communication channels open to the students. And I also needed to accept early that I would never actually make any meaningful connection with some number of students.

Video chat was the primary tool. Using the free version of Skype, combined with a projector and external speakers in the remote classroom, I was able to establish consistent telepresence on the remote end. But having one-on-one video chats with students on request and having regular and personalized email and message board interaction was the real foundation of creating the relationship. I learned each student's name and home nation, and in turn they started asking a lot of questions to me about living in the United States and my experiences in their various countries. I also made a point of performing my lectures from interesting locations, including from the eighth-grade classroom of my hometown middle school, in order to create a sense of humanity in what could be seen as just a cold, dry voice from the ether.

More so than with face to face classes, I found this class reluctant to ask questions in the classroom. Even though there was two-way communication via the telepresence, there was a notable lack of interaction. The classroom dynamic felt more like a live television broadcast than an immersive telepresence. The availability of a dedicated resource to handle the in-classroom communication was essential. Having this as a classroom experience as opposed to an individual subscriber model helped build the class's sense of teamwork and identity. The students having a trusted agent able to interpret instructor intent was also very useful.

Specific tools that helped this effort:

1. **The Initial Meeting:** The first week of physical presence set the standard that permeated the rest of the engagement.
2. **The TA:** With a room full of strangers with a wide range of cultural backgrounds, there's no substitute for having an agent in the room.
3. **The Technology:** Pushing telepresence into a room via the Skype tool made me "present" in a way that no other technology could have accomplished.
4. **The Personal Touch:** I made a regular point of going "off script" and sharing details of my own life and surroundings. It was important to become more than just a talking head on a screen, but to be a human on the other side of the technology.

This remote class experience was measured a success by both students and HiOA administration, although with some reservations. All agreed it wasn't "as good as" a full-time face to face instructor but all agreed that it was a viable approach, consistent with general perceptions in the academic and technical industries (Purvanova 2014). The key to this success, I believe, was in the first week of establishing an in-person relationship. By having met each student, and them having met me, we were able to establish a "trust baseline" that helped overcome or avoid communication problems (Covey, 2009).

The Manager of a Remote Employee on Short Term Rotating Assignment

I was the manager of a diverse team of system and database administrators on location with our data center and the majority of our customer base. The company also had some overseas operations, including one well-provisioned but lightly-staffed warm site. All services were expected to be available at the warm site, and we regularly used it as an alternative capability during break-fix problems or for scheduled upgrades or outages. The site consisted of 60% of the physical servers and infrastructure footprint compared to what we had at the primary location, but 90% of the services and capabilities were represented. The majority of the day-to-day work on the servers could be handled remotely, but a physical presence was required to maintain our standard for seamless service delivery in the event of an emergency.

I maintained a 90-day rotating assignment to the remote location, which was remote both in terms of being far away from our home office and also in a location that was remote from much of anything else. Everyone on the team had an opportunity to spend three months away from home, without much to see or do other than maintain servers and systems that didn't need much attention. Privately, the team considered this to be a mild punishment, so it wasn't really popular. I didn't get a lot of fight when I assigned someone, but I also didn't have a lot of volunteers.

Every three months, another team member would show up, get a day of turn-over, and then settle in to play games, read books, go to the gym, while trying to stay focused and actually provide operational value. The amount of operational mindset varied wildly from individual to individual.

The challenge with this situation was keeping people focused, motivated, and feeling like they were part of the overall team, when all the decisions and the majority of the work were being done by the team back home. A team member would arrive at the location and maybe be excited about making a difference, about doing something meaningful, but this energetic attitude would wear off in a week or two and the member would become just a warm body that would need to be kick-started when they returned to the home office.

Perhaps even worse were the few who would maintain high levels of professional motivation and would put pressure on to change and improve things constantly for the whole time they were at the remote site. The people who would never stop tinkering and adjusting were also the people who felt that the systems weren't in production so they could make changes without coordinating with the greater team. Not unlike Benetytė, and Jatuliavičienė's (2014) findings on the negative impact of "too much trust," too much motivation can have a negative impact on overall product.

The challenge is not just to keep one individual motivated and focused. In some cases it's to keep the individual from being too motivated. And in all cases, the challenge is to combine a sense of purpose and accomplishment with a smooth re-entry to the greater team.

Over time the cyclic nature of this assignment led to hot and cold areas. Some people gravitated more towards the assignment and some more away, which would naturally result in a subset of the team having most of the burden but also, conversely, actually doing the least amount of actual core work over the course of a year. At the start, I decided that I would force the assignment to be applied evenly across the whole team and not allow just one or two people to do it all the time. This decision was both counter-intuitive and unpopular, but I felt that the remote assignment presented too much opportunity for growing an unbalanced workforce, and too much opportunity for the remote member to drift from the team's core.

I attempted to generate and assign tasks that were evenly delivered and balanced and that would not be seen as busywork but would also maintain a baseline level of effort that would give the team member a sense of accomplishment. I encouraged innovation and ideas, but I was sure to keep an open pathway for ideas to be sent back home. I felt it was vital that actions taken at the remote site would always be the result of tasking from the team, but it was equally important that a remote team member's motivation and drive were not discouraged. Meetings for the sake of meetings were discouraged, but daily touch-points via telephone and email were important. Video teleconferencing was not an option due to the expense of that type of capability at the time.

After several years of managing the team this way, the successful tools I used were:

1. Keep a sense of fairness and balance. Don't let individuals shirk, don't let individuals do more than their share either. Maintain the sense of shared duty.
2. Avoid the appearance of busywork, which is demoralizing, but don't allow there to be no work at all, which is even more demoralizing.
3. Encourage people to feel a sense of pride in supporting the team by accepting the difficult assignment.
4. Always listen to and encourage new ideas from the remote employee, but never allow the remote employee to be a completely independent actor. Good ideas should be repackaged as tasks that come from the leadership team like any other task. This maintains control for the leader and also gives the employee a sense of pride in being able to influence, while limiting the dangers of uncoordinated activities.
5. Have regular touch points, but not so much that the employee feels babysat. The employee is entrusted with a difficult assignment, and should be made to feel that way. This is different than a full-time remote employee, who needs a lot more engagement to maintain a sense of belonging in the team.

It didn't always work out as planned, but it always worked. There's a lot of variability in people's personalities, especially when they're not accustomed to being away from the comforts of team and home. The overall concept was to motivate people just enough to stay focused, but to provide an easy outlet for overly-motivated ideas that would introduce problems in the environment. Create a set of boundaries for the employee that give a sense of control without the feeling that everything is controlled, and a sense of fairness and mutual support in a situation that may not be completely "fair."

The Virtual Guide

I have taught classroom and online Computer Science courses for the University of Maryland University College (UMUC) for over fifteen years. UMUC's student body has traditionally been made up of adult learners, and for the vast majority of my time teaching for UMUC's European Division the focus was on providing high-quality, affordable, and, ultimately, flexible higher education to U.S. service members stationed at overseas locations. In fact, not only was the focus on service members, but registrations were exclusive to this group and their immediate families.

Many enlisted people join the military in part to gain access to higher education both while in uniform and after completion of service, and in many situations a young enlisted troop actually gains some amount of credit towards awards and promotions as a result of taking college classes. For well over ten years I taught up to five course sections each year to these motivated students. While I still teach for UMUC today, the operating model has changed to open up enrollments to the general civilian population as well.

The online teaching dynamic has many potential approaches, and most everyone in a teaching role brings something unique to the experience. Teaching in an online environment was still very novel when I started working in the online space in 2002, though over the course of the following decade it grew from novelty to flexible tool in the student's arsenal to the primary approach for many students to complete an entire degree program. So too did teaching evolve from a novelty that added a section to a teaching load now and then to a core capability and a key skill that had special requirements.

Teaching is less about leading than it is about guiding. It's less about managing than it is about encouraging. But there remains a simple underlying dynamic: the teacher assigns tasks that the students perform. There is a feedback cycle with respect to assigned tasks. There are requirements for adhering to schedules and for communicating effectively. In the virtual sense of an online learning environment, these challenges are central. Even more so with this student body, where a single virtual classroom of thirty students may include people spread across Iraq, Afghanistan, Germany, Italy, Kuwait, Bahrain, England, Japan, Korea, Guam, Turkey, and the U.S. To add to the challenge, it was not uncommon to have students who were in permanent change of station (PCS) or temporary duty assignment (TDY) in the middle of class. And occasionally a student would be in a location that was actually under attack by hostile forces, which adds an intensity to the concept of "family emergency" that in other circumstances just means missing classes or assignments due to an ill or deceased family member.

My approach to providing leadership and guidance in this dynamic, diverse virtual environment is to clear away the noise and keep laser-focus on the things that really matter. My time of "supervising" a "team" of 30 students spread across the world and living in wildly-varied conditions is limited by the duration of a semester, which is currently eight weeks at UMUC. In this time, I need to teach a set of core skills and provide meaningful activities and evaluations that build new skills and provide the validation of learning, and I need to ensure that a student is ready for a follow-on class and that the university's general standards have been met. If I'm lucky, I'll have an opportunity to provide academic or career advice or mentorship, but as valuable as that is, teaching the core is the only thing that really "counts."

Our online learning environment is already disadvantaged compared to the effectiveness of a face-to-face experience because it is asynchronous and largely text-based. It's difficult to tell from a forum post if a student is struggling unless they explicitly state that they're indeed struggling. Establishing an early and meaningful bond with each student is a goal. Sharing personal and professional details, encouraging students to do the same, and having 100% positive interaction with each student from day one is essential. They must know that there's a live person in the leadership role and that this person is engaged and there for them.

Daily interaction helps remove the barrier that online learning creates. Regular touch points, an appearance of motion in a virtual classroom, and frequent feedback creates a sense of both motion and belonging. When these feelings take root, the student stops focusing on the downfalls of online learning and is better able to focus on the subject being taught.

Being available to be contacted for consultation on tasks or assignments is obvious, but to be really effective, the method of contact has to suit the student. Telephone and video chat have to augment email. Whenever possible, face-to-face meetings really bring the virtual environment to life for a student. Flexibility is key when the team is spread so far apart and in challenging environments. I have had telephone calls "patched through" to me from local military bases, where the origin of the call was both remote and undisclosed, just so we could work through a particularly difficult assignment.

With over a decade of near-continuous teaching, leading hundreds of remote students in a shifting online virtual environment to meet specific goals and learning outcomes, I can offer key approaches that work to keep the individual focused and lead them to success.

1. **Dependability:** Giving the appearance of being always-on.
2. **Flexible Communications:** Allowing and encouraging the remote individual to contact the leader in the way that works for them.

3.   Absolute focus on core requirements within maximum time boundaries, and complete forgiveness for anything else. Get everyone across the finish line.

These virtual-team classes strongly mirrored Berry's six attributes (2011): defined and limited membership, independent function with shared purpose, shared responsibility, span organizational boundaries, geographically dispersed, and reliant on technology for communications. Ultimately, the biggest factor for success in this virtual environment is flexibility. While many instructors may disagree with my approach, I have never, in over a decade of teaching in this environment, penalized for late work or denied an incomplete grade (Schimmer, 2011). What matters is the core skills being learned, and when everything else in a student's life is constantly changing, the due date of an assignment is far less relevant. Have I been taken advantage of along the way? Of course, but the occasional slacker taking advantage of me is OK when compared to the incredible results and products delivered by the vast majority.

The Manager on Extended Travel

I managed a team of around twenty technical people, from entry-level shift-working tier-one to late-career senior database administrators. We were all co-located in the same facility and had a strong relationship that included regular after-work get-togethers and in-office birthday parties. Throughout the organization it was known that this was the tightest-knit group, a point in which I took great pride.

One summer I had a unique opportunity. Our business had many subsidiary locations around the world and we dealt with many of them regularly over telephone and email, but we had rarely actually met the personnel and never visited their locations. One location in particular was having significant problems with their data processing and the on-site staff not only seemed unable to fix it, but they were having difficulty even recognizing the nature of their problem. This one site was the catalyst for a "world tour" of every subsidiary location to perform a "health check," which was not quite an "audit" but more than a social call. A senior technical manager would make week-long visits to review processes, interview staff, and help identify gaps and get-well plans for problem areas. I volunteered to visit ten locations over the course of three months.

But, of course, I still had a team to lead at the home office. While I had a team member designated as my deputy and he could perform many of my administrative functions, the general direction and tenor of the group was my responsibility and three months is a long time for a team to be rudderless. Nature abhors a vacuum, and this is also true for a leadership vacuum, especially in a stressful, high-paced workplace where competing and conflicting demands can render an individual contributor ineffective quickly.

My goal was to continue to provide general oversight and direction and to give a sense of presence and control, without actually spending the same amount of time and effort I would do while actually in the office. Consider what a team needs from a manager: A sense of cohesion and teamwork, guidance and prioritization, top-cover for external demands, direction for specific tasks, and administrative functions like hiring, firing, resigning, vacation, time cards, and benefits questions. Some of these things can be deferred, some delegated, but some could not be done by anyone else without changing the character of the team.

In preparation for the trip, I delegated the basic administrative functions down to my deputy and to the administration front-office. I delegated more serious administrative functions up, asking my manager to handle any personnel actions that couldn't wait. Day to day task management could be handled by my deputy, but bigger-picture guidance and direction should remain my responsibility, regardless of where I was. In this virtual team scenario, a shared leadership model was the best approach (Hoch & Kozlowski 2014).

With each week in a different location, each location providing a different set of communications tools, and with my own schedule at each site being driven by the site's own management team, it was not really possible to have a set meeting schedule for reach-back to my own team. I knew I would need to make touch-points with the team as frequently as possible, but I also didn't want to give any sort of impression that I didn't have trust in what they were doing while I was gone: I needed to be a remote manager, not a remote micromanager.

The approach I used:

1.   Delegate or defer everything but my core function.
2.   Create a feeling of stability and regularity even though the actual situation was volatile, and give the employees the key things they want from their manager (HR Specialist, 2011).
3.   Use every tool available, but don't get hung up on any specific tool.
4.   Don't be a remote micromanager.

The result was as expected. After a three month trip with a new location every week and with varied ability to communicate, and our sense of teamwork never wavered. A combination of work-day telephone, official email, personal email, after-hours home telephone, and occasional video, across multiple touch points in the organization gave the correct impression that I was still very much present. A self-governing approach to not call constantly and to not dig into details and to allow delegates to act independently helped prevent a sense of micromanaging, although it was probably not avoided entirely. The measures of the success of the activity were twofold: Despite a lot of volatility in our overall company, there were no problems in our team while I was gone, and when I returned nothing had changed – the team was still the high-functioning, tight-knit group it was when I left it.

SOLUTIONS AND RECOMMENDATIONS

I have described several different distributed team scenarios, long-term and short-term, one-to-many and many-to-one. In my experiences, I can account for one common lesson: Every situation presents a new challenge to overcome. In general, these are the principles I have collected over the years that I try to follow when approaching the task of managing a distributed team:

- To get the best communication, communicate in a way that is natural to the people with whom you're trying to communicate.
- Use the right tool, which is whatever tool works. Instant message, message boards, wiki, email, telephone, video chat, text message are all starting points.
- Understand the real work and prioritize activities, and focus on what really matters from a deliverable perspective
- Start out flexible, stay flexible. As hard as the distance is on the manager, it's even harder on the employee.

- Create a sense of normal, stable, predictable for the employee. Give the employee space to be able to focus on the task rather than worry about the job.
- Team identity and cohesion matter. From the manager's perspective, identity and cohesion help a team to perform better.
- Motivation should be governed as well as grown. Pride matters; give it a place to live and grow, or it will grow in unproductive ways.
- A sense of fairness helps people stay focused. A sense of unfairness gets amplified when relationships are stretched by time and distance.
- Have clear lines of authority, and delegate/deputize local, in-person trusted actors whenever possible, but only for the things that make sense. Clear understanding of authority needs to be balanced with allowing and encouraging autonomy
- Manage, don't micromanage, just like in a physical team
- Make it personal, because relationships matter most when there are problems

FUTURE RESEARCH

Virtual team leadership is an area of practice that will continue to benefit from the academic community. Continued study into the mechanisms of trust between individuals and organizations when cross-cultural teams are developed will be important for the future of knowledge work. In addition, creation and enhancement of formal academic and training programs for the development of both leaders and workers will deliver tangible results as workforces mature and employers consider virtual team arrangements to be normal and valuable rather than just cost-effective and sufficient.

CONCLUSION

This paper has presented several case studies and provided an overview of the current research into leading virtual teams. Solutions and recommended approaches have been discussed. Management is a skill that can be learned. Leadership is a character trait that can be grown. Applying management and leadership to a virtual leader situation where the team is spread out over space, time, and culture is a unique challenge that can be overcome with dedication, flexibility, and understanding of the challenges from the employee perspective.

REFERENCES

Benetytė, D., & Jatuliavičienė, G. (2014). Building and sustaining trust in virtual teams within organizational context. *Regional Formation and Development Studies*, *10*(2), 18–30. doi:10.15181/rfds.v10i2.138

Berry, G. R. (2011). Enhancing effectiveness on virtual teams: Understanding why traditional team skills are insufficient. *Journal of Business Communication*, 0021943610397270.

Covey, S. M. R. (2009). *How the Best Leaders Build Trust*. Retrieved February 1, 2015 from http://www.leadershipnow.com/CoveyOnTrust.html

Heim, M. H. (1993). *The Metaphysics of Virtual Reality* (1st ed.). New York, NY: Oxford University Press, Inc.

Hoch, J. E., & Kozlowski, S. W. (2014). Leading virtual teams: Hierarchical leadership, structural supports, and shared team leadership. *The Journal of Applied Psychology*, *99*(3), 390–403. doi:10.1037/a0030264 PMID:23205494

Huettner, B., Brown, M. K., & James-Tanny, C. (2006). *Managing Virtual Teams: Getting the Most from Wikis, Blogs, and Other Collaborative Tools*. Plano, TX: Wordware Publishing Inc.

O'Leary, M. B., Wilson, J. M., & Metiu, A. (2014). Beyond being there: The symbolic role of communication and identification in perceptions of proximity to geographically dispersed colleagues. *Management Information Systems Quarterly*, *38*(4), 1219–1243.

Purvanova, R. K. (2014). Face-to-face versus virtual teams: What have we really learned? *The Psychologist Manager Journal*, *17*(1), 2–29. doi:10.1037/mgr0000009

Sarker, S., Ahuja, M., Sarker, S., & Kirkeby, S. (2011). The role of communication and trust in global virtual teams: A social network perspective. *Journal of Management Information Systems*, *28*(1), 273–310. doi:10.2753/MIS0742-1222280109

Schimmer, T. (2011). *Enough with the Late Penalties!* Retrieved February 1, 2015 from http://tomschimmer.com/2011/02/21/enough-with-the-late-penalties/

Seely, A. (2014). /var/log/manager: Rock Stars and Shift Schedules. login:, *39*(3).

Specialist, H. R. (2011). *9 things employees want from their managers (and 5 things they don't)*. Retrieved February 1, 2015 from http://www.thehrspecialist.com/print.aspx?id=32033

Watkins, M. (2003). *The first 90 days: Critical success strategies for new leaders at all levels*. Boston: Harvard Business School Press.

Chapter 5
Skill Building for Virtual Teams

Amir Manzoor
Bahria University, Pakistan

ABSTRACT

Advances in Information and Communication Technologies (ICT) are creating new opportunities for organizations to build and manage virtual teams. Such teams are composed of employees with unique skills, located a distance from each other, who must collaborate to accomplish important organizational tasks. As such, it is very important for organizations to identify and develop skills that critical for virtual teams to succeed. Participation in and management of virtual teams comes with its own unique challenges and opportunities. This chapter explores virtual teams, their benefits and challenges to organizations, and provide ways to ensure that virtual team members and leaders in their organizations have the skills, competencies and tools needed to succeed. Specific recommendations to improve skills of virtual teams are also provided.

INTRODUCTION

During the decade of 90s a large number of public and private sector organizations and non-profit organizations implemented high performance, team-based management. The reason behind this move was the fact that organizational structures that consisted of high-performance teams continued to outperform organizations having conventional teams (Taha, Ahmed, & Ale Ebrahim, 2009). A properly implemented team-based approach helps to improve organizations in in virtually every measure—from productivity to morale; from quality to shareholder return (Beth Watson-Manheim, Chudoba, & Crowston, 2002).

In today's world, telework is gaining significance (Boell, Campbell, Cecez-Kecmanovic, & Cheng, 2013; Greer & Payne, 2014). According to the Telework Research Network, regular telecommuting grew by 61 percent between 2005 and 2009, and based on current trends, the organization estimates that the number of telecommuting workers will grow to nearly five million by 2016—a 69 percent increase (Lister & Harnish, 2011). Globalization requires employees and business partners to be geographically and temporally distant from one another, deploying information technologies within a virtual organization is an obvious choice for overcoming spatial and temporal boundaries (Boudreau, Loch, Robey, & Straud,

DOI: 10.4018/978-1-4666-9688-4.ch005

1998; Keller, 2014). Technology is changing fast and the number of collaborative software is growing. These collaborative software facilitate virtual work. These software have now become so comprehensive and easy-to-use that teams of people dispersed geographically are using them to perform their virtually and more efficiently. The recent downturn in global economy has forced businesses to cut travel costs. This phenomenon has also made virtual work more popular (Kam & Katerattanakul, 2014).

Virtual teams are groups of people who are geographically dispersed but leverage technology to work together virtually (Hinds & Bailey, 2003). The use of virtual teams is growing as the amount of virtual work continues to grow. Today, it is hard to find an organization that doesn't have one or more virtual workers and virtual teams. Virtual teams are now a reality in the workplace. If this trend in the workplace environment continues, virtual working will increasingly influence the way we operate, and the "effective virtual team worker" will be a valued asset. A key benefit to forming virtual teams is the ability to cost-effectively tap into a wide pool of talent from various locations (Crisp & Jarvenpaa, 2015).

Most project managers with a few years of experience or more are likely to have managed a project where some or even all of the project members were remotely located. Being a virtual team worker is not for everyone or every organization. A virtual team worker is more likely than the collocated worker to suffer from feelings of isolation if the set-up is not right, and they need to be more self-managing and focus their efforts in a particular way. According to experts, virtual teams are here to stay. For organizational success, it is necessary for organizations to learn how to work with them (Purdue University, 2008).

The global weakening of economies and increased realization by businesses that increased travel reduced productivity and increased has fueled the growth of virtual teams into corporate structures. Inclusion of virtual teams can result in an average productivity increase of 27% and in many instances the net productivity increase can surpass organizational real-estate cost savings. However, increase in productivity vary across organizations and industries (UNC Kenan-Flagler, 2010; Purvanova, 2014). There exist ample reports showing increased number of employers continuing to expand the number of virtual workers and team in their organizations (Leonard & Trusty, 2015). A 2009 survey found that more than 90% planned either to increase virtual work or keep it on the same levels (Leonard, 2011). Another survey conducted in 2010 found that more than 20% of organizations had planned to increase their virtual team members within one year and more than 75% planned to keep their number of virtual workers at the same level (Lockwood, 2010). Another poll of organizations in 2010 found a large proportion of organizations planning to increase their telecommuting workers within the next five years (Lockwood, 2010).

Various studies confirm virtual teams offer many positive benefits for both employers and employees. These benefits include increased flexibility and productivity, better work outcomes, increased knowledge sharing, and reduced time-to-market, attract talent from around the globe, reduced operating costs, and increased access to low-wage resources (Lockwood, 2010; Boutellier, Gassmann, Macho, & Roux, 1998; Chudoba, Wynn, Lu, & Watson-Manheim, 2005). Virtual teams are also eco-friendly (Mocanu, 2014). According to Lister and Harnish (2011), the existing telecommuting work in USA results in annual saving of 390 million gallons of fuel and annual decrease of around 4 million tons of greenhouse gases.

Management of and participation in virtual team pose some unique challenges and opportunities (McDonough, Kahnb, & Barczaka, 2001). The objective of this chapter is to explore the benefits and challenges of virtual teams to organizations and provide suggestions/recommendations that virtual team members and leaders of the organizations can use to acquire tools, skills, and competencies needed for the success of virtual teams.

CHALLENGES AND SUCCESS FACTORS OF VIRTUAL TEAMS

Trust building and conflict management are some in-built challenges of the virtual team concept (Jarvenpaa & Leidner, 1998). Communication among virtual team members becomes more difficult in cases of global virtual teams. In these global virtual teams, cultural barriers are often significant issues (Ale Ebrahim et al., 2009). A survey of more than 30,000 employees of multinational companies in USA (RW3, 2010) revealed that more than 45% of virtual team members never met each other and only 30% met each other once in a year. For 85% employees, the top challenge of working in virtual teams was the inability to read nonverbal cues. For 81% employees, virtual teams didn't provide cooperative interactions among team members and it was also difficult to build trust and rapport in virtual teams. A significantly large proportion of employees thought virtual teams didn't provide enough time to build relationships. More than 70% employees thought that conflict management and decision-making was more challenging in virtual teams than in conventional teams. For more than 60% employees, expressing opinion in virtual teams was challenging. Some other hurdles noted by the employees were different time zones, language, holidays, local laws and customs and technology (RW3, 2010; Hastings, 2010).

In global virtual teams, these challenges are more intense and cultural issues, such as "low context" culture (a culture messages are much more specific and importance put on what is said) and "high context" culture (a culture that relies on nonverbal cues and focus more on the relationship, the setting, and previous interactions to interpret what someone means) can significantly inhibit communication among team members (Hastings, 2008). Another stumbling block is to ensure that all members of virtual team have access to the technology selected for and used by virtual teams. Virtual team members need to be able to assess non-verbal cues of other team members (O'Leary & Mortensen, 2005a;M. O'Leary & Mortensen, 2005b). As such, conventional communication technologies, such as telephone and email, becomes less useful for virtual teams. Organizations need to pay attention while selecting technology for virtual teams because selection of wrong technology can negatively affect team members' communication, trust building and productivity (Lockwood, 2010).

These challenges to virtual teams are not insurmountable. Support and encouragement from organizational leadership and HR and their active involvement in selection and training of virtual team members, selection of technologies to be used by virtual teams, and training for use of the selected technologies can turn these challenges into opportunities. For virtual teams managing conflict, making decisions, and expressing opinions is more challenging as compared to conventional teams (Malhotra, 2010). If there exist no mechanism to express opinion or manage conflict among team members, no wonder decision making would be more challenging in virtual teams.

Generating quality output and innovative ideas from virtual teams is challenging mainly due to hurdles of language, time zone differences, and culture (Coronas, Oliva, Luna, & Palma, 2015). The lack of visual and tactile interaction further exacerbates these challenges. However, research shows that, despite these challenges, successful intercultural teams are able to produce quality output and more innovative ideas. This highlights the fact that, to be successful, virtual teams need to follow certain rules of interaction among team members (Malhotra, 2007).

General Challenges Faced by Virtual Teams

Some of the great challenges faced by virtual teams include different time zones, language, local laws and customs; and technology. It is interesting to note that technology is not a very significant challenge as compared with other challenges. Time zones represent a very significant challenge (Malhotra, 2006; Malhotra 2010; Wadsworth & Blanchard, 2015). To be managed properly, the challenges mentioned here need several compromises. For example, the schedule of virtual meetings can be prepared in such a way that the time zone issue is not borne by team members in a single country or time zone. Holiday schedules can distributed in advance and language difficulties can be addressed by follow up written or telephonic communication (Markus, Manville, & Agres, 2014).

Personal Challenges of Working in Virtual Teams

According to Malhotra et al. (2007), the greatest virtual challenge faced by team members is inability to read non-verbal cues (or body language) followed by absence of collegiality, difficulty establishing rapport and building trust, difficulty seeing the whole picture, reliance on email and telephone, and a sense of isolation. According to studies and anecdotal evidence, inability to read non-verbal cues is closely linked to the lack of face-to-face meetings as many virtual team members never meet in person (Glikson & Erez, 2015). Virtual team members also feel difficulty in increasing their productivity when there is no cooperative interaction among team members and establishing trust and rapport among team members is difficult. Summing it up, virtual teams need to establish clearly defined structures, procedures, and processes and arrange periodic face-to-face meetings. Virtual team leaders should allow time during these face-to-face meetings for building rapport, collegiality, and team trust. The lack of trust can be due to many reasons such as lack of trust in the perceived competence of team members or a lack of trust in their dedication and commitment to the team. These two aspects of lack of trust are directly related to cultural behaviors. To address these aspects of lack of trust, virtual teams need proper communication structures and establish dialog among team members (Peñarroja, Orengo, Zornoza, Sánchez, & Ripoll, 2015).

Face-to-Face Meeting with Virtual Team Members

The lack of face-to-face meetings with virtual team members, despite their desire, makes it more difficult to form a virtual team. Even more difficult is to provide energy and resources to sustain such a virtual team. The reasons described below illustrate why creating and sustaining a virtual team is so difficult.

Incomplete Communication

Studies show that 80% of all human communication is conveyed through factors other than the meaning of the words that are spoken (Thompson, 2011). Many factors help establish the meaning of a comment. These factors include tone of voice, facial expressions, speed, cadence, body language etc. Without this understanding we can get confused e.g. to understand whether an apology is cynical or sincere. Email is a communication media that is more susceptible to misinterpretation. In a telephonic conversation we can pick the cadence, speed, and meaning though we cannot pick the important nonverbal cues.

Limited Relationship Building

In 1980s, Hewlett-Packard coined a concept MBWA (Management By Wandering Around). The idea emphasized the importance of regular, informal contact between managers and employees. According to

MBWA concept, informal contact between team members and managers would be possible that would help surface issues and solutions normally not discussed in formal meetings. Further, managers can have opportunity to address concerns, dismiss rumors and forward company strategy. Even if MBWA is not part of organizational culture, it can be noted that causal interactions among team members and managers outside pre-planned meeting influence a significant proportion of decision making. Some studies estimated that as much as 30% of senior management time is spent in causal encounters (such as unplanned coffee break conversation) and is considered equally valuable by the senior management and team members. This informal contact can extend beyond the boundaries of workplace and lead to highly creative and innovative ideas. In a virtual team setting, these opportunities for relationship building and idea sharing are far more limited (Maznevski & Chudoba, 2000).

More Complex Interactions

While a conventional team members enjoy the proximity of team members and no restriction of a time zone, a virtual team may not have the same advantages. In fact, many mundane tasks, such as scheduling a meeting, can become a daunting task for a virtual team if team members reside at distant locations in very different time zones. In such cases, holding a real-time conference can be impractical and we need medium of communications where the presence of person receiving the information is not necessary. Electronic and voice mail could be very suitable in such cases. However, there exist a risk of misinterpretation by the receiver of the information sent. This misinterpretation could lead to some great erroneous conclusion because the receiver was not able to ask clarifying questions and receive feedback.

Challenges of Virtual Team Meeting

Meetings of virtual teams are faced with many challenges. Some significant challenges include insufficient time to build relationships among team members, pace of decision making, differences in leadership styles and methods of decision-making, and non-participation of team members. It is vital that well-defined structures and processes are in place that would be helpful not only enhance the comfort level of team members but also increase their performance.

CHARACTERISTICS OF EFFECTIVE VIRTUAL TEAMS

The support services required for virtual teams has not grown at the same pace as virtual teams. These support services are crucial for success of virtual teams. Erickson and Gratton (2007) identified some characteristics shared by successful virtual teams.

- **Executive Support:** Executive support of social relationships building in virtual teams is crucial for virtual team's success. This is because these social relationships help building trust among virtual team members. Executives in different organizations have their own ways of providing such support to build and support social relationships. However, the most successful executive employed "signature" practices. These practices are tailored to a particular organization, hard to replicate, and memorable.

- **Effective HR Practices:** Two HR practices are particularly effective for virtual team success. The first is skill building for collaborative behavior and the second is informal community building. Some significant traits of collaborative behavior include appreciation of others, engaging in meaningful conversations, creative and productive conflict resolution etc. Some prominent practices involved in informal community building include feedback, mentoring, and coaching. These practices help connecting virtual workers to the organization. One other area that needs HR attention is succession planning and promotions. HR need to track both in order ensure appropriate recognition and credit is given to the members of the virtual teams (Leonard, 2011).

- **Well-Structured Teams:** Right people for virtual teams are vital for team success and HR should make sure that the selected members of virtual teams are self-reliant, self-motivates, and able to succeed working independently. These members should be initiative takers, have strong communication skills, and tend to like or tolerate ambiguity (Leonard, 2011).

- **Strong Team Leaders:** Strong leadership forms the foundation of virtual team's success. While leaders of virtual teams share many characteristics of leaders of conventional team, there exist some key differences. Virtual teams lack face-to-face communication and as such it is difficult to build trust and rapport among team members. Virtual team leaders' need a strong focus on relationship building and demonstrate excellent communication skills and emotional intelligence in order build this trust and rapport. Virtual team leaders must also be process-oriented with a strong track record of producing results. This is because decision making in virtual team environments can be real challenge (Lockwood, 2010).

- **Practices of Effective Virtual Leaders:** According to Rosen et al. (2000), effective virtual leaders establish and maintain trust through the use of communication technology. These leaders have a clear understanding and appreciation of the diversity in the virtual team. These leader leverage this team diversity, enhance team visibility outside the organization, and ensure benefits of participating in virtual teams do reach the team members.

PRACTICAL/MANAGERIAL IMPLICATIONS AND RECOMMENDATIONS

The growing importance of global collaboration translates into virtual work and goes on to point out the key areas in which virtual teams represent unique challenges to the productivity of organizations and individuals. There is a need to recognize that virtual teamwork is different and requires specific training, tactics, and support. Furthermore, today's leadership needs to include specific competencies to structure and manage virtual teams that invariably comprise people from different cultures and work styles and who come together to meet at different times of the 24-hour work cycle. Virtual teams need to establish specific work rules (i.e., rules for respectful interaction) that are assumed in colocated teams. They also need to pay greater attention to team structure than co-located teams do. In addition, virtual teams must carefully monitor and adhere to the work rules they have created. Finally, they need to be aware of the influence of culture on work styles and to develop procedures to assure intercultural effectiveness.

Holding of regular team meetings, use of online chat, video conferencing, one-to-one conversations, e-mail conversations, an online directory of team members profiles, a track of team members participation in different team meetings held in different time zones, restricting multi-tasking during calls and meetings are some of the useful practices organizations can use improve the relationships among virtual team members (Hastings, 2010).

Following are some tips to help with the challenges of cultural diversity faced by many global virtual teams. The first word of advice for virtual meetings: set ground rules for team interactions. Some practical ideas to help set those ground rules include speaking slowly, avoiding interruption, listening to understand, speaking as though remote participants are in the room, avoiding use of a computer or text message during meetings, and setting/distributing agendas for meetings beforehand.

HR Support for Virtual Teams

HR should establish appropriate policies and procedure to support virtual work. HR should also ensure IT support is available to the virtual teams so that any query related to IT infrastructure is resolved on time. HR should also arrange e-learning sessions, and training sessions on cultural sensitivity and leadership styles for virtual team members.

Selection Process of Virtual Teams

HR should assist leaders of virtual teams at the team formation stage by helping them ensure the prospective team members possess the characteristics of successful virtual team participants. These characteristics include self-motivation, self-reliance, and ability to work independently and tolerate ambiguity. Since all organizations are expected to expand, this particular support from HR would help ensure organization is well-positioned for its future expansion.

Knowing the lack of face-to-face communication is a real obstacle in trust and rapport building among team members, successful leaders of virtual teams focus on building team norms and enhancing communication among team members. To enhance communication, team leaders often setup communication protocols, set explicit expectations and objectives, and establish clearly defined roles of team members. These leaders also try to balance the "suffering" of team members through rotation of timing of virtual meeting in different time zones. These leaders also offer frequent feedback, mentoring, and coaching.

It is highly likely for virtual team members to lose track of project deadlines. Successful virtual team leaders make use of technology to perform close tracking of progress and productivity of team members. These leaders ensure virtual team meetings provide ample time for social relationship building, ensure equal participation from team members, and provide conflict management mechanism. One common method of reporting project progress is use of balanced scorecard. Team leader post these balanced scorecard measurement in team's virtual workspace (Malhotra et al., 2005).

To enhance team visibility and let team members appreciate the value of their work, effective leaders of virtual teams regularly report virtual team activities and progress other managers and stakeholders. It helps foster team mentality and ensure recognition of team members for their participation. Some other ways that can be used to recognize team members' participation include virtual rewards, recognition of member participation at the start of virtual meetings, and informing the local supervisors of the team member of their achievements (Malhotra et al., 2007). HR needs to ensure that the selected leaders of virtual teams either possess the skills and competencies needed to become effective leaders or could develop them with additional training.

Virtual Team Technologies

Virtual team members should be able to assess facial expressions and nonverbal cues of each other. This is fundamental requirement to establish trust among team members and virtual team success. HR should pay special attention, during formation of virtual team, on technologies that would fulfill this fundamental requirement of virtual team success. Some technologies to be assessed by HR and IT include instant messaging and chat facility (such as different messenger services and Skype), collaborative technologies (such as IBM Lotus Notes), remote access technologies, video conferencing on the web (e.g. WebEx), file transfer software, email software, phone (either conventional or Voice over IP) (Malhotra, & Majchrzak, 2005). HR should ensure that all team members have access to the technologies provided to virtual teams. HR should also ensure that training should be provided to virtual team members on how and when to use these technologies. While implementing these technologies, HR should consider providing some sort of social networking platform within the organizational IT infrastructure. This social networking platform should be reserved for virtual teams. Provision of this platform would allow virtual teams to share their experiences, build trust among team members, and develop a sense of community. Some studies argue that employers should refrain from policing these social networking platforms because that may inhibit interaction among team members (Leonard, 2011, Malhotra, 2010).

Training

Members of successful virtual teams possess skills that are high level and complex. Developing team members with such skills is a long-term task. Organizations may find themselves in situations where a team member possess strong knowledge that is key to project's success. However, the same member lacks communication skills critical to perform in virtual team (Dorr, 2011). Or there could be a great communicator who lacks knowledge of technology used in virtual teams. It is imperative for HR to identify all such skill gaps and ensure virtual team members are provided training to close those gaps. According to Rosen et al. (2006), the key areas where virtual team members need training include virtual team leadership, coaching and mentoring team members, team progress monitoring, management of external relationships, and evaluating/rewarding team member contributions. There exist many good examples of best practices companies use to train their virtual team members. Sabre, Inc., a travel technology company based in Texas, held team-building sessions for their virtual team members. In these sessions, virtual team members were trained to develop mission statement, objectives, and individual roles. Dow Chemical, an American multinational chemical corporation headquartered in Midland, Michigan, arranged regular training sessions on etiquette and meeting management for their virtual teams. Rocketdyne, an American rocket engine design/production company headquartered in California, used information-sharing technologies to train their virtual teams. GlaxoSmithKline plc (GSK), a British multinational pharmaceutical company headquartered in Brentford, London, arranged training sessions to improve the communication skills of their virtual teams (Rosen et al., 2006). Rosen et al. (2006) provided a model virtual team training program. This model offered a comprehensive prototype for virtual team training that reflects the best practices of successful virtual teams. This model can be used as a starting point for training in any organization seeking to implement or improve virtual teams.. This program provides various training modules. The modules include fitting the technology to the task, setting expectations, measuring and rewarding team contributions, coaching and mentoring virtual team members, modeling desired virtual team behaviors, and managing external relations. Training modules for team leaders include mastering virtual team technology, communication skills, and team management.

CONCLUSION

Virtual teams are increasingly prevalent in today's world. Abundant high quality information is available on how to work effectively as a virtual team and the economies that can be achieved from virtual teams. Use of virtual teams offers great potential for harnessing talent from many locations. Managing a virtual team can be highly rewarding and many competencies required to manage a virtual team are the ones required to manage conventional teams. However, virtual teams are highly sensitive to communication styles. For their success, appropriate communication styles should be used depending on the occasion.

A trend towards global collaboration is evident and organizations have many fundamental areas where virtual teams can help organizations increase organizational and individual productivity. Challenges such as conflict management, decision making, time zone differences, and cultural and language barriers have such a strong impact on the success of virtual teams. Virtual team leaders need to structure and manage team members' relationships for optimal results. Specific and explicit work rules needs to be developed for virtual teams and there must be a mechanism in place to assure their observance and document their successful implementation. In order leverage the diverse talent available in their employee base distributed across the globe, organizations need to understand the cultural influences in their workplaces and develop appropriate procedures for sustainable intercultural effectiveness in their workplaces. For organization looking to leverage the strengths of their diverse workforces, virtual teams is an excellent option with a promising future. Successful virtual teams can increase productivity, lower operating costs and speed the time to market. Compared with traditional teams, virtual teams face some unique challenges that can be converted into opportunities with extensive support and encouragement from organizational leaders and HR. HR can foster success of virtual teams by participating in the selection process of virtual team members and leaders and making sure the team members and leader possess the right combination of communication skills and business acumen and offer training programs to bridge any identified skill gaps of team members and leaders. This chapter contributes to our understanding of the virtual teams. More specifically, this chapter provides a deep understanding of the issues of skill development of virtual teams.

REFERENCES

Ale Ebrahim, N., Ahmed, S., & Taha, Z. (2009). Virtual teams: A literature review. *Australian Journal of Basic and Applied Sciences*, *3*(3), 2653–2669.

Beth Watson-Manheim, M., Chudoba, K. M., & Crowston, K. (2002). Discontinuities and continuities: A new way to understand virtual work. *Information Technology & People*, *15*(3), 191–209. doi:10.1108/09593840210444746

Boell, S. K., Campbell, J., Cecez-Kecmanovic, D., & Cheng, J. E. (2013). *The Transformative Nature of Telework: A Review of the Literature*. Academic Press.

Boudreau, M.-C., Loch, K. D., Robey, D., & Straud, D. (1998). Going global: Using information technology to advance the competitiveness of the virtual transnational organization. *The Academy of Management Executive, 12*(4), 120–128.

Boutellier, R., Gassmann, O., Macho, H., & Roux, M. (1998). Management of dispersed product development teams: The role of information technologies. *R & D Management, 28*(1), 13–25. doi:10.1111/1467-9310.00077

Chudoba, K. M., Wynn, E., Lu, M., & Watson-Manheim, M. B. (2005). How virtual are we? Measuring virtuality and understanding its impact in a global organization. *Information Systems Journal, 15*(4), 279–306. doi:10.1111/j.1365-2575.2005.00200.x

Cisco. (2009). *Creating a Collaborative Enterprise.* Retrieved from http://www.cisco.com/c/dam/en/us/solutions/collateral/enterprise/collaboration-strategies/C11-533734-00_collab_exec_guide.pdf

Coronas, T. T., Oliva, M. A., Luna, J. C. Y., & Palma, A. M. L. (2015). Virtual Teams in Higher Education: A Review of Factors Affecting Creative Performance. In *International Joint Conference* (pp. 629–637). Springer. doi:10.1007/978-3-319-19713-5_55

Crisp, C. B., & Jarvenpaa, S. L. (2015). Swift trust in global virtual teams. *Journal of Personnel Psychology.*

Dorr, M. (2011). *Developing Real Skills for Virtual Teams.* Retrieved from http://onlinemba.unc.edu/wp-content/uploads/developing-real-skills.pdf

Glikson, E., & Erez, M. (2015). Emotion display norms in virtual teams. *Journal of Personnel Psychology.*

Gratton, L., & Erickson, T. J. (2007). Eight ways to build collaborative teams. *Harvard Business Review, 85*(11), 100. PMID:18159790

Greer, T. W., & Payne, S. C. (2014). Overcoming telework challenges: Outcomes of successful telework strategies. *The Psychologist Manager Journal, 17*(2), 87–111. doi:10.1037/mgr0000014

Hackman, J. R. (1987). The design of work teams. In J. Lorsch (Ed.), *Handbook of Organizational Behavior.* Englewood Cliffs, NJ: Prentice-Hall.

Hastings, R. (2008). *Set Ground Rules for Virtual Team Communications.* SHRM Online. Retrieved from http://www.shrm.org

Hastings, R. (2010). *Fostering Virtual Working Relationships Isn't Easy.* SHRM Online. Retrieved from http://www.shrm.org

Hinds, P. J., & Bailey, D. E. (2003). Out of sight, out of sync: Understanding conflict in distributed teams. *Organization Science, 14*(6), 615–632. doi:10.1287/orsc.14.6.615.24872

Jarvenpaa, S. L., & Leidner, D. E. (1998). Communication and trust in global virtual teams. *Journal of Computer-Mediated Communication, 3*(4).

Kam, H.-J., & Katerattanakul, P. (2014). Structural model of team-based learning using Web 2.0 collaborative software. *Computers & Education, 76,* 1–12. doi:10.1016/j.compedu.2014.03.003

Keller, M. R. (2014). *Effective global virtual teams: The impact of culture, communication, and trust.* University of Maryland University College.

Kim, D. H. (1998). The link between individual and organizational learning. *The Strategic Management of Intellectual Capital*, 41–62.

Leonard, B. (2011). Managing virtual teams. *HRMagazine, 56*(6), 38.

Leonard, E., & Trusty, K. (2015). *Supervision: Concepts and practices of management.* Cengage Learning.

Lipnack, J., & Stamps, J. (1997). *Virtual teams: Reaching across space, time, and organizations with technology.* Jeffrey Stamps.

Lister, K., & Harnish, T. (2011). *The State of Telework in the US.* Telework Research Network.

Lockwood, N. (2010). *Successfully Transitioning to a Virtual Organization: Challenges, Impact and Technology.* Alexandria, VA: SHRM Research Quarterly.

Malhotra, A. (2010). *Managing an A-Team of Far-flung Experts Requires Special Leadership Tactics.* Retrieved from http://www.kenan-flagler.unc.edu/~/media/files/documents/Malhotra-leadership-tactics

Malhotra, A. & Majchrzak, A. (2005). Virtual Workplace Technologies. *MIT Sloan Management Review, 46*(2), 11-16.

Malhotra, A., Majchrzak, A., & Rosen, B. (2007). Leading virtual teams. *The Academy of Management Perspectives, 21*(1), 60–70. doi:10.5465/AMP.2007.24286164

Markus, M. L., Manville, B., & Agres, C. E. (2014). What makes a virtual organization work: Lessons from the open-source world. *Image.*

Maznevski, M. L., & Chudoba, K. M. (2000). Bridging space over time: Global virtual team dynamics and effectiveness. *Organization Science, 11*(5), 473–492. doi:10.1287/orsc.11.5.473.15200

McDonough, E. F., Kahnb, K. B., & Barczaka, G. (2001). An investigation of the use of global, virtual, and colocated new product development teams. *Journal of Product Innovation Management, 18*(2), 110–120. doi:10.1016/S0737-6782(00)00073-4

Mocanu, M. D. (2014). Virtual Teams–An Opportunity in the Context of Globalization. *Business Excellence and Management, 4*(1), 47–53.

O'Leary, M., & Mortensen, M. (2005). *Subgroups with attitude: Imbalance and isolation in geographically dispersed teams.* Presented at the Academy of Management Conference, Honolulu, HI.

O'Leary, M. B., & Cummings, J. N. (2007). The spatial, temporal, and configurational characteristics of geographic dispersion in teams. *Management Information Systems Quarterly, 31*(3), 433–452.

Peñarroja, V., Orengo, V., Zornoza, A., Sánchez, J., & Ripoll, P. (2015). How team feedback and team trust influence information processing and learning in virtual teams: A moderated mediation model. *Computers in Human Behavior, 48*, 9–16. doi:10.1016/j.chb.2015.01.034

Purdue University. (2008). *Virtual teams are here to stay: to succeed, we must learn how to work with them.* Retrieved June 26, 2015, from http://www.purdue.edu/uns/x/2008b/080730O-BeyerleinTeams.html

Purvanova, R. K. (2014). Face-to-face versus virtual teams: What have we really learned? *The Psychologist Manager Journal, 17*(1), 2–29. doi:10.1037/mgr0000009

Rosen, B., Furst, S., & Blackburn, R. (2006). Training for virtual teams: An investigation of current practices and future needs. *Human Resource Management, 45*(2), 229–247. doi:10.1002/hrm.20106

Taha, Z., Ahmed, S., & Ale Ebrahim, N. (2009). Virtual teams: A literature review. *Australian Journal of Basic and Applied Sciences, 3*, 3.

Thompson, J. (2011, September 30). *Is Nonverbal Communication a Numbers Game?* Retrieved June 26, 2015, from http://www.psychologytoday.com/blog/beyond-words/201109/is-nonverbal-communication-numbers-game

Townsend, A. M., DeMarie, S. M., & Hendrickson, A. R. (1998). Virtual teams: Technology and the workplace of the future. *The Academy of Management Executive, 12*(3), 17–29.

Wadsworth, M. B., & Blanchard, A. L. (2015). Influence tactics in virtual teams. *Computers in Human Behavior, 44*, 386–393. doi:10.1016/j.chb.2014.11.026

Chapter 6
E–Tools for E–Team:
The Importance of Social Ties and Knowledge Sharing

Cathrine Linnes
Hawaii Pacific University, USA

ABSTRACT

Organizations are heavily investing in virtual teams to enhance their performance and competitiveness. These types of teams are made possible by advances in computer-mediated communication and software that allows people to work collaboratively on projects without being co-located or even working at the same time. Managing teams and collaborating online presents unique challenges. Maintaining a productive virtual team requires more than just the willingness of global participants, but even more so the tools to conduct and manage virtual projects. It is therefore important to incorporate online collaboration skills into the IT curriculum at the university level. This chapter provides a general overview of virtual teams; today's collaborative tools, and discuss expertise necessary for virtual teams to be successful.

INTRODUCTION

The globalization has changed the traditional workplace to a more modern workplace where the technology has created new opportunities that did not exist before (Cascio, 2015). The world has become more competitive because of the endless opportunities of information and resources due to the globalization. According to Townsend et al. (1998), an organization has no choice but to operate in a world shaped by globalization and the information revolution. There are only two options: adapt or die (Townsend et al. 1998). A solution to how an organization can embrace the technological revolution is through virtual teams. Virtual team members primary interact through electronic communication systems, and are often consisting of members that are dispersed both organizationally and geographically (Cascio, 2000). Teams used to be located in one place with face-to-face meetings and collaboration as the most efficient or sometimes only option. Now teams have become virtual, consisting of telecommuters and co-workers located around the globe collaborating on various jobs. In academia educators are struggling to get todays students feel comfortable collaborating on IT project solely online, it can even be hard

DOI: 10.4018/978-1-4666-9688-4.ch006

in the classroom sometimes. However, if combined with face-to-face meetings then the teams become more effective. There are many e-tools, which are helping to close this gap such as Bit bucket, Bitrix, Google Docs, Instant Messaging, Drop Box etc. Google Docs allows for multiple people modifying a document at the same time whereas Drop Box does not. Jing is an interesting program that allows you to take a screen capture or make a video and add a voice over. This is most beneficial when creating instructional videos or tutorials. Skype has been around for years and offers the most popular form of video conference calls. The biggest perk is that the service is free for mobile devices. On the other hand basecamp is an online collaboration tool that is used to manage projects to include task lists and team communication (Rawson, 2010).

The current work environment is quite complex with many changes and uncertainties. This has led to changes to the ways in which teamwork occurs. Virtual teams rely on information being communicated to all its members swiftly and team members have to feel comfortable and knowledgeable using various tools to increase accuracy and efficiency with the product being developed.

Factors that improve the quality of teamwork include the effectiveness of each individual, the information sharing process, project commitment, and joint responsibility for the milestones (Senge, 2006). Whether face-to-face or virtual, teams require parallel and distributive leadership (Andrews et al., 2004). Virtual teams are based on a leadership culture of learning, trust, and openness, which is communicated to all members that there are choices and that any individual can demonstrate leadership and thus influence direction and development (Walker & Shuangye, 2007). Virtual teams must have members who are capable of reflection, active and critical thinking, and who are capable of moving in unison even if individuals are separated geographically. Lister and Harnish (2011) reported there are about 2.9 million U.S. virtual workers since 2005. Between 2005 and 2009, it grew by 61%. Based on the current trends, it will grow to 4.9 million by 2016.

Remote teams in the workplace are becoming more commonplace. According to estimates from the American Community Survey, "telecommuting has risen 79 percent between 2005 and 2012, and now makes up 2.6 percent of the American workforce, or 3.2 million workers" (Tugend, 2014). With current advancing trends in mobile technologies, "managers must adjust their approach to achieving their business goals" and align mission accomplishment to support remote teams with the fundamentals of technology and consistent communication (McGannon, 2014). According to Reynolds (2014) forty-four percent of the respondents wanted to telecommute at least part of the time according to a survey of 1,500 people conducted by the website flexjobs. This drastic increase in virtual team members does not account for the part-time remote participants which is actually an even larger number than the full time employee. A strong indicator as to the upward spike in virtual/remote employees is the availability for utilization across a grand scope of business objectives. Supply, engineering, product development, information technologies, finance and customer service functions and initiatives only scratch the surface of the overall applicability of virtual environments (Kock, 2011), and in a global economy and internet based market this trend does not seem to be slowing down. In addition, more free-wifi is being provided in more business locations and by transportation agencies (airlines, bus, train, boat, etc.). According to a Boston Business Journal article dated February 3, 2015 a spokesman for Microsoft encouraged employees to "...[To] take advantage of the numerous productivity tools at the disposal including Microsoft Lync and Skype for online meetings, and other cloud and mobile work solutions that allow our employees to be productive in the office, at home, or in the local coffee shop" (Boston Business Journal, 2015). Businesses are making it very convenient for the user to stay connected while being away from the office.

Since technology is taking todays and tomorrow's works in this direction, educators must incorporate e-tools and project based learning into the curriculum this is especially crucial for IT students to ensure they will function effectively from day one when starting to work. We might have a situation where, Jim, John, Joe, Jane, Juliet, Eloise and Neil have been assigned a software engineering project together. Jim tends to take charge and makes sure everything gets done on time and delivered according to specifications. John is happy because he was on a team with Jim prior and received a high grade without putting in too much effort. Joe has been working on his degree one course at a time due to his hectic work schedule, and because of Eloise family obligations, she is only able to work on school projects in the evenings. Jane, Juliet and Neil is new to the field, and have limited programming skills and therefore not able to assist much. Social loafing can be a problem, therefore introducing team structure, establish proper criteria, prepare learners for team roles, and guide learners through team communication is a must.

BACKGROUND

Electronic collaboration (or e-collaboration) is defined as the process of collaborating in a project or program by using electronic technologies (BusinessDictionary.com, 2014). It is not long ago that collaboration was only possible face-to-face or through the old fashion telephone. Since then, technology has rapidly advanced and the workforce has changed accordingly. According to Kock (2005), e-collaboration has a promising future in academic research and software development. Regardless of these promising developments, e-collaboration has always been somewhat problematic.

Industries are more global than ever these days, which forces corporations to rethink how to conduct their business and operate smarter. Organizations can essentially either adapt to new technologies or opt to fall further behind their competitors. Global distributed teams (also referred to as virtual teams) are most common in IT (Carmel & Agarwal, 2002; Herbsleb & Mockus, 2003; Sarker & Sahay, 2004), but are also seen frequently in new product development and other areas of business (Malhotra, Majchrzak, Carman, & Lott, 2001).

It is difficult to imagine the world without the Internet. Taking a look at social network users around the world we can see that the increase has been very dramatic. In 2010 there were 970 million worldwide users of social networks. In 2015 the same amount was 1960 million and the statistics show that in 2018 the number is going to be 2044 million (Statista, 2010). Social media like Facebook has for example contributed to the rapid growth of virtual teams by allowing people to stay connected and seek out new relationships. Right now there are over 1.23 billion monthly active users, 945 million mobile users, and 757 million daily users (Protalinski, 2014). There are of course many discussion forums pertaining to software development where users can go to look for collaboration opportunities. Electronic tools have made it much easier to stay connected. However, electronic-collaboration is not a perfect solution to team work but if used in conjunction with traditional physical meetings or discussions it can further enhance the productivity of a group overcoming factors of time and place.

Furthermore, global distributed software teams when teams and members are located in more than one physical location. These teams work across boundaries of space using modern technologies to conduct their task. They can be faced with obstacles of time zones, language barriers, cultural differences, and perhaps regulations and legal differences (Carmel, 1999). "Near shoring" can therefore be a smarter choice to "offshoring" because near shoring is the transfer of business or IT processes to companies that share one or more of these characteristics. To collaborate successfully requires team members to be able to communicate effectively and efficiently.

Exploring various e-tools and looking at the various generations is important because the office of the upcoming generations may not look like our offices nowadays. Virtual team development is going to become a prevalent concept where work means logging into the companies' system from your home and collaborating with your colleagues who work for different teams. The physical location like buildings and offices become less important. Today, we find virtual teams in business, in between governments, education, but also socially.

The objective of this chapter is to provide a) general overview of virtual teams; b) look at today's collaborative tools, c) discuss expertise necessary for virtual teams to be successful, and d) discuss e-collaboration in higher education as it is important to prepare the next generation for the future. The solution and recommendation, future research directions and conclusion follow this section.

VIRTUAL TEAMS

Technology has continued to transform the way people interact with one another. As more and more businesses move toward globalization, there is a requirement to have virtual teams. People are no longer required to physically meet face-to-face to collaborate. Instead, they can now work with each other even if they are scattered in different countries around the world. It is also more common and expected that an employee is available online 24/7. A virtual team will have various benefits and challenges that determine its performance and outcome (Gera, 2013). Virtual teams allow for flexibility, but can also lead to miscommunication. There is no guarantee the virtual teams will be successful, but it provides an opportunity for innovative ways to complete projects. Today's workplace is continually evolving to keep up with rapidly changing and improving technology. For businesses to thrive in ever more competitive environments, they must seek out new ways of doing business. Virtual teams are becoming more and more prevalent as organizations recognize the numerous advantages of e-collaboration. This section is divided into five areas:

1.   Today's Collaborative Tools,
2.   Virtual Team Structure and Management,
3.   E-collaboration in Higher Education,
4.   Advantages of Virtual Teams,
5.   Disadvantages of Virtual Teams.

Today's Collaborative Tools

Teamwork plays an important part in the workforce. The use of virtual teams has increased dramatically in software development. New collaborative technologies are dramatically impacting the way we work and will work in the future. These technologies are now giving us the freedom and flexibility to work from anywhere, anytime, and on any device. Being able to find subject matter experts and connect with colleagues (known or unknown) across the globe is now a possibility that didn't exist in earlier years. With the spread of technology permitting the increase of virtual teams and global distributed workgroups, traditional workplace hierarchies are being flattened as virtually any employee can connect or communicate with anyone else at the company regardless of seniority. Information is being opened up and shared instead of being locked down or seen or shared only in person. Systems are being connected and opportunities for improvements to business processes, customer experiences, and overall efficiency abound. These new technologies are also allowing simple actions to have big impacts. For example a manager or the CEO of a company "liking" or commenting on an idea that an employee might post publicly inside of the company provides immediate opportunities for feedback or praise that used to have to wait for far less frequent in-person interactions. Research has clearly shown that reinforcement (or punishment) administered close in time to the behavior warranting it is optimal, so these technologies can help managers excel at this part of their work. Additionally, more employees now have a voice within their organizations and have the ability to become leaders without having to be managers. Thomas, Bostrom and Gouge (2007) found that virtual teams in software development on average used thirteen different technologies in their projects.

Several popular technologies are being used in today's software development projects such as cloud-based solutions, social media tools, version control software, document sharing video chat tools and instant messaging. These tools can help the users feel like they are working in a traditional team setting.

Cloud

Cloud collaboration is also gaining momentum as new revolution to enhance access service. A study reported by Forbes (2013) found that 64% of executives say that cloud-based collaboration tools accelerate business results. Also, about 93% of executive believe cloud based collaboration stimulate innovation. An example of cloud collaboration is Microsoft SharePoint, where teams can move unfinished documents into SharePoint document library to aid team members to have instant access to policies and procedures (Diffin, Chirombo, & Nangle, 2010). According to Chu (2011), working collaboratively is an essential skill necessary to stay relevant in today's global society.

Furthermore, mobility is not just about being able to work and get access to people and information from a mobile device. It's also about being a mobile worker which means you can work from anywhere, anytime, and on multiple devices. The idea of "connecting to work" has become more prevalent within organizations as they are starting to allow for more flexible work environments. With an Internet connection you can now access everything you need to get your job done. The advent of "cloud" storage and computing from anywhere frees files and data from the confines of file cabinets, hard drives, thumb drives, and "sneaker nets." The notion of having to work 9-5 and commuting to an office is no longer true for many. The smart and progressive organizations around the world have already been making the necessary investments to adapt to these changes. Some have been doing so for several years already whereas many organizations are still trying to educate themselves about what these changes mean. Regardless of where you are in this spectrum, the future of work is something that you must plan for and adapt to if you want your organization to continue to exist.

Virtually every collaboration platform today has a cloud-based deployment option. This means that the barrier to entry is virtually zero. Business units no longer need to wait for corporate approval or the blessing of IT to make investments in these areas. Anyone with a credit card and access to the Internet now has the ability to deploy whatever technology best fits their needs. This is a huge shift inside of an organization, which traditionally had to rely on IT to deploy any type of new technology, the cost used to be high and the deployments complicated. Individuals as well as teams opt to use cloud storage, which allow users access to all their files instantly as well as the ability to selectively or widely share their documents. The top ten mobile cloud storage are currently reported to be Justcloud.com, Zip Cloud, myPCBackup.com, SOS Online Backup, SugarSync, Mozy, BackupGenie, Dropbox, Box and CRASHPLAN. Just cloud.com and myPCBackup.com are increasing in popularity, SOS Online Backup and Dropbox are seeing a setback where the other six are remaining constant in popularity (TheTop-10BestOnlineBackup.com, 2014). Microsoft has also joined the crowd with a variety of free and paid individual and business versions of its OneDrive (formerly called SkyDrive).

Social Media

Lack of the right technology can also be an obstacle to the virtual team. There are a number of collaborative tools that can help virtual teams succeed and work more efficiently and effectively, such as web-based chat tools, web-based asynchronous conferencing tools, e-mail, list serves, collaborative writing tools, group decision support systems, teleconferencing suites, workflow automation systems, and document management technologies, etc. What we see inside an organization today stems from social media tools such as Facebook, LinkedIn, Foursquare, and Twitter.

The benefits of these social media tools are that they are free to use, students and workers are familiar with the tools. It makes it extremely easy to inform the pubic or members of the team regarding project status, system down time, upcoming training sessions as well as it forms a community among the members. The downside is that the information is stored on other companies' servers and can become their property. Twitter, Facebook, LinkedIn are great tools to use for academic projects as students are always online.

Version Control

Popular tools for software development teams to collaborate are tools such as Unfuddle, GitHub, JIRA, and Bitbucket among many others. These tools help track product releases and increase team productivity. In most cases, it is not optimal to operate without them. It can easily become quite chaotic where team members do not have the latest version of the application being developed, or team member deleting other member's work or making changes to the system causing other peoples code no longer to work. To better explain Unfuddle is a secure, hosted software development environment providing Git hosting, Subversion hosting, bug and issue tracking, milestones, time tracking, audit trails and more. It offers an all-in-one solution for ticket management and code repository. GitHub is a web-based hosting service for software projects that use the Git revision control system. GitHub offers both commercial plans and free accounts for open source projects. GitHub also operates a paste bin-style site at gist.github.com, wikis for the individual repositories, and web pages that can be edited through a Git repository. JIRA is currently the leading bug tracking, issue tracking, and project management tool. Trusted by more than 11,000 businesses, JIRA combines agile delivery of these features with a customizable workflow and a pluggable integration framework to increase the velocity of the software development team. Bitbucket is a code hosting site, for the Git and Mercurial version control systems. It provides a fully featured environment for managing development, including a wiki, a powerful issue tracker, and easy collaboration with others (alternativeTo, 2014). GitHub and Bitbucket are popular tools used among professionals and students. However, it takes some time to learn how to push and pull code efficiently. GitHub is an on-line workspace that works with Git. It's that simple. Nevertheless it can be used for so much more than simple version control. GitHub can be the place for you to share your resources with the world. Again, teaching students how to use these tools is a must. It is no way without it if working in a team as you will always be up-to-date. A pull request is truly the most useful aspect of GitHub. When a team member makes changes to a repo that you control, they can submit a pull request. You will then pull their changes into your repo and merge.

Track Document Edits

Technologies that support-sustaining productivity will ensure accountability are maintained where the lack of direct oversight is no longer valid. Dropbox, Evernote, Google Docs and Microsoft Office 365 are examples of the collaborative tools that track document edits and allow the team to understand the creation process quickly with minimal distractions (Lambotte, 2013). For example, a team could be working on a Microsoft Word document that is saved and accessible by the entire team on Dropbox. When the Dropbox application realizes a change to the file, a notification will be sent out to the whole group, and everyone will be able to view the changes. When work is completed in this manner, a manager can maintain oversight from afar by realizing file information such as last saved by, date and time. The tools mentioned here are tools used by all industries. Students are frequently using them on their class projects. Tools like this can really help to move a team forward more quickly.

Video Chat

Consistent and clear communication is critical when working with remote teams. Microsoft Skype, Google Hangout, Face Time and Wunderlist all are examples of currently supported technologies that support distant communications. These tools have been extremely popular video-chat products. When keeping in mind of time zone constraints, it is critical to use these services to ensure consistent communication is happening. For example, Wunderlist allows teams to "share specific lists with team members so everyone can see and update tasks, post files and add comments" (Olson, 2014). These tools make it possible to have a face-to-face conversation with a combination of high-speed Internet, webcams and smartphones. Despite the popularity, the literature indicates that face-to-face meetings are still favored and important for the performance of teams even though electronic tools are being used more and more frequently (Jarvenpaa *et al.*, 1998; Govindarajan & Gupta, 2001; Arvey, 2009). According to Elmore (2006) studies show that conference calls was the most frequently used tool among virtual team members, that worked for large American companies, followed by web collaboration tools such as e-mail, voice mail, video-conferencing, fax and telephone. The current video-chat products have come a long way to support the success of a team. However, there is still a valuable psychological effect of and desire for meeting in person for most people.

Being able to communicate real time can help to clear up any misunderstandings that you might not be able to catch via other means. Many organizations use messaging systems instead of telephones and e-mails to get ahold of other employees even when housed in the same building. One can set the status to available, busy or away. Anyone in the organization or team can then see if the person is there. This is very effective, but it also requires a person to respond quickly. The same goes in the academic setting, if you want to get a students attention send an instant message.

Virtual Team Structure and Management

Organizations are forced to develop a strategically flexible organization due to the global competition. Many organizations are asked to downsize and create leaner structures. These technologies allow organizations to be more flexible and put together global teams of the best employees possible. The team structure can be quite flat which is an advantage. Fewer layers of management mean fewer approvals, decisions can be made quicker, can provide better and more frequent communication among the team and management, and can give team members a better understanding of the goals. It is up to the virtual team leader to facilitate relationships normally created when a team communicates in person.

Therefore, it is important that the team leaders take time to set up the team effectively. A study conducted by O'Leary and Mortensen (2008) reported that when teams were unequal in size the smaller group ended up being disadvantaged as events unfolded. This can be detrimental to the success of a project. This is an easy fix. Additionally, in order to ensure successful collaboration team leaders and organizations must create an organizational mechanism to create social spaces among team members. For example, the team lead can create an "electronic water cooler", an environment where virtual team members can chat, e-mail, exchange information, or just keep in touch to see what team members are making progress on, perhaps on a popular social media platform such as Facebook or Instagram.

Furthermore, because virtual team members usually do not share a common physical workspace, the team leader should find a way to develop and maintain a sense of virtual team identity. Distance should not be a communication barrier; virtual team members should not feel a sense of "out of sight, out of mind" (Elvekrog, 2015; Gupta-Sunderji, 2013; Kostner, 1994). According to O'Hara- Devereaux and Johansen (1994) distance teams often fall into the trap of working in silos having relationships built only between the leader and individual members. Perhaps creating a website with the location, photograph, and personal information of each virtual team member can help make the work relationship more personable and shorten the distance. It is equally important for members try to establish good working relations with one another. Without these relationships, virtual teams may find they have no rapport without the team leader.

In addition, strong symbols are needed to unite people across time and space. Virtual team leaders must promise as much communication access for their virtual team members as they would offer to colleagues located in the same building. If possible, the best thing a virtual team leader can do is to bring all members of the virtual team together face-to-face at the very beginning to learn more about each other professionally and personally. A short, informal videoconference or virtual meeting would also allow the virtual team to build rapport. Perhaps a symbolic name for the team can be created at the start of the project along with a team t-shirt with the name and logo of the team. This can help to create some sort of belonging.

The opportunities for improved efficiency often drive many managers to consider and implement virtual teamwork. In fact, Boiney (2001) found that over two thirds of Fortune 1000 companies are using self-managed and virtual teams. Cascio (2000) states there are many good business reasons for using virtual teams including reduced real estate expense, increased productivity, higher profits, improved customer service, access to global markets, and environmental benefits. Team leaders needs to set the communication patterns from the start, be accessible for its team members, make the members feel they are part of the team, and focus on the end result. Effective leadership has always been about setting clear priorities, establishing strategic direction, and holding employees accountable (Gupta-Sunderji, 2013).

E-Collaboration in Higher Education

To better understand acceptance and use of technology among students, we need to look at some of the general characteristics of these different generations and how they may be changing the workforce. There are palatable differences between the generations showing up in university classes that must be understood to best meet their needs and preferences in order to help them succeed when entering the workforce.

Generation Z is the group of people born after the millennial generation. There are no precise dates but sources indicate 1995 to present. They are known as "Technoholics". They are entirely dependent on IT and have little grasp of alternatives. Their signature products may turn out to be Google Glass, Graphene, nano-computing, 3-D printing, and driverless cars. Their way of communicating is by hand-held communication devices, perhaps as fashion accessories or integrated into clothing. They prefer Face Time to face-to-face when communicating. When making financial and other decisions, their solutions will often be digitally crowd-sourced (Robertson Associates, 2013). Generation Y is the group of people following generation X. There are no precise dates but sources indicate 1981-1995. They are defined as digital natives. Their signature products are tablets and smart phones. Their way of communicating is by SMS and social media. They prefer online or mobile (text messaging) communications and face-to-face when making financial decisions. Generation Y has a clear passion for technology. They like using it and prefer to learn in team settings (Childs, Gingrich, & Piller, 2009; Coates, 2007; Dulin, 2005; Robertson Associates, 2013; Shih & Allen, 2007). Generation X is the group of people born after the generation referred to as Baby Boomers. There are no precise dates but sources indicate 1961-1980. They are defined as digital immigrants. Their signature product is personal computers. Their way of communicating is by e-mail or SMS. They prefer text messaging or e-mail over face-to-face meetings. When making financial decisions they prefer online meetings and face-to-face if time and circumstances permit (Robertson Associates, 2013). Baby Boomers are the group of people born after WWII until about the mid 1960s. They are early information technology adaptors. Their signature product is the television while their way of communicating is through the telephone. They prefer face-to-face meetings with e-mail or phone as a backup. When making financial decision they prefer face-to-face but increasingly will go online (Robertson Associates, 2013).

The generations listed above are quite different in the way they prefer working. Linnes (2014) reported that 55.9% of graduate students studying information systems preferred to read their class material online and 77% reported they felt comfortably reading online. All of the participants owned electronic devices, where 91% owned a smartphone and 95.8% owned a laptop. Majority of these students were generation Y. This means students should be open to try new technologies. A push for constant skill improvement should be part of an organizational culture. However, one cannot forget that organizational culture is based on experiences gained by its members even before they joined. The habit of improving ones skills should originate from at least the university level. Therefore, it is important for universities to inculcate their graduates not only with knowledge and actual experience in personal teamwork and collaboration, but e-collaboration as well. Just because they grew up with technology it does not mean they know how to use it effectively. It is important for a manager to understand the generation differences to make their team successful. Educators need to train their students to work with modern technologies, as well as to teach them what makes highly effective virtual teams.

Advantages of Virtual Teams

There are many benefits to working virtually. One of the benefit to virtual teams is that organizations can tap into resources rapidly to create a specialized work team that acts like a team, works like a team but doesn't look like a typical team because team members may not be co-located (Stough, Eom & Buckenmyer, 2000). Another significant advantage of virtual teams is that the business can hire an employee with the needed skill set, regardless of the employee's location. This drastically widens the pool of candidates for the company to choose from, as well as decreasing or eliminating relocation costs. The business is able to find employees whose skills match those needed, as opposed to being limited to potential employees in the geographic area of the business (Nydegger and Nydegger, 2010). It would not be possible to put together the same type of team in a traditional team structure. E-collaborative technologies such as computer-based conferencing systems are of critical importance to the success of a virtual team (Arnison & Miller, 2002).

In the absence of water-cooler socializing and philosophizing, virtual teams rely on technology to build trust between team members, resulting in greater synergy and ultimately team success in carrying out work tasks (Arnison & Miller, 2002; Stough et al., 2000). Virtual teams offer several benefits to the organizations and its employees. Virtual teams allows for employees to have greater control over their work schedules, which can lead to happier and more productive workers. Many employees seek to have a work-life balance. Virtual teams provide the opportunity to have a balance, as they may be able to work from home or are able to work flexible hours. Virtual teams have fewer office-related distractions, which can improve productivity (Lekushoff, 2012). Ciotti (2013) cited a 2009 study by Cisco in which they determined that about 69% of employees said their productivity was higher when teleworking versus when working in a traditional office. Employees can be more flexible, such as working with multiple virtual teams on various projects, or even for different companies simultaneously. In addition, to the intangible advantages of comfort and increased productivity, there are financial benefits for teleworkers. Brown (2013) estimates that teleworkers who work from home half the time can save between $2,000 and $7,000 a year, primarily in commuting costs. Additionally, teleworkers in the United States can claim tax deductions for using their home, personal computer, or other personally owned items for business purposes.

Virtual teams can help an organization cut costs by reducing travel or office costs. Since a for-profit business exists to make money, employing virtual teams must make financial sense. According to Brown (2013) a business can save as much as $11,000 dollars per year for each teleworking employee. These cost savings include not having to provide the square footage for the employees to work in, which saves on real estate expenses, property taxes, and utilities, such as electricity. Smaller buildings are cheaper to purchase and maintain, so employing virtual teams who do not require physical space at their place of employment can save the business money (Management Study Guide, n.d.). IBM for example reporting saving 40 to 60 percent per site annually by eliminating offices for all employees except those who truly needs them. Productivity increased with 15 to 40 percent by using virtual teams (Cascio, 2000).

Research has found that individuals interacting in virtual environments report feeling more comfortable expressing themselves (Kim, 2000). Similarly, a recent study showed that avatar realism increased individuals' willingness to share information with others (Bailenson, Yee, Merget, & Schroeder, 2006). Information sharing and disclosure are behaviors critical to both teamwork and organizational performance, especially in a virtual context that lacks sustained personal contact (Rafaeli & Ravid, 2003). At the team level, research has showed that VW use leads to higher group-oriented learning and process engagement (Nydegger & Nydegger, 2010; Jarmon, Traphagan, & Mayrath, 2009), in addition to spontaneous and opportunistic conversations that are integral to building more intimate social connections (Meyer & Swatman, 2009). Another significant advantage of virtual teams is that the business can hire an employee with the needed skill set, regardless of the employee's location. This drastically widens the pool of candidates for the company to choose from, as well as decreasing or eliminating relocation costs. The business is able to find employees whose skills match those needed, as opposed to being limited to potential employees in the geographic area of the business (Nydegger and Nydegger, 2010).

There are numerous advantages to employing virtual teams, which is why many organizations are relying on them for at least part of their workforce. The number of teleworkers, or employees who work from home, increased 73% from 2005 to 2011 (Brown, 2013). The decrease in the number of people commuting to work every day has one advantage that benefits everyone – decrease in pollution and consumption of non-renewable resources. This is better for the environment and society as a whole. In addition, there are benefits for both the businesses and the employees to embrace teleworking and being members of virtual teams.

Disadvantages of Virtual Teams

Despite having modern technology and tools, the individuals who have grown up with or into them still face communication and performance challenges. A reduction in or absence of face-to-face contact sometimes makes these challenges more difficult to handle especially where there is a lack of trust and/or when context or intention can be so easily misperceived. Managing remote teams is similar to managing regular teams, but will require greater emphasis on building trust, fostering communication, implementing team processes, and utilizing technology (Lemmex, 2005). There are several things to be aware off such as; trust, communication, knowledge sharing, social ties, and lack of engagement.

Communication, trust, and cohesion are essential in any team, but are more difficult to obtain in virtual teams compared to traditional teams (Nydegger and Nydegger, 2010). If team members have never met face-to-face, it can sometimes be hard to trust the person on the other side of a keyboard. Communication can be more difficult, but with today's technology, it can also be easier. According to one study, 83% of more than 2,000 employees responded, "their ability to communicate and collaborate with co-workers was the same as, if not better than, it was when working on-site" (Ciotti, 2013). Cohesion can be particularly difficult, especially if the teleworkers are geographically or culturally separated. Organizational culture can be a significant factor in determining how employees and teams are supposed to behave and interact; employees who have only e-collaborated for a company will likely not have the same level of indoctrination in the organizational culture as employees who have worked on-site (Nydegger & Nydegger, 2010). Thus, it is important for managers to focus on continually improving a virtual team's communication, trust, and cohesion in order to ensure success in their assigned task.

In addition, to a potential lack of organizational culture, virtual team members must also learn to cooperate with coworkers from different cultures (Ebrahim, Ahmed, and Taha, 2011). Employees who are not cognizant of coworkers' differing cultures and expectations can unintentionally offend one another or be less effective than if they understood each other completely. Supervisors or team leads should ensure that virtual team members are aware of cultural differences in order to maximize the team's communication and effectiveness. A major challenge is of course, the inability for teams to read nonverbal cues. It can also be hard to build relationships during the few hours when meetings are held.

According to de Pillis and Furumo (2007) lack of trust is fatal to the effectiveness of virtual teams. Human beings gain trust through body language, which is harder to maintain virtually, especially when it cannot be seen (Watson-Manheim & Belanger, 2002). Body language can be much more powerful than verbal communication. In addition, it is usually easier to negotiate or persuade an outcome face-to-face than through distance. Take for example a situation where you send a letter or e-mail. It is much easier for a person to turn down your request than if you show up in person. Trust takes time to build up in any situation, but in virtual situations, it can take even longer.

Knowledge sharing, brainstorming, persuasion, negotiating alternatives, or decision-making take on a complete new form that excludes the use of body language when done virtually. Face-to-face meetings are strong vehicles for participants to learn the relative norms of the organization as well as its culture. Individuals learn about the various ways things "operate" in organizations by observing how others behave and display emotions. Information such as the value and meaning of time (i.e. showing up on time), who has power in the organization, what is reinforced and punished, etc. are all things people learn in face-to-face meeting which otherwise might not be observable in electronically based communication devices (Arvey, 2009). Despite this, there appears to be no going back. Virtual communication is increasing and slowly replacing face-to-face collaboration and meetings.

There are several key areas that are deemed challenging while working in a virtual team environment, such as creating a sense of community and differences in time zone, language, and cultural (Jones, Oyung, & Pace, 2005). The first of these challenges, building a sense of community and shared purpose can be achieved through use of instant messaging, video interaction such as with Skype or Face Time, and an upfront information session meeting where team members get to meet each other in person to reduce the barriers. The second challenge is differences in time zone. However, this is an issue that can be overcome by recording meetings to allow for viewing or listening at member's own convenience. In addition, multiple training sessions to accommodate different time zones can be held. The third challenge relates to language and cultural differences. If a company choses to outsource, which is frequently done in IT, it may possible and preferable to select vendors or partners from areas that have the same social ties or speak the same language. For example, in parts of the U.S. where the population speaks Spanish, it may make sense to collaborate with companies in South/Central America. In Europe, English-speaking Ireland has served as a hub for many years to companies in the United States. Teams also work best with other teams who have the same understanding and background. Where this is not an option, it is crucial that virtual team members who collaborate and communicate with others of different cultures be as educated as possible to these possible or actual differences. It is important that these issues are resolved early on to avoid a devastating project. A final concern may be the individual differences in things such as attention span and stamina team members may have. Reducing the length of virtual meetings and using video and images to maintain the attention of the audience can help.

Working exclusively in a virtual environment has its disadvantages and challenges. One important aspect of virtual collaborative enviroments is the lack of physical, social interactions where behavioral cues are important in gauging feedback from other members. In a study conducted by Ku, Tseng, and Akarasriworn (2011); some comments from their subjects included:

- I prefer to meet my peers and instructors face to face.
- I feel I learn better when in a traditional classroom setting. Sometimes the information can become confusing and if that happens we can stop the instructor right then and there to clear up the situation.

- If we are working on group projects I believe it helps if we can meet face to face to work on our project.
- By just relying on email or blackboard to communicate can sometimes slow the feedback time down.
- If we could meet, we would set a time to meet, be there and finish the task.

Another interesting, but related finding in another study showed that subjects preferred time slots that accommodated real-time conversations over leaving a dialogue open for another 24 hours (Zaugg, 2011). While online, virtual workspaces are convenient, and in some cases cost-efficient; team members require a certain amount of human interaction to maintain a healthy synergistic work environment. The lack of social cues inhibited collaboration (Ann, 2010). Technology such as video teleconferencing offers a means of member interaction that is comparable to physical presence. While video teleconferencing does have some technological challenges such as latency and high bandwidth considerations, it can mitigate the lack of social cues that are missing from other tools.

While there are several significant disadvantages to virtual teams, the advantages often outweigh them. More and more companies and employees are moving to teleworking and virtual team structures and this trend will likely continue for the foreseeable future. As virtual teams become more numerous, it is important for employees to adapt to this new working environment. The freedom and flexibility working at a distance brings is a huge plus for any IT worker.

There are challenges to any team structure. Free riding, where individuals are perceived not to complete their fair share creates challenges and slows down the team. It is also likely far easier to hide in a virtual team than in a face-to-face team structure. Free riding is less visible online which can lead to team members contributing even less (Chidambaram & Tung, 2005). Classic psychological research shows that the mere presence of others (perceived or real) facilitates performance on a wide variety of tasks (Bond & Titus, 1983). In line with this, a study conducted by de Pillis and Furumo (2007) indicated that virtual teams yield significantly lower performance, lower satisfaction and a lower result-to-effort ratio than face-to-face teams.

Other issues workers face, are for example that Face Time happens not to be available in all countries. The United Arab Emirates and Saudi Arabia are one of the countries that do not allow Face Time due to regulations on IP-based communications (Ray, 2010). Internet connections can be slow or weak in certain parts of the world making it hard to collaborate online. Internet services are being built out every year to handle the heavy usage by individuals and organizations. Most video tools do not allow unlimited number of concurrent users, except for Skype offers a beta release version. The technology can of course get in the way, I can see you but I cannot hear you type of a deal. Of course, it is not just to hold up a document during a meeting and say take a look at this when online. Underdeveloped countries can also be a problem, as they will fall further and further behind in the technical environment. When jobs are being outsourced to other parts of the world, it can drive wages down. Security and privacy issues can be a dilemma for organizations. Information transferred via the Internet can of course get in the wrong hands but so can information shared via face-to-face meetings. Lastly, generation Z has entered college and some started to enter the workforce therefore many of today's issues will eventually dissolve by itself. They have grown up with technology; they feel much more comfortable using it, but the school system still plays a major part in preparing the students. By partnering up with corporations educators can learn first hand what the current problem is. Also the industry might be using certain technology that is currently not used in the classroom. Technology will only get better making it more convenient and efficient to work from a distance. This is only the beginning for a great technology future.

SOLUTIONS AND RECOMMENDATIONS

Technology is only one part of the equation. It is important to remember that a virtual team is a social system. The team's success will depend on the quality information received, how they apply, and how they communicate this information. It can be difficult enough to work on a project with a person sitting right next to you who is on all the same wavelengths as you. For a virtual team to be successful, there need to be executive support, effective HR practices in place, team needs to be structured well and there needs to be a strong team leader.

Starke-Meyerring and Andrews (2006) lay out the following recommendations and learning environment needed for virtual teams to ensure success. Ensure that communication tools such as calendars, blogs and other tools are incorporated through the entire organization (and in the case of educational settings, within the entire curriculum). Arrange for various communication channels to be available such as voice, video, and/or chat, synchronous or otherwise. Ensure team members have equal access to information and resources and that the team can control their own online workspace. Use technologies to track changes to documentation or code as well as technologies that alert users of changes made. There should be ample opportunity to determine the impact of various technologies used. If teams from different organizations are working together, ensure that all teams have and are using the same tools. Ensure that privacy is protected.

It is important that communication guidelines are created to create a solid learning culture to increase performance of virtual teams. Other researchers have indicated the same importance. For example Watkins and Marsick (1993) described six requirements as the framework for creating and supporting a learning environment: create continuous learning opportunities; promote dialogue and inquiry; encourage collaboration and team learning; establishing systems to capture and share learning; inspiring people toward a collective vision, and connect the organization to its environment. Herrington (2004) repeated these requirements and added additional conditions such as equality in power and access, both linguistically and technologically; synchronous communication; reliable, fast connections; robust visual and audible information for interaction; equal access to information; and additional actual international exchange. This is vital for maintaining a sound technological environment for learning in virtual teams. Educators must reach out to local businesses and government agencies frequently to find out what the current status is so the instructions can be altered to better fit the need of the community. The generation shift will also change the landscape moving us toward a more digital work force.

FUTURE RESEARCH DIRECTIONS

There seems to be a need for more studies on e-collaboration in higher education. We have a new generation of students (generation Z) just entering college and some entering the workforce for the first time. But, our institutions are still filled with generations X and Y who are still finding their way through technologies new to them. Higher education has a responsibility to prepare students for successful professional careers with all the tools necessary to perform the job. This can be done by introducing the students to more modern ways of working and making them highly efficient in using the tools.

It would be interesting to study information technology students to see how comfortable they are using various collaboration tools and how successful the project ended up being. This will give degree programs an idea of what needs to be done to prepare students better. Further, it would be interesting to see how the different generations of IT students are handling the online collaboration tools. Generation Z might not be any more tech savvy than generation X and Y even though they are born with technology. As technology advances new issues and opportunities will arise.

CONCLUSION

Being able to collaborate effectively in modern distributed systems is extremely important in today's globalization. The lack of physical presence is challenging despite brand new technologies on the market. Most of us already in business or in higher education grew up communicating face-to-face when making important decisions, so changing what feels natural to us is extremely hard. Team leaders need to be global thinkers, to listen and to keep communication simple and clear. It is easy to be misunderstood when not in the same room. Use new and different technologies and work on building relationships and trust all take time, which can be an extra challenge in an age delivering instant effects and gratification. Trust can be gained by showing respect for others and showing an understanding of others cultures (Gross, 2002; Maynard & Mehrtens, 1993).

For universities to be successful in their efforts they need to have students collaborate with corporations, governmental agencies, or other organizations in hands-on service learning projects. These types of projects can play an invaluable role educating students and preparing them to be ready to enter the workforce and be effective contributors. Whenever possible, it is also recommended that students are exposed to and gain experience in working not only in direct collaborative teams, but also in distributed e-collaborating teams. This is a joint effort where faculty needs to better understand the newer generation as well new technology and business needs to incorporate technology into the curriculum. Businesses and governments need to form close partnerships with universities. Working virtually is becoming the norm. As technologies evolve new and better practices will emerge.

REFERENCES

alternativeTo. (2014). *Unfuddle*. Retrieved November 20, 2014 from: http://alternativeto.net/software/unfuddle/

Andrews, D., Conway, J., Dawson, M., Lewis, M., McMaster, J., Morgan, A., & Starr, H. (2004). *School revitalization: The IDEAS way (ACEA Monograph Series No. 34)*. Winmalee, Australia: Australian Council for Educational Leaders.

Ann, D., Lenore, N., & Chris, L. (2010). Facilitating transdisciplinary sustainable development research teams through online collaboration. *International Journal of Sustainability in Higher Education*, *11*(1), 36–48. doi:10.1108/14676371011010039

Anthony, T., DeMarie, S., & Anthony, H. (1998). Virtual teams: Technology and the workplace of the future. *The Academy of Management Perspectives, 12*(3), 17–29. doi:10.5465/AME.1998.1109047

Arnison, L., & Miller, P. (2002). Virtual teams: A virtue for the conventional team. *Journal of Workplace Learning, 14*(4), 210427294. doi:10.1108/13665620210427294

Arvey, R. (2009). *Why face-to-face business meetings matter.* Retrieved November 20, 2014, from: http://www.iacconline.org/content/files/WhyFace-to-FaceBusinessMeetingsMatter.pdf

Bailenson, J. N., Yee, N., Merget, D., & Schroeder, R. (2006). The effect of behavioral realism and form realism of real-time avatar faces on verbal disclosure, nonverbal disclosure, emotion recognition, and copresence in dyadic interaction. *Presence (Cambridge, Mass.), 15*(4), 359–372. doi:10.1162/pres.15.4.359

Boiney, L. G. (2001). Gender impacts virtual work teams. *Graziadio Business Report, 4*(4), 5.

Bond, C. F., & Titus, L. J. (1983). Social facilitation: A meta-analysis of 241 studies. *Psychological Bulletin, 94*(2), 265–292. doi:10.1037/0033-2909.94.2.265 PMID:6356198

Boston Business Journal. (2015). *How Boston-area companies and hospitals are dealing with the snow and their employees.* Retrieved February 28, 2015 from http://www.bizjournals.com/boston/blog/start-ups/2015/02/how-boston-area-tech-companies-are-dealing-with.html?page=all

Brown, J. (2013, February 12). *Virtual Team Management Trends and Telecommuting.* Retrieved on February 24, 2015 from http://blog.hubstaff.com/virtual-team-management-trends/

Business Dictionary. (2014). *Electronic collaboration.* Retrieved November 20, 2014, from: http://www.businessdictionary.com/definition/electronic-collaboration.html

Carmel, E. (1999). *Global software teams: Collaborating across borders and time zones.* Upper Saddle River, NJ: Prentice-Hall.

Cascio, F. W. (2015). *The Virtual Workplace: A Reality Now.* SIOP Retrieved from: http://www.siop.org/tip/backissues/tipapril98/cascio.aspx

Cascio, W. F. (2000). Managing a Virtual Workplace. *The Academy of Management Executive, 14*(3), 81-90.

Chidambaram, I., & Tung, I. (2005). Is out of sight out of mind? An empirical study of social loafing in technology supported groups. *Information Systems Research, 16*(2), 27–39. doi:10.1287/isre.1050.0051

Childs, R., Gingrich, G., & Piller, M. (2009). The future workforce: Gen Y has arrived. *Engineering Management Review, 38*(3), 32–34.

Chu, S. K., & Kennedy, D. M. (2011). Using online collaborative tools for groups to co-construct knowledge. *Online Information Review, 35*(4), 581–597. doi:10.1108/14684521111161945

Ciotti, G. (2013). *Why Remote Teams Are the Future (and How to Make Them Work).* Retrieved February 28, 2015 from: http://www.helpscout.net/blog/virtual-teams/

Coates, J. (2007). *Generational learning styles.* LERN Books.

de Pillis, E., & Furumo, K. (2007). Counting the cost of virtual teams. *Communications of the ACM*, *50*(12), 93–95. doi:10.1145/1323688.1323714

Diffin, J., Chirombo, F., & Nangle, D. (2010). *Cloud Collaboration: Using Microsoft SharePoint as a Tool to Enhance Access Services*. Retrieved February 24, 2015, from http://contentdm.umuc.edu/cdm/ref/collection/p15434coll5/id/1046

Dulin, L. (2005). *Leadership preferences of a Generation Y cohort: A mixed methods investigation*. (Ph.D. dissertation). University of North Texas. Retrieved August 15, 2014, from ABI/INFORM Global. (Publication No. AAT 3181040).

Ebrahim, N., Ahmed, S., & Taha, Z. (2011). Virtual Teams and Management Challenges. *Academic Leadership Journal, 9*(3), 1-7. Retrieved on February 24, 2015 from https://www.academia.edu/302000/Virtual_Teams_and_Management_Challenges

Elmore, B. (2006). It's a SMALL world after all. *Baylor Business Review, 25*(1), 8-9. Retrieved May 14, 2015 from: http://web.a.ebscohost.com.ezproxy.hpu.edu/ehost/pdfviewer/pdfviewer?vid=6&sid=6c324351-f088-43c9-ab84-c188741ab7b0%40sessionmgr4003&hid=4101

Elvekrog, J. (2015). *5 Ways to Ensure Remote Employees Feel Part of the Team*. Entrepreneur. Retrieved May 11, 2015 from: http://www.entrepreneur.com/article/243795

Forbes. (2013). *Collaborating in the Cloud*. Retrieved February 28, 2015 from https://www.cisco.com/c/dam/en/us/solutions/collateral/collaboration/hosted-collaboration-solution/forbes_cisco_cloud_collaboration_business.pdf

Gera, S. (2013). Virtual teams versus face to face teams: A review of literature. *Journal of Business and Management, 11*(2). Retrieved from http://www.academia.edu/4858172/Virtual_teams_versus_face_to_face_teams_A_review_of_literature

Govindarajan, V., & Gupta, A. K. (2001). Building an effective global business team. *MIT Sloan Management Review, 42*(4), 63–71.

Gross, C. (2002). Managing communication within virtual intercultural teams. *Business Communication Quarterly, 65*(4), 22–38. doi:10.1177/108056990206500404

Gupta-Sunderji, M. (2013). *Long-Distance Leadership: Managing Virtual Teams*. Retrieved May 14, 2015 from http://www.hrvoice.org/long-distance-leadership-managing-virtual-teams/

Herbsleb, J. D., & Mockus, A. (2003). An empirical study of speed and communication in globally-distributed software development. *IWWW Transactions on Software Engineering, 29*(6), 1–134.

Herrington, T. (2004). Where in the world is the Global Classroom Project? In J. Di Leo & W. Jacobs (Eds.), *If classrooms matter: Progressive visions of educational environments* (pp. 197–210). New York: Routledge.

Jarmon, L., Traphagan, T., Mayrath, M., & Trivedi, A. (2009). Virtual world teaching, experiential learning, and assessment: An interdisciplinary communication course in Second Life. *Computers & Education, 53*(1), 169–182. doi:10.1016/j.compedu.2009.01.010

Jarvenpaa, S. L., Knoll, K., & Leidner, D. E. (1998). Is anybody out there? Antecedents of trust in global virtual teams. *Journal of Management Information Systems, 14*(4), 29–64.

Jones, R., Oyung, R., & Pace, L. (2005). *Working virtually: Challenges of virtual teams.* Hershey, PA: Cybertech Publishing. doi:10.4018/978-1-59140-585-6

Kim, J. Y. (2000). Social interaction in computer-mediated communication. *Bulletin of the American Society for Information Science and Technology, 26*(3), 15–17. doi:10.1002/bult.153

Kock, N. (2005). What is e-collaboration? *International Journal of e-Collaboration, 1*(1), i–vii.

Kock, N. (2011). *E-Collaboration Technologies and Organizational Performance: Current and Future.* New York: Information Science Reference. doi:10.4018/978-1-60960-466-0

Kolowich, L. (2014, October 22). *Will Telecommuting Replace the Office? How Technology Is Shaping the Workplace.* Retrieved February 28, 2015 from http://blog.hubspot.com/marketing/technologyre-moteworkstatsinfographic

Kostner, J. (1994). *Virtual leadership: Secrets from the Round Table for the multi-site manager.* New York, NY: Warner Books.

Lambotte, F. (2013). *Managing & Working in Virtual Teams.* Retrieved February 28, 2015 from http://www.academia.edu/4415499/How_to_manage_a_virtual_team_How_to_communicate_in_virtual_teams

Lekushoff, A. (2012). Lifestyle-driven virtual teams: A new paradigm for professional services firms. *Ivey Business Journal, 76*(5). Retrieved February 28, 2015 from http://iveybusinessjournal.com/publica-tion/lifestyle-driven-virtual-teams-a-new-paradigm-for-professional-services-firms/

Lemmex, S. (2005). *Successfully Managing Remote Teams.* Retrieved February 28, 2015 from ftp://ftp.software.ibm.com/software/emea/dk/frontlines/SuccesfullyManagingRemoteTeams-S.pdf

Linnes, C. (2014). College Student Perception of Electronic Textbook Usage. *American Journal of Information Technology, 4*(2), 21–39.

Lister, K., & Harnish, T. (2011, June). *The State of Telework in the U.S.* Retrieved February 24, 2015, from http://www.workshifting.com/downloads/downloads/Telework-Trends-US.pdf

Malhotra, A., Majchrzak, A., Carman, R., & Lott, V. (2001). Radical innovation without collacation: A case study at Boein-Rocketdyne. *Management Information Systems Quarterly, 25*(2), 229–249. doi:10.2307/3250930

Maliniak, D. (2001). Design Teams Collaborate Using Internet Fast Track. *Electronic Design, 49*(11), 69.

Management Study Guide. (n.d). *Advantages and Disadvantages of Virtual Teams.* Retrieved on February 24, 2015 from http://www.managementstudyguide.com/virtual-teams-advantages-and-disadvantages.htm

Maynard, H., & Mehrtens, S. (1993). *The fourth wave: Business in the 21ˢᵗ century.* San Francisco, CA: Berrett-Koehler Publishers.

McGannon, B. (2014, June 23). *Managing Remote Employees.* Retrieved February 28, 2015 from http://www.lynda.com/Business-Skills-tutorials/Managing-remote-employees/156090/179147-4.html

Meyer, P., & Swatman, P. (2009). *Virtual worlds: The role of rooms and avatars in virtual teamwork.* Paper presented at the 2009 Americas Conference on Information Systems (AMCIS), San Francisco, CA.

Nydegger, R., & Nydegger, L. (2010, March). Challenges in Managing Virtual Teams. *Journal of Business & Economics Research, 8*(3), 69-82. Retrieved on February 24, 2015 from http://www.cluteinstitute.com/ojs/index.php/JBER/article/view/690

O'Hara-Devereaux, M., & Johansen, R. (1994). *Globalwork: Bridging distance, culture, and time.* San Francisco, CA: Jossey-Bass.

O'Leary, M. B., & Mortensen, M. (2008). A surprising truth about geographically distributed teams. *MIT Sloan Management Review, 49*(4), 5–6.

Olson, D. (2014, October 2). *Manage Projects with Wunderlist.* Retrieved February 28, 2015 from http://moultriejournal.net/2014/10/02/manage-projects-with-wunderlist/

Protalinski, E. (2014). *Facebook passes 1.23 billion monthly active users, 945 million mobile users, and 757 million daily users.* Retrieved February 28, 2015 from http://thenextweb.com/facebook/2014/01/29/facebook-passes-1-23-billion-monthly-active-users-945-million-mobile-users-757-million-daily-users/

Rafaeli, S., & Ravid, G. (2003). Information sharing as enabler for the virtual team: An experimental approach to assessing the role of electronic mail in disintermediation. *Information Systems Journal, 13*(2), 191–206. doi:10.1046/j.1365-2575.2003.00149.x

Rawson, R. (2010). *The 8 Best Collaboration Tools for Virtual Teams.* Time Doctor. Retrieved May 4, 2015, from http://blog.timedoctor.com/2010/12/03/the-8-best-collaboration-tools-for-virtual-teams

Ray, B. (2010). Apple wipes smile off FaceTime in the Middle East. *The Register.* Retrieved November 20, 2014, from: http://www.theregister.co.uk/2010/10/19/facetime/

Reynolds, B. (2014, October 1). *Survey: People Who Want Flexible Jobs and Why.* Retrieved February 25, 2015, from http://www.flexjobs.com/blog/post/survey-people-who-want-flexible-jobs-and-why/

Robertson Associates. (2013). *Which generation are you? X/Y/Z? Lost?* Retrieved April 10, 2014 from: http://www.robertson-associates.eu/blog/2013/11/29/which-generation-are-you-xyz-lost

Sarker, S., & Sahay, S. (2004). Implications of space and time for distributed work: An interpretive study of US-Norwegian system development teams. *European Journal of Information Systems, 13*(1), 3–20. doi:10.1057/palgrave.ejis.3000485

Senge, P. (2006). *The fifth discipline: The art and practice of the learning organization* (2nd ed.). Melbourne, Australia: Random House.

Shih, W., & Allen, M. (2007). Working with Generation-D: Adopting and adapting to cultural learning and change. *Library Management, 28*(1/2), 89–100. doi:10.1108/01435120710723572

Smith, C. (2013). *Working in a Virtual Team Using Technology to Communicate Collaborate.* Retrieved February 28, 2015 from www.mindtools.com/pages/article/working-virtual-team.htm

Starke-Meyerring, D., & Andrews, D. (2006). Building a shared virtual learning culture. *Business Communication Quarterly, 69*(1), 25–49. doi:10.1177/1080569905285543

Statista. (2010). *Number of global social network users 2010-2018*. Retrieved February 28, 2015 from: http://www.statista.com/statistics/278414/number-of-worldwide-social-network-users/

Stough, S., Eom, S., & Buckenmyer, J. (2000). Virtual teaming: A strategy for moving your organization into the new millennium. *Industrial Management & Data Systems, 100*(8), 370–378. doi:10.1108/02635570010353857

TheTop10BestOnlineBackup.com. (2014). *Review of the best cloud storage services*. Retrieved November 20, 2014, from: http://www.thetop10bestonlinebackup.com/cloud-storage

Thomas, D., Bostrom, R., & Gouge, M. (2007). Making knowledge work in virtual teams. *Communications of the ACM, 50*(1), 85–90. doi:10.1145/1297797.1297802

Tugend, A. (2014, March 7). It's Unclearly Defined, but Telecommuting Is Fast on the Rise. *The New York Times*. Retrieved February 28, 2015 from http://www.nytimes.com/2014/03/08/your-money/when-working-in-your-pajamas-is-more-productive.html

Walker, A., & Shuangye, C. (2007). Leader authenticity in intercultural school contexts. *Educational Management Administration & Leadership, 35*(2), 185-204.

Watkins, K., & Marsick, V. (1993). *Sculpting the learning organization*. San Francisco, CA: Jossey-Bass.

Watson-Manheim, M., & Belanger, F. (2002). Support for communication-based work processes in virtual work. *e-Service Journal, 1*(3), 61–82. doi:10.2979/ESJ.2002.1.3.61

KEY TERMS AND DEFINITIONS

Asynchronous Communication: Communication occurring independent of time or location.

Communication: The exchange of ideas, messages, or information through the use of technologies.

E-Collaboration: The process of collaborating in a project or program by using electronic technologies.

Knowledge Sharing: An activity where information, skills, or expertise is exchanged among individuals.

Offshoring: Relocation, by a company, of a business process from one country to another.

Social Ties: Connection between people.

Telecommuters: People working from home using phone, Internet, or other communication devices.

Trust: Refers to a sense of reliability and confidence necessary to complete tasks in a working environment.

Virtual Teams: A group of people who rely on communication technologies to achieve their goals. Virtual teams is also known as Global Distributed Teams.

Chapter 7
Knowledge Transfer and Knowledge Creation in Virtual Teams

Nory Jones
University of Maine, USA

ABSTRACT

Tacit knowledge transfer and knowledge creation represents perhaps the best means of sustainable competitive advantage through continual innovation. As organizations become more distributed in their different offices, virtual teams become more common and valuable. The question of how these virtual teams can effectively transfer tacit knowledge and create new knowledge thus becomes of importance to organizations. This chapter focuses on this issue and presents supporting evidence related to tacit knowledge transfer and creation, virtual teams, and how businesses can effectively harness capacity of virtual teams to transfer valuable tacit knowledge and create new knowledge.

I. INTRODUCTION

Teams are used in virtually every organization to improve productivity, responsiveness and effectiveness because of the synergies of bringing different expertise, talent and perspectives together. Virtual teams represent a growing part of the business environment given the globalization of business. They are normally defined as "groups of geographically and/or organizationally dispersed coworkers that are assembled using a combination of telecommunications and information technologies to accomplish an organizational task" (Berry, 2011, Page 187). A recent survey of businesses by the Society for Human Resource Management (SHRM) (Minton-Eversole, 2012) demonstrated the pervasiveness of virtual teams in the workplace. In this study, they surveyed "379 randomly selected HR professionals from SHRM's membership" and showed that almost half (46%) of the organizations surveyed used virtual teams. It was not surprising that their findings also showed multinational organizations were "more than twice as likely (66%) to use virtual teams compared with those with U.S.-based operations (28 percent)."

DOI: 10.4018/978-1-4666-9688-4.ch007

Some other interesting results from this study included: • Why they used virtual teams: 53%- "the need to include talent in different locations because their work is taking on a more global focus." 59%- to increase collaboration throughout the organization, 39%- to improve productivity, 39%- to reduce travel time and costs. • Of the respondents, 50% of the public for-profit companies were likely to use virtual teams as well as 49% of private for-profit companies. Only 9% of government agencies were likely to use virtual teams. If we assume that this survey is representative of most businesses, we can infer that virtual teams are very important to businesses and probably increasing in use and importance as business continues to extend its reach in terms of customers, suppliers, and collaborative activities with other businesses. This form of virtual collaboration provides an organization with competitive advantage by combining the best core competencies of expertise across an organization (Romero et al., 2011). They further provide organizations with the ability to adapt quickly to changing environmental and competitive conditions by assembling virtual teams with specific expertise to respond rapidly to these dynamic markets (Klotz-Young, 2012). However, it is not enough to create virtual teams and assume that team members will effectively communicate, collaborate, share knowledge, innovate and achieve goals. Rather, there are many challenges associated with virtual teams. This chapter explores tacit knowledge transfer in virtual teams including the impact of culture and leadership, and concludes with some lessons learned from the literature on how to effectively manage virtual teams for effective tacit knowledge transfer. Why is this important? With business operating on a global scale, expertise and talent is usually not local. Therefore, businesses should learn how to leverage the knowledge and expertise of people anywhere any time for sustainable competitive advantage and continual improvement and innovation. One way to do this is to learn to harness the power of virtual teams in an effective way. This chapter attempts to share knowledge on this topic by searching the literature and presenting research on the topic in a reasonably comprehensive manner. The chapter concludes with an analysis of trends from this literature review.

II. VIRTUAL TEAMS

According to Montoya et al (2011), a virtual team can be thought of as "a group of geographically dispersed people who interact through interdependent tasks guided by a common purpose with the support of communication technology" (Page 451). In their research, they suggest that several major factors can impact effective or ineffective virtual team communication and collaboration including trust, team composition, and cultural differences. However, they also recognize that since people on virtual teams must interface with technologies to communicate and collaborate, these technologies can also play a pivotal role in the success or failure of the team initiative. They suggest that the current technologies that are widely used can be improved upon with new, emerging technologies to enhance collaboration and communication. Specifically, they suggest that 3-dimensional (3D) technologies can provide a much richer environment than the commonly used technologies such as web sites, email, document repositories, wikis, and blogs. We infer that commonly used video and audio web-based technologies such as Google Hangout (free) and the similar technologies are included in the category of current 2- dimensional technologies. An example they provide of an emerging 3D environment is "Lenovo operates a 3D virtual showroom using the web. Alive platform where prospective customers can interact with products and salespeople" (Page 452). These 3D technologies use simulations to provide more of a real world experience to the users. Some of these environments use "avatars" such as the Second Life program where users create representations of themselves that are immersed in these simulated environments. This enables team members to simulate face-to-face interaction by using expressions, body language, audio and visual communication with their avatars.

In their interesting study about virtual team communication and collaboration in these 3D verses regular environments, they drew upon the theory of Rousseau et al. (2006) in examining several major team-related behaviors. These included: "(i) information exchange or communication behavior, (ii) cooperation or contribution behavior, and (iii) integrating or coordination behavior" (page 455). They suggest that normal virtual teams require more communication than face-to-face teams because of the lack of non-verbal communications, visual cues, feedback and nuances, and other behaviors that contribute to a richer communication environment in face-to-face teams. The ease and effectiveness of communication in teams is important because people need to establish trust, relationships and shared meaning in order to communicate and collaborate effectively.

Another interesting dimension of team behavior is the contribution of each team member. These authors suggest that in normal virtual environments, it is easier for "loafers" to not contribute as much as more productive, motivated team members. In contract, in face-to-face team interactions or possibly using avatars, it is easier to see the people and what they are doing, thus making each team member more accountable to contributing their fair share of the work. The findings from their study suggest that using 3D environments can mimic face-to-face interactions resulting in more effective, higher-performing teams.

Daim et al; (2012) examined distributed virtual teams in the high tech industry. These authors explored global virtual teams (GVTs), suggesting that most high tech companies use their talent as productively as possible by creating on-demand global teams for specific projects. Therefore, employees can be assigned to multiple teams for different projects with different virtual team members, across very different geographic areas and cultures, with responsibilities to different team leaders or managers. Therefore, in this matrix structure, employees report to different virtual team leaders for each specific project as well as to their functional managers.

One interesting observation from these authors was that a key component of success in these virtual global teams was to send team members to the different countries to meet and work with their team members on a face-to-face basis for a period of time. This allowed them to understand the different cultures and to develop a relationship and level of trust and communication with the different team members before they began the virtual team project. Trust and shared goals were shown to be key components of effective communication in the virtual teams once the project began.

In their study, they explored issues of communication in these different virtual global teams and found that culture differences exerted a huge influence, creating many communication problems. In terms of national culture, language represented a major barrier in terms of understanding and problems with shared meaning in translation. In addition, national culture exerted influences on perceptions of work time and expectations such as cultures where religious beliefs dictated expected times to work verses pray. In contrast, other cultures created expectations that people should be on-call to work whenever needed. Another cultural dimension that influenced communication was organizational and professional culture. For example, while the organization might attempt to create a standard organizational culture stressing innovation, risk taking and knowledge sharing, the different professions of people on the team might create major barriers. For example, engineers on the virtual team could have expectations of communicating only hard data after a thorough analysis verses a marketing person on the team who preferred exploratory communication of ideas. Finally, generational differences led to communication problems as Millennials had different priorities and communication styles than Baby Boomers. All of these differences were further manifest in the preferred technologies used. For example, Millennials would feel more comfortable using social media and mobile type technologies than Baby Boomers.

Another study explored the impact of culture on global virtual teams (Dekker et al, 2008). These authors developed a framework of factors critical to effective virtual team performance; shown in Table 1. In their research, they found that while there is a small amount of overlap with face-to-face teams, most of the factors are unique to virtual teams.

These authors posit that people in global virtual teams do not fully recognize the huge cultural differences in both language, norms and values until the team is formed and work processes begin which can create conflict and poor team performance. From their research, they suggest that a "fusion model, in which culturally diverse teams have to accept and respect the coexistence of differences and utilize the unique qualities of those differences, produces the best team outcomes" (Page 442-443). They also explored the impact of Hofstede's major cultural dimensions on virtual team performance. These include power-distance (the degree to which managers and subordinates prefer a hierarchy vs. a more egalitarian system), uncertainty-avoidance (the degree to which people prefer strict rules and procedures vs. more emphasis on developing relationships), individualism-collectivist (the degree to which people are integrated into groups vs. being individualistic), masculinity (degree to which men and women differ on their roles). Their study found that cultures with low power-distance tend to include team members more and are less concerned with power and status of the team members. They found that people from high individualist cultures tend to be more responsible and feel more accountable in their individual performance to the team. However, people from countries low on the masculine scale tend to be more caring and concerned about the team members. We would infer that this would be true for collectivist cultures also. Finally, they uncovered another dimension stressed by the virtual team members; the need for respectfulness of each other. This is consistent with their theory that fusion involves understanding

Table 1. Categories of interaction behavior in virtual teams and how team members should behave per category

No.	Category Label Interaction Behavior
1	**Media Use:** Effectively matching the media to the task and effective use of media.
2	**Handling Diversity:** Taking into account language-, time zone-, and cultural differences when nteracting and behaving accordingly.
3	**Interaction Volume:** Communicating short, to the point, and only when necessary.
4	**In-Role Behavior:** Taking task and goal of the team seriously and complying with obligations.
5	**Structuring of Meeting:** Planning and structuring of meetings.
6	**Reliable Interaction:** Being predictable in behavior and responsive to messages of team members.
7	**Active Participation:** Showing active participation in meetings by contributing and listening.
8	**Including Team Members:** Including and inviting team members for contribution.
9	**Task-Progress Communication:** Communicating deadlines, actions, and progress of a task to the team.
10	**Extra-Role Behavior:** Showing pro-social behavior towards team members.
11	**Sharing by Leader:** Sharing of information and decisions with the team by team leader.
12	**Attendance:** Being involved in the meeting and not showing up late or not at all. No multitasking.
13	**Social-Emotional Communication:** Talking about non-task-related subjects.
14	**Respectfulness:** Behaving in accordance with the hierarchy of the team.

and respecting each other's different cultures. Salminen-Karlsson (2013) supported this in her research showing that virtual teams were successful if they were supported by trust, an attitude of respect for each other's cultures and a willingness to learn.

Another review of cultural impacts on virtual teams by Symons, and Stenzel (2007) provided similar accounts of the benefits of global virtual teams. However, they also shared some of the downsides of virtual teams including "low individual commitment, role overload, role ambiguity, absenteeism, lack of synergy amongst team members, conflicts due to dual or multiple reporting lines, different holidays and working hours, communications breakdowns due to unreliable technology and cultural variances" (page 2). In contrast to the prior study, these authors also suggest that the two major categories that influence the success or failure of virtual teams include

1. Leadership and culture, and
2. Communication and technology.

In terms of technology, the authors remind us that different countries have different levels of sophistication in terms of the technologies available as well as those that are commonly used. Therefore, managers should include assessment of needed technologies and training for virtual team participants.

They also suggest that in virtual teams, leadership is a results-oriented process that is mitigated by different cultural norms, but in the end, relies on a trust-based virtual culture that values diversity and respects differences in cultural norms. They reiterate the basic premise that trust is normally considered the most critical element in developing an effective virtual team. They also suggest that the "Cultural Orientation Model, which combines key dimensions-based cultural concepts `into ten cultural dimensions" created by the Training Management Corporation (TMC), represents a good assessment of cultural diversity in virtual teams. They provide an interesting example, "American culture is considered to be control focused, flexible in terms of structure, present and doing oriented, low context, direct, instrumental and informal in communication, where linear inductive thinking dominates and where people that believe in equality, individualism and competition prefer a certain distance between themselves and others. Communication is relatively impersonal, fact-centered, and informal. If it is an accurate representation (and there is a danger of stereotyping), this characterization of achievement (rather than relationship-orientation) would help a virtual team with a US participant to recognize cultural differences. It is not surprising that the internet and working formats based on information technology have been led from the USA where progress, drive and discovery are highly valued" (page 13).

An article by Kayworth and Leidner (2002) explored the role of leadership in virtual teams in greater depth. These authors suggest that communication in virtual teams may be severely compromised by the computer technologies used which do not provide the visual cues (body language, tone, expertise, social status, facial expressions needed to communicate effectively. In addition, computer interactions often do not provide the level of trust and relationship building as in face-to-face teams. In asynchronous environments, discussion threads can be disjointed and often overwhelming. As discussed in prior sections, different cultures also create problems in shared meaning via different cultural nuances, language and norms. In addition, the need to learn the different technologies itself can present barriers to effective team communication and performance. Finally, in global teams, different time zones and busy work schedules make communication and performance even more challenging. In their review of leadership theories in general, these authors "suggest that effective leadership may be a function of the manager's ability to display a varied and complex set of behavioral repertoires in response to complex organizational circumstances" (Page 12).

They further contend that because of the increased complexity of virtual teams, that effective virtual team leaders must be able to handle even more complexity and diversity than traditional team leaders. In their study, they found that effective virtual team leadership was associated with team member's perceptions of communication effectiveness, mentoring capabilities, empathy, positive attitude, and leader's ability to establish role clarity in the team. The team members wanted leaders who communicated promptly, frequently, directly, clearly and unambiguously. Using effective collaboration tools appeared to improve trust and the development of relationships. Effective virtual leaders exhibited care, concern and understanding, but also provided clear structure on goals, tasks and expectations. While communication is important in all teams, it appears to be more important in virtual teams as well as the ability of the leader to develop a climate of trust and collaboration via social facilitation, mentoring and a culture of caring and shared vision and goals with an assertive task orientation. The use of appropriate technologies also influences the effectiveness and performance of virtual teams.

El-Sofany et al (2014) surveyed participants in virtual IT project teams and provide consistency in the findings of the prior authors. Their results showed that the major factors for success in virtual teams included:

- The team has to be able to see the whole picture/Clarity of objectives,
- Leadership/Project management,
- Communication,
- Reliability/Efficiency,
- Teamwork,
- Quality Management.

Another study (Politis & Politis, 2011) examined the role of the "big five personality traits" on leadership on virtual teams. These traits include "extraversion, negative affectivity, (neuroticism or emotional stability), agreeableness, conscientiousness and openness to experience" (page 342). They further define these traits as "extraversion refers to people (extraverts) who are outgoing, talkative, sociable and assertive. Introversion and passivity refers to individuals (introverts, the opposite of extraverts), who are less sociable, shy, and cautious and they feel comfortable being alone. Negative affectivity (known as neuroticism or emotional stability) is the tendency to experience negative emotions, feel distressed, hostile and critical of oneself and others. In contrast, people with low negative affectivity are poised, secure, and calm and governed by emotional control. Agreeableness, refers to people who get along well with other people – some scholars label agreeableness with 'friendly compliance'. Conscientiousness refers to people who are careful, dependable and self-disciplined and openness to experience is the tendency to be original, flexible, creative, risk taking and be open to a wide range of stimulus" (page 343). The results of their study show that leaders who are more likeable, cooperative, trusting and helpful will be more effective with virtual teams. They tend to be more empathetic and are more conscientious in providing feedback and support. Therefore, leaders who have "high agreeableness, high conscientiousness, high openness to experience and high extraversion in order to implement virtual offices and manage a more flexible workforce consisting of remote and mobile workers" (page 348).

Hertel et al, (2004) explored the impact of task interdependence and goal setting on virtual team performance and motivation of the team members. Specifically they examined three elements relating to team performance: "goals (goal interdependence), task behavior (task interdependence), and evaluation of behavior outcomes (outcome interdependence)" (page 4). In the first element, they speculate that getting buy-in to the goals of the task would more important in virtual teams. In task interdependence, they speculate that if the team members must work closely together with interconnected tasks, they will be more involved, engaged and motivated. With outcome interdependence, they found studies to support their contention that team-based rewards tend to motivate teams to work together more productively. Finally, they incorporate trust as a crucial element in effective virtual team performance. Their "results suggest that high quality of goal setting, high task interdependence, and the use of team-based rewards might be independent ways to enhance the effectiveness of virtual teams" (page 22).

Similarly, Massey et al. (2003) explored the impact of time (temporal coordination mechanisms) on the development of relationships in virtual teams and their ultimate performance and effectiveness. They suggest that a major problem in virtual team is coordinating time schedules which results in sub-optimal work flow and communication. This is because asynchronous communication often lacks the nuanced feedback required for effective tacit knowledge transfer. Therefore, they hypothesize that coordinating communications in virtual teams will lead to higher team performance and quality. Their results supported this hypothesis, showing that real time (synchronous) communication provided the ability to discuss issues in greater depth, with more shared meaning, thus resulting in higher quality interactions and team performance.

III. TACIT KNOWLEDGE TRANSFER

Tacit knowledge is defined as "Unwritten, unspoken, and hidden vast storehouse of knowledge held by practically every normal human being, based on his or her emotions, experiences, insights, intuition, observations and internalized information." http://www.businessdictionary.com/definition/tacit-knowledge.html#ixzz3NPVqctSB Polyani (1983) famously noted, "We know more than we can tell. This fact seems obvious enough; but it is not easy to say exactly what it means. For example, we know a person's face, and can recognize it among a thousand, indeed among a million. Yet we usually cannot tell how we recognize a face we know. So most of this knowledge cannot be put into words."

Smith et al (2007) did an interesting study of conducting a focus group with knowledge management experts from around the world. They started with the assumption that "knowledge transfer occurs when knowledge is diffused from an individual to others. Thus, knowledge transfer is a function of the interactions between people and its effectiveness is largely dependent on factors that encourage or inhibit interpersonal relationships" (Page 28). In their focus groups, they looked at different types of knowledge transfer and concluded that different methods were effective with different types of tacit knowledge. They describe best practices as knowledge that has worked well and should be transferred to other people. A key component in facilitating this was a clear relative advantage of the knowledge perceived by people as well as an easy way to document, archive and retrieve it. Expertise was the second type of knowledge examined. One way this tacit knowledge was successfully shared at an engineering company was the use of web-based videos with live Q&A sessions. The content was valuable and the experts were rewarded for sharing. Another successful approach was "fireside chats" where experts had online discussion groups that provided valuable knowledge and insights to people throughout the organization who could benefit from that knowledge. Another successful knowledge transfer program involved proactive mentoring which paired a knowledgeable person with someone who wanted or needed that knowledge. Experiences learned was facilitated by "project debriefs" where the team would meet and share their experiences after the project was complete. These sessions were recorded and available to others. Storytelling was used effectively by some companies to explain concepts and situations in context, which were also videoed and available online. Finally, some companies used social events to invite different people from the organization with different expertise to meet, share ideas, and develop relationships and trust.

Goffin and Koners (2011) explored the challenges of tacit knowledge transfer in teams. While reiterating the challenges of real tacit knowledge transfer, they suggest that this normally occurs via close interactions with people through shared experiences. They also reiterate the value of metaphors and storytelling in transmitting tacit knowledge outside the local group. This last finding may be relevant to virtual teams who often do not have the rich face-to-face experiences and nuances and could instead benefit from context-rich stories and metaphors.

Knockaert et al. (2011) suggest that since tacit knowledge resides in the knower, it requires extensive interpersonal communication and interaction in order to effectively transfer tacit knowledge to another person. Often this is done through mentoring or apprenticeships with great personal interactions over a long time period. However, in virtual teams, there are several potential barriers to this. First, since the team members may not meet in person, the communication is usually less rich than in face-to-face teams and the team members may not develop the long term relationships and trust to communicate effectively for tacit knowledge transfer. In addition, as mentioned in the prior section, virtual teams are often composed of members from very different backgrounds, occupations and cultures, which creates additional significant barriers to shared meaning, communication and tacit knowledge transfer. While these researchers studied tacit knowledge transfer in traditional face-to-face teams, we can make a few inferences from their research findings. First, they found that a team needed to have a critical mass of expertise (tacit knowledge) for effective team performance. Second, the team members worked and communicated best if they were motivated by intrinsic factors such as a belief in the importance of the mission and vision of the team as well as a sincere desire to collaborate and communicate within the team. To overcome the cognitive dissonance of people from different backgrounds and cultures, they found that having boundary-spanning individuals on the teams to act as translators / liaisons was very helpful. They also found that having team members who had worked together before and were familiar with each other or having orientation / bonding sessions to build relationships and trust were also helpful in tacit knowledge transfer.

Bertels et al. (2011) studied the impact of communities of practice and organizational climate (culture) on the effectiveness of dispersed (virtual) teams. They define communities of practice as "groups composed of members who share information, insight, experience, and tools about an area of common professional discipline, skill, or topic (McDermott, 2000; Wenger, 1998). The purpose of communities of practice is to build and exchange knowledge among self-selected members, who are glued together by a passion, commitment, and identification with the group's expertise" (page 761). They further suggest, "It is in our tacit knowledge that our intuition, insight, and "gut feel" originate" (page 759). They reiterate the difficulty of sharing tacit knowledge in these dispersed teams with people potentially from different countries, cultures, and fields of expertise, but suggest that developing effective communities of practice with positive organizational climate (culture) can facilitate this tacit knowledge transfer. They also reiterate that communities of practice and cultures that promote "risk taking, trust, and open interaction improves the creation of new knowledge" and by inference, tacit knowledge sharing (page 758). They suggest that this works because tacit knowledge can be shared via "situated learning" in communities of practice. That is, in a "a situated theory of learning, true understanding involves ''living it,'' i.e., it requires the learner to be situated within the context because learning is a social process which will be affected by social and cultural contexts" (page 760). Therefore, in their study, they explored the concept that "Proficiency of dispersed collaboration is related to the ability of employees active in the front end of innovation to overcome barriers related to distance, culture, and IT-enabled communication" (page 764). Their findings suggest that improved performance in dispersed collaboration (tacit knowledge transfer in virtual teams) is improved with communities of practice that are supported with a positive organizational climate (culture).

Harris (2009) explored tacit knowledge transfer in the context of learning and collaboration among small and medium size businesses (SMEs). He suggests, "Effective collaboration offers a framework by which tacit knowledge can be shared amongst participants through a process of socially constructed learning. The perspective is characterised by the way in which language, in the form of conversations, discourses, narratives and stories provide the means by which the participants share experiences and construct a shared "reality", in which renewed understanding and actions create new enacted opportunities. The social nature of these collaborative communities gives rise to learning opportunities amongst participants and for members to obtain exposure to wider network of communities" (page 217). He also reiterates Nonaka's (1991) 3-step process for knowledge transfer:

1. *Individual learning to develop localised knowledge;*
2. *Tacit knowledge transfer to disseminate knowledge across the organisation; and*
3. *Organisational learning to embed the new knowledge into the organisation's procedures and unconscious actions, thus developing competitive advantage (Page 218).*

The results of his research with focus groups showed that SME's could benefit from collaborative e-learning networks in the form of communities of practice (COP). What SME's are interested in is action-oriented tacit knowledge transfer that will help all of the in win-win situations. Therefore, the virtual collaborative COP's could be valuable if the participants had a common goal and shared their expertise to create new knowledge and helped everyone gain a competitive advantage via innovative and knowledge creation.

Klitmøller and Lauring (2013) examined the interaction of media type with culture/language and knowledge sharing in virtual teams. They suggest in theory that complex or ambiguous tasks or projects are more successful using "rich media", such as video-conferencing. In contrast, they suggest that "lean media", such as e-mail, can lead to poor communication with misunderstandings and problems with the effectiveness of the virtual teams. However, they also suggest that virtual team members can use "lean media" for simple, explicit ("canonical") information effectively. They further contend that cultural differences make the need for rich media even more important to allow for observation and transfer of cultural differences and nuances. In addition, different languages further contribute to the difficulties in knowledge transfer, which would be exacerbated by virtual media. Their study of virtual teams involving people from India and Denmark confirmed these propositions. Specifically, virtual teams with cultural and language differences require rich media to effectively communicate and share knowledge, especially tacit knowledge. Simple explicit knowledge can be shared via lean media such as email or discussion forums.

IV. CONCLUSION: HOW VIRTUAL TEAMS CAN EFFECTIVELY TRANSFER TACIT

Virtual teams continue to grow in number and importance as organizations recognize their value in an increasingly interconnect world. The issue is how to leverage the knowledge and expertise of the virtual team members so that the team works effectively and synergistically to solve problems, attain goals, and accomplish tasks.

In this chapter, we reviewed the literature on virtual teams and tacit knowledge transfer. Several themes emerged from this review:

1. The effective performance of virtual teams is influenced by:
    a. Trust and relationships,
    b. Culture and language,
    c. Effective leadership,
    d. Common goals and effective team structure.
2. Effective tacit knowledge transfer within virtual teams is influenced by:
    a. Trust and relationships,
    b. Appropriate media,
    c. Use of appropriate knowledge sharing strategies.

The literature consistently shows that people on virtual teams need to first develop solid trusting relationships. In an ideal environment, an organization should devote resources of time and money to bring people on virtual teams together at the beginning to bond and develop trust and long term relationships. However, given organizational constraints, if this is not possible, then the use of rich media such as video-conferencing and synchronous tools (chat, social media, etc.) can compensate to some degree for face-to-face relationship building.

Culture involves both national, regional and organizational cultures which can be vastly different among the virtual team members. Different languages among the team members can also cause major problems in the team. Therefore, the literature suggests that managers devote resources in time and funding to explore these cultural and language differences and provide team members with education and training as well as time to learn from each other. In addition, when managers model attitudes of tolerance and appreciation for differences in culture and language, this tends to motivate the team members to similarly value these differences as value added.

Effective leadership is also a key factor in successful virtual teams. The literature suggests that effective leaders of virtual teams should be both task and people-oriented. They need to put in more time providing structure and feedback to compensate for the lack of face-to-face interactions in the team. This includes great, consistent communication, mentoring, empathy, and providing specific roles and tasks for the team members. Effective leaders should also recognize that in forming the teams, people who are open, agreeable, likable, trusting and helpful will make better virtual team members. Effective leaders also need to develop team reward structures and temporal scheduling to bring the team members together synchronously at times.

Similarly effective tacit knowledge transfer is more challenging in virtual teams since the team members often do not have the opportunity to develop trust and long-term relationships that are usually necessary for effective tacit knowledge transfer. Therefore, in addition to the factors mentioned in the prior paragraph including trust and relationship, several factors can help facilitate tacit knowledge transfer. First, use appropriate technologies / media. For tacit knowledge transfer, people often need to simulate face-to-face interactions to understand complex problems, issues of tasks. This often involves using video-conferencing, real-time interaction media like chat, 3D virtual environments or social media. These provide the opportunity for questions, feedback, nuances in body language, tone and much richer discussions among participants. Similarly, developing effective communities of practice (CoPs) helps people develop rapport in support of common goals and a common vision. Other tacit knowledge sharing strategies in these media can include the use of stories, mentors, videos or discussion forums.

In conclusion, it is advisable to take a serious look at the factors that can impede effective tacit knowledge transfer in virtual teams. The factors discussed demonstrate their complexity. However, with serious research, education and planning, organizations can use virtual teams to help them achieve a competitive advantage via the diversity of expertise and perspectives that they provide.

REFERENCES

Berry, G. (2011). Enhancing Effectiveness on Virtual Teams: Understanding Why Traditional Team Skills Are Insufficient. *Journal of Business Communication, 48*(2), 186–206. doi:10.1177/0021943610397270

Bertels, H., Kleinschmidt, E., & Koen, P. (2011). Communities of Practice versus Organizational Climate: Which One Matters More to Dispersed Collaboration in the Front End of Innovation? *Journal of Product Development and Management, 28*, 757–772.

Daim, T., Ha, A., Reutiman, S., Hughes, B., Pathak, U., Bynum, W., & Bhatla, A. (2012). Exploring the communication breakdown in global virtual teams. *International Journal of Project Management, 30*(2), 199–212. doi:10.1016/j.ijproman.2011.06.004

Dekker, D., Rutte, C., & Van den Berg, P. (2008). Cultural differences in the perception of critical interaction behaviors in global virtual teams. *International Journal of Intercultural Relations, 32*(5), 441–452. doi:10.1016/j.ijintrel.2008.06.003

Denison, D. R., Hooijberg, R., & Quinn, R. E. (1995). Paradox and performance: Toward a theory of behavioral complexity in managerial leadership. *Organization Science, 6*(5), 524–540. doi:10.1287/orsc.6.5.524

El-Sofany, H., Alwadani, H., & Alwadani, A. (2014). Managing Virtual Teamwork in IT Projects: Survey. *Journal of Advanced Corporate Learning, 7*(4), 28–33. doi:10.3991/ijac.v7i4.4018

Goffin, K., & Koners, U. (2011). Tacit Knowledge, Lessons Learnt, and New Product Development. *Journal of Product Innovation Management, 28*(2), 300–318. doi:10.1111/j.1540-5885.2010.00798.x

Harris, R. (2009). Improving tacit knowledge transfer within SMEs through e-collaboration. *Journal of European Industrial Training, 33*(3), 215–231. doi:10.1108/03090590910950587

Hertel, G., Konradt, U., & Orlikowski, B. (2004). Managing distance by interdependence: Goal setting, task interdependence, and team-based rewards in virtual teams. *European Journal of Work and Organizational Psychology, 13*(1), 1–28. doi:10.1080/13594320344000228

Hofstede, G. (1980). *Culture's consequences: international differences in work-related values*. Beverly Hills, CA: Sage.

Hofstede, G. (2001). *Cultures consequences: comparing values, behaviors, institutions, and organizations across nations* (2nd ed.). Thousand Oaks, CA: SAGE Publications.

Kayworth, T., & Leidner, D. (2002). Leadership Effectiveness in Global Virtual Teams. *Journal of Management Information Systems, 18*(3), 7-40.

Klitmøller, A., & Lauring, J. (2013). When global virtual teams share knowledge: Media richness, cultural difference and language commonality. *Journal of World Business, 48*(3), 398–406. doi:10.1016/j.jwb.2012.07.023

Klotz-Young, H. (2012). *The Virtual' Marketing Team*. Security Distributing & Marketing.

Knockaert, M., Ucbasaran, D., Wright, M., & Clarysse, B. (2011). The Relationship Between Knowledge Transfer, Top Management Team Composition, and Performance: The Case of Science-Based Entrepreneurial Firms. *Entrepreneurship Theory and Practice, 35*(4), 777–803. doi:10.1111/j.1540-6520.2010.00405.x

Massey, A., Montoya-Weiss, M., & Hung, Y. (2003). Because Time Matters: Temporal Coordination in Global Virtual Project Teams. *Journal of Management Information Systems, 19*(4), 129–155.

Minton-Eversole, T. (2012). *Virtual Teams Used Most by Global Organizations, Survey Says*. Accessed on December 10, 2014 from http://www.shrm.org/hrdisciplines/orgempdev/articles/pages/virtualteams usedmostbyglobalorganizations,surveysays.aspx

Montoya, M., Massey, A., & Lockwood, N. (2011). 3D Collaborative Virtual Environments: Exploring the Link between Collaborative Behaviors and Team Performance. *Decision Sciences, 42*(2), 451–476. doi:10.1111/j.1540-5915.2011.00318.x

Nonaka, I. (1991). The knowledge-creating company. *Harvard Business Review, 69*(6), 96–104.

Politis, J., & Politis, D. (2011). The Big Five Personality Traits and the art of Virtual Leadership. In *Proceedings of the European Conference on Management. Academic Conferences, Ltd.*

Romero, D., & Molina, A. (2011). Collaborative networked organisations and customer communities: value co-creation and co-innovation in the networking era. *Production Planning & Control: The Management of Operations, 22*(5-6), 447-472.

Rousseau, V., Aube, C., & Savoie, A. (2006). Teamwork behaviors: A review and an integration of frameworks. *Small Group Research, 37*(5), 540–570. doi:10.1177/1046496406293125

Salminen-Karlsson, M. (2013). Swedish and Indian Teams: Consensus Culture Meets Hierarchy Culture in Offshoring. In *Proceedings of the European Conference on Information Management & Evaluation. Academic Conferences & Publishing International Ltd.*

Smith, H., McKeen, J., & Singh, S. (2007). Tacit Knowledge transfer: Making it Happen. *Journal of Information Science and Technology, 4*(2), 23–44.

Symons, J., & Stenzel, C. (2007). Virtually borderless: An examination of culture in virtual teaming. *Journal of General Management, 32*(3), 1–17.

Chapter 8
Electronic Collaboration in Organizations

William Dario Avila Diaz
Independent Researcher, Colombia

ABSTRACT

Electronic collaboration was born with the new technologies, which establish a more harmonious balance of organizations in an increasingly global, open and competitive digital economy, called, nowadays, "the economy of the crowds". This economy has caused changes in the organizations of the century, as new administrative principles. In this context, organizations use new business models to achieve its objectives to a meager cost. Similarly, they have managed the integration of different levels and optimizing performance of the entire organization together through electronic media and online collaboration. This work shows the areas of the different levels and forms of organizational electronic collaboration.

INTRODUCTION

Electronic collaboration of organizations has facilitated the construction of robust models that have been developed on the basis of a literary growth of global experiences, such as YouTube and Facebook. These may involve, in the levels of organization, such as integration and cooperation of workers, a joint task with the use of information and communications technology. The reality of the organizations of XXI century is reflected in the broad collaboration of professional support in the social, cultural and economic boundaries, worldwide, in terms of partners in line with common interests and the possibility that they can work deeper and with faster joint.

The members of collaborative teams share ideas and experiences and solve problems through social networks using the Internet as a form of communication that transcends geographical environment. Such geographic expansion of organizations can be beneficial, since the diversity of online collaboration could contribute significantly to the growth, development and integration of the organization with the uptake of new ideas, new knowledge transfer and possible solutions.

With respect to these issues, questions arise, among which we highlight the following: electronic collaboration can contribute to the harmonious balance between all functions of the organization? will be

DOI: 10.4018/978-1-4666-9688-4.ch008

worth the effort to make information technology and communication constitute into a support in electronic collaboration or online?, online collaboration wins on access to new sources of production and services to reduce cost?, with the arrival of the twenty-first century, organizations have entered the economy of crowds to reduce costs and obtain solutions to answers in such a short time? In other words, the main question might be this: electronic collaboration will contribute to the modernization of organizations?

This article is structured as follows: the first section provides a brief history and the significant role played by in electronic collaboration within the organization and beyond it, plus how it has helped set out the integration of the different functional levels of the organization; then, modes of online collaboration that have contributed to the strengthening of the strategic objectives of the organizations of XXI century are presented and, finally, concludes with the most important and transcendental aspects that may result from use of electronic collaboration in the organizations, according to the questions asked.

CONTEXT

Historically, the concept of online collaboration itself arose in 2006, when the Internet became Web 2.0. This caused a change in the world of business and in the global economic system. Canadian writers Don Tapscott and Anthony Williams, authors of *Wikinomics: the new economics of intelligent masses*, said at the time that new technologies, demographics and the global economy contribute to the emergence of new modes of production or services and will influence the economic paradigm based on mass collaboration and the intensive use of technologies that will grow even more in the future.

It is worth saying that the intensive use of technology (Freire, 2014) provides a more harmonious balance between all functions and facilitates compliance with the requirements of innovation and the impact of design changes caused by the organizations in the century XXI, as quoted in the book *Build innovation* of Mexican engineer Enrique Alberto Cabrera Medellin (Medellin, 2013: 39-41).

The content of the book *Wikinomics: the new economy of intelligent masses* has as a supporting base four innovative ideas for future organizations, namely: openness, interaction, exchange and overall performance (Wikinomics, 2014). In connection with the opening, this makes organizations rely solely on internal resources and capabilities. The use of open standards is the first start. Business information ceases to be secret, reducing transaction costs, accelerate business networks and promotes trust and loyalty of their environment. With regard to peer interaction, it is a new form of horizontal organization that is emerging and competing with the hierarchical organization. Workers at all levels, with the collaboration of technological tools, are self-organizing to design and produce products or services. As for the exchange, these involve working with digital creations as well as to reproduce and mix a zero marginal cost. Finally, overall performance, where the impact of the consequences of innovation and wealth creation occurs. In short, one can say that global online collaborative platforms allow businesses, workers and, in general, society a globalized way activity.

It should be noted that the operation of online collaboration is an approach to integrate the organization, so that appropriate services are provided in appropriate places and at the right time, in order to minimize costs and achieve specific objectives, to forge more productive tools such as information technology and communication. Note that, today, with the help of these tools, it has increased levels of coordination and collaboration among all stakeholders. So interpersonal knowledge gaps are confronted more spontaneously (Jones, 2008: 318). This leads to a management approach called "Chain integration of products and services" as a key element for achieving the strategic objectives.

Referring to the support of information technology and communication in the process of integration of the products or services in the organization, are based on four dimensions: first, the integration of information, which is to share information between all members of different levels; the second is the synchronized planning, where plans are designed together for the introduction of new products or services in order to reduce costs, optimize capacity and improve services; the third is the coordination of shared workflow, where the sequence of activities is formalized to perform and comply with the objective to integrate and automate business processes and coordination of plans and operations, and finally, It is the implementation of new business models in which combinations of multiple patterns that help the organizational management, such as the following:

- Construction of markets that connect buyers and sellers. This facilitates transactions between them; eg B2B, B2C or C2C.
- Adaptation of traditional business rules that provide content and services that benefit the activities and management of the organization, such as portals, online classified ads, advertising content and incentives for attention.
- Setting patterns, where information about users and their habits are extremely valuable for use as *marketing* campaigns.
- Accommodation services available only on the web and in virtual stores.
- Installation Guidelines to provide services directly; thus reducing costs, the customer service is improved and understanding of customer preferences is optimized.
- Feasibility of modules based on customer loyalty set in access to services.
- Accommodation subscription in order to provide high value-added content.
- Implementation of a utility model of pay per view.

The performance of online collaboration within an organization achieves its competitive edge for access to the conquest of new sources of production and services through incremental innovations in continuous improvement, adaptation, improvement and upgrading of such products or services. As for the main innovations bring these entire categories of new services such as wireless communications and architectural innovations, which refer to the reconfiguration of the system components that constitute services; for example, virtual stores where services are offered at lower cost and are oriented to a critical and defined mass of customers.

It should be noted that innovation models of online collaboration are complex linear processes, iterative and where almost always given a series of systematic steps that influence the innovative performance, making it easier for new management principles of the organization. These models range from brochures, advertisements, catalogs and services.

In particular, online collaboration is a design that facilitates the integration of the various levels and functions and optimizes performance across the organization as a whole. This phenomenon explains why organizations today are gradually moving from the old notion of a mechanistic structure to a systemic structure, whose essential properties arise from the relationship between the parties and the understanding of a phenomenon that can be understood in the context of a larger set of basic and essential components of organizations (Capra, 1999: 46-49; Jones, 2008: 318).

So there are interests at different levels of the organization and there are four types of collaboration online such as operative or technical, tactical, strategic and directive regardless of the structure that is working whether this hierarchy for efficiency and scalability or "redarquia" (collaboration, diversity, adaptability, self-organization, etc.) that allows coordinating human efforts without sacrificing the initiative and creativity of the people (Cabrera, 2013: 27; Cabrera, 2013b: 5).

- The electronic collaboration of operational or technical support monitoring activities, where people are integrated, media monitoring tools and processes of planning, decision making and monitoring. It should be noted that online collaboration or electronic offers a unique view of the situation of the organization at the level of implementation and the benefits to be gained both internally and externally. One aspect of online collaboration at this level is the ability to offer products or services on the exact measure, strongly linked to the production with precision and without additional cost.

- With regard to online collaboration in the tactical side, it helps to integrate and innovate in knowledge, and in services, implying significant changes in the characteristics of the existing product or service; in the process, causing significant events in the methods of distribution or delivery of the service; in the company, as to the application of new methods of organization, such as changes in business practices and workplace sorting and in trade, which involves the application of new marketing methods. These methods include changes in the design, promotion, placement and on ways to fix the prices of products or services offered.

Successful relationships are based on creating an inclusive environment for all business processes that span more than one area, such as production, processing, marketing and organization. These functional areas are electronically coordinated to increase efficiency and responsiveness of the organization and it is also possible to connect the business processes of other organizations in collaboration.

At this level, internet technologies allow all members of the functional areas to communicate with each other instantly, by using updated information to adjust purchasing, logistics, manufacturing, *software*, etc. The functional areas can use a web interface to enter provider systems and see if the inventory and production capacities are sufficient to meet the demand for products or services.

Is worth mentioning that the information technologies of the twentieth century are based on closed models of vertical integration, but the advent of the Internet, especially Web 2.0 tools, changed the landscape: they are now more efficient, effective and cheap when you outsource. Exploiting the opportunities created by technology necessarily imply transparency; bearing uncertainty and accept reduced control, while traditional companies always work with a strict control over the processes and workers and kept secret the details of the organization (Freire, nd: 4).

One of the tools used today to reduce costs are the contents of the public cloud, which allows a high level of cooperation, through a platform where systems can exchange information transparently. Public clouds set in internet, to coordinate innovation in production, processing, marketing and organization, provide a framework for trading activities business collaboration.

Organizations that survive over time become very efficient because they produce a limited number of services using standardized routines in collaboration.

- Online collaboration, at the strategic level, facilitates the control and decision-making, increasing profit and overcome competitors. In this regard, electronic collaboration can contribute to the value of organizations, such as increasing the performance of accounting, improving its strategic position and improving the implementation of organizational processes. This broader understanding online collaboration covers an understanding of the administrative, technical and organizational dimensions, called basic knowledge of collaboration systems. These skills include both behavioral approaches as a technique for the study of them, ranging from the organization and management to the technological. As for the organization, allowing the support for collaboration of the strategic, political, cultural and structural procedures; in connection with the administration, help make decisions and formulate action plans to solve the problems of organization and, with respect to technology, allow the construction of specific information systems, carefully designing and creating the set of services needed to work.

The most common of this service, at the strategic level, is electronic commerce, which represents new forms of digital management, both within organizations and outside them. Increasingly, the Internet provides a better core to technology, as you can connect to thousands of organizations in one network and create the foundation for a great digital market, while it is also connected to all organization members; and the way we work is changed, redefined business models, marketing processes reinventing, changing of organizational culture is encouraged and created closer relations with customers and suppliers.

Online collaboration helps managers to confront and resolve strategic issues and identify long-term trends for both the organization and the external environment. Its main function is to support the changes in the external environment with the existing capacity of the organization. Similarly, it also supports key business functions related to innovations in production and commercial and organizational processes.

It is noteworthy that online collaboration of strategic kind, requires an environment of trust where all members agree to cooperate and respect the mutual commitments. They must be willing and able to work together on the same goal to redesign some of their business processes so they can coordinate activities more easily.

We must emphasize that today's organizations rely on these new partnerships to enhance their planning, production and distribution of products and services. Also worth mentioning that the use of internet technologies, in this aspect of collaboration, provides a platform on which systems from different companies can exchange information transparently. Also, the networks established through the web, to coordinate inter-agency business processes, provide a framework for developing collaborative activities of online trading.

There the fact that the incorporation of new Internet technologies transforms the management of organizations in most developed new models of success stands. The emergence and spread of blogging, social networking and operating system, *free software* and digital communities are identified as technology changes; as well as clear indicators of social and cultural change and not only, or even primarily, new technologies, which allow you to do more efficiently. It should be noted that social networks make possible new organizations more quickly, participatory, open, transparent and truly focused on talent adaptation (Cabrera, 2013: 25).

In relation to our societies and cultures of the XXI century, these undergo a radical transformation, especially organizations in the strategic aspect, to reduce or eliminate barriers through technology. This actually makes new practices not only be between the levels of the organization, but they are also interfering internally in the corporation, such as collaboration, sharing and remixing of content as an essential part of creative process and the use of flexible licensing of intellectual property (Freire, nd: 4).

- Online collaboration in the management field helps solve strategic issues, combining the external environment with existing organizational capacity. It is noted that it plays an important role in the administration to create, identify and use data mining in applications based on communication, in order to strengthen the ability to perceive and respond to their environment. It also notes that electronic collaboration to an organization perceives and responds to its environment and promotes learning by identifying, capturing, encoding and the distribution of explicit and tacit knowledge. Electronic collaboration in the management field is critical to sponsor a collaborative work environment to share ideas and exchange documents and generate ideas in order to make good decisions. Today, a growing number of companies using intelligent tools such as balanced scorecard, in order to analyze the fulfillment of its mission of organization to achieve its vision, according to the principles of unity, loyalty, fairness, relevance and transparency and values of honesty, respect, tolerance, coexistence, solidarity and organizational justice. All this, to make decisions and formulate action plans in order to solve problems of organization. Managers perceive business challenges in the environment, set the strategy for solving problems and allocate human and financial resources to implement the strategy and coordinate collaborative work. They must exercise responsible leadership from beginning to end. Online collaboration reflects the hopes, dreams and reality of the directors of the real world in the twenty-first century.

MODALITIES OF ONLINE COLLABORATION

Over the past few years it has seen an enhancement in electronic collaboration within organizations and outside them, as a means of competition for survival. So says the Russian thinker Piotr Kropotkin Alexeyevich (2014) in his book *Mutual aid*: *UN factor in evolution*, in which exposes investigative results after exploring and researching the inclination of men to mutual aid. According to this researcher, mutual help has a very long history and is deeply intertwined with the whole development of mankind and men have preserved this trait to this day, despite all the vicissitudes of history. Kropotkin also states that cooperation leads humanity to a higher and harmonious economic relations, to meet the material needs without worrying about all the profit, product development and intellectual faculties and moral qualities ("mutual support", 2014). Therefore, we can assume that with the pass of time it has been tilting the functioning of organizations into a new ecosystem of organization, collaboration between users, customers, partners, employees, competitors, suppliers, etc. These partners or smart and organized crowds have an increasingly active role and, sometimes, inclusive and decisive. For this reason, organizations have been incorporated in its market, the optic of crowds at each of the levels, phases of design, development and marketing of products or services (Gutierrez-Rubi and Freire, 2013).

The specialist in strategy and direction of the organization, the Spanish Juan Freire, provides, in its article "Cultures of innovation and design thinking, new paradigms of business management", the basic features that make organizations build a new paradigm known as design thinking. These are:

- **Collaboration:** Especially those who have a different and complementary expertise to improve work processes and building consensus and coalitions.
- **Abductive:** By creating new options for finding new and better solutions to new problems.
- **Experimental:** Prototyping and increased hypotheses; contrast and cyclically repeating this process to find what works and what does not work, to manage risk.
- **Personal:** Consider the unique context of each problem and the users involved.

- **Integrative:** Perceiving the entire system and its connections.
- **Interpretation:** The emergence of a problem and possible solutions.

To Juan Freire, this design thinking is an emerging model approach in many different areas, ranging from the design of products / services or strategies and business models of conflict resolution. The design appears as a philosophy and a strategy for solving problems.

With regard to the process of solving problems and developing strategies for implementing the ideas of design thinking, they require a collaborative effort involving all levels of the organization to display prototypes with ideas and those ideas are turned into action quickly. It is a convenient convergence of strategies to address complex problems that affect large and diverse groups. These problems are not solved with simple solutions without the direct and active participation of customers or future users.

Thus, according to Freire (nd: 1-3), design thinking emphasizes process over the final results. The design of the contexts in which individual and collective labor and social is developed, interactions, particularly are the best way to achieve positive results and build innovative organizations. So organizational models should be the main objective of innovation aimed at a specific purpose. Creativity and innovation in services and products will be largely an indirect consequence of previous decisions and actions.

As regards to the new century, an environment of increasing complexity and uncertainty, it has risen economic analysis and business based on the study of business models originated. This led to a transformation of technological and innovative market, which occurs simultaneously, without neglecting the changes wrought in the markets of products or services or redefining the competitive field of global organizations, which have contributed to the so-called business models. They could see the light with open innovation, which is presented as a panacea for many, but, usually, is included in the same period of the processes, while sharing many similarities, they represent different organizational models: the widespread adoption of information technology and communication at all levels of the organization, organizational commitment to open cooperation between different levels of processes, as well as the implementation and exploitation transverse or longitudinal organizing their capabilities and technological knowledge controlling (Morcillo, 2011).

It should be noted that open innovation (Freire, nd: 9-10) is crucial for the two profiles and allows on one hand, reduce costs for the internal model, and, on the other hand, accelerate the innovation cycles in building new products or services, marketing and obsolescence, which are shorter. It is noteworthy that open innovation can accelerate the process and keep an organization in the global competitive market rate:

- **Crowdsourcing:** A model of organizational innovation which intends to use an external network. Therefore, broader, that could only be achieved with internal resources at all levels of talent and ideas to power a competitive organization that is unchanged from the forms of exploitation and commercialization of intellectual property.
- **An Open Platform:** In this case, the organization creates a platform to individual workers at different levels, to develop ideas and products or services. This model seeks to maximize creativity, and is organized, usually in modular processes.

Significantly, both models are not incompatible, quite the contrary; They can be mobilized in the same environment and share three characteristics: collaboration, networking and shared ownership; however, they can also vary greatly, as the first, which is the *crowdsourcing*, reduces costs and accelerates, while the second, which are open platforms, the goal is to manage and define intellectual property.

Regarding to *crowdsourcing*, subject marking the central focus of this article, it is a practice closely associated with the Internet and the emergence of a new digital culture in the development of cooperation projects between organizations (Gutierrez-Rubi and Freire, 2013). This concept was coined by wired magazine editor Jeff Howe, in June 2006, in an article entitled "Increased *crowdsourcing*". This author believes that it is a mechanism to openly call for an indeterminate group of people who are fit to perform the tasks in order to respond to the response to a complex issue and thus contributing to the cooperation of the freshest and most relevant ideas ("crowdsourcing", 2014).

In particular, *crowdsourcing* allows access to the execution of works or actions in the group of internet and linked to the business communities selflessly. This implies a change in values and culture; also, in the way we work, interact and communicate (Marquina, 2013). In his book, *The Fifth Discipline*, Peter Senge American engineer, ensures that collaborative work is vital to the organization because it provides the focus and energy to expand the capacity of creating and its effect awakens the commitment of the community formed online (online). Keep in mind that the collaborative work becomes part of a greater purpose, which is embodied in the products or services of the organizations. Accelerates learning with the help of information and communications technology linking the world through social networks and promotes freedom of digital connection. It also encourages collaborative work, experimentation and willingness to take risks. When we dive into a collaborative work, we know what to do, but often we have no idea how. The experiments are conducted because we think they will lead us where we want to (Cabrera, 2013A, Peter, 1995: 260-267).

Within the movement of the collaborative work, depending on the scope or the problem to be solved, with input from others, you can find different models of collaboration, which have clear components of commitment and objectives through the use of social networks and communities online and are focused on the connection and communication (Coleman, 2012). It is about new ways of doing business, which allows changing the culture of organizations towards a more humane behavior of its employees, which correspond to different levels, and by extension, to be more effective (Pesquera, 2013: 6):

- **Crowdfunding (Micromecenazgo):** Dates back to 2004, when French producer Guillaume Colboc and Benjamin Pommeraud launched an online fundraising campaign to finance the film *Demain la veille* (*Place of yesterday*). In just three weeks they managed to collect the money, With Which They Could shoot the film ("Micromecenazgo", 2015). This, Basically, Certain is to collect a sum of money to fund a private initiative; Until completion of the same, the employee Receives The acknowledgments or buy the product or service at a price much lower before it is released to the market. ("Emprendelandia.es", 2015).
- **Crowdcreation:** Basically involves the creation, creatively, an idea, however small, from a user community with the aim of collaborating. Current technologies have provided the basis for developing cooperation tasks and are a trend in which all participants are building a viable project ("Escritor de soja", 2014; EOI, 2012; Morcillo, 2011: 73).
- **Crowdvoting:** Its use originated in 1969, when a German television program promoted crowd voting ("Crowdvoting", 2014). It consists above all in the massive vote of contests of ideas or large surveys lasting a short time (Velasco, 2013).

The *crowdvoting* can range from the simple answer "like", "yes / no" to the extensive evaluations. Now it is very common to find sites like YouTube, Facebook or digital where surveys as contributions in accordance with the results of the vote, users dictate their preferences ("Crowdvoting", 2014).

The *crowdvoting* is a method used by organizations to know the point of view of the community on various topics. It also provides an accurate and cost effective to select the best concepts, ideas and directions as part of an evaluation process form ("Qmarkets", 2014).

- **Crowdwisdow:** This term appeared in 2004 in the book *The Wisdom of Crowds*, written by the American journalist James Surowiecki. But before that, in 1906, the English statistician and psychologist Francis Galton performed an experiment in a cattle trade show: there, asking the audience to estimate the weight of an open animal. Among the participants were skilled farmers and the general public. In that place, on the basis of knowledge and experience, there was a successful outcome, almost with little deviation from the true weight of the meat. Galton concluded that collective reliability criteria were more optimal than the individual arguments; therefore called *crowdwisdow* (wisdom of crowds). It is a statistical phenomenon by which individual knowledge or biases cancel estimate thereof in any direction, so that hundreds or thousands of individual cases mean surprisingly precise answers (Iuisi, 2012: 6-8). The effect could be further defined as described for decades: as a combination of information on groups, which ends in decisions that often are better than by a single member of the organization could have been achieved ("The wisdom of the crowd", 2015; "Wisdom of the groups," 2014). For proper operation, it should generate diversity of opinion, independence (there is no influence over the other), decentralization and aggregation; that is, you can move from personal opinions to collective decision (Gutierrez-Rubi and Freire, 2013).

In their studies, Surowiecki was able to identify three categories of economic problems (Iuisi, 2012: 8-9): the first is related to the objectives inherent to cognitive problems; these have or will have only definitive solutions, but are finite or multiple and some are preferable to others, such as the winner of a sports league or the most appropriate place to build a supermarket. The latter are recorded as coordination problems, where members of a group must coordinate their behavior with the occasions where everyone individually has the same objectives as the pricing of financial assets in the stock market. The third refers to cooperation where group members must cooperate even against their individual interests.

Crowdwisdow success is based on the diversity of the group formed, since the results are effective; independence, by providing new information most likely to succeed; decentralization motivated by the desire to work in a decentralized manner in trouble, so best collective solutions are obtained, and coordination that produce results of coordinated decision. Today, the importance of information and communications technology, facilitating largely the solution of the problems of coordination. It is noteworthy that the coordination of group decisions is much more feasible when the culture, norms, customs and social practices are shared by its members and the ongoing cooperation based on mutual trust (Iuisi, 2012: 13-19).

CONCLUSION

On the one hand, electronic or online collaboration, in recent years, has led to a change in the organizations of XXI century, has become a way to produce or provide services to challenge the economy which is increasingly global and participatory. Therefore, organizations are increasingly dependent on its technological and human resources, as well as its internal capabilities to respond in a coordinated manner to the priority needs of the global market through online collaboration.

As for information technology and communication, this has helped in the process of organizational integration, because for the innovations, have held continuous improvement, with offering products or services at a lower cost to a critical mass of defined customers.

Furthermore, electronic or online collaboration has conquered new sources to produce or deliver new services with minimal investment, thanks to the widespread use of information and communications technology at all levels of the organization and opening processes towards international cooperation, in order to keep pace with competitive market and achieve the strategic objectives set by the organization. This has been achieved with the *crowdsourcing* model. This model has served as preferred mechanism by the organizations of XXI century, and that helped, at minimal cost, to openly call a corporate group (as volunteers) of experts online to intervene to solve complex ideas innovative.

Within the model of *crowdsourcing*, there are variants for ideas they have achieved, and aims to achieve the objectives, as a mechanism to ensure that the organization has better support. Of these variants, we highlight the following: *crowdfunding, crowdcreation, crowdvoting* and *crowdwisdow*.

Finally, as a contribution that leaves this paper, XXI century organizations are increasingly open to electronic collaboration or online, thanks to the intensive use of information and communications technology and the advantages of using the internet. Within organizations, as time passes, a state of greater cohesion between all workers is experienced through online collaboration and aims to seek targets in an increasingly competitive and global economy. As it is well known, and there are no geographical boundaries, cultures and languages, and global dissemination of knowledge is a fact, thanks to electronic collaboration or online. Now, the challenges that will face the organizations of the future are related to processes, management strategies and human resources, because time and space are volatile, uncertain, complex and ambiguous. It notes that, in the future of organizations exist some changes to make good decisions, such as *crowdsourcing*, where new environments will be contemplated; technological leaps; changing preferences; changes in laws or regulations; changes in the economic, social, political and cultural context; renovations in communications and ecological transformations.

Also, the strength of the future of organizations will boost how to unlearn in order to learn new styles of online collaboration to make successful decisions in a short time, so is that true visionary become champions to face the fear of the various economic factors, habits, security, information and knowledge.

Thus, *crowdsourcing* is a new global challenge where the epicenter generates a diversity of cultures and knowledge and, most importantly, professional experts on specific topics will agglutinate. Note that these future communities will deploy advanced technologies that will require online work for the benefit of organizations.

REFERENCES

Cabrera, J. (2013, March 15). *Redarquía. Change management in the era of collaboration* [Slideshare]. Retrieved from http://es.slideshare.net/fullscreen/jcabrera/redarqua-gestin-del-cambio-en-la-era-de-la-colaboracin/1

Cabrera, J. (2013a, July 19). *Redarquía* [Slideshare]. Retrieved from http://es.slideshare.net/jcabrera/redarquia-el-orden-emrgente-en-la-era-de-la-colaboracin

Cabrera, J. (2013b). *Redarquía and organizational change* [Web log post]. Retrieved from http://cabreramc.files.wordpress.com/2013/02/11_redarquc3ada-y-cambio-organizacional.pdf

Capra, F. (1999). *The web of life*. Barcelona, Spain: Anagram.

Coleman, D. (2012, June 20). *The remains of social enterprises* [Web log post]. Retrieved from http://www.cmswire.com/cms/social-business/the-challenges-of-the-social-enterprise-016147.php#null

Collaborative Work. (2014, December 15). In *Wikipedia, the free encyclopedia* [Electronic version]. Retrieved from http://es.wikipedia.org/wiki/Trabajo_colaborativo

Crowdsourcing. (2014, November 24). In *Wikipedia, the free encyclopedia* [Electronic version]. Retrieved from http://es.wikipedia.org/wiki/Crowdsourcing

Crowdvoting. (2014, July 3). In *Wikipedia, Die freie Enzyklopädie* [Electronic version]. Retrieved from http://de.wikipedia.org/wiki/Crowdvoting

Emprendelandia.es. (2015). *Crowdfunding reward (reward-based crowdfunding): Collective funding for job creation*. Retrieved from http://www.emprendelandia.es/que-es-el-crowdfunding

Eoi. (2012). *Crowdcreating: conceptualization and production of the product or service (Campus EOI Sevilla)*. Retrieved from http://www.eoi.es/portal/guest/evento/1994/crowdcreating-conceptualizacion-y-produccion-del-producto-o-servicio-campus-eoi-sevilla

Freire, J. (2014, April 22). *Nomad* [Web log post]. Retrieved from http://nomada.blogs.com/

Freire, J. (n.d.). *Cultures of innovation and design thinking, new paradigms of management*. Universidade da Coruña and EOI School of Industrial Organization. Retrieved from http://laboratoriodetendencias.com/wp-content/uploads/2011/09/DesignThinking.pdf

Gutierrez-Rubi, A., & Freire, J. (2013). *Manifesto crowd. The company and the intelligence of the crowds*. Retrieved from https://books.google.com.co/books?id=7XIkJWE7qeoC&printsec=frontcover&vq=Crowd+wisdom&hl=es&source=gbs_ge_summary_r&cad=0#v=onepage&q=Crowd%20wisdom&f=false

I'm a Writer. (2014). *What is crowdcreation?* [Web log post]. Retrieved from http://www.soyescritor.com/?q=crowdcreation

Iuisi. (2012). *The use of the collective wisdom and operational means of intelligence analysis*. Retrieved from http://www.iuisi.es/15_boletines/15_ISIe/doc_ISIe_13_2012.pdf

Jones, G. (2008). *Organizational theory. Design and organizational change*. Mexico City, Mexico: Prentice Hall.

Joyanes, L. (1997). *Cibersociedad*. Madrid, Spain: McGraw-Hill.

Laudon, K., & Laudon, J. (2004). *Management Information Systems*. Mexico City, Mexico: Prentice Hall.

Marquina, J. (2013, April 18). *Collective intelligence: crowdsourcing* [Message in a blog]. Retrieved from http://www.julianmarquina.es/tag/web-social/

Medellin, C. (2013). *Build innovation*. Mexico City, Mexico: Fese.

Micromecenazgo. (2015, January 6). In *Wikipedia, the free encyclopedia* [Electronic version]. Retrieved from http://es.wikipedia.org/w/index.php?title=Micromecenazgo&oldid=79243220

Morcillo, P. (2011). *Natural innovating. The pass says it all.* Retrieved from https://books.google.com.co/books?id=9Fj3pIPqmzgC&pg=PA73&dq=Crowdcreating&hl=es&sa=X&ei=xyCgVMa8IZHUgwTGoILgAQ&ved=0CDIQuwUwAw#v=onepage&q=Crowdcreating&f=false, 73.

Mutual Support. (2014, December 11). In *Wikipedia, the free encyclopedia* [Electronic version]. Retrieved from http://es.wikipedia.org/w/index.php?title=Apoyo_mutuo&oldid=78703967

Pesquera, M. (2013, September 3). *The programmable university in scale-free networks.* [Slideshare]. Retrieved from http://es.slideshare.net/mapesquera/la-universidad-programable-en-redes-libres-de-escala-map-2103?next_slideshow=1

Peter, S. (1995). *Fifth Discipline.* Barcelona, Spain: Granica SA.

Piotr Kropotkin. (2014, December 30). In *Wikipedia, the free encyclopedia* [Electronic version]. Retrieved from http://es.wikipedia.org/w/index.php?title=Piotr_Kropotkin&oldid=79108122

Qmarkets. (2014). *Learn from the wisdom of the crowd* [Web log post]. Retrieved from http://www.qmarkets.net/additional-products/crowd-voting

Velasco, J. (2013, February 7). *Think Big* [Web log post]. Retrieved from http://blogthinkbig.com/crowdsourcing-colaboracion-motor-ideas

Wikinomics. (2014, August 31). In *Wikipedia, the free encyclopedia* [Electronic version]. Retrieved from http://es.wikipedia.org/wiki/Wikinom%C3%ADa

Wisdom of Crowds. (2014, August 5). In *Wikipedia, the free encyclopedia* [Electronic version]. Retrieved from http://es.wikipedia.org/wiki/Sabidur%C3%ADa_de_los_grupos

Wisdom of the Crowd. (2015, January 3). In *Wikipedia, The Free Encyclopedia* [Electronic version]. Retrieved from http://en.wikipedia.org/wiki/Wisdom_of_the_crowd

Section 2
Information System:
Design, Methodologies, and Strategic Thinking

Chapter 9
Developing Project Team Cohesiveness in a Virtual Environment

Lisa Toler
Ashford University, USA

ABSTRACT

As more projects require the specialized technical skills of those who work in virtual environments due to dispersed geographic locations, project managers of these distributed virtual teams (DVT) must gain insight into achieving project success amongst team members who hold varying operational and world perspectives. When organizational managers decide to implement virtual teams (VT), can they develop strategies to overcome the lack of social interaction, cultural differences, and preconceived notions that can hinder the development of a collaborative and cohesive team? In addition, leading DVTs in a manner that encourages collaboration, diversity, competency building, open communication, and overcoming feelings of isolation must be met in this technology-based environment. This chapter addresses the dilemma of managers in which they must have a clear understanding of what communication and relationship-building techniques and management systems are best suited.

INTRODUCTION

The emphasis of this chapter is on dispersed project management team members who rarely meet face-to-face and how cohesiveness in performance and expectations can be achieved. In addition, a key discussion in this chapter is to gain knowledge that can assist virtual team leaders and organizational project managers in understanding individual perceptions of project success for virtual teams, based on the individual perceptions of team members. The chapter provides insights into what behaviors, tools, and techniques are indicators to virtual team members that they are effectively attaining project objectives. The chapter may help organizations determine where to focus training initiatives for project managers and virtual team members and what communications systems project managers need to use to achieve a cohesive understanding of project objectives for effective performance in the absence of face-to-face communication.

DOI: 10.4018/978-1-4666-9688-4.ch009

To achieve these objectives, seminal research on different aspects of the dispersed virtual project management team were reviewed. Specifically, beginning with Eveland and Bikson (1988) as to what defines the virtual team and to distinct elements identified by Jarvenpaa and Leidner (1999) that help or hinder the virtual teams' ability to achieve project success and trust effectively has been evaluated. The literature reviewed in this chapter has also included scholarly works by Jarvenpaaa and Leidner (1999) and Kayworth and Leidner (2000) that examine communication and the role of a leader in the virtual team.

The original research conducted and the studies reviewed for this book chapter show that virtual teams are central to the success of today's organizations that must compete in a worldwide marketplace to understand how the global virtual team is affected by technology. However, scholars examining the changeable nature and individual perceptions of the dispersed virtual project team itself suggest that there is a need to understand how to effectually encourage a collaborative and collectively minded environment. This is key to gauging the future success of the team and the health of the project in which the members contribute to.

BACKGROUND

Project management as a profession caught on in the mid-1950s and the discipline, methods, principles, and tools have evolved over the ensuing years and decades (Padar, Pataki, & Sebestyen, 2011). As more organizational leaders see the need to develop higher benchmarks for implementing strategic planning and tactical execution of projects, senior leaders are endorsing the standardized tools and techniques espoused through professional project management guiding principles. The key benefits of utilizing project management strategies include not only focusing on measuring project success through schedule, budget, and scope components, but additionally by focusing on developing the practitioners' level of insight into effective leadership and communication skills.

More clarity is needed to explain how effective project managers use team communication and relationship-building and management elements in dispersed virtual project teams to influence the overall project success when team members have less face-to-face interaction. In the 1990's, Hallam (1997) conducted a quantitative study of 2,000 team members from all across the United States and from various industries in an effort to understand the complexities of work team structures and dispel common beliefs about team effectiveness.

Hallam (1997) illustrated one such commonly held belief that as individuals work together in teams, they develop one general idea about what is going well and what is not going well on a team. The ability to maintain independent thought was believed to hinder the work teams' capacity to come together in a collective effort to achieve project success. Findings showed that each member has their own experience and perception as to what is working successfully and what is not in team projects (Hallam, 1997). However, what we need to understand thoroughly is if this same finding holds true on a dispersed virtual project team where constant communication is lacking and an interrelationship exists in which one change is likely to influence multiple factors and end results.

In addition, another belief examined by Hallam (1997) was that people typically dislike working on teams. This myth too has been dispelled as research findings indicated that surveyed team members enjoyed the sense of community and excitement of sharing ideas with others in a team (Hallam, 1997). Yet, the research has not gone far enough into the exploration of the dispersed project team environment, which does not have the benefit of that same sense of community.

Studies from the early 2000s that explored the dispersed virtual team's ability to collaborate and communicate identified several challenges. Previous articles (Hayes et al., 2000, Feinberg et al., 2004) tend to support the findings that in order to enable collective performance of virtual teams, the sole contributing factor is a higher level of team member effort. However, in more recent articles from 2009 to 2011, topics surface from the standpoint that varying types of team member effort facilitate different outcomes such as technical assistance, team skills, organizational structure, time in meetings, extent of virtualness, and benefits for the virtual team members (Berry, 2011; Stark & Bierly, 2009, Turel & Zhang, 2010; Wells et al.).

According to Sivunen (2006) individuals who are part of the dispersed virtual team do not spend time developing social or interpersonal skills because most of their time and effort is spent working tasks through computer-mediated methods. Sivunen (2006) investigated through a qualitative study focusing on four virtual team leaders how they attempted to strengthen team identification through computer-mediated communication. Due to individual members having different cultural backgrounds and working often times in different countries, the role that technology plays in team identity and developing common goals becomes key to the virtual team leaders' ability to provide feedback and promote a cohesive atmosphere (Sivunen, 2006).

Likewise, in a quantitative study on virtuality and technology's impact on team performance in a globalized business world, Ahuja (2010) found virtual teams faced complex issues in work coordination, trust, and team participation. The reliance on technology to become successful is only effective if virtual team members can overcome working in a less personal atmosphere. This can be achieved by developing high levels of trust, learning to communicate clearly, having strong leadership in place, and the application of appropriate levels of user-friendly technology (Ahuja, 2010). Since teams working in a virtual environment must contribute collectively to these areas on a continual basis, Ahuja (2010) found that the lack of user hardware skills, software training, and computer knowledge could hinder the team's ability to overcome the drawbacks that may impact virtual teams.

In general, the bulk of the studies on virtual teams have been quantitative and the focus has been on leadership and information technology (Eveland & Bikson, 1988; Zigurs et al., 1988; Sprague & Greenwell, 1992; Cross & Rieley, 1999' Markus et al., 2006; Daim et al., 2012). In these cases, the sample of participants in the research has been work teams or organizations. Conversely, the focus of studies that have been qualitative in nature has been on communication and trust in the virtual environment (Jarvenpaa et al., 1998; Ahuja & Galvin, 2003; Ardichvili et al., 2003).

Though these previous studies provide the foundation for this chapter, the relationships and roles of the virtual team as it relates explicitly to the organization or to technology have been explored. This chapter provides an in-depth and critical review of the literature that is germane and contributes to the understanding of how different individual perceptions of project effectiveness of virtual project management teams can lead to collective project success when teams are distributed and have limited face-to-face interaction. In addition, the findings from the author's own qualitative research study will be brought into the discussion. The chapter explains these findings and explores various bodies of literature that study and makes clear the dispersed virtual project management team's ability to effectively communicate and work together in a cohesive manner to achieve mutual and collective project goals.

MAIN FOCUS OF THE CHAPTER

Issues

Many project management elements can have a negative or positive influence on the dynamics of the project management team and the final outcome of the project itself. The consequence of the role that the stakeholder plays is an important concept to appreciate when examining the significance of the interaction within the dispersed project team. This is due to the reality that the constraints of the dispersed team are magnified by the fact that not only do the individual members have to achieve cohesion amongst themselves, but they must also proactively manage and communicate the needs of each key stakeholder with limited face-to-face interaction.

When organizational managers decide to implement virtual teams within the company, can they develop strategies to overcome the lack of social interaction, cultural differences, and preconceived notions that can hinder the development of a collaborative and cohesive team? In addition, leading dispersed virtual teams in a manner that encourages collaboration, diversity, competency building, open communication, and overcoming feelings of isolation is a challenge that must be met in the technology-based environment of the virtual team. This research addresses the dilemma of managers and leaders in which they must have a clear understanding of what communication and relationship-building techniques and management systems are best suited to meet the needs of the dispersed virtual project team to achieve both project and organizational objectives.

As Table 1 demonstrates, a limited degree of theoretical research of virtual teams began in the late-1980's defining virtual teaming and applying quantitative methodology to assess virtual teams against face-to-face work groups (Hollingshead, Mcgrath, and O'Connor, 1993). However, the exploration and bulk of research on this subject matter increased during the mid-1990s and continues to be an active area of research today due to the cost benefits and the growth in internationally based workforces.

The focus of these earlier investigations of the topic of study in the field of Organization and Management mainly encompassed two foci: (i) communication and trust in global virtual teams (Jarvenpaa & Leidner, 1998) and (ii) effectiveness of the virtual team (Furst, Blackburn, & Rosen, 1999. These studies provide the foundation for investigations into the definition and viability of the virtual team and the management of these dispersed teams.

The researcher conducted a qualitative single case study. The purpose of this study was to broaden the project management knowledge base about dispersed virtual project teams toward an improved understanding of individual perspectives of effective project team performance. Specifically, the researcher sought to understand through interviews and surveys with virtual project team members and leaders how

Table 1. Categories of prior researcher on virtual teams

Dates	Categories
1988-1999	Defining and Developing the Virtual Team
1988-Current	Leadership and the Virtual Team
1988-Current	Information Technology and the Virtual Team
1998-Current	Communication and Trust in the Virtual Team

individual perspectives of project effectiveness can lead to a cohesive understanding of project objectives and a collective understanding of success when members are not co-located. To achieve this goal, an examination of the relevant literature throughout all phases of this study was conducted.

Literature Review

This section provides an in-depth and critical review of the literature that is germane and contributes to this research study. It explores various bodies of literature that study and explain the dispersed virtual project management team's ability to effectively communicate and work together in a cohesive manner to achieve mutual and collective project goals. A review of the literature indicated several key areas that were relevant to the foundation for conducting this study in addition to providing supporting material for key constructs. The ensuing sections will critically review studies that explore positive and negative affect on how an individual perceives the success of a project. Pertinent bodies of literature include the following:

1. Defining and developing virtual teams,
2. Group dynamics and leadership of the virtual team,
3. Developing trust and communication in the virtual environment.

During the course of the literature review, it has become evident that there continue to be multiple definitions of the *virtual team* and an indeterminate set of models for assessing the dynamics within the group that affects project performance and dispersed virtual project team effectiveness. Therefore, the literature on defining and developing virtual teams became necessary for laying the history of and foundation for the theoretical views that have formed the definition and classification taken for this study.

The phenomenon investigated in this study is a relatively new one, and thus, the critical review of literature examined studies conducted from the earliest relevant studies conducted in 1988 on information technology available at the time to the most current research dynamics and leadership of the virtual team as a group. The researcher identified in the review of the literature gaps in the knowledge base on providing answers to these issues that confront the dispersed project team. For example, both seminal and contemporary virtual project team research has focused primarily on a quantitative methodology for assessing trust and communication (Jarvenpaa and Leidner, 1998); information technology for global virtual work teams (Eveland and Bikson, 1988); and development of a virtual office (Davenport & Pearlson, 1998). While other research in this area has produced quantitative literature that has mainly been centered on how technology affects the operational environment of the global virtual team (Townsend, DeMarie, & Hendrickson, 1998). These issues and gaps are identified and discussed within each segment of the literature review.

Defining and Developing the Virtual Team

To acquire a better appreciation of how the virtual team is defined in the literature, an examination of the various perceptions and characterizations in the scholarly writings on the virtual project management team was made. As the research on distributed virtual teams is a fairly new phenomenon, a specific delimiting time frame from 1990 to 2013 was used to conduct the literature review in this section. In order to determine which literature should be retained for a pertinent discussion on progression of this idiom, the researcher established criteria that: the literature communicated the specific differences between traditional teams and the virtual team, social and cultural concepts and dynamics, and specific characteristics of the virtual team.

In order to assess the scholarly interpretation of the virtual team, Powell, Piccoli, and Ives (2004) conducted a review of the current literature between 1998 and 2002 on virtual teams. They uncovered several definitions in the current literature of virtual teams and arrived at their definition through an analysis of 43 papers. The most common and preferred definition of the virtual team is that of a group of separated individuals that are physically, organizationally, and time-dispersed workers brought together to achieve one or more tasks by information and through telecommunication technologies (Powell et al., 2004).

According to this definition, the virtual team is brought together in response to a specific need or specialized function in which the scope of work provides a beginning and end date. At the core of the dispersed virtual team is the use of technology for communication. The traditional team and team member can successfully and effectively communicate because they are typically located near one another. This provides repeated opportunity for frequent feedback and coordination of tasks between individual members. However, the researchers found that because the virtual team encompasses members who are disbursed, face-to-face meetings do not occur frequently or more often than not do not take place at all.

Powell et al. (2004) posit that this lack of face-to-face communication makes it complicated and difficult for distributed teams to exchange information in a timely manner. In addition, cultural differences can foster an inability for the virtual team to develop a concrete social structure. This, in conjunction with time-dependent issues, different social schedules, and the lack of formalized training can create difficulties in the virtual team that traditional teams do not come upon (Powell et al., 2004). This can include conflict resolution, cohesive understanding of project objectives, risk response, and overcoming challenges to the project by the influence of the team members' organization or stakeholders. However, as Powell et al. (2004) uncovered, because the virtual team represents a progressive form of the workgroup for organizations that is flexible and responsive in a global environment, understanding the concept, theory, and nature of the dispersed virtual team is critical to team viability and effectiveness.

Additional research found in the literature that examines and defines virtual teams that rely on technology-mediated communication is that of Martins, Gilson, and Maynard (2004). Noted by Martins, et al. (2004), earlier definitions of dispersed virtual teams only sought to compare and contrast virtual teams and face-to-face teams. However, research conducted in the early 2000's to date focuses specifically on defining the virtual team in a *real world* organizational setting by examining team processes and team outcomes while faced with time, location, social network and organizational boundaries (Martins et al., 2004). Thus, the evolving definitions include the previous dimensions of the virtual team but also highlight the fact that virtual teams are a form of team first while integrating *virtualness* as a team characteristic. Therefore, Martins et al. (2004) focused their research on member satisfaction and the effects of virtual interaction through team composition.

Most of the literature on virtual teams prior to this such as that of Weisband (1992); Lee and Spears (1992); Daly (1993); Straus and McGrath (1994); Bouas and Arrow (1996); Tanm Wei, Watson, et al. (1998); examined the effects of virtual interaction on team effective outcomes and performance outcomes such as effectiveness and decision quality. This earlier research posited that in order for member satisfaction to occur on the virtual team the virtual interaction was reliant on the nature of the task and on team make-up. Martins, et al., (2004) found that in order to encourage progress in the understanding of dispersed virtual teams and in order to move the research forward, studies need to be more focused on *team-ness* in conjunction with *virtualness*.

Based on a compilation of the research articles cited thus far, is applied to identify the key differences between the dispersed virtual team and the traditional team (Table 2).

This guiding theory of the virtual team as a complex and evolving group enabled researchers such as Kayworth and Leidner (2002); Huang, et al. (2010), Wei, et al., (2002); and Martins, et al., (2004) to better understand the relationship between virtual team performance and outcomes. For example, Martins, et al., (2004) found that prior research conducted by Jarvenpaa, Knoll, and Leidner (1998) on virtual teams ignored the role of time on group processes and outcomes. However, upon closer examination, more recent studies such as that of Schmidt, Montoya-Weiss, and Massey (2001) have found that virtual team members' satisfaction with the teams processes and results increased with time. In addition, social context also affected virtual interaction on team outcomes. Groups whose members were more judgmental of one another produced the greatest amount of innovative solutions. However, the more supportive virtual team groups had the most fulfilled members and their perceptions of the levels of effectiveness were greater than those of non-supportive groups or face-to-face groups.

In referencing and advancing the findings of Powell et al. (2004) on virtual team member commitment, Kuruppuarachchi (2009) led a qualitative case study to demonstrate the function of virtual team models within an organization and reviewed the benefits and drawbacks of the virtual team versus the face-to-face or traditional team. Like Powell et al. (2004), Kuruppuarachchi (2009) defined virtual teams as those that exist in various time zones, cultural boundaries and spatial limitations; while contributing to a common intent and through the use of technologies to communicate effectively. The benefits outlined in the course of this research were the possibility to recruit talented employees, incite creativity and resourcefulness among individual team members, and create equal opportunity in the workplace.

The research of Kuruppuarachchi (2009) found that virtual teams also create the possibility of having more flexibility in resource allocation, creativity due to diversity, and taking advantage of the accessibility of a large pool of expertise irrespective of location or distance from team member to team member. In addition, the researcher posits that employees experience more flexibility with working hours and the organization can have financial gains through reduce costs and improve productivity. However, the researcher also identified drawbacks of the virtual project team that were uncovered in implementing a project as a case study.

During the case study, Kuruppuarachchi (2009) found that linked to difficulties in overcoming problems was the lack of direct interaction and personal contact with immediate managers. These dispersed virtual project team drawbacks for the organization were found to be ineffective communication due to the lack of face-to-face communication, loss of vision or common understanding of project objectives, structure may not fit the organizational or operational environment, lack of permanent reports, supervision and

Table 2. Virtual team vs. traditional team

Virtual Team	Traditional Team
Meets face-to-face infrequently	Meets face-to-face regularly and as needed
Distributed organizational, nationally, globally	Co-located near one another
Relies heavily on electronic means of communication	Uses electronic means of communications but does not rely on it
Individual team members work independently on project tasks that have a formal project beginning and project end and closeout. Leaders and members utilize formal project management tools and techniques	Individuals may or may not work on formal projects and may or may not use formal project management tools and techniques

monitoring and performance management or quality control can be difficult, and the requirements of training and developing skills of individual members to work in the virtual team (Kuruppuarachci, 2009). All of the above studies are key to laying the foundation for defining how we look at the dispersed virtual team, and how we currently view the importance of these types of teams in today's global marketplace (Powell et al., 2004; Martins et al., 2004; Kuruppuarachci, 2009).

Dynamics and Leadership of the Virtual Team as a Group

As technology developed to enhance the global communities' ability to communicate more effectively, scholars such as Zigurs, Poole, and DeSanctis (1998) and Cross and Rieley (2009) began to examine the sense of community within the global virtual environment and assess the rapid change in the ability or inability of dispersed virtual teams to exchange information. Thus, when examining the literature on the dynamics of the dispersed virtual project team as a group, empirical research on computer technology and the amount of behavioral influence it has on these types of groups must be discussed. Seminal research conducted by Zigurs, Poole, and DeSanctis (1988) examined effects of computer-supported group decision-support systems (GDSS) and its relationship to influencing behavior. The researchers defined influence behavior as verbal, nonverbal, and group-messaging actions that attempt to control the direction of group behavior (Zigurs et al., 1988).

Zigurs et al., (1988) conducted a quantitative study analyzing 32 U.S. work groups. In the course of this study, influence behavior was measured by summing both verbal acts and non-verbal procedural acts. Zigurs et al. (1988) hypothesized that due to the anonymity of computer technology, groups spent significant effort on issues and ideas rather than the individual who presented those ideas to the group thus having a positive outcome on project effectiveness and performance. In addition, group dynamics were found to be more collaborative and cohesive due to the use of computer-aided technology. This is because external status characteristics are more neutralized through this form of communication. However, these findings also indicate that computer support did not result in higher quality decisions being made. The research also uncovered that computer-supported groups spent substantive time communicating goal-oriented procedures and tasks in an effective manner.

In order to further explore the initial findings of Zigurs et al. (1988) on the effects of technology on team dynamics and team learning, an additional search of the literature resulted in review of a conceptual article written on team learning best practices and tools. Cross and Rieley (1999) conducted research that hypothesized that by improving team learning on how to tackle important tasks on a project basis, the team's ability to succeed synergistically would increase and lead to project success. Since organizations have turned more toward team-based structures to improve collaboration within their companies, project and organizational leaders need to implement practices that support learning best team practices (Cross & Rieley, 1999).

Training

A major component of this research study is to uncover what management skills and tools can help the virtual project team collectively reach project success. There is an assumption that the lack of sufficient formalized training in the area of project teams and project management lead to stress and frustration for project team members. Further hypothesized is that this may also lead to the project team's inability to work cohesively especially in cases where the team members are not co-located or dispersed and lack the day-to-day and face-to-face communication that can help maintain collaboration and cooperation for the team to achieve project success.

Sprague and Greenwell (1992) suggest that traditional employees require more frequent and regular training to work in virtual project teams. Though managers recognize the value of project team and project management education because of the increased usage of project management within their organizations, the results of their survey indicated that over 40% of participants had no previous training with project teams or project management. In addition, participants who had had previous training had not been provided updated training or education in four to seven years. This lack of formalized training in specific project management areas of focus such as project planning, project controls systems, and reporting practices, for example, decreases the organization's ability to become and remain competitive in a global marketplace.

Team Learning

The qualitative research of Cross and Rieley (1999) found that there were several practices that inhibited teams from learning and negatively impacted the team's dynamics and performance on projects. A survey of the members of 22 project teams in professional services, financial services, and manufacturing organizations was conducted. These members had recently completed projects and were asked what they learned from their experiences and what either prohibited their learning or facilitated it (Cross & Rieley, 1999). The researchers found that if leaders and managers established a strict division of labor, individual opportunities for learning and interaction with other team members became restricted. In addition, team members who became less engaged with task assignments ultimately became less attentive to their jobs. Therefore, leaders must find ways to stimulate the interests of individual team members. This can be accomplished through individual learning through development of other important skills that are important to the success of the project as a team (Cross & Rieley, 1999).

Cross and Rieley (1999) recommend that in order to optimize team or group learning and thus increased potential for project success especially within the dispersed virtual team, team members must understand how they fit into a team. Having an understanding of how the job they are doing is related to the overall team's mission increases the team's and the individual's ability to accomplish the project's mission (Cross & Rieley, 1999). However, the researchers caution that in order for the individual team member to have an accurate view of how they fit within the team, they must feel empowered with an awareness of control, encouraged by knowledge, and a sense of self-efficacy.

In a separate study conducted that same year on virtual team member empowerment, Kirkman and Rosen (1999) linked this phenomenon to higher levels of productivity, quality, job satisfaction, and organizational commitment. Empowerment in work teams is defined as having four critical dimensions. The first dimension is *potency* or competency in performing specific tasks for generalized project effectiveness. The second dimension is *meaningfulness* or the ability for team members to collectively share the team's experiences as having value and worth. *Autonomy,* or the individual's ability to freely and independently make choices that are executed by the team, is the third dimension of team empowerment. Finally, *impact,* or the team member's ability to produce work that is viewed as significant to an organization, enables a collective understanding for teams on both individual and team project success (Kirkman & Rosen, 1999). The quantitative field study conducted by Kirkman and Rosen (1999) examined the multidimensionality of team empowerment and the relationship between empowerment and project outcomes. The researchers found that both training and higher levels of empowerment in decision making increased team member communication, motivation, and collective coordination of activities.

To further understand the phenomenon of multidimensionality on individual perceptions of time and its effects on meeting deadlines, a review of the literature uncovered the research of Waller, Conte, Gibson, and Carpenter (2001). Waller et al, (2001) postulate the need for dispersed teams to make effective decisions in a timely manner in order to achieve project success. However, their review of the literature found that there are two time-oriented individual differences that influence team members' perceptions of deadlines that could hinder a successful outcome for the project. These two differences involve *time urgency* and *time perspective*, which influences the ability of the team to work under time pressure and meet deadlines (Waller et al., 2001).

The Waller et al. (2001) literature review of existing theories posited by Hambrick and Mason, 1984; Gersick, 1989; Conte et al., 1995; Waller et al., 1995b; Waller et al., 1999a, discovered that time urgency is associated with task prioritization and scheduling of tasks within and allotted timeframe. Individuals who are time-urgent tend to schedule more activities than can fit into the available amount of time. These individuals often set their own goals and use deadlines to measure remaining time resources. Individuals who are time perspective concentrate more on either past, present, or future time (Waller et al., 2001). Culture, religion, family, work background, and education all may influence in individual's time perspective. If an individual has a more present-time perspective, he or she will focus less on future-oriented pre-emptive thinking than other individual team members. The differences between these types of individuals' behavior can lead toward miscommunications, conflict, and ineffective performance to collectively meet deadlines (Waller et al., 2001). These differing perceptions of deadlines and time emphasizes that in order to increase the distributed teams' ability to be effective, especially in terms of the decision-making process, team autonomy that satisfies all team members' feelings of authority, responsibility, and accountability must be *formally* incorporated into the management and leadership of the dispersed virtual team's daily practices.

Ellemers, Gilder and Haslam (2004) conducted research on motivating individuals and groups at work as it relates to social identity and group performance. Their research referenced the work of Kirkman and Rosen (1999) and sought an understanding of motivation through individual social identity processes and how those individual and group processes interact to determine work motivation. In examining several individual and group motivational theories, the researchers discovered that the more contemporary work situations such as we find with dispersed virtual project management teams, a situational and dynamic identity exists for individuals. Understanding and anticipating individual behavioral shifts and how individuals perceive themselves as a collective or work team and their social identity can positively or negatively impact the group's performance as a whole. Learning to understand the circumstances in which group and project team members sustain their efforts and motivation on behalf of the group can raise project team leaders' and project managers' ability to gauge project performance and establish a common goal for the workgroup. However, the virtual team must also learn to function within the larger organization.

Markus, Manville, and Agres (2006) researched how performance and behavior in an environment that is becoming more *open source* or virtual can be successful. Their findings posit that in order to ensure success through organizational order, there must be intrinsic motivation and self-management of the organization's virtual employees. This can be accomplished through economic rewards that emphasize benefits to the individual and emphasize collective successful performance. In addition, key rules that define appropriate conduct that benefits the collective effort must be maintained within the virtual group. In this regard, communication technology that is required to gauge performance and enforce rules must also be used to influence the level and quality of work and the projects direction (Markus et al., 2006).

Decision-Making Process

Bourgault, Drouin, and Hamel (2008) conducted an empirical study using a quantitative methodology to examine the importance of a formal decision-making process within highly dispersed distributed project management teams and its effect on collaboration at the project level. Using an online questionnaire given to project managers based in Canada, 149 responses were reviewed. Historically, the researchers found that alliances at the project level mainly involved a network of dispersed teams and individual team members who were enthusiastically involved in shared endeavors and resources (Bourgault et al., 2008). However, new challenges for project managers have been created with the expansion of competition to a global marketplace to facilitate a collaborative and effective work environment for distributed project management teams. Bourgault et al. (2008) found that collaborative relationships and joint ventures provide project managers and organizational leaders with a competitive advantage if they can understand the distinct dimensions and challenges that exist within the dispersed virtual team. The project management perspective on decision-making processes includes planning and executing projects within a clearly defined time frame (Bourgault et al., 2008).

Bourgault et al. (2008) posit that the individual's and group's ability to make sound and effective decisions in a dispersed project management team is difficult for several reasons other than stakeholder buy-in and directives. First, traditional interpersonal cues that can help to facilitate communication are limited for the virtual team by the use of communication technology. Second, team effectiveness can be hindered by the fact that there is no past history amongst the team members. This diminishes the individual team member's ability to interpret other team member's thought processes that lie behind decisions that are made for the team (Bourgault et al., 2008).

Global competitiveness can be strengthened through telecommunications technology. Virtual teams rely heavily on this type of technology because it makes it possible to bring together individuals on the project team who are separated across the globe. Brandt, England, and Ward (2011) examined the cultural and process dynamics of the virtual project management team through best telecommunications practices. The authors discovered through a review of the literature on the topic of virtual teams that there are several common themes and key differences that drive the need for virtual teams to work in different ways.

The first of these common themes includes tools and techniques for communication. Communication issues for the virtual project management team such as through e-mail and videoconferencing required establishing explicit rules of engagement (Brandt, England, & Ward, 2011). Unlike project team members who are co-located, it is highly beneficial for the dispersed team member to establish what will and will not be done when collaborating with one another so that the team remains effective during the course of their electronic communications. The main impetus for establishing these rules upfront stems from cultural and language differences that are found in the virtual project management team.

Cultural differences, especially those language and behavioral differences that are unintended to be interpreted as non-inclusive and rude, break down the virtual project team's ability to be effective and cohesive (Brandt, England, & Ward, 2011). The effective virtual team learns to successfully interpret social behaviors based on the varying degree of cultural norms that can exist within a global virtual team.

Similarly, project managers and team leaders should consider along with knowledge and availability when assembling their virtual teams the ability for individual virtual team members to exercise good social skills. This establishes the foundation for effectual exchange of information and a higher degree of effective team performance.

Virtual Team Leadership Effectiveness

The effect of leadership on team collaboration in virtual teams is another key construct when reviewing literature on virtual project management teams and project effectiveness. Adaptive abilities in advanced information technology and leadership of dispersed virtual project teams is the subject of research conducted by Avolio, Kahai, and Dodge (2001). They conducted a literature review of research to uncover how advanced information technology (AIT) could influence and is influenced by leadership. The authors found that the ability for team members to work together along with team leaders to accomplish project goals was enhanced by the level of access to new technology in order to develop knowledge. Understood by Avolio et al. (2001) is the need for both leadership systems and AITs to *co-evolve* over time in order to elevate the team's development and performance. The authors also reviewed the literature on collaborative group technology that could be used to enhance trust and interaction in virtual teams thereby leading to project success. They also came to the conclusion as with Huang et al. (2010) that a transformational leader who has the ability to inspire and motivate through their communication abilities creates an atmosphere that results in an increase in information exchange among virtual team members. The effective virtual team leader, according to Avolio et al. (2001), has the ability to instill confidence in the individual's ability to perform tasks, and by implementing the use of a media rich environment for the virtual team, augments the teams' intellectual stimulation, which maintains focus and promotes motivation to achieve success.

Kayworth and Leidner (2002) inquired through their qualitative study what the role and nature of team leadership is in the virtual setting. Their conclusions, based on the creation of 13 virtual teams who each were assigned mandatory tasks to complete, postulate that the virtual team leader who exhibited a mentoring role and articulated role relationships among the virtual team members were highly effective at achieving project success. In addition, the researchers found that the nature of the technology used has significant influence on linking team members together. Thus, it is crucial for the team members' to adapt to advances in information technology. This finding is key to further exploration of the connection between a collective achievement of project success, project planning, and team communication to leadership that is part of the subject of this thesis.

Similarly to and advancing the research conducted by Kayworth and Leidner (2002), Huang, Kahai, and Jestice (2010) explored the interaction between leadership styles and the influence on virtual team performance in decision-making tasks. The more recent findings of Huang et al., (2010) substantiate the previous findings of Kayworth and Leidner (2002). The researchers conducted a laboratory study of 485 undergraduate students and grouped them to mirror real-world, short-term virtual teams that did not know each other. Huang et al., (2010) define a cooperative climate as one in which there is a feeling of supportiveness from team members and leaders and that participation in discussions was safe. The researchers examined two types of leadership behaviors in virtual teams and how they relate to and influence team collaboration. This included the transactional leader and the transformational leader. Their results indicated that certain leadership behaviors could improve and sustain a cooperative climate within the virtual team.

The research showed that the transactional leader influences followers by exchanging rewards for performance. This type of leader clarifies roles and expectations and provides team members with rewards based on meeting tasks goals. The behavior of the transactional leader tends to improve cohesiveness of task objectives within the virtual team. On the other hand, according to Huang et al. (2010) the transformational leader does not focus on individual performance. The transformational leader helps team members evolved past their own self interests and connect to the group's mission by emphasizing the identity of the collective in order to achieve project success effectively. Therefore team members are more likely to cooperate with one another in the virtual environment through the transformational leadership that help to improve a cooperative climate and group cohesiveness (Huang et al. 2010).

Based on a summary of characterizations found in the above studies in the literature, provides a comparison of the effective virtual team leader characteristics with the ineffective virtual leader (Table 3).

Communication and Trust in Global Virtual Teams

In 1998, Jarvenpaa and Leidner conducted research on communication and trust in global virtual teams. Concurrently that same year, Jarvenpaa and Leidner, along with Knoll (1998) conducted a similar study on the concept of trust in global virtual teams. In both studies, the objective was to explore the precursors to trust in a global virtual team setting. Jarvenpaa, Leidner, and Knoll (1998) created a two-week trust building exercise with 75 teams consisting of four to six members in order to examined any significant effects on the team members' perceptions of the other members' ability and propensity toward trust. In this qualitative study, the researchers found that the more frequent and substantive communication between the team members, the higher the level of trust and engagement. These findings indicate that teams that suffer from lapses in communication were more reluctant to trust other team members and were less likely to feel motivated and committed to their virtual team. In practical terms, this study is important for understanding how to improve virtual team process outcomes through proactive communication and frequent interaction between virtual team leaders and amongst the dispersed team members (Jarvenpaa, Leidner, & Knoll, 1998).

The secondary research by Jarvenpaa and Leidner (1998) added an additional component of this research on trust and perceived member ability. They incorporated communication behavior that might facilitate trust in the global virtual team. Through their exploratory case study, the absence of face-to-face interaction by the global virtual team was examined to understand what communication behaviors might facilitate the development of trust. Due to the fact that virtual teams are understood to be an improved

Table 3. Effective virtual team leader vs. ineffective leader

Traits	Effective VT Leader	Ineffective VT Leader
Ability to inspire	Inspires through frequent communication	Creates atmosphere of disagreement and confusion through lack of communication
Creates sharing environment	Empowers members to contribute	Tells members what to do
Offers intellectual stimulation	Encourages development of skills and knowledge	May offer development and training but does not allow members to exercise that knowledge and skills
Motivation	Promotes Motivation through independent thought and ownership of project goals	Members are not involved with the planning or decision-making process so they tend to focus on themselves rather than the good of group

resource in a dynamic global business environment, the concept of reliance on computer-mediated communication technology in order to create a collaborative work environment is extremely important to examine and understand. Participants of this case study included 350 students from 28 universities organized over a period of 6 weeks. Students were assigned to teams of 4 to 6 people ideally from different countries and unfamiliar with one another. Students communicated solely through electronic mail to reach their other team members. Questions were geared toward measuring levels of trust within each team as well as the amount and type of communication behaviors (Jarvenpaa and Leidner, 1998).

The development of trust appeared to occur early on in the team's existence when there was extensive social discussion. However, Jarvenpaa and Leidner (1998) found that social communication was not sufficient enough to maintain trust over a longer period of time. The more successful teams in building trust where those that encouraged each other on project tasks with optimism in addition to developing some level of social communication. In addition, in concurrence with their own findings along with Knoll (1998), a frequent and regular pattern of communication developed in higher levels of trust and abilities to contribute to achievement of goals. Though these findings encourage leadership of virtual teams to actively promote these types of actions on the part of the virtual team members, neither study could definitively conclude that trust was possible to maintain in the global virtual environment purely through the use of computer-mediated technology.

Socialization and Motivation

A generally accepted assumption behind the understanding of distributed virtual teams is that they operate differently from face-to-face or traditional groups. This understanding is based on the findings of several formative research articles in the literature. Amongst these articles includes the study conducted by Ahuja and Galvin (2003) on socialization in virtual teams versus the traditional team. A content analysis of 673 e-mail communications among members of virtual groups during a three-month time period was conducted.

Ahuja and Galvin (2003) specifically studied how newcomers of virtual teams sought information and how established team members provided that information and compared the findings against how traditional teams carryout the same information sharing capability. Because group information-sharing communication processes are key to developing socialization and knowledge, proving the similarities and differences in how this occurs in the virtual team versus traditional ones is highly important to further advancing awareness of the dispersed virtual team.

The researchers concluded that similar to traditional groups, socialization begins to occur for new members of the team because they need to make sense of their environment and assigned tasks. This helps them to reduce their level of uncertainty in their new virtual environment and establish team members are eager to assist in this matter. However, individuals in traditional teams use more passive means to acquire information then virtual teams. In cases of the traditional team member, actions and behaviors were observed rather then directly asking others about group expectations and procedures. This passive approach cannot be conducted in a virtual team environment since members are not meeting face-to-face. All team members must be equally participative and more proactive when acquiring information on group activities and project objectives or risk the possibility of not achieving project goals.

Ardichvili, Page, and Wentling (2003) conducted research on impediments to participation in virtual knowledge-sharing communities. The researchers conducted a qualitative study of motivation and obstacles to employee participation in a virtual community. Three virtual communities were used to interview 30 members by phone and through face-to-face communications using open-ended questions. Their findings presented suggestions on reasons behind barriers to employees contributing their knowledge in virtual settings and their implications. Interview respondents believed that barriers to sharing recreated because people are afraid might be viewed as either in accurate or not relevant to the subject at hand. In addition, the research showed that in the virtual environment clear direction and communication is needed so that team members understood specifically what information should be shared electronically.

Furthermore, use of communication technology and how it is was organized and managed by virtual team leaders also created barriers in information sharing if not developed with the team to address specific issues. These findings support the conclusions drawn and discussed earlier in this section by Eveland and Bikson (1988). Virtual team members are more willing to use electronic means of communication and share knowledge in a virtual community if they are given clear guidance on what should be communicated and also managed in a way that lessons their perception of feeling intimidated to post ideas electronically. Finally, confidentiality and security concerns when relying on electronic means of communication must be proactively managed to strengthen team member fears of knowledge sharing in the virtual environment.

Daim, Ha, Reutiman, Hughes, Pathak, Bynum, and Bhatla (2012) conducted a more recent study exploring and advancing previous findings for the reasons for communication breakdown within global virtual teams that can jeopardize project success. The researchers conducted a literature review examining five specific criteria that were judged to be responsible for having the greatest impact on communications breakdown in dispersed teams. In addition, subject matter experts were interviewed through questionnaires and face-to-face interviews while mapping the weight of these criteria based on the different statements given. Daim et al. (2012) used the generally accepted management stages of team development; i.e. forming, norming, reforming, and performing as the model for interpreting group development.

The five criteria examined by Daim et al. (2012) included cultural differences, interpersonal relationships, leadership abilities, technology, and trust. The researchers supported concepts in the literature that cultural differences arise in dispersed virtual teams because members are typically located in different countries. Language barriers, cultural expectations, and differences in understanding what is being communicated impacts the virtual teams overall level of performance. Ability to reason collectively is also another negative impact that cultural differences can bring to the virtual team.

As the researchers further examined, ineffective ability to manage cultural differences alone does not lead to the breakdown in communications in virtual teams. The second criteria, interpersonal relations, affects the virtual team's ability to work effectively through electronic communication because elements such as body language and tone of voice are absent when there is limited or no face-to-face communication. This finding is important to the research of distributed virtual project management teams because it reinforces the knowledge that a constant reliance on electronic means of communication for the virtual team can lead to misunderstandings that can negatively impact team communication and productivity. Therefore, virtual teams rely heavily upon strong leadership to monitor and manage their team's interactions and electronic communications regularly.

Daim et al. (2012) examine the critical importance of strong leadership for distributed virtual teams to advance and have project success. Their research showed that the leader's ability to encourage constant and positive communication amongst the team members and proactively establish proper team communication guidelines reduced greatly the potential for miscommunication and conflict within the team. Key to these findings is the fact that again the transformational leader is acknowledged to have the strongest effect on team performance in the virtual environment because of their ability to influence positive behavior. The research also indicates, though, that strong leadership must include the ability to clearly define authorities and responsibilities amongst individual team members on the dispersed virtual team. However, ways of ensuring that this is accomplished through technical means is not discussed and may be an significant area of study on the subject of distributed virtual teams for future research.

As we have seen postulated in other literature discussed here, today's global marketplace demands that managers have the ability to implement technological tools for the distributed virtual project management team to collaborate with. Daim et al. (2012) examined how an inability for managers to use the best products on the market that best meets the needs of their virtual team members can result in a negative impact to project results. In addition, the researchers further explained that because no one single electronic collaboration tool possesses all the functionality that users need to exchange information in the absence of face-to-face communication, managers must further have the ability to tailor products and create solutions that will help achieve project success Daim et al. (2012). Furthermore, by choosing the best-suited electronic communication tools for their team, virtual team managers can help their team build a high level of trust early on in the team's developmental stage.

The literature critically reviewed in this section provided the background for what and how other researchers have examined this phenomenon (Table 4). Perspectives from research reviewed have mainly been from a descriptive and quantitative approach rather than the intuitive and qualitative approach taken by this researcher discussed in the next section.

Table 4. Comparison of literature review methodologies and methods

Author(s)	Quantitative	Qualitative	Sample	Support Dissertation
Defining and Developing the Virtual Team				
Powell, Piccoli, and Ives (2004)			Issues Paper / Lit Review	Need for more research on leadership role
Martins, Gilson, and Maynard (2004)			Lit Review	Cost/Benefit of VTs & Direction of Research
Kuruppuarachchi (2009)		X	Organization	Communication and Team Structures
Dynamics and Leadership of the Virtual Team				
Zigurs, Poole, and DeSanctis (1988)	X		Three and four-member groups	Group decision making using computer technology
Sprague and Greenwell (1992)	X		Manufacturing companies	Employee training to work on virtual teams
Cross and Rieley (1999)		X	22 Virtual Project Teams	Learning in virtual team setting. Shared project vision development
Kirkman and Rosen (1999)	X		Work Teams/ Organizations	Virtual Team Empowerment
Waller, Conte, Gibson, and Carpenter (2001)			Lit Review	Virtual Team member perceptions of time
Ellemers, Gilder and Haslam (2004)			Issues paper	Social identity, motivational processes and work motivation
Markus, Manville, and Agres (2006)		X	Organizations/Lit Review	Work motivation/networked organizational forms
Bourgault, Drouin, and Hamel (2008)	X		Project Management Professionals	Decision-making in distributed teams, collaboration

continued on following page

Table 4. Continued

Author(s)	Quantitative	Qualitative	Sample	Support Dissertation
Brandt, England, and Ward (2011)			Issues paper	Differences between virtual teams and traditional teams
Avolio, Kahai, and Dodge (2001)	X		Issues Paper/Lit Review	Advanced information technology and e-leadership
Kayworth and Leidner (2002)		X	Undergrads and Grad Students	Leadership roles and communication effectiveness
Huang, Kahai, and Jestice (2010)		X	Undergrads	Leadership styles and effects on virtual team collaboration
Information Technology and the Virtual Team				
Eveland and Bikson (1988)	X		Two task forces	Overcoming physical and social barriers to interact through computer support
Townsend, DeMarie, and Hendrickson (1998)	X		Organizations	Virtual teams and information technology, team building
Communication and Trust in the Global VT				
Jarvenpaa, Leidner, and Knoll (1998)		X	Grad Students	Team building and Quick Trust development
Jarvenpaa and Leidner (1998)		X	Grad Students	Social Coping Skills
Ahuja and Galvin (2003)		X	Organizational members and Academics	Information exchange and communication
Ardichvili, Page, and Wentling (2003)		X	Multinational Corporation	Motivation and barriers to participation in virtual environments
Daim, Ha, Reutiman, Hughes, Pathak, Bynum, and Bhatla (2012)	X		Matrix Organization/ Virtual Project Team Members	Communication breakdown in the virtual team
Performance Measures of the Virtual Team				
Reed and Knight (2010)	X		Information technology practitioners	Project risk virtual vs. traditional teams

SOLUTIONS AND RECOMMENDATIONS

This section provides a summary of this researcher's own analysis of the findings that came out of this investigation. This was a qualitative case study that broadened the project management knowledge base about dispersed virtual project teams toward an improved understanding of individual perspectives of effective virtual project team performance.

A review of the literature for this research uncovered numerous topics that were concerned with virtual teams though the use of various approaches was uncovered. By reviewing all relevant material in the literature for their application to this study, the researcher builds upon this area of inquiry by explaining how actual virtual team members perceive effective project management approaches, how these virtual project team members achieve collective success and maximize project effectiveness, and how individuals on virtual project teams feel about expressing themselves openly about the project's condition in the virtual environment.

This study examined virtual teams and the individual perceptions of effective project management that contribute to a collective effort in project success. The expectation was to provide the stimulus for the development of a more novel model for assessing distributed virtual project team effectiveness and recommendations for employing good project management professional practices on these types of virtual teams.

The specific questions that the researcher wanted to answer were:

1. How do virtual team members perceive effective project management approaches that contribute to the collective success of the project, such as through project planning and project risk and change management practices?
2. How does project team management ensure commonality in perspectives and maximize project management effectiveness in virtual project teams where various perceptions of effectiveness and success exist?
3. How do individual project team members feel about freely sharing their perspectives about the project's condition when there is limited face-to-face interaction? The findings for each question are shown in Figure 1 and discussed in the subsequent paragraphs.

Research Question 1

This initial inquiry into Research Question 1 encompassed the responses of 21 individual dispersed virtual project management team members to an online open-ended questionnaire and those of five virtual project management team leaders through semi-structured interviews again using open-ended questions. The first two sections of both questions were aimed at answering this first research question. This research question was asked in order to understand how virtual team members perceive project planning, project risk, and change management as well as any other project management practices that contribute to collective project success.

Phase 1 continued over the course of three months in order to collect enough data from virtual team member participants in the online survey while in parallel gathering data through Phase 2, the inter-

Figure 1. Mind map of research findings

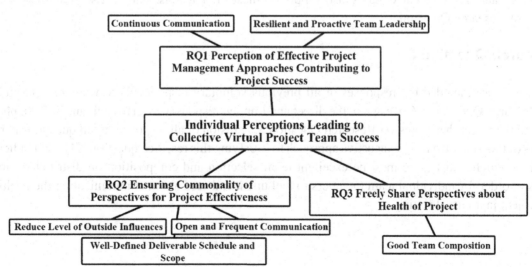

views. Both the direct and implied responses from participants in both phases allowed the researcher to answer Research Question 1. Participant responses allowed the researcher an in depth understanding of the feelings of the individual virtual team member and virtual project team leader on dispersed teams in regard to effective project management of virtual projects. Respondents perceived continuous communication amongst the team members throughout the life of the project a necessity when applying project management techniques in order to communicate and minimize project risk. Additionally, a resilient and proactive project team leader is a necessary component for creating and maintaining a cohesive, effective dispersed virtual project management team for communicating project risk and maintaining project objectives. The next section discusses the results of Research Question 2.

Research Question 2

This discussion involved the responses of all previous voluntary respondents to Research Question 1 as well as document reviews. Again, both the candid and inferred responses from participants in both phases allowed the researcher to answer Research Question 2. The second half of both questionnaires was directed at answering this research question in addition to any ancillary responses to all inquiries in both phases. Respondents perceived a well-defined deliverable schedule and scope of work as well as open and repeated communication are uniformly responsible for achieving commonality amongst individual virtual project team members. Equally so, there are strong feelings for what could hinder this achievement, i.e., outside influences, poor communication, and team member qualification. This researcher, in order to find supporting papers and citations for the findings uncovered in Research Question 2, applied the document review phase.

Document reviews were conducted by analyzing findings in the literature and in project management records that might support the researcher's above interpretation. The review of six PWPs of dispersed virtual project management teams deemed successful by program mangers and the review of peer-reviewed literature on ensuring project success did indeed uncover support for this analysis. The PWPs of successful virtual projects all communicated clear goals, strategies, and a well-defined scope of work. Critical success factors pinpointed in the literature for ensuring project success of both co-located and non-co-located teams also supported these conclusions. The next section uncovers the results of Research Question 3.

Research Question 3

This final case included the responses of all previous voluntary respondents to Research Question 1 and Research Question 2. Once again, the direct and implied responses from participants in both phases allowed the researcher to answer Research Question 3. A fusion of all responses to all questions in both phases of this study led to an exhaustive approach to satisfying this research question. The data indicated to this researcher that there must be excellent team selection and composition on distributed virtual teams in order for individual team members to feel unencumbered about communicating the health of the project in a virtual setting.

FUTURE RESEARCH DIRECTIONS

Studies on virtual teams have focused on leadership effectiveness and managing virtual teams, the advantages and disadvantages for organizations of establishing virtual offices and virtual teams, information technology and its influence on virtual teams. However, there still remains this unanswered question on enhancing dispersed virtual project management team effectiveness from the individual perspective and data collected from actual dispersed virtual project management teams. There are two areas that are recommended by this researcher for future research, namely,

1. International and organizational opportunities, and
2. Socialization.

The next section will run through suggestions for each of these areas in detail.

International and Organizational Opportunities

Disparate opportunities across international and organizational boundaries in the use of technology to enhance information sharing on dispersed teams have been studied in the literature as an antecedent to determining project success. However, the influence of the larger organization on project success of virtual teams has not been specifically addressed to learn how this may affect the cohesion of the virtual team. Views of virtual project team members versus their home organization and country may cause an adverse impact on the team's overall performance to collectively achieve original project objectives due to divergences surrounding project issues and priorities. Current research can be enhanced by focusing on organizational inferences that create an atmosphere for the virtual team members where decisions cannot be made due to the lack of empowerment stemming from differing policies by the home country and/or organization. Managers can learn how to work with the virtual project team members in understanding how to empower its members to achieve project success while still protecting organizational and governmental doctrines and policies. The next section discusses socialization as an opportunity for future research.

Socialization

There are still several unanswered questions regarding how the virtual team members socialize in a dispersed environment, and is this socialization an important factor in achieving project success. Do individual dispersed virtual project management team members socialize differently from when they are on face-to-face teams? Do these individuals perceive the need for socialization differently in a virtual environment vs. the traditional environment? If so, how is this difference overcome by team members and team leaders or project managers of the virtual project management team in order to achieve cohesion and collective project effectiveness? How does that ultimately impact the achievement of project success on global virtual teams? These are some of the questions surrounding the socialization and communication aspects of the dispersed virtual project management team that need to be examined.

By further examining socialization of virtual project management teams and international perspectives of achieving collective project success from the individual's point of view the organizational and management communities gain a better understanding of the main issues. Furthermore, by engaging divergent methodologies, specifically more qualitative studies to appreciate underlying individual perceptions, this area of research will be advanced and taken in a new direction.

CONCLUSION

This research study uncovered several key elements that led to a construct of individual perceptions of effective project management that contribute to a collective effort and project success of virtual teams. This understanding was gained through the perceptions and expressed feelings of actual dispersed virtual project team members and leaders. In particular, the following findings bore corroboration: the need for continuous communication amongst the team members when applying project management techniques in order to minimize project risk; a resilient, proactive project team leader for maintaining a cohesive, effective dispersed virtual project management team for sustaining original project objectives; a well-defined deliverable schedule and scope of work as well as open and repeated communication are responsible for achieving commonality amongst individual virtual project team members; outside influences and lack of team member qualification can hinder reaching commonality; and there must be good team composition first in order for individual team members to feel open about communicating in the virtual environment.

REFERENCES

Ahuja, J. (2010). A study of virtuality impact on team performance. *The IUP Journal of Management Research, 9*(5), 27–56.

Ahuja, M., & Galvin, J. (2003). Socialization in virtual groups. *Journal of Management, 29*(2), 161–185. doi:10.1177/014920630302900203

Ardichvili, A., Page, V., & Wentling, T. (2003). Motivation and barriers to participation in virtual knowledge-sharing communities of practice. *Journal of Knowledge Management, 7*(1), 64–77. doi:10.1108/13673270310463626

Avolio, B. J., Kahai, S., & Dodge, G. E. (2001). E-Leadership: Implications for theory, research, and practice. *The Leadership Quarterly, 11*(4), 615–668. doi:10.1016/S1048-9843(00)00062-X

Bourgault, M., Drouin, N., Daoudi, J., & Hamel, E. (2008). Understanding decision making within distributed project teams: An exploration of formalization and autonomy as determinants of success. *Project Management Journal, 39*(S1Suppl.), S97–S110. doi:10.1002/pmj.20063

Brandt, V., England, W., & Ward, S. (2011). Virtual teams. *Research & Technology Management, 54*(6), 62–63.

Cross, R., & Rieley, J. (1999). Team learning: Best practices and tools for an elusive concept. *National Productivity Review, 18*(3), 9–18. doi:10.1002/npr.4040180303

Daim, T., Ha, A., Reutiman, S., Hughes, B., Pathak, U., Bynum, W., & Bhatla, A. (2012). Exploring the communication breakdown in global virtual teams. *International Journal of Project Management, 30*(2), 199–212. doi:10.1016/j.ijproman.2011.06.004

Davenport, T., & Pearlson, K. (1998). Two cheers for the virtual office. *Sloan Management Review, 39*(4), 51–65.

Ellemers, N., Gilder, D., & Haslam, S. A. (2004). Motivating individuals and groups at work: A social identity perspective on leadership and group performance. *Academy of Management Review, 29*(3), 459–478.

Eveland, J. D., & Bikson, T. K. (1988). *Work group structures and computer support: A field experiment.* Retrieved at: http://tuiu.academia.edu/JDEveland/Papers/1207189/Work_group_structures_and_computer_support_A_field_experiment

Furst, S., Blackburn, R., & Rosen, B. (1999). Virtual team effectiveness: A proposed research agenda. *Information Systems Journal, 9*(4), 249–269. doi:10.1046/j.1365-2575.1999.00064.x

Hackman, J. R. (1983). *A normative model of work team effectiveness.* New Haven, CT: Yale School of Organization and Management, Research Program on Groups Effectiveness.

Hallam, G. (1997). Seven common beliefs about teams: Are they true? *Leadership in Action, 17*(3), 1–4.

Hollingshead, A., Mcgrath, J., & O'Connor, K. (1993). *Group task performance and communication technology: A longitudinal study of computer-mediated versus face-to-face work groups.* Thousand Oaks, CA: Sage.

Huang, R., Kahai, S., & Jestice, R. (2010). The contingent effect of leadership on team collaboration in virtual teams. *ScienceDirect: Computers in Human Behavior, 26*, 1098–1110.

Jarvenpaa, S., Knoll, K., & Leidner, D. (1998). Is anybody out there? Antecedents of trust in global virtual teams. *Journal of Management Information Systems, 14*(4), 29–64.

Jarvenpaa, S., & Leidner, D. (1998). Communication and trust in global virtual teams. *Journal of Computer-Mediated Communication, 3*(4), 1–53.

Kayworth, T., & Leidner, D. (2002). Leadership effectiveness in global virtual teams. *Journal of Management Information Systems, 18*(3), 7–40.

Kirkman, B., & Rosen, B. (1999). Beyond self-management: Antecedents and consequences of team empowerment. *Academy of Management Journal, 42*(1), 58–74. doi:10.2307/256874

Kuruppuarachci, P. (2009). Virtual team concepts in projects: A case study. *Project Management Journal, 40*(2), 19–53. doi:10.1002/pmj.20110

Markus, M. L., Manville, B., & Agres, C. (2000). What makes a virtual organization work? *Sloan Management Review,* (Fall): 13–26.

Martins, L. L., Bilson, L. L., & Maynard, M. T. (2004). Virtual teams: What do we know and where do we go from here? *Journal of Management, 30*(6), 805–835. doi:10.1016/j.jm.2004.05.002

Mihalic, J. (2013, October). Leading the way with thought leadership in the project management community. *PMI Today,* 3-5.

Mihhailova, G. (2007). *From ordinary to virtual teams: A model for measuring the virtuality of a teamwork.* Retrieved at: http://managementstudyguide.com/degree-of-virtuality-in-teams.htm

Padar, K., Pataki, B., & Sebestyen, Z. (2011). A comparative analysis of stakeholder and role theories in project management and change management. *International Journal of Management Cases, 13*(4), 252–260. doi:10.5848/APBJ.2011.00134

Powell, A., Piccoli, G., & Ives, B. (2004). Virtual Teams: A review of current literature and directions for future research. *The Data Base for Advances in Information Systems, 35*(1), 6–36. doi:10.1145/968464.968467

Reed, A., & Knight, L. (2010). Project risk differences between virtual and co-located teams. *Journal of Computer Information Systems,* (Fall), 19–30.

Schmidt, J., Montoya-Weiss, M., & Massey, A. (2001). New Product Development Decision-Making Effectiveness: Comparing Individuals, Face-to-Face Teams, and Virtual Teams. *Decision Sciences, 32*(4), 575–600. doi:10.1111/j.1540-5915.2001.tb00973.x

Sivunen, A. (2006). Strengthening identification with the team in virtual teams: The leaders' perspective. *Group Decision and Negotiation, 15*(4), 345–366. doi:10.1007/s10726-006-9046-6

Sprague, D., & Greenwell, R. (1992). Project management: Are employees trained to work in project teams? *Project Management Journal, 23*(1), 22–26.

Townsend, A., DeMarie, S., & Hendrickson, A. (1998). Virtual teams: Technology and the workplace of the future. *The Academy of Management Executive, 12*(3), 17–29.

Waller, M., Conte, J., Gibson, C., & Carpenter, M. (2001). The effect of individual perceptions of deadlines on team performance. *Academy of Management Review, 26*(4), 586–600.

Yoo, Y., & Kanawattanachai, P. (2001). Developments of transactive memory systems and collective mind in virtual teams. *The International Journal of Organizational Analysis, 9*(2), 187–208. doi:10.1108/eb028933

Zigurs, I., Poole, M. S., & DeSanctis, G. (1988). A study of influence in computer-mediated group decision making. *Management Information Systems Quarterly, 12*(December), 625–644. doi:10.2307/249136

KEY TERMS AND DEFINITIONS

Communication Media: Any technological or computer-mediated system that assists in communication.

Distributed/Dispersed Project Teams: Work teams that are not co-located or can be separated by geography, time, and organization and rely on technology to achieve project goals while they accomplish interdependent tasks.

Project Management: A discipline that is at the foundation of promoting knowledge and practices that lead to project success through the application of "generally recognized" processes, skills, tools, and techniques (Project Management Institute, 2008, p. 4).

Project Risk Management: The process of determining what risks may affect the project and documenting these points. A plan of action in response to these risks should be developed in order to mitigate risks during the course of the project (Project Management Institute, 2008).

Project Team Effectiveness: Measured by the level of effort group members collectively expend carrying out their work, the joint amount of knowledge and skill members bring to the group tasks, and the appropriateness to the task of the performance strategies used by the group in its work (Hackman, 1983).

Team Performance: Measured through communication, level of trust, team participation and coordination, and work outcomes (Ahuja, 2010).

Virtual Project Teams: "Use technology to varying degrees in working across locational, temporal, and relational boundaries to accomplish an interdependent task" (Martins, Gilson, and Maynard, 2004, p. 808).

Chapter 10
Designing and Managing ERP Systems for Virtual Enterprise Strategy:
A Conceptual Framework for Innovative Strategic Thinking

Yi Wan
Aston Business School, UK

ABSTRACT

The business environment today is transforming towards a collaborative context compounded by multi-organizational cooperation and related information system infrastructures. This chapter aims to examine Enterprise Resource Planning (ERP) systems development and emerging practices in the management of multi-organizational enterprises and identify the circumstances under which the so-called 'ERPIII' systems fit into the Virtual Enterprise paradigm; and vice versa. An empirical inductive study was conducted using case studies from successful companies in the UK and China. Data were collected through 48 semi-structured interviews and analyzed using the Grounded-Theory based Methodology (GTM) to derive a set of 29 tentative propositions which were then validated via a questionnaire survey to further propose a novel conceptual framework referred to as the 'Dynamic Enterprise Reference Grid for ERP (DERG-ERP)'; which can be used for innovative decision-making about how ERP information systems and multi-organizational enterprises – particularly the Virtual Enterprise may be co-developed.

INTRODUCTION

Enterprise Resource Planning (ERP) systems have developed extensively over the last decades in response to changing business requirements, technological developments, and new organizational strategies. According to the *APICS Dictionary* (11th Edition) (Blackstone & Cox, 2005), ERP is defined as a "framework for organizing, defining, and standardizing the business processes necessary to effectively plan and control an organization so the organization can use its internal knowledge to seek external ad-

DOI: 10.4018/978-1-4666-9688-4.ch010

vantage" (p. 38). This definition also indicates that ERP can be viewed as an information management strategy which enables the integration of various business units through a common system platform; this is echoed by other scholars (Beheshti, 2006; Johnson et al., 2004; Klaus et al., 2000).

It has been noted that most extant research on ERP systems design and management focuses on improvements in ERP functionality within a single unitary organization (Chen, 2001; He, 2004; Michel, 2000). Nevertheless, it is generally acknowledged that manufacturing and service companies today are facing a dynamic turbulent business environment, and therefore, can be encouraged to think differently and move beyond traditional single organizational boundaries whilst becoming involved in multi-organizational collaborations (Hoffmann, 2007; Rayport & Sviokla, 1995). This has stimulated the emergence of a new operations strategy in which competitive advantage is based on the development of relationships with partners (Walters, 2004). This chapter follows this premise and thereby uses the European Commission's definition of an *enterprise* to explore how ERP systems can be designed and managed to effect changes in multi-organizational enterprise structures and *vice versa*; in turn, identify the circumstances under which the Virtual Enterprise paradigm can be realized by using the next generation ERP systems coined in this chapter as 'ERPIII'. The EC's definition of an *enterprise* is, "… an entity including partnerships or associations that can be made up of parts of different companies" (European Commission, 2003). This chapter builds on this definition and *does not* therefore consider manufacturing and service operations to be single legal entities operating in isolation, but instead embodies the (multi-organizational) *enterprise management* concepts (European Commission, 2003), where parts of companies work with parts of other companies to deliver complex product and service systems.

Some operations management researchers already realize that multi-organizational enterprises – particularly the Virtual Enterprise strategy cannot be described through simple contractual exchanges; but are better thought of as operational interdependencies based on complex interactive of operations and information technology (IT) (Banker et al., 2010; MacBeth, 2002). Likewise, information systems (IS) researchers suggest that integrated technical solutions – particularly ERP systems, which could make the multi-organizational enterprise management concept a full technical reality, are not far away (Chorafas, 2001). These works emphasize the fact that successful multi-organizational (virtual) enterprise strategy relies on the correct type of ERP information systems being used, as well as highlighting the importance of investigating how an ERP system fits into the multi-organizational operation and structure, in order to properly pursue the Virtual Enterprise strategy.

There is an emerging body of studies beginning to advocate the inter-organizational information systems (IOIS) (Saeed et al., 2011; Vathanophas, 2007). There is however a perennial pressing challenge for alignment between multi-organizational (virtual) enterprise management thinking and ERP systems design, adoption and development; which is imperative to provide a useable decision-making framework for thinking innovatively about co-development of ERP systems and multi-organizational collaboration – particularly the Virtual Enterprise paradigm. Thus this chapter aims to empirically examine ERP systems development and emerging practices in the management of multi-organizational enterprises and identify the circumstances under which the so-called 'ERPIII' systems fit into the Virtual Enterprise strategy; and *vice versa*. This aim is fulfilled by achieving three research objectives:

1. Summarize developing trends in ERP systems;
2. Describe the principles of Virtual Enterprise (VE) paradigm whilst confronting it with the Extended Enterprise (EE) and Vertically Integrated Enterprise (VIE) forms; and

3. Propose and describe a new conceptual framework known as the Dynamic Enterprise Reference Grid for ERP (DERP-ERP) to improve the concomitance between strategic operational thinking and ERP systems design and management within the context of multi-organizational (virtual) enterprises.

The remainder of the chapter is structured as follows. The next section critically review the literature related to ERP systems development and multi-organizational enterprise paradigms – particularly the Virtual Enterprise. This is followed by a description of the grounded theory-based research methodology. The findings and theoretical discussion are reported in the subsequent section; leading from this, a useable framework is then proposed. Finally, the chapter concludes by highlighting the contributions to the body of ERP-Virtual Enterprise knowledge and identifying the implications for future work.

LITERATURE REVIEW AND THEORETICAL BACKGROUND

ERP Evolutionary Trend: From ERP to ERPII and On towards ERPIII

Traditional ERP systems are internally integrated information systems which are used to gain operational competitive advantage (Blackstone & Cox, 2005, p. 38; He, 2004) by primarily supporting core internal functions such as operations and production, and which may be extended to include other closely related functions such as sales and distribution, and accounting and finance (Al-Mudimigh et al., 2001; Davenport, 1998). These traditional ERP system types (sometimes also referred to as ERPI) typically have a high degree of proprietary in-house development requiring considerable financial commitment to implement and integrate with other organizational applications; such as Product Data Management (PDM) and Decision Support System (DSS) (Stevens, 2003; Themistocleous et al., 2001).

The origins of ERP systems are firmly based in manufacturing and their fundamental structure built upon Material Requirements Planning (MRP) (Harwood, 2003; Shehab et al., 2004), Manufacturing Resource Planning (MRPII) (Wight, 1984) and later Computer Integrated Manufacturing (CIM) (Jacobs & Weston Jr., 2007; Rashid et al., 2002). Apparently, traditional ERP does not necessarily support the increasing scope of future business requirements for Internet based commerce (Bond et al., 2000; Moller, 2005; Songini, 2002; Vazquez-Bustelo & Avella, 2006). In response, further functional modules are developed as 'add-ons' to form ERPII systems and the mantra of "ERP is dead – long live ERPII" is often used by contemporary systems developers (Eckartz et al., 2009). Thus, traditional ERP systems are slowly usurped by ERPII (sometimes also known as 'XRP' – eXtended Resource Planning); as ERPII is recognized as an integral part of business strategy enabling multi-organizational collaborations through extension of operations to close and trusted partners (Bagchi et al., 2003). Modules such as Advanced Planning and Scheduling (APS), Supply Chain Management (SCM), Customer Relationship Management (CRM), Demand Chain Management (DCM), Vendor Managed Inventory (VMI), Business Intelligence (BI), and Data Warehouse (DW) are all parts of ERPII systems (Kumar & van Hillegersberg, 2000); giving the potential for multi-organizational operations and Internet based commerce (Davenport and Brooks, 2004). One might say that the first generation of ERP primarily supported and enhanced *single* organizational operations (Akkermans et al., 2003) whilst ERPII supports "… resource planning co-operations *between* different organizations at a meta-level" (Daniel & White, 2005).

Currently ERPII is the dominant type of system to support modern manufacturing enterprises. However as competition increases and markets become even more turbulent, many manufacturers are trying to re-design their operations and ERP systems to have even greater flexibility (Anussornnitisam and Nof, 2003). As a result information systems solutions based on technologies such as Enterprise Application Integration (EAI), Service-Oriented Architecture (SOA), SaaS (Software as a Service) (Bass and Mabry, 2004; Sharif et al., 2005), utility and cloud computing technologies (Maurizio et al., 2007; Rappa, 2004; Sharif, 2010) and open-sources applications (Benlian and Hess, 2011) are becoming more prevalent. These technologies bring with them further flexibility, agility, efficiency, scalability and re-configurability for ERP systems and operations; because they provide the potential for multi-organizational connectivity (Torbacki, 2008; Wilkes and Veryard, 2004) – particularly for the Virtual Enterprise structure.

The future for ERP systems is still uncertain though as SOA, SaaS, Utility and openly-sourced enterprise applications bring new challenges around granularity of data-sharing, business privacy and decentralization of strategic objectives (Candido et al., 2009; Xu et al., 2002). Despite these new challenges one can observe these emerging technologies changing the way that ERP systems are currently being perceived and developed. For instance one can find 'Virtual Enterprise Resource Planning (VERP)' and 'Federated ERP' concepts being deployed using cloud computing, SOA, SaaS and PaaS (Platform as a Service) technologies (Cummins, 2009; Pal and Pantaleo, 2005). These new technical and conceptual information systems developments may provide more sustainable competitive advantage and make the (multi-organizational) enterprise management concept – particularly the Virtual Enterprise a future reality. For managers who may be seeking to temporize their structure and operations strategy in response to economic turbulence and uncertainty, this is an important trend to be aware of.

In this chapter the author refers to the *next generation* of Enterprise Resource Planning systems as 'ERPIII'. The author defines ERPIII as a *flexible, powerful information system incorporating web-based technology which enables enterprises to offer increasing degrees of connectivity, collaboration and dynamism through increased functional scope and scalability*. This definition considers contemporary management thinking about multi-organizational enterprise concepts (e.g. Virtual Enterprise) brought out by academic literature cited in this chapter. Table 1 summarizes recent ERP systems development trends outlined above; from traditional ERP to ERPII, and on towards ERPIII on which the new contingency framework described towards the end of this chapter is partially founded.

The Multi-Organizational Enterprise Management: VE, EE, and VIE

The concept of applying (multi-organizational) enterprise strategy is important because it is widely accepted that embracing new business partnerships and collaborative arrangements (e.g. virtual enterprise) can contribute to the sustainability of a business (Achrol & Kotler, 1999). For instance, Tencati and Zsolnai (2009) state that the 'enterprise' concept helps a business fit better within its business environment, social, and culture contexts. Likewise Binder and Clegg (2006) claim that, "… the success of collaborative enterprise management depends on the ability of companies to intermediate their internal core competences into other participating companies' value streams and simultaneously outsource their own peripheral activities …". Similarly Li and Williams (1999) indicate that "firms should focus on their core competences and share expertise and risks with each other in order to develop inter-firm collaboration in strategic processes …" This thinking indicates that competitiveness relies on the overall performance of all partners in an 'enterprise' rather than just one company's internal operations. This chapter herein focuses on the three main types of multi-organizational enterprises: the Vertically Inte-

Table 1. Summary of ERP trends: ERP to ERPII, and on towards ERPIII

Key Element	ERP	ERPII	ERPIII
Role of system	Single organization optimization and integration (Akkermans et al., 2003; Park and Kusiak, 2005; Scott and Vessey, 2000)	Multi-organisation participation with some collaborative commerce potential (Bagchi et al., 2003; Daniel and White, 2005; Zrimsek, 2003)	Multi-organisation, Internet based, with full collaborative commerce functionality (Hauser et al., 2010; Ponis and Spanos, 2009; Torbacki, 2008)
Business scope	Manufacturing and distribution, automatic business transactions (Al-Mudimigh et al., 2001; Chen, 2001)	Often sector-wide offering upstream and downstream integration (Bendoly et al., 2004; Bond et al., 2000)	Facilitating cross sectors strategic alliances (Muscatello et al., 2003; Wilkes and Veryard, 2004; Wood, 2010)
Functions addressed	Manufacturing, product data, sales and distribution, finance (Davenport, 1998; Monk and Wagner, 2009)	Most internal organisational functions supported with some limited supplier and customer integration (Li, 1999; "Ted" Weston, 2002, 2003)	All internal functions supported plus core inter-company processes (Hauser et al., 2010; Wood, 2010)
Processes supported	Internal, hidden, with an intra-company boundary (Al-Mashari et al., 2003; Markus and Tanis, 2000)	Externally connected with intra-enterprise (i.e. inter-company) focus (Bond et al., 2000; Moller, 2005; Songini, 2002; Tapscott et al., 2000)	Externally connected, open network to create borderless inter-enterprise/industry-wide focus (Muscatello et al., 2003; Ponis and Spanos, 2009; Wood, 2010)
Information system architecture	Web-aware closed and monolithic (Hicks and Stecke, 1995; Stevens, 2003; Themistocleous et al., 2001)	Web-based, componentized, non-proprietary (Callaway, 2000; Monk and Wagner, 2009) Internally and externally available, often subscribed to by joint ventures (Ericson, 2001; Li, 1999; Moller, 2005)	Web-based communication, service-oriented architecture (Hofmann, 2008; Ponis and Spanos, 2009) External exchange via open source and cloud computing (Buco et al., 2004; De Maria et al., 2011)

grated Enterprises (VIE), the Extended Enterprises (EE), and the Virtual Enterprises (VE) to illustrate multi-organizational enterprise management behavior, as well as confronting the VE – as the targeted strategy with the VIE and EE forms.

Vertically integrated enterprises (VIE) operate as large single well-integrated multi-functional firm striving for scales of economy, they typically have bureaucratic reporting hierarchies (Lynch, 2003) which evolve as, "a response to pre-existing market power problems or as a strategic move to create or enhance market power in upstream and downstream markets" (Joskow, 2003, p. 25). A VIE will typically process ultraraw materials through to end-consumer products and services to embed a firm within an industry (Harrigan, 1985; Vallespir & Kleinhans, 2001). A classic example is the Ford Motor Company is in its 20th century heyday (Monteverde & Teece, 1982; Crandall, 1968). As a result competitiveness maybe gained through reduced transaction costs (Harrigan, 1984, 1985; Mahoney, 1992), strong quality control, higher barriers to new entrants (Rothaermel et al., 2006), and rapid response to volume changes (Richardson, 1996). However, the competitive damage created by excessive vertical integration can be substantial, as in the examples of the U.S. automobile and steel industries in 1983. Hence, instead of building VIE, quasi-integration and joint ventures should be formed to obtain strategic flexibility. Firms could have components engineered to their tight and highly specific instructions by outsiders rather than fully own and control adjacent business units in the vertical chain, as do Japanese automobile manufacturers, for instance. In turn, some research suggests that 'make-or-buy' decisions (Anderson and Weitz, 1986; Vallespir & Kleinhans, 2001); strategic outsourcing or global sourcing (Chung et al.,

2004) and alliances make further enhancements to a VIE set-up (Arya & Mittendorf, 2008). Therefore, the downside to VIEs (Argyres, 1996) is that their structure and size can inhibit engagement with other organizations – particularly within the virtual business environment; hence the rate at which changing market requirements are addressable in collaboration with other organizations is reduced. To combat the downsides of VIEs – the extended enterprise structure and strategy should be used instead.

The 'extended enterprise' (EE) concept, in contrast to the VIE, is defined by Davis and Spekman (2004, p. 20) as "… the entire set of collaborating companies … which bring value to the marketplace …" and by Lyman et al. (2009) as "… a business value network where multiple firms own and manage parts of an integrated enterprise". This allows practices such as just-in-time (JIT) supply chain logistics (Sutton, 2006), collaborative innovation (Owen et al., 2008), and data warehouse interoperability (Triantafillakis et al., 2004) to be deployed more easily across company boundaries (Childe, 1998; Jagdev & Browne, 1998). This is because an EE structure allows organizations to focus on their core business and technical activities whilst outsourcing non-core activities to other members in their extended enterprise (Stalk et al., 1992; Thun, 2010). Thus extended enterprises are deemed to be more agile than vertically integrated enterprises. But despite reduced cross-company boundaries (O'Neill & Sackett, 1994), even EEs cannot manage to follow very highly economic turbulence and unpredictability because they operate in a partially restricted environment operated by known, trusted and willing members.

Highly turbulent and unpredictable market behaviors are best coped with by 'virtual enterprise' (VE) (Byrne & Brandt, 1993; Katzy & Dissel, 2001) rather than an EE or a VIE as virtual enterprises (VEs), in contrast to the EEs and VIEs, are the most agile type of enterprises. VEs are best thought of as a jigsaw of operations and information systems from more than one business entity loosely governed by decentralized specific objectives which delivers value in an agile manner towards its market opportunities (Goldman et al., 1995; Martinez et al., 2001). Virtual inter-organizational relationships like these can facilitate innovative agile manufacturing or supply chain more easily (Cho et al., 1996; Sharp et al., 1999) and deal with dramatic dynamic market changes (Madu & Kuei, 2004) through Internet based information and communication technologies (ICTs) (Hyvonen et al., 2008; Jagdev et al., 2008; Lipnack & Stamps, 1997). This is because firms' tendencies towards temporizing structure and strategy are more easily addressed. For example the book publishing business is constantly changing due to newly emerging digital technologies (e.g. Lightning Source's Internet based 'print-on-demand' (POD) publishing service is able to integrate hundreds of thousands of suppliers and buyers rapidly into a 'cost effective' deliver system; see lightningsource.com/process).

Browne and Zhang (1999) summarize that the EE and VE can be seen as two complementary enterprise strategies as their similarity lies in the fact that they both pursue multi-organizational partnerships in order to achieve business success in a very competitive environment. The main different is represented by the 'temporary' and 'dynamic' nature of the VE in comparison to the EE. Similarly, Jagdev and co-workers (1998; 2001) unveil that unlike EE, VE is a manifestation which is inherent in agile manufacturing, and which is made possible by heavily utilizing ICT systems; therefore, EE can be considered as a special case of the VE. Moreover, as manager seek to re-engineer companies – the SMEs in particular (Hanna & Walsh, 2000; Jagdev et al., 2008; Kaihara & Fujii, 2002) in response to uncertain business environment, the VE tends to replace the VIE (Daniels, 1998) and the EE because virtual enterprises are more suitable as they are, "opportunistic aggregations of smaller (business) units that come together and act as though they were a larger, longer-lived enterprises" (Goranson, 1999).

Table 2 summarizes the comparison between vertically integrated enterprises (VIE), extended enterprises (EE) and virtual enterprises (VE) types as discussed above using key elements which both characterizes and differentiates them on structural, strategic operations and IS bases. The multi-organizational enterprise types in Table 2 are used as partial bases for the new contingency framework given towards the end of this chapter.

RESEARCH METHODOLOGY

Considering the nature of the research subject and the above theoretical debate an exploratory and qualitative empirical approach was used based on inductive Grounded Theory-based methodological approach (Glaser and Strauss, 1967; Strauss and Corbin, 1990). This was generally structured into three phases: data collection, data analysis, and data validation. It should be noted that data collection and analysis were not conducted sequentially but iteratively until theoretical saturation was achieved.

Data Collection Phase (Choosing Interviewees)

Data collection was deployed by conducting 48 semi-structured face-to-face interviews from a variety of industries (construction, printing, electronics, logistics and banking) in the UK and China; covering 8 companies who deliver complex products and services across organizational boundaries whilst using

Table 2. Comparison between VIE, EE, and VE

Key Element	Vertically Integrated Enterprise (VIE)	Extended Enterprise (EE)	Virtual Enterprise (VE)
Characteristic of core competencies	Mature and well accepted Large scale of economies	Semi-mature with pilot experience Ideal for production ramp-up scenarios	Quick respond to the changing market and environment Low overheads
Strategic aims	Long term objectives	Medium-long term objectives	Short-term objectives
Partnership purposes	Long-term indefinite co-operation	Medium-long-term collaboration on variety of projects and products	Temporary team-working for single project or products
Organization stability	Stable hierarchy and inflexible structure	Relatively stable across the product value chain	Dynamic organizations with core competences
Organization type	Command & control unity Concern more on scales of economies	Product/service value chain based	Frequently project or niche market based
Co-ordination of partnership	Original equipment manufacturer supervises relationship with the partners	Manufacturer or prime contractor supervises the partnership	The most strategically influential member ('orchestrator') supervises the co-operation
Operational challenges	Legacy system transferring approaches (e.g. big bang vs. incremental ways)	Synergistic among complementing core competencies Compatibility around partners and IS/IT	Dynamic operating and unpredictable business environment Psychological issues
Risk degree	Comparative low	Moderate	Intensely high
IS/IT facilitators	In-house development of proprietary systems with traditional ERP system for intra-integration	Advanced IS/IT ERP merged with other new functional modules (e.g. SCM, CRM, VMI)	Sophisticated Web-based technologies (e.g. SOA, cloud computing, SaaS)

ERP systems to support their operational strategies. Interviewees were from operations, manufacturing, supply chain, IT, client service and finance functions. All the interviews took place between March 2011 and August 2011, lasting between 1 – 1.5 hours (producing 53 hours and over 800 pages of validated transcript). Key characteristics of the interview and background information on each of the case study sites are given in Table 3.

Table 3. Overview of the case companies and interview sample

Company	Industry Sector	Number of Interviewees	Role of Interviewees	Management Level	ERP Systems
Print-on-demand-Co (UK)	Printing manufacturer	6	Managing director	Senior	Content management system (CMS) Oracle (and PeopleSoft)
			Operations director	Senior	
			Manufacturing manager	Middle	
			Client service manager	Middle	
			IT system manager	Middle	
			Supply chain manager	Middle	
Printing-Co (UK)		5	Managing director	Senior	Print-Pack MIS systems
			Client service manager	Middle	
			Account director & sales manager	Middle	
			Production & administration manager	Middle	
			Studio manager	Middle	
Electronic-Co (UK)	Semiconductor manufacturer	6	Supply chain programme manager	Senior	SAP (ERP) systems
			Supply chain technologist	Middle	
			Finance manager	Senior	
			Logistics & manufacturing manager	Middle	
			B2B technologist	Junior	
			Supply planning & customer manager	Middle	
Logistics-Co (UK)	Transport and logistics service	9	Group service director	Senior	Sage (ERP) systems SAP (ERP) systems
			Head of sortation auto	Senior	
			Operations director	Senior	
			Group commercial director	Senior	
			Senior financial controller	Junior	
			IT director	Senior	
			Head of transport	Senior	
			Head of human resource	Senior	
			Operations control team manager	Middle	

continued on following page

Table 3. Continued

Company	Industry Sector	Number of Interviewees	Role of Interviewees	Management Level	ERP Systems
Zoomlion (China)	Crane manufacturer	7	Executive manager	Senior	SAP (ERP) systems
			Chief information officer	Senior	
			Logistics manager	Middle	
			Regional marketing & sales manager	Middle	
			Regional director	Middle	
			Credit manager	Junior	
			Business sales assistant	Junior	
Lanye (China)	Concrete and mixer manufacturer	5	General manager	Senior	Alutex (and GPS) ERP systems
			Chief information officer	Senior	
			Logistics director	Middle	
			Production manager	Senior	
			Chief executive officer	Top/executive	
Wanghai (China)		4	Chief executive officer	Top/executive	Three Prosper Technology
			Human resource manager	Senior	
			Inventory manager	Middle	
			Chief information manager	Senior	
Metrobank (China)	Banking	6	Chief executive manager	Top/executive	SAP (ERP) systems
			Head of human resource	Senior	
			Compliance manager	Senior	
			Chief finance officer	Senior	
			Chief operation officer	Senior	
			Chief information officer	Senior	

Data Analysis (Grounded Theory-Based Coding)

The textual analysis of over 800 transcribed pages of data via codification was done using the QSR NVivo 9.2 software tool based on the constant comparative method of Grounded Theory-based Methodology (Glaser and Strauss, 1967). Strauss and Corbin's (1990) hierarchical coding paradigm was applied. The author use *open*, *axial* and *selective coding* in order to reach the necessary conceptual density. It was applied at the intra- and inter-case level (each interview reflecting one individual case) as suggested by Strauss (1987).

Firstly, codes and categories were identified in an unrestricted *open coding* of the empirical data – during the coding process, memos were created that explained how the data were opened up to get a greater understanding of the responses, and 1367 free nodes were extracted. Secondly, *axial coding* of these provisional categories gave further insight into the inter-relationships of these categories; this technique revolves around the axis of core category at a time (Strauss, 1987) (giving 133 useable codes,

23 analytical categories, and 19 sub-categories). Finally, top-down *selective coding* was used to develop seven high level *core categories* or 'themes' that pulled together all the other detailed categories conceptually. These themes are

1.   Industrial impact,
2.   Enterprise structure and strategy design,
3.   Enterprise structure and strategy governance,
4.   ERP systems design,
5.   ERP systems management,
6.   Competence and competitiveness as main contingency factors, and
7.   Organization and people management. A generic overview of the final coding diagram is presented in Figure 1.

The coding process produces datum types and records the frequency of their occurrence. It also aggregates the analysis, reduces researcher bias and highlights the main issues from which propositions can be written. The codification process is therefore the provenance for the 29 tentative propositions; these parsimoniously summarize all the main issues contained the interview transcriptions and tie the theoretical debate to industrial practice. The resulting tentative propositions are presented in a theoretical narrative later in this chapter and can be seen in full in Table 4.

Data Validation (Questionnaire Survey)

The tentative propositions were constructed into a self-administered questionnaire survey and then validated using 116 industry experts (with backgrounds in purchasing, R&D, quality assurance, production

Figure 1. Generic coding diagram

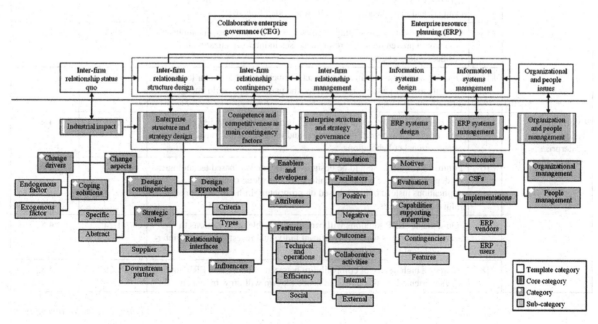

Table 4. Validating propositions relating to ERP systems and inter-organizational enterprise collaboration (N = 116)

Core Category	No.	Propositions Relating to	Mean Agreement	Mean Importance
Inter-firm relationship status quo (Industrial impact)	#1	Change in the manufacturing and service-driven industries is driven by a combination of dynamic globalization, internal organizational issues and general industrial forces	1.76	5.40
	#2	Increasing business complexity, cost-effectiveness and shorter turnaround time requires organizations to move towards more collaborative strategies	1.83	5.45
Inter-firm relationship structure design (Enterprise structure and strategy design)	#3	Inter-organizational relationships change over time, which is dependent upon individual core competencies	1.04	4.89
	#4	Inter-organizational relationships change over time, which is dependent upon the end product or service being delivered	1.39	4.99
	#5	Types of inter-organizational relationships and collaborative practices are determined by an industry-specific context	1.34	4.93
	#6	Service based inter-organizational collaborations have greater propensity to become virtual than product based inter-organizational collaborations	0.60	4.30
	#7	Organizations could use different approaches to inter-organizational collaboration, structure and strategy within different supply networks simultaneously	1.51	5.03
Inter-firm relationship structure management (Enterprise structure and strategy governance)	#8	Responsibilities and functional roles of each different organization needs to be clearly defined within the supply network	1.99	5.54
	#9	Collaboration with new external organizations requires internal business processes to be reengineered to accommodate new changes	1.38	5.20
	#10	In the context of inter-organizational collaboration, product-based organizations predominantly focus on the portfolio and quality of products, and the standardization of business processes	1.35	4.95
	#11	In the context of inter-organizational collaboration, service-oriented organizations predominantly concentrate on consumers' experiences	1.64	5.32
	#12	There is need for a leader or a 'broker' organization within the supply network who has core competencies and responsibilities to supervise, evaluate and manage cooperation between other organizations	0.91	4.89
	#13	Organizations are more willing to collaborate with other organizations who have a proven track record of successes in inter-organizational business collaborations	1.79	5.23
	#14	Once organizations obtain a similar set competences at a similar level of maturity as their partner organizations, the partnerships could change as a result	0.98	4.68
Information systems design (Enterprise resource planning systems design)	#15	The role of ERP systems in supporting operational business has evolved from intra-organizational optimization and integration into multiple inter-organizational collaborations	1.58	4.93
	#16	Future ERP systems should be designed based on web-based technologies by deploying service oriented architectures and cloud computing applications instead of being based on proprietary in-house enterprise information systems	1.47	4.90
	#17	'On-demand' ERP solutions will benefit and enable organizations to access technologies without significant individual investment cost in inter-organizational systems integration	1.24	4.87
	#18	There is a high degree of compatibility between 'cloud-based ERP' and service oriented architectures and hence the two will grow in unison	1.09	4.82

continued on following page

Table 4. Continued

Core Category	No.	Propositions Relating to	Mean Agreement	Mean Importance
Information systems management (Enterprise resource planning systems management)	#19	Information security and flexibility of ERP systems will be key determinants in their adoption and use in inter-organizational collaborations	1.94	5.62
	#20	Inter-organizational integration requires different organizations within the same collaborative supply network to use ERP system(s)	1.02	4.71
	#21	Inter-organizational integration requires ERP systems within the same collaborative supply network to use the same ERP system to become highly integrated	0.61	4.42
	#22	The tighter inter-organizational collaborative structures and strategies become; the more integrated and flexible ERP systems also need to become	1.33	5.04
	#23	Third-party consulting organizations are becoming increasingly responsible for handling web-based ERP system implementations, which could make non-web-based ERP vendors lose their influential positions over end-users	0.41	4.24
	#24	Inter-organizational collaboration can be facilitated best by integrating 'best of breed' functional modules from different ERP solutions, rather than customizing a single 'one-size-fit all' solution	0.66	4.44
Inter-firm relationship contingency (Competence and competitiveness as main contingency factors)	#25	Initial motives for inter-organizational collaboration are based upon the attractiveness of an organization's core competences	1.37	4.97
	#26	Collaboration between different organizations can create new meta core competencies and specific systems resulting in 'end-to-end' product-service solutions	1.50	4.98
	#27	Building inter-organizational collaboration is an effective way to reduce cost and lead time, increase the efficiency, improve flexibility and reactivity to demand; and encourage innovation	1.77	5.31
Organizational and people issues (Organization and people management)	#28	Organizational cultural diversity, trust issues and resistance to change have to be managed when adopting ERP systems, especially in inter-organizational collaboration	2.00	5.63
	#29	Organizational behavior is a key challenge when adopting and managing web-based ERP systems in inter-organizational collaborations	1.64	5.26

& manufacturing, logistics, marketing & sales, inventory management, IT and strategy development) from 16 different companies who were asked to assess each tentative proposition on two dimensions of perception "agreement" and "importance" using 7-point Likert scales as follows:

1. Agreement (strongly agree = 3, agree = 2, mildly agree = 1, neutral = 0, mildly disagree = -1, disagree = -2, strongly disagree = -3); positive scores indicate agreement and negative scores indicate disagreement.
2. Importance (extremely high importance = 7, very high importance = 6, high importance = 5, medium = 4, low importance = 3, very low importance =2, extremely low importance = 1); all positive scores were used as this was a weighting factor.

The validating ratings for the 29 final tentative propositions are given in Table 4.

FINDINGS AND THEORETICAL DISCUSSION

Data analysis and validation showed that enterprise resource planning systems design and management and multi-organizational enterprise governance is generally regarded as an effective perspective to maintain and achieve competitiveness for the whole 'enterprise' (e.g. virtual enterprise) and its individual value members as well as the 'enterprization of operations' (Clegg and Wan, 2013) with long-term and short-term effects (cf. propositions #26 and #27). Each member of the multi-organizational enterprise is affected by a variety of industrial forces; thus different multi-organizational enterprise structures and strategies may change over time (cf. propositions #3 and #4) and should be supported by different ERP information system types under different circumstances (cf. propositions #15, #20, #21, #22, and #24), in order to cope with the challenges of dynamic globalization, complex industrial changes, and shorter turnaround time required by the end consumers (cf. propositions #1 and #2).

In respect to the *enterprise structure and strategy design and governance*, managing core competencies is considered as a principal factor when making decisions to achieve the multi-organizational relationships and collaboration successfully as the competencies determine the role of the individual partners within the collaborative venture (e.g. virtual enterprise) (cf. propositions #3, #14, and #25) via the value or competitiveness they are creating for the entire virtual value chain. Also, becoming more influential within multi-organizational enterprise requires (the most influential or focal firms) managing competencies belonging to other member-companies. In addition, multi-organizational enterprise design may also be affected by the end (collaborative) products or service solutions being delivered by the 'enterprise', as well as different industry-specific contexts (cf. propositions #4, #5, #10, and #11). This is reflected by the fact that any type of multi-organizational collaboration in the service industries (e.g. 'print-on-demand' and logistics) will have greater propensity to become 'virtual enterprises' by using web-based enterprise information systems than those production-based strategic alliances (e.g. concrete manufacture) (cf. proposition #6). This is because most service-oriented business (e.g. parcel delivery and banking) require more flexible and agile operational performance with quicker and more accurate responsiveness to unpredictable market demands, owing to their inherent nature.

Furthermore, the existence of multiplicity of dynamic multi-organizational relationships within an 'enterprise' requires a differentiated management based on the respective relationship characteristics (cf. propositions #3, #4, #7, and #14). It also requires a leader or a 'broker' organization (e.g. the most influential or focal companies) that has core competencies and responsibilities to clearly define each functional roles and boundaries within an 'enterprise', as well as supervising, evaluating, and managing cooperation between the partners. This allows for a certain degree of autonomy within the collaborative venture (cf. propositions #8 and #12) and the ability to deploy or even create (new) competencies (e.g. 'end-to-end' product-service solutions/systems) through effective multi-organizational collaboration (cf. proposition #26). Besides, in the opinion of most interviewees, any effective multi-organizational collaboration with external organizations would require internal business processes of each individual (enterprise) member to be re-engineered to accommodate new changes (cf. proposition #9), which is a big challenge for multi-organizational enterprise management – particularly for the virtual enterprises which is highly dynamic and reconfigurable and aims for short-term business objectives. Hence, organizations – particularly the most influential or focal firms are more willing to collaborate with other organizations that have a proven track record of success within the multi-organizational enterprise business collaborations (cf. proposition #13).

With respect to *enterprise resource planning systems design and management*, firstly it was observed that the strategic role of ERP information systems in supporting operational business have evolved from intra-organizational optimization and integration into multiple inter-organizational collaboration (cf. proposition #15). This consequently gives birth to new IT technologies including SOA, cloud computing, and web services applications instead of traditional proprietary or monolithic in-house enterprise systems, which promise to provide quicker and less expensive cloud-based ERP services – as the next generation ERP systems – in order to establish and sustain new business partnerships and network structures (cf. propositions #16, #17, and #18). Specifically, the configuration and development of ERP systems supporting multi-organizational enterprises are expected to be linked to the adoption and spread in the use of service oriented architectures, with the uptake and increased maturity in one driving increased uptake and maturity in the other in a virtuous cycle, on the one hand (cf. proposition #18); and the 'on-demand' solutions based on web-based technologies could benefit and enable organizations – particularly (virtual) small and medium-sized companies to access innovative ERP systems without significant individual investment cost – in comparison to the 'on-premise' ERP solutions, in multi-organizational systems integration, on the other hand (cf. proposition #17).

In addition, it has been widely accepted that within the context of multi-organizational collaboration, different organizations (e.g. the most influential or focal companies) are requested to not only use ERP information systems but also use the *same* ERP information systems to become highly integrated and flexible via real-time information exchange (cf. propositions #20, #21, and #22). This may be facilitated by integrating 'best of breed' functional modules form different ERP solutions offered by different ERP vendors rather than customizing a single 'one-size-fit all' solution (cf. proposition #24) which is more time and cost-consuming. Moreover, information security, cost, and flexibility of ERP systems are considered as the most critical determinants in their adoption and use in multi-organizational enterprise collaborations (cf. proposition #19). Thus, sophisticated third-party consulting organizations are becoming increasingly responsible for handling web-based ERP system implementation, which could potentially make non-web-based ERP vendors lose their influential positions over end-users (cf. proposition #23).

The empirical findings of this research study also indicate that organizational cultural diversity, trust issues, and people's resistance to change have to be managed properly when adopting ERP systems in the context of multi-organizational enterprise (cf. proposition #28). Particularly organizational behavior is regarded as a key challenge for the web-based ERP systems use (cf. proposition #29), because members within an 'enterprise' might not be ready for the next generation of ERP information systems on the novel strategic concepts, i.e. the extended enterprises, virtual enterprises, and cloud-ERP information systems or SOA-based ERP infrastructure (a.k.a. ERPIII (Wan and Clegg, 2010)).

The New ERP Matrix

Successful ERP systems enabled multi-organizational enterprise design and management not only needs tools (e.g. Enterprise Matrix (Binder and Clegg, 2006)) to regulate collaborative activities (cf. propositions #12 and #25); but also requires tools to determine how enterprise information systems and technologies (e.g. ERP systems) are being used in different functional areas which make up the whole 'enterprise', i.e. connecting ERP functional modules within the multi-organizational enterprise (cf. propositions #22 and #24). Hence, drawing on the basics of virtual value chain (Rayport and Sviokla, 1995) and IT and business alignment concepts the new 'ERP Matrix' tool was developed to illustrate the capabilities of different ERP systems (or functional modules) to accommodate varying multi-organizational enterprise structures and strategy in a systematic manner; this is shown in Figure 2.

Multi-organizational relationships and collaboration are based on transactions between heterogeneous value members (e.g. publishers, book printers, and channel distributors) that traditionally pursue diverse strategies but try to fulfill a common task (e.g. joint products or collaborative complex service solutions development and completion) by establishing a mutual *modus operandi* through sharing real-time knowledge and information, technical know-how, and core competencies (cf. propositions #3 and #4). Such integration of collaborative activities will require supporting enterprise information systems and technologies (e.g. inter-connected ERP systems, web-based EDI, and electronic portals) which are the greatest enablers towards forming an e-integration among value members (cf. proposition #22).

In this sense, the new ERP Matrix tool helps to optimize the ERP information systems configuration and adoption within the whole 'enterprise' operation (represented by the respective collaborative activity) through the allocation of the most suitable *ERP modules* to support the operational requirements in different stages of the value stream based on their capabilities; which are determined by their targeted enterprise paradigms, strategic roles, deployment approaches, and systems advancement (cf. propositions #15, #16, #17, #19, #22, and #24). Therefore, this kind of allocation bridges 'structural holes' between the value members' information systems in the multi-organizational enterprise through the establishment of common ground based on the ERP modules that consist of capabilities catering to the core competence-based unique tasks (in different stages of the value stream) and adjunct functionalities facilitating interface connections (cf. propositions #3, #15, #22, and #24). In other words, the ERP Matrix tool can be seen as an artifact that helps people better understand multi-organizational ERP systems design, implementation, and landscape transformation through providing an architecture along which the key capabilities of ERP modules can be placed into the required functional units to fundamentally

Figure 2. The ERP matrix: a tool for determining how ERP systems capabilities supporting collaborative activities in enterprises by linking process, enterprise structure, and ERP systems use

ERP capabilities supporting inter-firm relationships and collaboration					
Collaborative activity:		Value stream			
		Process start ⟶			Process end
ERP functional module classification		Stage 1	Stage 2	...	Stage n
High usage rate ⬆ ⋮ ERP module(s) ⋮ ⬇ Low usage rate	ERP module 1	Operational requirements in 'stage 1' of the value stream is supported by 'ERP module 1'			
	ERP module 2				
	ERP module 3				
	...	Interface connection between 'stage 1' and 'stage 2' is supported by adjunct portals or links ⟷			
	ERP module n				Operational requirements in 'stage n' of the value stream is supported by 'ERP module n'

support the entire virtual value stream; which, in turn, achieves the most optimized 'enterprization'. It is thereby important to realize that the value stream, as shown by Figure 2, is supported *only* by the (sub-) information systems (i.e. the ERP modules) of the value members that can *actually* facilitate or add value to the collaborative activities (cf. propositions #3, #17, #24, and #25); whilst the parts of the entire (virtual) value chain are actively managed by the multi-organizational enterprise leader or governor (e.g. the most influential or focal firms) (cf. proposition #12).

A collaborative activity (see Figure 2) is a joint business activity in the multi-organizational enterprises and can involve collaborative products (e.g. crane, smart phones production), complex service solutions (e.g. print-on-demand realization), or a joint project (e.g. a construction project) that should be reasonably defined and circumscribed. This task should be conducted by a distinct leader or 'broker', e.g. focal manufacturer, the most influential service provider, or the joint project owner, who has the competence to i) evaluate the specific competencies of the value members, ii) allocate suitable core competencies to respective stages and tasks of the value stream, and iii) define the responsibilities of the boundaries between the value members (cf. propositions #8 and #12). Meanwhile, the corresponding ERP modules – as a set of powerful strategic weapons – need to be properly selected and implemented to facilitate multi-organizational (virtual) communication and collaboration needs, laying the foundation for external integration (e.g. supply network connectivity), allow simultaneous access to same data, as well as automating value stream processes. To be specific, in an 'enterprise' ERP modules can be effective means of optimizing planning applications, monitoring production constraints, managing demand forecasting, and keeping order delivery promises. In the cases of innovative print-on-demand service delivery, the value stream can be described as collaborative activities or processes right from the customer order placement (stage 1) to printing and packing books (stage 2), on towards the books distribution (stage 3), and ending up with establishing and managing relationships with the customers and end consumers (stage 4). In turn stage 1 may be supported by ERP module 1 with capabilities of 'electronic book storage'; stage 2 may be supported by 'content management and manufacturing' modules (i.e. ERP module 2); stage 3 may be supported by 'distribution management' module (i.e. ERP module 3) that could be adopted by *another* value member (e.g. downstream channel partner); and stage 4 may be supported by 'customer relationships management' systems (i.e. ERP module 4). However, in some cases, more than two value stream stages could be supported by a comprehensive ERP module or package (e.g. ERPII systems covering Data Warehouse, SRM, CRM, DSS, and e-business functionalities); and this indicates the necessity to rank the importance of different ERP modules by critically evaluating their *usage rate* (see Figure 2), i.e. how well the (ERP) module capabilities perform to support multi-organizational relationships and collaboration (cf. propositions #17, #19, and #22).

In addition, it is argued that the responsibilities of configuring and managing ERP systems within the context of (virtual) multi-organizational collaboration do not necessarily need to be occupied by a single value member (e.g. the focal firm or ERP vendor) but can involve various partners. In the semiconductor manufacturing industry it is, for example, often the case that the focal manufacturers define the overall multi-organizational ERP systems infrastructure and implementation approach; and select the suitable functional modules and the adjunct systems such as web portals, electronic hubs, and EDI technologies to form external linkages between trading partners. Their strategic choices on 'enterprise-wide' ERP design and management can be affected by targeted (multi-organizational) enterprise types (cf. proposition #22), IS deployment approach (cf. propositions #17 and #24), systems advancement (e.g. the intensity in use of web-based technologies) (cf. proposition #16), collaborative product attributes (cf. proposition #4), and partners' capabilities of using enterprise systems (cf. propositions #3, #20, and

#21). The multi-organizational enterprise governor of this e-integration project that is developed on, for instance, RosettaNet EDI connections or SAP i6 ERPII architecture, will then delegate the actual systems implementation to ERP vendors or third-party consulting companies (cf. proposition #23); whilst the significant value members (e.g. 1ˢᵗ tier OEMs and ODMs) might get involved in setting up the whole ERP platform. This, however, requires 'enterprise' leaders or facilitators to move away from their traditional roles as *tertius gaudens* and move towards *tertius iungens* (Obstfeld, 2005) or *primus inter pares* (Binder and Clegg, 2005b) (cf. proposition #29) whereas other key value members need to take more responsibilities for planning and managing multi-organizational enterprise-wide ERP project as well as establishing and integrating ERP systems (cf. propositions #20, #21, and #23).

As can be seen from Figure 2, the level of importance of ERP systems capabilities in the multi-organizational (virtual) enterprise, described by the *usage rate of ERP functional module* (see vertical axis in Figure 2) in the *collaborative activity* (see horizontal axis, i.e. value stream processes in Figure 2), can range from a high usage rate with the most strategic IS roles and effective IS capabilities in supporting multi-organizational relationships and cooperation (i.e. core modules) through the integrated backbone (e.g. SOA) (cf. propositions #16 and #24) to some sort of ancillary tools such as adjunct web-based portals and linkages for connecting the interfaces, which typically have less effect in facilitating multi-organizational enterprises integration. Therefore, the usage rate level of ERP modules is not only dependent upon their intrinsic IS functionalities and advancement (cf. propositions #16, #17, and #19) but also the stages of the value stream of the collaborative activity along with the value members' (core) competencies are delivered to (cf. propositions #3), as well as the targeted (multi-organizational) enterprise structures and strategy. For instance, during the concept phase of product research and development (R&D) ERP modules such as Product Content Management (PCM) and Product Lifecycle Management (PLM) that focus on centrally managing information about (joint) products and speeding up development processes can gain more influence within the 'enterprise' by contributing highly to the multi-organizational cooperation than ERP modules (e.g. MES, VMI, and EDI) that *only* deliver capabilities to the later stages of value stream (e.g. production and distribution) or allow simultaneous interface connections. Additionally, once the overall strategic orientation of the whole enterprise structure and strategy changes, e.g. moving from vertically integrated and extended enterprise paradigms towards virtual enterprise paradigm, a more integrated and flexible information systems infrastructure (e.g. SOA or cloud-based ERP information system) will be adopted at the highest (usage) rate, in order to connect (or even replace) the previous diverse and dispersed ERP modules (cf. propositions #16, #17, #19, and #22). Thus the new ERP Matrix is a vehicle for mapping and linking the architecture of ERP modules with capabilities in supporting different collaborative activities/stages of the value stream and the structure of the multi-organizational enterprises (cf. proposition #15).

The ERP Reference Grid

It has been observed in this research that traditional ERPI, ERPII, and ERPIII are *not*, as some would believe, enterprise information system types resulting from completely different information management strategies. This research study suggests that they are better thought of as a closed loop continuum of the same IS strategy focused on ERP systems enabled multi-organizational relationships and collaboration. In addition, the number and usage rate of different ERP systems types (or functional modules) for any one company participating in an 'enterprise' is closely aligned with the capabilities of supporting targeted (multi-organizational) enterprise structure and strategy and the feasibility of deploying their functionalities within the collaborative activities of the 'enterprise' (cf. propositions #16, #17, #19, #20, #21, #22, #24, #28, #29). This is referred to as 'enterprise supporting ERP capability of ERP information systems in the multi-organizational enterprises with regard to the respective collaborative activity (see Figure 2), i.e. the ability of an ERP module to be involved in the value stream due to its specific information systems competences (e.g. full collaborative commerce functionality, all internal functions supported plus core inter-company processes, and open network). In alignment with aspects of contingency theory, competence theory, and IT and business alignment view, the determination of an appropriate ERP systems design and management for the resulting multi-organizational enterprise-wide ERP information systems governance was identified to be dependent upon four main dimensions; these are

1. (Targeted) multi-organizational enterprise types,
2. Deployment approach,
3. (ERP) strategic roles in supporting enterprises, and
4. (ERP) systems advancement that are influenced by various technological and managerial factors.

In other words, the selection of an appropriate governance mode for ERP IS design and management within the multi-organizational enterprise is dependent upon various factors that influence the strategic capabilities (embedded in the ERP module) and implementation of ERP systems within the collaborative activity and ultimately in the 'enterprises'. The four identified dimensions, their related factors, and their impact on the 'enterprise supporting ERP capability' – reflecting in two key aspects, i.e. the intensity in use of web-based ERP information systems and the rate of change frequency of enterprise structure supported by ERP – are outlined in Table 5; which, in turn, can be linked up with the corresponding multi-organizational enterprise forms (i.e. VIE, EE, and VE).

For example, if the value members choose an *on-demand* ERP solution such as Software as a Service (SaaS) the enterprise systems deployment (approach) will be *simpler* than traditional ERP solution (i.e. *proprietary* or *on-premise*) since they do not have to purchase expensive equipment or make sure that they have sufficient infrastructure to handle the system. Rather, they just simply download a software application onto the computers and allow a hosting ERP vendors (or third-party consulting companies) to provide services (cf. propositions #16 and #17). This, therefore, gives stronger *flexibility*, *agility*, and accessibility for value members to adopt, access, and integrate different ERP modules within the multi-organizational enterprises resulting in *high* intensity in use of web-based ERP information systems that best serve the (multi-organizational) enterprise structure which has *high* rate of change frequency, i.e. the suggested 'enterprise' paradigm is dynamic *virtual enterprise*. In contrast, if the value members aim at *mature and well-integrated* (multi-organizational) enterprise type and *large scale of economies*, they aspire to deploy an ERP solution by hosting it internally on their own servers (i.e. *proprietary*) with great concerns about *internal operational integration and optimization*, as well as *security issues* (e.g. data protection), in order to have total control. Consequently, this requires *how* intensity in use of web-based ERP information systems because they want to keep the business data close to the source (with the central control in hands) instead of relying too much on an external Internet connection; this can best serve the (multi-organizational) enterprise structure which has *low* rate of change frequency, i.e. the suggested 'enterprise' paradigm is fully linked *vertically integrated enterprise*.

These examples show that the determination of an appropriate ERP systems design and management for the governance of enterprization (i.e. DERG-ERP) should not only based on the intensity in use of web-based ERP information systems but also on the rate of change frequency of enterprise structure

Table 5. Four dimensions influencing the enterprise supporting ERP capability in the enterprise

Four Key Dimensions	Related Factors	Impact on Enterprise Supporting ERP Capability (Correlation)		Suggested (Corresponding) Enterprise Strategy (Correlation)
		The Intensity in Use of Web-Based ERP IS	The Rate of Change Frequency of Enterprise Structure Supported by ERP	
(Targeted) enterprise types	• Mature and well-integrated • Relatively stable across the product/service value chain • Large scale of economies • Strategic outsourcing • Dynamic and temporary co-operation	Low Medium Low Medium High	Low Medium Low Medium High	VIE EE VIE EE VE
Deployment (approach)	• On-premise (a.k.a. proprietary) • On-demand (a.k.a. SaaS) • Hybrid ERP solution • Feasibility and simplicity	Low High Medium High	Low High Medium High	VIE VE EE VE (or EE)
Information systems strategic roles	• Internal operational integration and optimization • Multi-organization participation • Internet-based full collaborative commerce	Low Medium High	Low Medium High	VIE EE VE
(ERP) systems advancement	• Flexibility and agility • Security (a.k.a. data protection) • Technological compatibility between different systems	High Low Low	High Low Low	VE VIE VIE (or EE)

supported by ERP systems. Figure 3 summarizes the findings in a concise reference grid which shows four prevailing current and future ERP information system types and their enterprise supporting capability (ranked simply as 'high' or 'low' in terms of the two key aspects). In each of the quadrants the best suited ERP system type (or generation) (i.e. ERPI, ERPII, and ERPIII) depending on the intensity in use of web-based technologies and the rate of change frequency of the targeted (multi-organizational) enterprise structure is given with some of its key characteristics. Additionally, each quadrant of the ERP Reference Grid will be characterized in more detail in Table 6.

The Evolutionary Multi-Organizational ERP Configuration

The empirical findings show that, once established, multi-organizational relationships and their related governance (i.e. design and management) structures, as well as the supporting ERP information systems will and have to change over time (cf. propositions #3, #4, #5, #15, and #16) depending on the varying significance of contingency factors acting upon it (e.g. core competence, delivered products, strategic roles of ERP systems) (cf. propositions #3, #4, #5, #6, #7, #17, and #22). This is in order to stay adaptive to constantly and rapidly changing industrial, multi-organizational relationships, and information systems management requirements which reflects basic ideas of contingency and configuration theories in the sense that 'enterprise' structures and strategies and the supporting ERP information system types are complex adaptive systems that evolve within the 'ecosystem'. For instance, web-aware closed and monolithic (traditional) ERPI systems could be used to enable fully linked and stable multi-organizational

Figure 3. The ERP reference grid: determining appropriate ERP system types

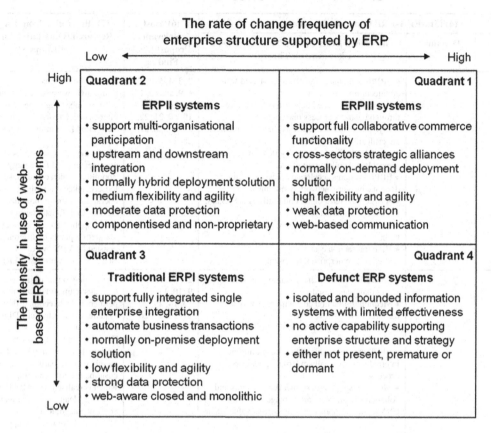

enterprise optimization and integration whilst automating business transactions (Chen, 2001; Stevens, 2003). Alternatively, componentized web-based ERPII systems could be used to facilitate more flexible multi-organizational cooperation with some collaborative commerce potential whilst focusing on integrating upstream and downstream of value stream (Bendoly et al., 2004; Daniel and White, 2005; Monk and Wagner, 2009).

These changes of multi-organizational enterprise-wide ERP information system design and management types seem to be constantly reiterating and evolving, and occur partially (i.e. based on the reconfiguration of ERP functional modules) leading to a closed loop continuum of information systems strategy focused on ERP enabled multi-organizational relationships and collaboration. Figure 4 suggests the evolutionary configuration that ERP information system (IS) types may go through within the context of multi-organizational enterprises. According to the above discussion, when ERP IS type evolves from traditional ERPI to ERPII and towards ERPIII, the value members engaging in the multi-organizational enterprises are required to increasingly adopt web-based information technologies to support more flexible and cloud-based enterprise systems. On the other hand, when the rate of change frequency of 'enterprise' structure becomes higher (i.e. transforming from stable and inflexible enterprise structure to dynamic temporary enterprise structure), the corresponding ERP systems design and management strategies will be developed from (traditional) ERPI systems to ERPII systems and on towards web-based ERPIII architecture.

Table 6. An illustration of the new DERG-ERP concept using empirical examples and links to literature

(4) (5) DERG-ERP Conceptual Element			(6) Most Relevant Propositional Finding	(7) Illustration from Empirical Research (Qualitative Empirical Examples)
Static	**Dynamic**	**Theoretical Description (Provenance from Literature on Theory)**		
Quadrant 1 Virtual Enterprise (VE) with ERPIII		• ERPIII contains a flexible, agent-based ICT architecture • Quick and dynamic inter-firm collaboration through business process management • Psychological issues such as trust and conflict are critical success factors • Flexible, agility, loose, temporary, and dynamic project based collaborative venture • ERPIII systems accelerate quicker and more dynamic business network communication • Assisted by SOA, cloud computing, PaaS, SaaS, and other web-based tools • Potential high risk with fragmented resource base • High transaction cost • High inter-enterprise integration	2, 3, 4, 5, 6, 7, 8, 9, 11, 12, 13, 14, 16, 17, 18, 19, 20, 21, 22, 23, 25, 26, 27, 28, 29	Both print-on-demand-Co and printing-Co were setting up on a small venture embracing large amount of inter-firm collaboration supported by web-based ERP information systems or EDI. This is the future enterprise management and IS strategy for Zoomlion, which could make them loosely linked with other partners' operations through more mature and flexible ERP functionalities. Lanye applied the VE strategy for integrating plants in different locations assisted by VPN (Virtual Private Network) and web-based ERP systems.
Quadrant 2 Extended Enterprise (EE) with ERPII		• Enterprise strategy changes into goal seeking rather than issue based • Medium transaction cost with relatively lean resource base • BPR for medium degree of intra-enterprise integration • ERPII can enable high level integration of internal and potentially external operational processes • Moderate supplier-customer relationships and collaborative alliances are managed by SCM/CRM systems approaching the virtual value chain concept • More stable, strategic, close, and permanent collaborative venture focused	2, 3, 4, 5, 7, 8, 9, 10, 11, 12, 13, 14, 19, 20, 21, 22, 25, 26, 27, 28	Electronic-Co was trying to integrate with upstream and downstream partners by connecting different ERP systems via RosettaNet EDI and B2B connections. Zoomlion adopted a new business strategy to re-position its value members: joint partners, suppliers, customers, and even competitors; which is realised by SAP ERPII systems. Meanwhile, lean management concept and strategic outsourcing from CIFA and Powermole is applied.
Quadrant 3 Vertically Integrated Enterprise (VIE) with traditional ERPI		• Proprietary ERP supposedly built upon real-time information • High degree of functional units integration • Involving predominantly production processes • Potentially permanent with high degree of intra-integration • Promotes business process re-engineering • Extensive internal resource and low transaction cost • ERP used reactively • Business strategy is driven by 'top-down' approach	2, 3, 4, 5, 7, 8, 9, 10, 12, 13, 14, 20, 21, 22, 24, 25, 26, 27, 28	After ERP systems launch Zoomlion had a high level of intra-integration. Also, large contributions are noted from value members who engaged within intra-enterprise activities. Wanghai had fully achieved an internal resource integration by adopting a full ERPI system package (e.g. Yonyou ERP systems) and ancillary tools such as RFID technology.
Quadrant 4 Defunct Enterprise (DE) with limited IT/IS efficiency		• No profits achievable • Rare IT/IS implementation or no ERP • Fixed single company configuration • No active engagement in a current collaborative activity • IT driven strategy via 'bottom-up' approach • Company focuses on solving 'issue-based' problems	3, 4, 5, 7, 9, 13, 15, 20, 21, 22, 25, 29	Zoomlion was initially founded on a high-tech academic institution without any explicit profitable or commercial purposes. Wanghai was a scrap recovery plant without any enterprise management and ERP IS strategy.

continued on following page

Table 6. Continued

(4) (5) DERG-ERP Conceptual Element			(6) Most Relevant Propositional Finding	(7) Illustration from Empirical Research (Qualitative Empirical Examples)
Static	**Dynamic**	**Theoretical Description (Provenance from Literature on Theory)**		
	Quadrant 1 to Quadrant 2 From VEs to EEs by changing ERPIII into ERPII	• Strategic move for successful joint ventures depending on the existing mutual relationships and experiences • Effective partnership along with expertises, technology, and knowledge management is critical to establish common enterprise strategies regarding the culture, trust, and advanced IT/IS issues • Changing ERPIII to ERPII for better governing medium-long term relationships with suppliers whilst predicting customer's demands	1, 2, 3, 4, 5, 7, 9, 12, 14, 20, 21, 22, 25, 26, 28	In order to offer a complete printing solution, printing-Co moved from VE to EE based on its existing and successful partnerships whilst applying EDI with its trust partners. Lanye intends to apply EE to achieve a more stable organisational structure with medium-long term inter-firm relationships. In this enterprise context, ERPII could be used based on strategic alliances instead of web-based architecture.
	Quadrant 2 to Quadrant 1 From EEs to VEs by developing ERPII to ERPIII	• Transformation of EE to VE can be adopted incrementally • Upgrading from ERPII to ERPIII would increase the companies' flexibility and adaptability for coping with a quick response to the business environment • ERPIII, SCM, CRM, and e-business applications merged with SOA, SaaS, cloud computing, etc. can optimise global supply network integration • Successful stable ventures trigger the creation of new temporary, agile, and dynamic ventures • Requires open minded management with proactive IT/IS strategies • Focus on temporary market opportunity through short-term collaboration • Enterprise strategies shift from company centric into 'borderless enterprises'	1, 2, 3, 4, 5, 6, 7, 9, 11, 12, 14, 15, 16, 17, 18, 19, 20, 21, 22, 23, 25, 26, 28, 29	Electronic-Co planned to design and implement the SOA-based ERP systems to become more agile, flexible, and responsive to the customers. In the future Zoomlion may develop from EE into VE by upgrading ERPII to ERPIII to address cost-effectiveness, product uniqueness, business network optimisation, and short-temporary seamless issues with industrial third parties. Metrobank endeavoured to be more responsive to dynamic market conditions; whilst new legal and regulatory requirements demanded greater transparency and more accurate and timely information. Thus it has transformed from EE to VE by upgrading ERPII to ERPIII NetWeaver.
	Quadrant 2 to Quadrant 3 From EEs to VIEs by changing ERPII into traditional ERPI	• The enterprise with predominantly medium asset specific content and information systems move to adopt 'lock-in' tactics to gain industrial dominance and market share • For the purpose of achieving economies of scale; known as the 'shake-out' stage • Shifting ERPII systems into traditional ERPI but still keep the intelligent ICT applications such as SCM, CRM, DSS, DW, etc.	1, 2, 3, 4, 5, 7, 9, 10, 12, 14, 19, 20, 21, 22, 24, 25, 26	Print-on-demand-Co has gained a large scale of economies by integrating and cooperating with different functional legal entities such as channel distributors and logistics (e.g. Amazon), publishers, and IT providers in a whole.
	Quadrant 3 to Quadrant 2 From VIEs to EEs by developing traditional ERPI to ERPII	• Business processes are re-engineered and lean thinking must be adopted in parallel • The most valuable members who engaged in the entire value chain have transferred from outside the company boundary to inside the enterprise boundary • A new strategic partnership has revived an existing and proven enterprise module by deploying it in an EE context • ERPII replaces traditional ERPI with SCM and CRM tools to gain medium inter-integration rather than merely intra-integration • Shifting from 'issue-based' problem solving into goal seeking strategy formulation via business driven 'top-down' approach	1, 2, 3, 4, 5, 7, 9, 12, 14, 15, 20, 21, 22, 25, 26	Electronic-Co has developed its ERP systems by extending the functional modules to include SCM, CRM, and EDW to address the real business-to-business integration, as well as manage and control suppliers better. By re-classifying the value members and re-designing business processes, Zoomlion's new production line is based on collaborative alliances with ERPII information systems.

continued on following page

Table 6. Continued

(4) (5) DERG-ERP Conceptual Element			(6) Most Relevant Propositional Finding	(7) Illustration from Empirical Research (Qualitative Empirical Examples)
Static	**Dynamic**	**Theoretical Description (Provenance from Literature on Theory)**		
	Quadrant 3 to Quadrant 1 From VIEs to VEs by developing traditional ERPI to ERPIII	• Traditional VIE or M&A strategies try to seek new innovative ventures to remain competitive • ERPIII replaces traditional ERPI towards a more flexible and agile information systems • Web-based technologies and other ICT tools will assist this new enterprise management pattern	1, 2, 3, 4, 5, 6, 7, 9, 11, 12, 14, 15, 16, 17, 18, 19, 20, 21, 22, 23, 25, 26, 28, 29	By re-classifying the value members, Lanye has transformed from VIE into VE by setting up its own 'Virtual Private Network' (VPN) and ERP-GPS infrastructure for achieving agile or even the leagile manufacturing in response to the dynamic complex marketing demands.
	Quadrant 1 to Quadrant 3 From VEs to VIEs by changing ERPIII into traditional ERPI	• In the case of highly asset specific can be controlled or influenced by former partners internally • Try to extend business portfolio and product/service differentiation to cover whole supply chain cycle via 'forward (vertical) integration' or 'backward (vertical) integration' strategies • Changing ERPIII to traditional ERPI aiming at in-house IT/IS development, in order to reduce the transaction cost	1, 2, 3, 4, 5, 7, 9, 10, 12, 14, 20, 21, 22, 24, 25, 26	As soon as completing the virtual business network across intra- and inter-organisational scopes, Lanye gradually changed its enterprise structure from VE into a more stable and fully linked VIE to gain more market profits and bargain power against its competitors within the same industry; whilst web-based ERP solutions need to be replaced by in-house ERP solutions.

Figure 4. The Evolutionary configuration of multi-organizational ERP information systems

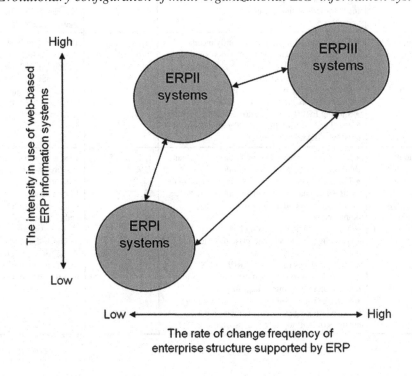

This kind of adaptive IS strategy paradigm can be regarded as a 'dynamic information systems community'; each of these ERP system types (i.e. ERPI, ERPII, and ERPIII) is considered to be a 'dynamic equilibrium' within the ecosystem 'multi-organizational enterprise-wide ERP strategy' around which one ERP system type consisting of ERP functional modules is configured and implemented for a certain period until flipping over to another ERP type (bifurcation), in order to best serve the targeted multi-organizational enterprise structures and strategies (e.g. virtual enterprise). However, as opposed to the assumptions of mere quantum change of the (ERP) design and management applied in complexity theory this chapter also reveals evidence for step-by-step adaption and reconfiguration of ERP information system types to balance emergence and control between different ERP functional modules more in line with the argumentation of contingency and configuration theory (see examples for both in Table 6). Moreover, these examples show that the bifurcation from one design and management type to another can follow a two-way pattern (hence the double sided arrows in Figure 4) although the clockwise cyclical pattern from ERPI systems through ERPII systems to ERPIII systems is the most common and likely evolution (or IS development) to be observed in practice.

A PROPOSED CONCEPTUAL FRAMEWORK

Figure 5 is a summary of the generalizable findings from the empirical studies presented as the final contingency framework known as the Dynamic Enterprise Reference Grid for ERP (DERG-ERP) which demonstrates how to guide the interactions between ERP information systems and the management of multi-organizational enterprises; and how to rightly pursue the virtual enterprise paradigm supported by the correct type of ERP systems to make a significant contribution to knowledge in the fields of information systems and multi-organizational enterprise management as well as for the application of this knowledge to practice. Thus the author believes it is a valuable and significant generalizable conceptual deliverable from this research.

The DERG-ERP as shown in Figure 5 is now described generically quadrant by quadrant; whilst the pairing VE-ERPIII in Quadrant 1 – the one that is mainly concerned by this chapter (in comparison to other two pairings VIE-ERPI and EE-ERPII) – is highlighted and suggested.

Traditional ERPI Systems Use in VIEs

In Quadrant 3 of the DERG-ERP in Figure 5 a VIE would be the most appropriate multi-organizational enterprise form using a traditional ERPI system which can support all core processes and provide some inter-departmental integration (within a single legal entity). Such system are relatively good at long term issue based (or detailed problem solving) tasks and help accomplish business driven top-down goals, although they do not contribute directly towards the strategic forward vision of a company because they are usually operational and transactional in nature; and so therefore tend to entrench current practice and become relatively reactive to strategic and environmental business changes, rather than being the driver of flexibility or change. Traditional ERPI system performs best when core competencies of strategic partners (a.k.a. value members) – particularly the most influential/focal firms in the multi-organizational enterprise are currently highly engaged, e.g. due to their mature, well-established, and widely useable

Figure 5. Dynamic Enterprise Reference Grid for Enterprise Resource Planning (DERG-ERP) contingency framework

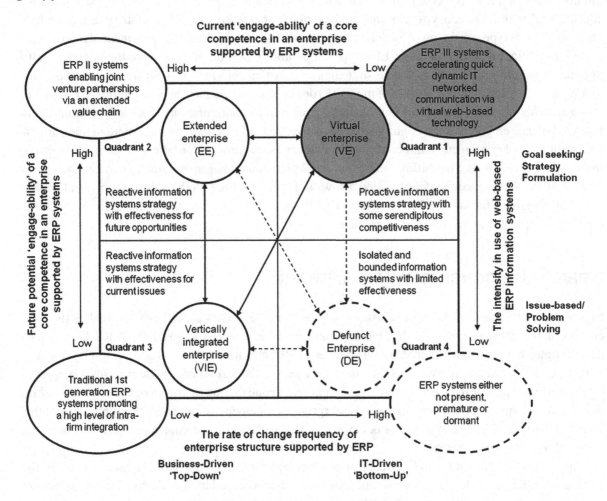

capabilities, but could decline in attractiveness in the future, e.g. because of fears that profit margins are eroding or that their technologies may become obsolete; thus allowing transaction costs to be minimized and scale of economy to be maximized.

ERPII Systems Use in EEs

In Quadrant 2 of the DERG-ERP in Figure 5 an EE is the most appropriate multi-organizational enterprise form. The EE best serves medium-to-large sized operations aspiring to form closer (joint venture) partnerships within an extended value chain. ERPII systems are able to extend ERPI capabilities to cover SCM, customer relationship functions, and some collaborative commerce potential to encourage active participation from other legal entities. ERPII systems can therefore drive business driven top-down tasks which can be directly used for achieving goals and formulating strategy across company boundaries (e.g. supply chain policies and collaborative forecasting with suppliers). ERPII is most effective when core competencies of strategic partners in the multi-organizational enterprise are currently, and in the near future, highly engaging, e.g. owning to their relatively mature nature and market success; this makes them highly attractive to other multi-organizational enterprise members, and therefore highly likely to be needed in new collaboration, with new *modus operandi*.

ERPIII Systems Use in VEs (The Mainly Concerned Pairing)

In Quadrant 1 of the DERG-ERP in Figure 5 a VE is shown. The VE best serves organizations (participating in the multi-organizational enterprise) which have aspirations for rapid growth (and so are likely to be relatively small) and see themselves as innovative and likely to be serial and parallel innovators or collaborators. ERPIII systems are able to facilitate temporary and highly agile operations using non-proprietary web-based technology for computer integrated manufacturing systems with decentralized operational control on a global scale and scope. ERPIII systems can therefore be used strategically to achieve strategic goals whilst still incorporating incremental IT driven changes required by bottom-up idiosyncrasies. ERPIII systems are considered to be pro-active IS with some almost serendipitous qualities (e.g. cloud-sourcing of innovative ideas) which fit well to the virtual enterprise type as long as the required security and trust-levels can be attained. Simultaneously, ERPIII is most effective when core competencies of strategic partners in the multi-organizational enterprise are currently lowly engaged but highly engaged in the future, e.g. because they usually have many newly emerging (core) competencies.

ERPIII applications are best used in multi-organizational (virtual) enterprise-wide operations within and across different legal entities (i.e. parts of companies). Based on traditional ERPI and ERPII principles, ERPIII based (virtual) enterprises will probably achieve the next level of business integration; namely to enable a strategic-level dialog between customers (or potential customers), an 'enterprise' integrator, and the extended supply chain using SOA, PaaS, SaaS technologies and Service-Level Agreement (SLA) management tools; and will most likely be maintained by a strategic IT/IS partner. Moreover, ERPIII type solutions could create truly integrated and borderless (virtual) enterprises; thus reaching near utopian levels of multi-organizational enterprise consciousness bringing about the simultaneous strengthening of operations, strategy, and IT interactivity, which the author refers to as the 'enterprization of operations".

Des and IS Misuse

Quadrant 4 of the DERG-ERP in Figure 5 shows a Defunct Enterprise (DE). Des occur when operations strategy, structural thinking, or information system policy have gone wrong or are premature; the challenge for operations and strategist in this business environment is to move to another more suitable types of multi-organizational enterprise (e.g. virtual enterprise) supported by corresponding ERP information systems as quickly as possible. In DEs ERP is often not widely used, used inappropriately or without any great effectiveness. Tasks are normally driven by bottom-up IT initiatives lacking strategic congruence.

Putting It All Together: Theory and Practice into a Useable Concept

To illustrate the implications of the developed concept of 'enterprization of operations' and the DERG-ERP contingency framework a structured recapitulation of the research presented above is given in tabular format in Table 6 which describes the 'static' typologies of (multi-organizational) enterprises and the supporting ERP systems, 'dynamic' changes they may undergo, theoretical description (a.k.a.

provenance from literature), relevant propositional findings, and qualitative empirical examples derived from 8 cases (i.e. an empirical illustration). Thus an illustration of the new DERG-ERP is based upon

1. A combination of conceptual interpretation of the empirical data (column 1 in Table 6) based on the characteristics of ERP and enterprise types (see Tables 1 and 2) combined with Figures 2 and 3;
2. The most relevant propositional findings (column 2 in Table 6); and
3. Empirical examples derived from the interviews (column 3 in Table 6).

Thereby, different examples are used to explain the 'static' and 'dynamic' components of the concept because this has been a cross-sectional study and not a longitudinal one. However, at an aggregated level Table 6 demonstrates the connection between the concept of 'enterprization of operations' (i.e. DERG-ERP) and the empirical examples which is representative of inductive grounded reasoning. In particular, the 'static' conceptual elements of 'VE-ERPIII' and the 'dynamic' transformational routes moving from VIE-ERPI, EE-ERPII, and DE with limited IT/IS efficiency to VE-ERPIII (i.e. Quadrants 2, 3, 4 to Quadrant 1) are highlighted and suggested.

CONCLUSION

This chapter summarizes recent trends in ERP systems development and examines the emerging practices in the management of multi-organizational enterprises. From 8 empirical cases 29 theoretical propositions were formed using Grounded Theory-based methodology. The new Dynamic Enterprise Resource Grid for ERP (DERG-ERP) conceptual framework is shown in Figure 5 which distills the generic principles from research into a single 'decision-making' framework, in order to assist managers in identifying the circumstances under which the so-called 'ERPIII' systems fit into the Virtual Enterprise structure and strategy; and *vice versa*.

The author claims that the generic DERG-ERP in Figure 5 is used to explain correlations between ERP system types and collaborative enterprise structures and strategy, from both manufacturing and service perspective. In all 8 cases it was observed that traditional ERPI was associated with VIEs, ERPII with EEs, ERPIII with VEs and limited IS/IT was observed in DEs. Therefore the author claims that there is a correlation between each of these pairings; which further explains and describes how and why ERP system develop from traditional ERPI and ERPII types to ERPIII type as well as how and why multi-organizational enterprise structures and strategy transforms from VIE and EE types to VE type – the one that is mainly concerned by this chapter.

The empirical findings specifically indicate that the core competence, expected competitiveness, and ERP information systems strategic capability (referred to as 'enterprise supporting ERP capability' in this chapter) are significant contingency factors that influence the design and management of multi-organizational enterprise structure and the supporting ERP systems. The DERG-ERP conceptual framework also suggests the evolutionary configuration that ERP information system types may go through within the ecosystem of 'multi-organizational enterprise-wide ERP strategy', which is determined upon two key aspects: (a) the intensity in use of web-based ERP information systems and (b) the rate of change frequency of multi-organizational enterprise structure supported by ERP information systems.

The new DERG-ERP framework gives some practical decision support and serves as a guideline to practicing information systems and multi-organizational enterprise managers. This study is also important to those companies grappling with the 'right' approach to steer their collaborative agile enterprise patterns such as virtual enterprise strategy and improve their company performance by adopting ERP systems, whilst seeking greater profits and efficiency by increasing their levels of multi-organizational collaboration. Besides this, the research will be of interest to those interested in the development of inter-organizational information systems and application of the new IT platforms and services designed to extend ERP modules and functionalities within the context of virtual operational environment (or e-collaboration).

DERG-ERP is limited by being based on 8 cases; and so is currently being tested and applied on other companies. This work contributes to a gap in extant literature about the correlation between ERP systems and multi-organizational innovation – particularly for the design and management of ERP systems supporting for virtual enterprise structure and strategy.

REFERENCES

Achrol, R., & Kotler, P. (1999). Marketing in the network economy. *Journal of Marketing, 63*(Special Issue), 146–163. doi:10.2307/1252108

Akkermans, H., Bogerd, P., Yucesan, E., & Van Wassenhove, L. (2003). The impact of ERP on supply chain management: Exploratory findings from a European Delphi Study. *European Journal of Operational Research, 146*(2), 284–294. doi:10.1016/S0377-2217(02)00550-7

Al-Mashari, M., Al-Mudimigh, A., & Zairi, M. (2003). Enterprise resource planning: A taxonomy of critical factors. *European Journal of Operational Research, 146*(2), 352–364. doi:10.1016/S0377-2217(02)00554-4

Al-Mudimigh, A., Zairi, M., & Al-Mashari, M. (2001). ERP software implementation: An integrative framework. *European Journal of Information Systems, 10*(4), 216–226. doi:10.1057/palgrave.ejis.3000406

Anderson, E., & Weitz, B. (1986). spring). Make-or-buy decisions: Vertical integration and marketing productivity. *Sloan Management Review, 27*(3), 3–19.

Anussornnitisam, P., & Nof, S. Y. (2003). E-work: The challenge for next generation ERP systems. *Production Planning and Control, 14*(8), 753–765. doi:10.1080/09537280310001647931

Argyres, N. S. (1996). Capabilities technological diversification and divisionalization. *Strategic Management Journal, 17*(5), 395–410. doi:10.1002/(SICI)1097-0266(199605)17:5<395::AID-SMJ826>3.0.CO;2-E

Arya, A., & Mittendorf, B. (2008). Pricing internal trade to get a leg up on external rivals. *Journal of Economics & Management Strategy, 17*(3), 709–731. doi:10.1111/j.1530-9134.2008.00192.x

Bagchi, S., Kanungo, S., & Dasgupta, S. (2003). Modeling use of enterprise resource planning systems: A path analytic study. *European Journal of Information Systems, 12*(2), 142–158. doi:10.1057/palgrave.ejis.3000453

Banker, R. D., Chang, H., & Kao, Y. (2010). Evaluating cross-organizational impacts of information technology – an empirical analysis. *European Journal of Information Systems, 19*(2), 153–167. doi:10.1057/ejis.2010.9

Bass, T., & Mabry, R. (2004). Enterprise architecture reference models: A shared vision for Service-Oriented Architectures. In *Proceedings of the IEEE MILCOM* (pp. 1-8).

Beheshti, H. M. (2006). What managers should know about ERP/ERPII. *Management Research News, 29*(4), 184–193. doi:10.1108/01409170610665040

Bendoly, E., Soni, A., & Venkataramanan, M. A. (2004). *Value Chain Resource Planning (VCRP): Adding value with systems beyond the enterprise.* Retrieved January 17, 2010, from www.fc.bus.emory. edu/~elliot_bendoly/VCRP_BH.pdf

Benlian, A., & Hess, T. (2011). Comparing the relative importance of evaluation criteria in proprietary and open-source enterprise application software selection – a conjoint study of ERP and Office systems. *Information Systems Journal, 21*(6), 503–525. doi:10.1111/j.1365-2575.2010.00357.x

Binder, M., & Clegg, B. T. (2005b). Partial evolutionary multiplicity: An approach to managing the dynamics of supply structures. In *Proceedings of the 18th International Conference on Production Research.* Universita di Salerno.

Binder, M., & Clegg, B. T. (2006). A conceptual framework for enterprise management. *International Journal of Production Research, 44*(18/19), 3813–3829. doi:10.1080/00207540600786673

Blackstone, J. H., Jr., & Cox, J. F. (2005). APICS Dictionary (11th ed.). APICS: The association for Operations Management.

Bond, B., Genovese, Y., Miklovic, D., Wood, N., Zrimsek, B., & Rayner, N. (2000). *ERP is dead - long live ERPII.* Retrieved November 8, 2009, from www.pentaprise.de/cms_showpdf.php?pdfname=infoc_report

Browne, J., & Zhang, J. (1999). Extended and virtual enterprises-Similarities and differences. *International Journal of Agile Management Systems, 1*(1), 30–36. doi:10.1108/14654659910266691

Buco, M. J., Chang, R. N., Luan, L. Z., Ward, C., Wolf, J. L., & Yu, P. S. (2004). Utility computing SLA management based upon business objectives. *IBM Systems Journal, 43*(1), 159–178. doi:10.1147/sj.431.0159

Byrne, J. A., & Brandt, R. (1993, February 8). The virtual corporation. *Business Week,* 36-41.

Callaway, E. (2000). *ERP – the next generation: ERP is Web Enabled for E-business.* Charleston: Computer Technology Research Corporation.

Candido, G., Barata, J., Colombo, A. W., & Jammes, F. (2009). SOA in reconfigurable supply chain: A research roadmap. *Engineering Applications of Artificial Intelligence, 22*(6), 939–949. doi:10.1016/j.engappai.2008.10.020

Chen, I. J. (2001). Planning for ERP systems: Analysis and future trend. *Business Process Management Journal, 7*(5), 374–386. doi:10.1108/14637150110406768

Childe, S. J. (1998). The extended enterprise – a concept of co-operation. *Production Planning and Control, 9*(4), 320–327. doi:10.1080/095372898234046

Cho, H., Jung, M., & Kim, M. (1996). Enabling technologies of agile manufacturing and its related activities in Korea. *Computers & Industrial Engineering, 30*(3), 323–334. doi:10.1016/0360-8352(96)00001-0

Chorafas, D.N. (2001). *Integrating ERP, CRM, Supply Chain Management, and Smart Materials.* New York, NY: CRC Press LLC and Auerbach Publications.

Chung, A. A. C., Yam, A. Y. K., & Chan, M. F. S. (2004). Networked enterprise: A new business model for global sourcing. *International Journal of Production Economics, 87*(3), 267–280. doi:10.1016/S0925-5273(03)00222-6

Clegg, B., & Wan, Y. (2013). ERP systems and enterprise management trends: A contingency model for the enterprization of operations. *International Journal of Operations & Production Management, 33*(11/12), 1458–1489. doi:10.1108/IJOPM-07-2010-0201

Crandall, R. (1968). Vertical integration and the market for repair parts in the United States automobile industry. *The Journal of Industrial Economics, 16*(3), 212–234. doi:10.2307/2097561

Cummins, F. A. (2009). *Building the Agile Enterprise with SOA, BPM and MBM.* Burlington, VT: Morgan Kaufmann Publishers and Elsevier Inc.

Daniel, E. M., & White, A. (2005). The future of inter-organizational system linkages: Findings of an international delphi study. *European Journal of Information Systems, 14*(2), 188–203. doi:10.1057/palgrave.ejis.3000529

Daniels, S. (1998). The virtual corporation. *Work Study, 47*(1), 20–22. doi:10.1108/00438029810196685

Davenport, T. H. (1998). Putting the enterprise into the enterprise system. *Harvard Business Review*, 121–131. PMID:10181586

Davenport, T. H., & Brooks, J. D. (2004). Enterprise systems and the supply chain. *Journal of Enterprise Information Management, 17*(1), 8–19. doi:10.1108/09576050410510917

Davis, E. W., & Spekman, R. E. (2004). *Extended enterprise: Gaining competitive advantage through collaborative supply chains.* New York, NY: Financial Times Prentice-Hall.

De Maria, F., Briano, C., Brandolini, M., Briano, E., & Revetria, R. (2011). Market-leader ERPs and cloud computing: A proposed architecture for an efficient and effective synergy. In *Proc. of the 10th WSEAS Conference on Applied Computer and Applied Computational Science.* Madison, WI: WSEAS.

Eckartz, S., Daneva, M., Wieringa, R., & Hillegersberg, J. V. (2009). Cross-organizational ERP management: How to create a successful business case? In *SAC'09 Proceedings of the 2009 ACM Symposium on Applied Computing.* Honolulu, HI: ACM.

Ericson, J. (2001). *What the heck is ERPII?* Retrieved May 27, 2012, from http://www.line56.com/articles/default.asp?ArticleID=2851

European Commission. (2003). Commission recommendation of 6 May 2003 concerning the definition of micro, small and medium sized enterprises. *Official Journal of the European Union, L, 124*(1422), 36–41.

Glaser, B. G., & Strauss, A. L. (1967). *The discovery of grounded theory: Strategies for qualitative research.* New York, NY: Aldine.

Goldman, S., Nagel, R., & Preiss, K. (1995). Agile Competitors and virtual organizations. New York, NY: van Nostrand Reinhold.

Goranson, H. T. (1999). *The agile virtual enterprise: Cases, metrics, tools.* Westport, CT: Quorum Books and Greenwood Publishing Group, Inc.

Hanna, V., & Walsh, K. (2000). Alliances: The small firm perspective. In *Proceedings of 4th International Conference on Managing Innovative Manufacturing* (pp. 333-340), Aston University.

Harrigan, K. R. (1984). Formulating vertical integration strategies. *Academy of Management Review, 9*, 638–652.

Harrigan, K. R. (1985). Vertical integration and corporate strategy. *Academy of Management Journal, 28*(2), 397–425. doi:10.2307/256208

Harwood, S. (2003). *ERP: The implementation cycle.* Burlington: Butterworth-Heinemann.

Hauser, K., Sigurdsson, H. S., & Chudoba, K. M. (2010). EDSOA: An event-driven service-oriented architecture model for enterprise applications. *International Journal of Management & Information Systems, 14*(3), 37–47.

He, X. (2004). The ERP challenge in China: A resource-based perspective. *Information Systems Journal, 14*(2), 153–167. doi:10.1111/j.1365-2575.2004.00168.x

Hicks, D. A., & Stecke, K. E. (1995). The ERP maze: Enterprise resource planning and other production and inventory control software. *IIE Solutions, 27*, 12–16.

Hoffman, W. J. (2007). Strategies for managing a portfolio of alliances. *Strategic Management Journal, 28*(8), 827–856. doi:10.1002/smj.607

Hofmann, P. (2008). ERP is dead, long live ERP. *IEEE Internet Computing, 12*(4), 84–88. doi:10.1109/MIC.2008.78

Hyvonen, T., Jarvinen, J., & Pellinen, J. (2008). A virtual integration – the management control system in a multinational enterprise. *Management Accounting Research, 19*(1), 45–61. doi:10.1016/j.mar.2007.08.001

Jacobs, F. R. (2007). Enterprise resource planning (ERP) – a brief history. *Journal of Operations Management, 25*(2), 357–363. doi:10.1016/j.jom.2006.11.005

Jagdev, H., Vasiliu, L., Browne, J., & Zaremba, M. (2008). A semantic web service environment for B2B and B2C auction applications within extended and virtual enterprises. *Computers in Industry, 59*(8), 786–797. doi:10.1016/j.compind.2008.04.001

Jagdev, H. S., & Browne, J. (1998). The extended enterprise – a context for manufacturing. *Production Planning and Conitrol, 9*(3), 216–229. doi:10.1080/095372898234190

Jagdev, H. S., & Thoben, K. D. (2001). Anatomy of enterprise collaboration. *Production Planning and Control, 12*(5), 437–451. doi:10.1080/09537280110042675

Johnson, T., Lorents, A. C., Morgan, J., & Ozmun, J. (2004). A customized ERP/SAP model for business curriculum integration. *Journal of Information Systems Education, 15*(3), 245–253.

Joskow, P.L. (2003). *Vertical integration, Handbook of New Institutional Economics.* Boston, MA: Kluwer.

Kaihara, T., & Fujii, S. (2002). IT based virtual enterprise coalition strategy for agile manufacturing environment. In *Proc. of the 35th CIRP Int. Seminar on Manufacturing Systems*, (pp. 32-37).

Katzy, B. R., & Dissel, M. (2001). A toolset for building the virtual enterprise. *Journal of Intelligent Manufacturing, 12*(2), 121–131. doi:10.1023/A:1011248409830

Klaus, H., Rosemann, M., & Gable, G. G. (2000). What is ERP? *Information Systems Frontiers, 2*(2), 141–162. doi:10.1023/A:1026543906354

Kumar, K., & van Hillegersberg, J. (2000). ERP experiences and evolution. *Communications of the ACM, 43*(4), 23–26.

Li, C. (1999). ERP packages: What's next? [electronic version]. *Information Systems Management, 16*(3), 31–36. doi:10.1201/1078/43197.16.3.19990601/31313.5

Li, F., & Williams, H. (1999). Interfirm collaboration through interfirm networks. *Information Systems Journal, 9*(2), 103–115. doi:10.1046/j.1365-2575.1999.00053.x

Lipnack, J., & Stamps, J. (1997). *"Virtual teams", reaching across space, time, and organizations with technology.* New York, NY: Wiley.

Lyman, K. B., Caswell, N., & Biem, A. (2009). Business value network concepts for the extended enterprise. In P. H. M. Vervest, D. W. Liere, & L. Zheng (Eds.), *Proc. of the Network Experience.* Berlin: Springer. doi:10.1007/978-3-540-85582-8_9

Lynch, R. (2003). *Corporate strategy* (3rd ed.). Harlow: Prentice-Hall Financial Times.

MacBeth, D. K. (2002). Emergent strategy in managing cooperative supply chain change. *International Journal of Operations & Production Management, 22*(7), 728–740. doi:10.1108/01443570210433517

Madu, C. N., & Kuei, C. (2004). *ERP and supply chain management.* Fairfield, CT: Chi Publishers.

Mahoney, J. T. (1992). The choice of organisational form: Vertical financial ownership versus other methods of vertical integration. *Strategic Management Journal, 13*(8), 559–584. doi:10.1002/smj.4250130802

Markus, M. L., & Tanis, C. (2000). The enterprise system experience – from adoption to success. In R. W. Zmud (Ed.), *Framing the domains of IT management: Projecting the future through the past* (pp. 173–207). Cincinnatti, OH: Pinnaflex Educational Resources, Inc.

Martinez, M. T., Fouletier, P., Park, K. H., & Faurel, J. (2001). Virtual enterprise: Organization, evolution and control. *International Journal of Production Economics, 74*(1-3), 225–238. doi:10.1016/S0925-5273(01)00129-3

Maurizio, A., Girolami, L., & Jones, P. (2007). EAI and SOA: Factors and methods influencing the integration of multiple ERP systems (in an SAP environment) to comply with the Sarbanes-Oxley Act. *Journal of Enterprise Information Management, 20*(1), 14–31. doi:10.1108/17410390710717110

Michel, R. (2000). *The road to extended ERP*. Retrieved May 8, 2009, from www.manufacturingsystems. com/extendedenterprise

Moller, C. (2005). ERPII: A conceptual framework for next-generation enterprise systems? *Journal of Enterprise Information Management, 18*(4), 483–497. doi:10.1108/17410390510609626

Monk, E. F., & Wagner, B. J. (2009). *Concepts in enterprise resource planning* (3rd ed.). Cambridge, MA: Course Technology, Cengage Learning.

Monteverde, K., & Teece, D. J. (1982). Supplier switching costs and vertical integration in the automobile industry. *The Bell Journal of Economics, 13*(1), 206–213. doi:10.2307/3003441

Muscatello, J. R., Small, M. H., & Chen, I. J. (2003). Implementing enterprise resource planning (ERP) systems in small and midsize manufacturing firms. *International Journal of Operations & Production Management, 23*(8), 850–871. doi:10.1108/01443570310486329

O'Neill, H., & Sackett, P. (1994). The extended manufacturing enterprise paradigm. *Management Decision, 32*(8), 42–49. doi:10.1108/00251749410069453

Obstfeld, D. (2005). Social networks, the tertius iungens orientation, and involvement in innovation. *Administrative Science Quarterly, 50*(1), 100–130.

Owen, L., Goldwasser, C., Choate, K., & Blitz, A. (2008). Collaborative innovation throughout the extended enterprise. *Strategy and Leadership, 36*(1), 39–45. doi:10.1108/10878570810840689

Pal, N., & Pantaleo, D. C. (2005). *The agile enterprise: Reinventing your organization for success in an on-demand world*. New York, NY: Springer SciencetBusiness Media, Inc.

Park, K., & Kusiak, A. (2005). Enterprise resource planning (ERP) operations support system for maintaining process integration. *International Journal of Production Research, 43*(19), 3959–3982. doi:10.1080/00207540500140799

Ponis, S. T., & Spanos, A. C. (2009). ERPII systems to support dynamic, reconfigurable and agile virtual enterprises. *International Journal of Applied Systemic Studies, 2*(3), 265–283. doi:10.1504/IJASS.2009.027664

Rappa, M. A. (2004). The utility business model and the future of computing services. *IBM Systems Journal, 43*(1), 32–42. doi:10.1147/sj.431.0032

Rashid, M. A., Hossain, L., & Patrick, J. D. (2002). The evolution of ERP systems: A historical perspective. In L. Hossain, J. D. Patrick, & M. A. Rashid (Eds.), *Enterprise Resource Planning: Global opportunities and challenges* (pp. 1–16). Hershey, PA: Idea Group Publishing. doi:10.4018/978-1-931777-06-3.ch001

Rayport, J. F., & Sviokla, J. J. (1995). Exploiting the virtual value chain. *The McKinsey Quarterly, 1*, 21–36.

Richardson, J. (1996). Vertical integration and rapid response in fashion apparel. *Organization Science, 7*(4), 400–412. doi:10.1287/orsc.7.4.400

Rothaermel, F. T., Hitt, M. A., & Jobe, L. A. (2006). Balancing vertical integration and strategic outsourcing: Effects on product portfolio, product success, and firm performance. *Strategic Management Journal, 27*(11), 1033–1056. doi:10.1002/smj.559

Saeed, K. A., Malhotra, M. K., & Grover, V. (2011). Interorganizational system characteristics and supply chain integration: An empirical research. *Decision Sciences, 42*(1), 7–42. doi:10.1111/j.1540-5915.2010.00300.x

Scott, J. E., & Vessey, I. (2000). Implementing enterprise resource planning systems: The role of learning from failure. *Information Systems Frontiers, 2*(2), 213–232. doi:10.1023/A:1026504325010

Sharif, A. M. (2010). It's written in the cloud: The hype and promise of cloud computing. *Journal of Enterprise Information Management, 23*(2), 131–134. doi:10.1108/17410391011019732

Sharif, A. M., Irani, Z., & Love, P. E. D. (2005). Integrating ERP with EAI: A model for post-hoc evaluation. *European Journal of Information Systems, 14*(2), 162–174. doi:10.1057/palgrave.ejis.3000533

Sharp, J. M., Irani, Z., & Desai, S. (1999). Working towards agile manufacturing in the UK industry. *International Journal of Production Economics, 62*(1-2), 155–169. doi:10.1016/S0925-5273(98)00228-X

Shehab, E., Sharp, M., Supramaniam, L., & Spedding, T. (2004). Enterprise resource planning: An integrative review. *Business Process Management Journal, 10*(4), 359–386. doi:10.1108/14637150410548056

Songini, M. L. (2002). J.D. Edwards pushes CRM, ERP integration. *Computerworld, 36*(25), 4.

Stalk, G., Evans, P., & Shulman, L. E. (1992). Competing on capabilities: The new rules of corporate strategy. *Harvard Business Review*, (March-April), 57–69. PMID:10117369

Stevens, C. P. (2003). Enterprise resource planning: A trio of resources. *Information Systems Management, 20*(3), 61–71. doi:10.1201/1078/43205.20.3.20030601/43074.7

Strauss, A. (1987). *Qualitative Analysis for Social Scientists*. Cambridge: Cambridge University Press. doi:10.1017/CBO9780511557842

Strauss, A., & Corbin, J. (1990). *Basics of qualitative research: Grounded theory procedures and techniques*. Newbury Park, CA: Sage.

Sutton, S. G. (2006). Extended-enterprise systems' impact on enterprise risk management. *Journal of Enterprise Information Management, 19*(1), 97–114. doi:10.1108/17410390610636904

Tapscott, D., Ticoll, D., & Lowy, A. (2000). *Digital capital*. Boston, MA: Harvard Business School Press.

"Ted" Weston, F.C., Jr. (2002). *A vision for the future of extended enterprise systems*. Presentation, J.D. Edwards FOCUS Users Conference, Denver, CO.

"Ted" Weston, F.C., Jr. (2003, November/December). ERPII: The extended enterprise system. *Business Horizons*, 49-55.

Tencati, A., & Zsolnai, L. (2009). The collaborative enterprise. *Journal of Business Ethics, 85*(3), 367–376. doi:10.1007/s10551-008-9775-3

Themistocleous, M., Irani, Z., & O'Keefe, R. (2001). ERP and application integration: Exploratory survey. *Business Process Management Journal, 7*(3), 195–204. doi:10.1108/14637150110392656

Thun, J. H. (2010). Angles of integration: An empirical analysis of the alignment of internet-based information technology and global supply chain integration. *Journal of Supply Chain Management, 46*(2), 30–44. doi:10.1111/j.1745-493X.2010.03188.x

Torbacki, W. (2008). SaaS – direction of technology development in ERP/MRP systems. *Archives of Materials Science and Engineering, 31*(1), 57–60.

Triantafillakis, A., Kanellis, P., & Martakos, D. (2004). Data warehousing interoperability for the extended enterprise. *Journal of Database Management, 15*(3), 73–82. doi:10.4018/jdm.2004070105

Vallespir, B., & Kleinhans, S. (2001). Positioning a company in enterprise collaborations: Vertical integration and make-or-buy decisions. *Production Planning and Control, 12*(5), 478–487. doi:10.1080/09537280110042701

Vathanophas, V. (2007). Business process approach towards an inter-organizational enterprise system. *Business Process Management Journal, 13*(3), 433–450. doi:10.1108/14637150710752335

Vazquez-Bustelo, D., & Avella, L. (2006). Agile manufacturing: Industrial case studies in spain. *Technovation, 26*(10), 1147–1161. doi:10.1016/j.technovation.2005.11.006

Walters, D. (2004). New economy – new business models – new approaches. *International Journal of Physical Distribution & Logistics Management, 34*(3/4), 219–229. doi:10.1108/09600030410533556

Wan, Y., & Clegg, B. T. (2010). Enterprise management and ERP development: Case study of Zoomlion using Dynamic Enterprise Reference Grid. In *CENTERIS 2010 – Conference on ENTERprise Information Systems* (pp. 191-198). Springer Verlag.

Wight, O. (1984). *Manufacturing Resource Planning: MRPII. Williston: Oliver Wight Ltd.* Publications.

Wilkes, L., & Veryard, R. (2004, April). Service-oriented architecture: Considerations for agile systems. *Microsoft Architect Journal.* Retrieved May 16, 2010, from www.msdn2.microsoft.com

Wood, B. (2010). *ERP vs ERPII vs ERPIII future enterprise applications.* Retrieved October 3, 2010, from www.r3now.com/erp-vs-erp-ii-vs-erp-iii-future-enterprise-applications

Xu, W., Wei, Y., & Fan, Y. (2002). Virtual enterprise and its intelligence management. *Computers & Industrial Engineering, 42*(2-4), 199–205. doi:10.1016/S0360-8352(02)00053-0

Zrimsek, B. (2003). *ERPII vision.* Paper presented at US Symposium/ITxpo, Gartner Research (25C, SPG5, 3/03), San Diego, CA.

KEY TERMS AND DEFINITIONS

Enterprise: An entity, regardless of its legal form, including partnerships or associations regularly engaged in economic activities. Practically this means parts of companies working with parts of other companies to collectively deliver complex product service systems.

Enterprise Resource Planning: An electronic information system includes a set of business applications or modules, which links various business units of an organization (or multi-organizational enterprises) into an integrated system with a common platform for flow of information across the entire business.

Enterprise Supporting ERP Capability: Enterprise Supporting ERP Capability determines the type and ability of ERP systems to be applied in the multi-organizational enterprise due to its specific information systems competences, capabilities supporting the targeted enterprise structure and the feasibility of deploying their functionalities within the enterprise.

ERPI: Internally integrated information systems used to gain operational competitive advantage by primarily supporting core internal (operational) functions.

ERPII: An enterprise information system recognized as an integral part of business strategy enabling multi-organizational collaboration through extension of operations to close and trusted partners.

ERPIII: A flexible, powerful information system incorporating web-based technology which enables (multi-organizational) enterprises to offer increasing degrees of connectivity, collaboration and dynamism through increased functional scope and scalability.

Extended Enterprise: Parts of companies working with parts of other companies to collectively deliver complex product service systems. This is a semi permanent multi-organizational enterprise structure designed to be flexible and agile.

Vertically Integrated Enterprise: Parts of companies working with parts of other companies to collectively deliver complex product service systems. A multi-organizational vertically integrated enterprise operates almost as large single well-integrated multi-functional firm striving for scales of economy.

Virtual Enterprise: Parts of companies working with parts of other companies to collectively deliver complex product service systems. A multi-organizational virtual enterprise is designed to be short term and highly agile.

Chapter 11
Design of Information Spaces and Retrieval of Information using Electrostatics in Virtual Spaces

Vaibhav Madhok
University of British Columbia, Canada

Navin Rustagi
Baylor College of Medicine, USA

ABSTRACT

Humans have a rich awareness of locations and situations that directs how we interpret and interact with our surroundings. The principle aim of this paper is to create 'Information Spaces' where people will use their awareness to search, browse and learn. In the same way that they navigate in a physical environment, they will navigate through knowledge. An information space is a type of design in which representations of information objects are situated in a principled space. In this chapter we present an architecture based on the principles of electrostatistics, which presents a model for design of information spaces. Our model gives an easy conceptual framework to reason about how information can be represented as well as secure ways of extracting and storing information leading to a design which are easily scalable in virtual team environments.

1. INTRODUCTION

The design and representation of information has been studied extensively in the recent past [1,2]. In particular, networks representing information occur naturally as part of human design efforts. Many properties of modern day networks, like the internet topology, road networks exhibit properties close to that of networks occurring naturally in nature(Barabasi et al.,

2004). Insights from physical systems have been of great use in designing the structural properties of modern day synthetic networks built out of the human interaction graph (Barabasi et al, 2004;

DOI: 10.4018/978-1-4666-9688-4.ch011

Dorogovtsev et al., 2013). With the advent of modern telecommunication infrastructure, it is increasingly possible to connect as virtual teams in an adhoc, spontaneous and time bound manner, with constraints on time to live, fault tolerance and reliability. The design of the information spaces underlying a virtual team, can have implications to the robustness and efficiency of human-to-human interactions. Also with the advent of several data collection initiatives (Reips et al., 2014), with most of the data unstructured it is increasingly important to analyze how information is retrieved in this environment. Past work in this area has relied on extracting relational features to optimize extraction algorithms (Korfhage, 2008).

One source of design principles is to employ insights from the fields of professionals who construct spatial information design, museum exhibit designers, librarians. Another source is to get inspiration from nature and let the natural laws of physics work for us. This is precisely the objective of this paper. The Laws of Physics, electrostatics in this case, will work for us.

In this paper we present a design of information spaces based on fundamental principles of electrostatistics. Humans have a rich awareness of locations and situations that directs how we interpret and interact with our surroundings. The principle aim of this paper is to create 'Information Spaces' where subjects will use their awareness to search, browse and learn. The information objects are put in principled space, which can be regarded as the surface of a metallic conductor. This vectorization of information is the basis of many modern information retrieval systems (Singhal, 2001). Now the charge density on the surface will decide up to what 'depth' a certain object can be accessed. The basic idea is to put information objects on the principled space and let the subject control the electrostatic field around the conductor so that the redistribution of charges will provide different 'depths' or 'levels' to which a certain information object can be accessed. This approach lends itself to a mathematical treatment of Information retrieval techniques, which have been discussed in detail in (Manning et al, 2008) and (Berry et al., 1999). In their work, the authors use orthogonality of representations of data as matrices to simplify the task of resolving ambiguous queries. In our work we use the underlying physical laws to simplify the task of information retrieval.

Many other methods of information retrieval use indexing as a method of fast information retrieval (Frakes, 1992). While this makes information access fast, indexing can still be a time consuming process for potentially petabytes of data in typical search queries (Lai, 2008). A probabilistic architecture will alleviate this problem to some extent (Maron, 2008), but our architecture can give an O(1) access time in the best case scenario. Another advantage of our architecture over traditional information retrieval architectures is the inherent design scalability and parallelization afforded by using user specific query infrastructure.

2 . BACKGROUND

Information design and retrieval involves representation of information in an abstract form and its retrieval when the user enters a query. Queries are formal statements which specify what specific information is being requested. Several objects of information might match a particular query and they can be ranked by quantifying the degree of relevancy. Information is mapped in a database which can be used to store audio, video or images (Goodrum, 2000, Jonathan 1999).

Research in the design of information spaces and retrieval of information spans almost hundred years. It started with cards, key punches and tabular data for census. Progress was made by employing statistical methods in information retrieval in the 1950's and continued in the next few decade. In the last quarter of the 20th century, with the advent of computers, artificial intelligence and Bayesian methods became important components of information retrieval systems. In particular, Saltons's vector space model (Salton, 1975) and Kent's statistical sampling method to determine the relevancy of a query became cornerstones of this field.

Models of information retrieval can be classified on the basis of the mathematical basis (Singhla, 2001). For example, set theoretic models represent information stored as text and data. Depending on the nature of information, they can be based on Boolean or Fuzzy retrieval. The other type of mathematical models are the algebraic models. Here the information is represented as a vector space or matrices. A query in such a model is a vector in this vector space. Lastly, probabilistic models, based on Bayes theorem, involve information retrieval as a process of probabilistic inference. Various metrics are used to quantify the quality of information retrieval in the models. Precision, the fraction of documents relevant to user information, and recall, fraction of the documents relevant to the query are important metrics used to benchmark the efficiency of information architectures.

In this work, we propose a new model for the design of information spaces and also propose how this can help managers and leaders in effective decision making, organization, direction and control. We propose an interdisciplinary approach involving information technology, artificial intelligence and management paradigms to address the challenges faced by managers in diverse settings. Our goal is two fold: Use of information technology to solve business problems and more importantly, proposing a new design for information spaces which might have a wide applicability. Our approach can be used to design information systems to meet the needs of a wide variety of organizations and managers and help them be more efficient in their ability to harness vast amounts of information. In the next section, we describe our model for information design and retrieval that employs concepts from electrostatics and show how this design is useful for managers and leaders in diverse scenarios.

3. METHODS

This paper gives an analogy from nature in the design of information spaces. If the information space is considered as to be the surface of a metallic conductor, then we can develop a method to represent and extract information using the principles of electrostatics. The basic idea behind the use of a metallic conductor is that it has a lot of free electrons and thus on the application of an electric field there will be a redistribution of charges every time.

In our model each point in the space inside the volume of the conductor can be regarded as a 'Query'. Thus the volume of the conductor is our 'Query Space". Also by choosing a point inside the conductor, we can generate a query that cannot be projected through normal information retrieval. Not only can the information objects be accessed, but new queries can be generated by just selecting an arbitrary point inside the conductor. This query will have a unique answer by our information architecture. Most importantly we argue that a machine can be instructed to learn from this information architecture.

We now present a rule which correlates the electrostatic charge potential to the extent of data which can be accessed. A metallic conductor has a lot of free electrons and thus on the application of an electric field there will be a redistribution of charges every time. Greater the charge density at any given point, greater will be the magnitude or the amount to which the corresponding information object will be accessed. It can be regarded as the granularity or depth to which the information object can be accessed.

Let the information density be referred to as **I.** Let the surface charge density be referred to as **σ.** The rule is as follows.

Rule: $I = K\sigma^{\alpha}$,

where k is the constant of proportionality.

Therefore, greater the charge density at any given point, greater will be the magnitude or the amount to which the corresponding information object will be accessed. It can be regarded as the granularity or depth to which the information object can be accessed.

4. EXAMPLES AND DISCUSSIONS

As a proof of principle of our formulation of an information space we demonstrate how electrostatic charge potential over conductors can be harnessed for design of information spaces.

1. **Representation of Information:** We refer to the coordinate system in this section to be the usual x, y, z coordinate system, with the usual notation of ΔA for the infinitesimal change in area A. Initially the charge density on the inner surface of the conductor (whenever we will talk about the charge density that decides the extraction of information we will talk about the charge density on the inner surface of the conductor) is zero. Therefore according to our rule, though the information is present it cannot be extracted. (see Figure 1a)
2. **Query:** A point charge of magnitude -Q is placed at the center of the spherical shell as shown below in Figure 1b. This charge is placed for the purpose of information abstraction. We call it the 'query charge'. The free electrons inside the spherical shell will tend to go towards the outer surface and there will be residual positive charge on the surface. The inner surface charge density can be calculated using Gauss's Law for electric flux.

    According to the Gauss's Law:

$$\oint E \cdot ds = q / \varepsilon_{0}$$

where 'q' is the net charge enclosed within the surface S. As the electric field at any point of the surface S is zero, the net charge (q) enclosed by the surface will be zero. i.e. in the above formula q = 0. Thus there is some positive charge on the inner surface. Because of the symmetry of the figure this charge will distribute itself uniformly. Now according to our rule, the information can be extracted. Information density I, is given by the formula:

$$I = K\sigma^{\alpha}$$

Note that the expression for I has spherical symmetry. This means that the amount of information corresponding to each and every information object on the spherical shell will be extracted at a uniform

Figure 1. a) Representation of information object; b) a basic query represented by a negative charge inside the sphere; c) navigating through information space and querying in depth: As we move the negative charge -Q, we can obtain information in depth.

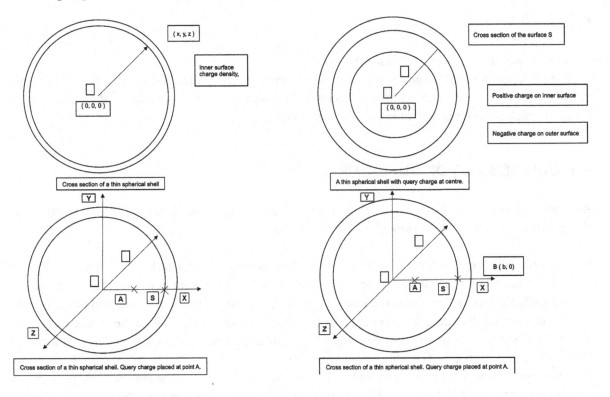

'granularity' or 'depth'. Now we will consider the case when we want a particular information object in more detail than the others.

3. **Querying in Depth:** Our 'query charge', of magnitude -Q has been placed at a point A in the Figure 1c . Let OA = (r^2/b). We are required to find I at points on the inner surface of the spherical shell on the line joining O and A. We report the result here, using the "method of images" in electrostatics. Therefore I = K [(Q_b/4πr(b-r)^2)[1+b/r]] $^\alpha$. Thus the information object corresponding to the point S or the region around S will be extracted in greater depth.

4.1 Discussions

4.1.1 Interpretations Arising from Information Architecture

This information architecture lends itself to certain interpretations of concepts.

1a: Each point in the space inside the volume of the conductor can be regarded as a 'Query'. Thus the volume of the conductor is our 'Query Space'. This is in fact the basic underlying philosophy of this paper: A charged particle inside a conductor is actually a Query in nature, and the laws of physics solve it.

1b: By choosing a point inside the conductor, a query can be generated that is novel. Not only can the information objects be accessed, but new queries can be generated by just selecting an arbitrary point inside the conductor. This query will have a unique answer by our information architecture.

2a: A machine can be instructed to learn from this information architecture. In Figure 2a the information is mapped on a spherical shell and in the process of subsequent retrieval: the charged particle describes the shown curve.

Figure 2b shows the section of the curve described by the 'query charge' infinitesimally divided into combination of steps in radial and tangential direction. The journey of charge in the radial direction (B to C) gives the detail of certain section of information and the journey in the tangential direction (A to B) can give access to some other information object without increasing the depth with which they are to be extracted. If we can instruct our charged particle to follow a certain curve, we will get the extraction of the different information objects according to our curve. This is an alternative way of getting queries. The machine can deal with such curves more easily rather a certain query language having complicated syntax.

2b: The previous bullet introduces the feature of scanning the information. Suppose we have to scan the information with a certain given 'depth' or detail. Figure 3a and b describes the information extraction features in detail.

In the curve describing the motion of the charged particle, we can give a 'spike' at the required point. In the figure, the information object corresponding to the region in the neighborhood of the point (r, Θ) (r is the radius of our spherical shell) will be extracted in greater detail. This is because it is this information object, which corresponds to the 'spike'.

Figure 2. a) Curve described by query charge; b) query charge in tangential and radial directions

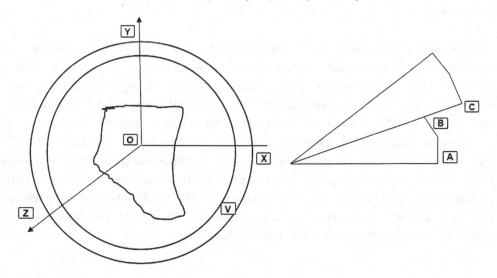

Figure 3. a) The path taken by the query charge. It represents basic scanning of "all" information at a constant depth. b) Obtaining information at differential depths

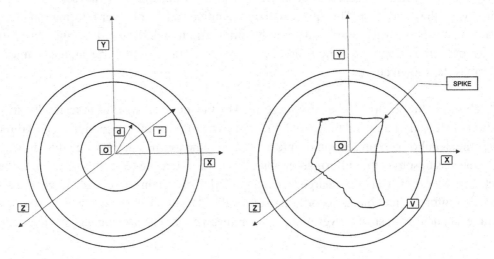

3: Consider the problem: Which curve should the 'query charge' describe to extract the maximum amount of in formation from the surface of the spherical shell? The answer to this problem will be a concentric spherical shell of radius d, such that d>r, where r gives the radius of the conductor. Thus we can infer that the maximum possible amount of information extracted from the system corresponds to the maximum possible volume that is enclosed by the 3-dimensional curve described by our 'query charge'. In more general terms, consider the volume covered by the curve describing the motion of the 'query charge'. Let it be a sphere of radius d. Now a) the greater the volume of this sphere, more the information that can be extracted from the system. b) Also, any point inside the volume of this smaller sphere won't yield any extra information. In other words: The information that can be extracted by moving the charge inside this spherical volume of radius d is a subset of the information extracted by the 'query charge' when it described the surface of this sphere.

4.1.2 Applications

1. **Representation of Information in Multimedia:** In this section we describe some applications of this concept. On a computer, the information objects can be mapped on the surface of the conductor, like a database. The mouse pointer can act as our 'query charge'. Depending upon the way user moves the mouse, the electrostatic field will get modified and there will be redistribution of charge and hence different degrees of information extraction as shown in the Figure 4. This is in line with existing hierarchical clustering models for extraction of data (Jardine et al, 1971) .

2. **Connectedness of Related Information:** Consider the case of our spherical shell as shown in the Figure 5. The points O, P and Q' are collinear. Let P (r',θ) be the location of the generator charge. Therefore the maximum charge density will at the point Q'(r, θ). But notice that the region near the point Q' will also have relatively high charge density. The charge density function σ will decrease gradually as θ deviates from its value corresponding to the point Q'.

Figure 4. Information extraction at different levels of resolution

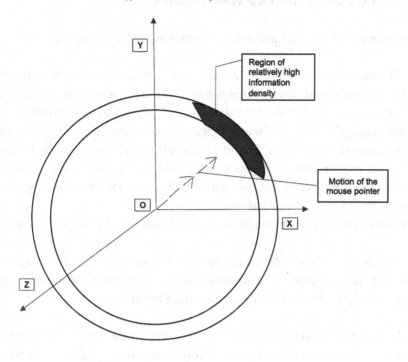

Figure 5. Clustering of related information

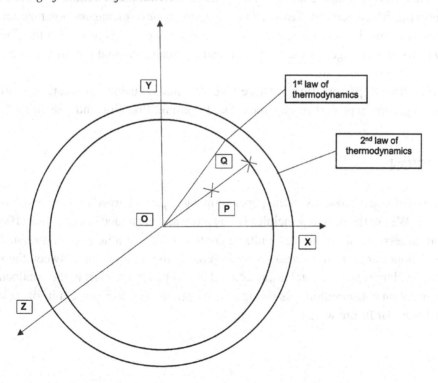

5. INFORMATION DESIGN AND MANAGEMENT

Design of information spaces has a central role in various aspects of management.

- **Planning:** The manager needs to have an access to information at various levels on demand to make good decisions. Our design facilitates this as the end user (the manager in this case) determines at what detail he needs the information. Therefore, embedding the information in this way helps him/her manage information in an efficient way. In our model, navigation through information space is geometrical in nature. The information is mapped n the surface of a conductor. This facilitates human-machine interaction in a very intuitive way. For example, the manager can ask queries relevant to his project by navigating a multidimensional information space instead of navigating through information mapped in tabular form.
- **Organizing:** Organizing information in our model is inherent by way of its construction.

Information is mapped geometrically on the surface of a conductor and subsequently retrieved by employing the laws of electrostatics. As we discuss later, the shape of the conductor (e.g, an ellipse, star shaped, a sphere) can help managers organize information better.

- **Decision Making and Directing:** One of the most important aspects of our design is that it has the potential for automated decision making. Since most of the information retrieval is done by letting the laws of electrostatics work for us, it will be interesting to explore if this can be extended to automated decision making as well. This is a question we plan to address in our future work.
- **Restructuring Management Teams:** Our design can help managers organize and restructure management teams. For example, if the information space is designed such that the information mapped is about a management team, it will result in managers making informed decisions about the restructuring of the team.
- **Process Groups:** Representation and accessing information using our algorithm will help in various process groups in project management like Initiating, Planning and Executing.

6. CONCLUSION

As stated, the underlying philosophy of the paper is the fact that a charged particle inside a conductor is a query in nature. We use the design principles from nature and get added benefits like efficient retrieval of information, accessing information at multiple depths and being able to obtain related information as well. Several important questions arise as well. What is the relationship between the shapes of the conductors, their volumes and the total information that can be retrieved from the conductor? Is there a notion of "information conservation"? Is this the most optimal way to represent information? We hope to answer all this in our future work.

REFERENCES

Barabasi, A. L., & Oltvai, Z. N. (2004). Network biology: Understanding the cell's functional organization. *Nature Reviews. Genetics, 5*(2), 101–113. doi:10.1038/nrg1272 PMID:14735121

Berry, M. W., Drmac, Z., & Jessup, E. R. (1999). Matrices, vector spaces, and information retrieval. *SIAM Review, 41*(2), 335–362. doi:10.1137/S0036144598347035

Dorogovtsev, S. N., & Mendes, J. F. (2013). *Evolution of networks: From biological nets to the Internet and WWW.* Oxford University Press.

Feynman, R. P., Leighton, R. B., & Sands, M. (2013). *The Feynman Lectures on Physics, Desktop Edition* (Vol. 1). Basic Books.

Foote, J. (1999). *An overview of audio information retrieval. In Multimedia Systems.* Springer.

Frakes, W. B. (1992). Introduction to information storage and retrieval systems. *Space, 14,* 10.

Goodrum, A. A. (2000). Image Information Retrieval: An Overview of Current Research. *Informing Science, 3*(2).

Jardine, N., & van Rijsbergen, C. J. (1971). The use of hierarchic clustering in information retrieval. *Information Storage and Retrieval, 7*(5), 217-240.

Lai, E. (2008). *Size matters: Yahoo claims 2-petabyte database is world's biggest, busiest.* Retrieved from http://www.computerworld.com/article/2535825/business-intelligence/size-matters--yahoo-claims-2-petabyte-database-is-world-s-biggest--busiest.html

Manning, C. D., Raghavan, P., & Schütze, H. (2008). *Introduction to information retrieval* (Vol. 1). Cambridge, UK: Cambridge University Press. doi:10.1017/CBO9780511809071

Maron, M. E. (2008). An historical note on the origins of probabilistic indexing. *Information Processing & Management, 44*(2), 971–972. doi:10.1016/j.ipm.2007.02.012

Reips, U. D., & Matzat, U. (2014). Mining "Big Data" using Big Data Services. *International Journal of Internet Science, 9*(1), 1–8.

Salton, G., Wong, A., & Yang, C. S. (1975). A vector space model for automatic indexing. *Communications of the ACM, 18*(11).

Singhal, A. (2001). Modern information retrieval: A brief overview. *IEEE Data Eng. Bull., 24*(4), 35–43.

Chapter 12
A New Formalism for Diagnosis and Safe Development of Information Systems

Calin Ciufudean
Stefan cel Mare University, Romania

ABSTRACT

Failure diagnosis in large and complex information systems (LCIS) is a critical task due to respect the safe development of these systems. A discrete event system (DES) approach to the problem of failure diagnosis of LCIS is presented in this chapter. A classic solution to solve DES's diagnosis is a stochastic Petri nets. Unfortunately, the solution of a stochastic Petri net is severely restricted by the size of its underlying Markov chain. On the other hand, it has been shown that foraging behavior of ant colonies can give rise to the shortest path, which will reduce the state explosion of stochastic Petri net. Therefore, a new model of stochastic Petri net, based on foraging behavior of real ant colonies is introduced in this paper. This model can contribute to the diagnosis, the performance analysis and design of information systems.

INTRODUCTION

As modern world evolves, so does the technique, economy, the life style, and unfortunately, the information avalanche many times deformed by the sources, often deformed and delayed by transmission channels and media. Therefore, nowadays mainly due to stress, low infrastructure facilities and rush life style we confront with a new problem: the diagnosis of multiple informational news and data releases of mutant informational transmission infrastructure. Failure diagnosis in large and complex information systems (LCIS) is a critical task due to respect the safe development of these systems. A discrete event system (DES) approach to the problem of failure diagnosis of LCIS is presented in this chapter. A classic solution to solve DES's diagnosis is a stochastic Petri nets. Unfortunately, the solution of a stochastic Petri net is severely restricted by the size of its underlying Markov chain. On the other hand, it has been shown that foraging behavior of ant colonies can give rise to the shortest path, which will reduce the

DOI: 10.4018/978-1-4666-9688-4.ch012

state explosion of stochastic Petri net. Therefore, a new model of stochastic Petri net, based on foraging behavior of real ant colonies is introduced in this paper. This model can contribute to the diagnosis, the performance analysis and design of information systems.

The property of diagnosis is introduced in the context of the availability problem of the DES. We propose a systematic procedure for diagnosis implemented with a new class of stochastic Petri nets, i.e. Ant Colony Decision Petri Nets (ADPN), and related models (e.g., stochastic reward nets stochastic activity networks) are gaining increased acceptance as tools for analyzing complex systems. The acceptance of such high-level formalism is due to their ability to represent LCIS in a compact and convenient way, while still describing an underlying discrete-time Markov chain (DTMC), (Campos, s.a., 1991). Our approach proposes non-expanded dimension of DTMC so we afford to model LCIS and take advantage on a finite horizon of their behavior, due to DTMC properties. The prediction and the optimization of the performance of information systems represent a must for the modern socio-economic systems. We consider an information system as coupled processing elements working co-operatively and concurrently on a set of related tasks. We notice that a information system consists of discrete and continuous parts, and according to the chart level, these parts may have different significations; for example a discrete part seen from a macro-level looks like a continuous one, and a continuous part of the system seen from a macro-level may be analyzed as a discrete one. Various algorithms have been used to perform this task. In general, there are two approaches for the diagnosis of complex systems (Bacelli & Zin, 1992): deterministic models and probabilistic models. In deterministic models, it is usually assumed that the task arrival times, the task execution times, and the synchronization involved are known in advance to the analysis. This approach is very useful for the performance evaluation of real-time control systems with hard deadline requirements. In probabilistic models, the task arrival rates and the task service time are usually specified by probabilistic distribution functions. Probabilistic models usually give a gross prediction on the performance of a system and are usually used for the early stages of system design when the system characteristics are not well understood. Various algorithms like evolutionary computations (Michalewicz, Z. & Michalewicz, M, 1997), genetic algorithms, adaptive cultural models etc. have been used to perform this task. Swarm intelligence (Kennedy & Eberhart, 1995) links artificial intelligence to the concept of fish shoaling or swarming theory. It is based on the social behavior of flocks of birds/shoals of fish and the success of the group is due to the communication established between them. Ant colony optimization algorithm proposed by Marco Dorigo in 1992 in his PhD thesis (Dorigo, 1992) is based on this simple concept and the paradigms can be easily implemented. It is relatively a new concept and is used for target tracing by autonomous communicating bodies. This paper presents the application the ant colony optimization (ACO) for searching targets in a given scheduling problem using a number of tasks referred to as *agents* in the rest of the chapter.

This chapter has three main parts according to the formalisms used to optimize the diagnosis of a hybrid system:

1. Ant colony optimization (ACO) for scheduling the tasks of complex systems,
2. Petri nets model for modeling the tasks synchronization, and
3. Henstock-Kurzweil integral introduced in a new technique for determining the reliability of information systems in the presence of perturbations with parameters described by highly oscillating functions.

The complexity of constructing the diagnosis and testing the diagnosis is exponential in the number of states of the system and double exponential in the number of failure types. Ant Colony Optimization (ACO) is an intuitive developed approach that takes inspiration from the behavior of real ant colonies to solve NP - hard optimization problems. It has been successfully applied to various hard combinatorial optimization problems. In this chapter we present the first application of ACO to Petri nets formalism, in order to simplify the models achieved with GSPN for solving the diagnosis of LCIS.

It has been shown that distributed-hybrid systems can be modeled as event graphs (Laftit, s.a., 1992), (Proth & Xie, 1994). When the information times are deterministic (respectively stochastic), the cycle time (respectively the mean cycle time) of the model is the period (respectively the mean period) required to manufacture a given set of parts which fits the required ratios. The smaller the cycle time (respectively the mean cycle time) the higher the productivity of the system. When the firing times of transitions are deterministic, it is possible to define the cycle time of an elementary circuit. This is given by the ratio of the sum of the firing times associated with the transitions of the circuit and the number of the tokens in the places which belong to the circuit, which is constant (we address strongly connected graphs with the number of tokens in any elementary circuit, constant):

$$C_i = \frac{T_i}{N_i} \tag{1}$$

where i = 1, 2, ..., n number of elementary circuits of the graph; C_i = cycle time of elementary circuits i; $T_i = \sum_i^n r_i$ = sum of the execution times of the transition in circuit i; $N_i = \sum_i^n M_i$ = total number of tokens in the places in circuit i.

In this case it has been proven that the cycle time of a strongly connected event graph is equal to the greatest cycle time of all elementary circuits. Furthermore, given a value C* greater than the largest firing time of all transitions, an algorithm has been proposed in (Campos, s.a., 1991) to reach a cycle time less than C*, while minimizing a linear combination of the number of tokens in the places. The coefficients of the linear combination are the elements of a p-invariant. When the event graph is the model of a ratio-driven distributed system (such as information system), C* has to be greater than the largest cycle time of all command circuits (Zurawski & Zhon, 1994). A command circuit is an elementary circuit, which joins the transition that models the operations performed on the same machine. Such a circuit contains one token to prevent more than one transition firing at any time in each elementary circuit. In other words, C* must be greater than the time required by the bottleneck machine to perform a sequence of parts which fits with the production ratios. In the case of random firing times, it is no longer a task for the elementary circuits to evaluate the behavior of the event graph and to reach a given performance. Thus, the results presented in this chapter, which aim at reaching a given mean cycle time in a steady state while minimizing a linear combination of the place markings, are particularly important at the preliminary design level of information systems working on a ratio-driven basis. This applies in particular to distributed systems in flexible information systems. In section 2 we briefly introduce the Ant Colony Optimization (ACO) algorithm (Clerc, 1999), (Clerc, 2006), (Nananukul & Gong, 1999) in the section 3 we frame the mean cycle time of event graphs, in the section 4 we introduce a new technique for diagnosis of event graphs. In section 5 based on an example we propose a new method for determining the availability of information systems in the presence of highly oscillating perturbations, and some concluding remarks are given in section 6.

ANT COLONY DECISION PETRI NET DIAGNOSER

The complex system, e.g., a flexible information system (LCIS) to be diagnosed is modelled as a finite state machine of DES's formalism:

$$W = \left(S, E, t, m_0 \right) \tag{2}$$

where S is the state space, E is the set of events, t is the partial transition function and m_0 is the initial state of system. The model W accounts for the normal and failed behaviour of the system. Let $E_f \leq E$ denote the set of failure events which are to be diagnosed. Our objective is to identify the occurrence of the failure events. Therefore we partition the set of failure events into disjoint sets corresponding to different failure types:

$$E_f = E_{f_1} \cup E_{f_2} \cup \ldots \cup E_{f_m} \tag{3}$$

This partition is motivated by the following considerations (Kennedy & Eberhart, 1999):

1. Inadequate instrumentation may render it impossible to diagnose uniquely every possible fault;
2. It may not be required to identify uniquely the occurrence of every failure event. We may simply be interested in knowing whether failure event has happened as the effect of the same failures in the system.

So, when we say that "a failure of type F_i has occurred", we mean that some event from the set E_{fi} has occurred.

In (Kennedy & Eberhart, 2009) the diagnosis is defined as follows: A prefix-closed and live language L is said to be I-diagnosis with respect to the projection P, the partition E_f, and the indicator I if the following holds:

$$\left(\forall i \in E_f \right) \cdot \left(\exists n \in N \right) \cdot \left(\forall s \in E_{f_i} \right) \cdot \left(\forall t \in \frac{L}{s} : st \in I\left(E_{f_f} \right) \right) \left[\|t\| \geq n; \Rightarrow D \right] \tag{4}$$

where the diagnosis condition D is:

$$\omega \in P_L^{-1} \left[P\left(st \right) \right] \Rightarrow E_{f_i} \in \omega \tag{5}$$

Note that I (E_{fi}) denotes the set of all traces of L that end in an event from the set E_{fi}. The behaviour of the system is described by the prefix-closed live language L(A) generated by A (see relation (2)). L is a subset of E*, where E* denotes the keen closure of the set E (Michalewicz, Z. & Michalewicz, M., 1997). ‖s‖ denotes the length of trace s∈E. L/s denote the post language of L after s, i.e.

$$\frac{L}{s} = \left\{ \begin{matrix} t \in E * \\ st \in L \end{matrix} \right\} \tag{6}$$

We define the projection P:E*→E in the usual manner:

$$P\left(\varepsilon\right) = \varepsilon \text{ and } P\left(s_1 \cdot s_2\right) = P\left(s_1\right) \cdot P\left(s_2\right),\ s_1 \in E * \text{ and } s_2 \in E \tag{7}$$

where ε denotes the empty trace.

The above definition, e.g. relations (4) and (5), means the following: Let s be any trace generated by the system that ends in a failure event from the set E_{fi}, and let t be any sufficiently long continuation of s.

Condition D then requires that every trace belonging to the language that produces the same record of observable events, and in which the failure event is followed by certain indicator, should contain a failure event from the set E_{fi}. This implies that on some continuation of s one can detect the occurrence of a failure of the type F_i with a finite delay, specifically in at most n_i transitions of the system after s. To summarize, here diagnosis requires detection of failures only after the occurrence of an indicator event corresponding to the failure. In this paper we improve this approach by according a gradual importance of failure indicators, in correspondence with the availability of the system. In our assumption the diagnoser is a stochastic Petri net (SPN), where the places are marked with the availability of the correspondent production cell. The availability of a production cell is calculated with a Markov chain, where the transitions reflect the gradual importance of the failures in the cell. We may say that the diagnoser is an extended observer where we append to every state estimate a label. The labels attached to the state estimates carry failure information and failures are diagnosed by checking these labels. We also assume the system W is normal to start. A diagnoser is a deterministic finite state machine whose transitions correspond to observations and whose states correspond to the set of system states and failures that are consistent with the observations. The transitions of the diagnoser are labelled with observable events, and the states of the diagnoser are labelled with sets of pairs (v, l) denoting a state and a failure label of the abstracted model. In our approach, the diagnoser efficiently maps observations to sets of possible system states and failures, and it is modelled with a new class of Petri nets, called here Ant Colony Decision Petri Nets (ADPN), which are an extension of our previous work (Ciufudean, s.a., 2005) where we introduced a class of Stochastic Coloured Petri Nets (SCPN). Here, the colour of tokens in ADPN, represents the colour of the ants, grouped in families. We suppose that in our model there are different ant families (e.g., red ants, black ants, s.a.), each kind of ant has a specific pheromone; an ant will sense the pheromone in the nodes of the net and will follow only the specific path that was marked with the pheromone of its family. In the initial marking of the Petri net we know the number of the test ants, by colour. Considering that after firing a transition in the net, the ant leaves its pheromone in the control place of the respective transition (see Figure 1), and then dies, after the first ant reaches the end of the

Figure 1. The basic structure of ADPN

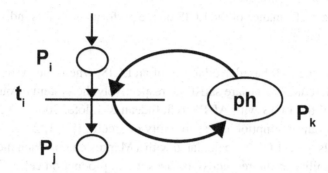

graph we count the number of the ants remained in the first place of the net. We conclude which is the shortest way in the net, i.e. which family of ants found the optimum path, considering that a family of ants will never follow the same way as another ant family.

In Figure 1 we may see that the control place, p_k, of transition t_i memorize the pheromone ph_k of the ant which burns first the transition t_i. We say that transition t_i will be fired only by ants with colour ph_k, where ph_k has the same signification as that given in relation (8), (Ciufudean, s.a., 2007). The firing rates of transitions in ADPN are given by the next relation:

$$f_i = \frac{\left(ph_i\right)^{\alpha_i} \cdot \left(h_i\right)^{\beta_i}}{\left(ph_i\right)^{\alpha_i} + \left(h_i\right)^{\beta_i}} \tag{8}$$

In relation (8) ph_k is the pheromone dropped in the control place by the first ant, that burns the transition t_i; h_i is the classic exponential firing rate of a transition in a stochastic Petri net; probabilities α_i and β_i control the failure rate, respectively the repair rate of elements (machines, electronic devices, etc) of a complex system, such as a Flexible Information System (LCIS). We define our ADPN as follows:

An ADPN is a fire-tuple (P, T, k, m, V), where:

P = {p_1, p_2, ..., p_n}, n > 0, and is a finite set of places;
T = {t_1, t_2, ..., t_s}, s > 0, and is a finite set of transitions with P∪T ≠ Ø, P∩T = Ø;
K = {Pk_1, Pk_2, .., Pk_s}, s > 0, and is a finite set of pheromone - control places;
m = P → N, and is a marking whose i[th] component is the number of tokens in the ith place. An initial marking is denoted by m_0;
V = T → R, is a vector whose component is a firing time delay with an ant decision function.

In our work we assumed that when a device, sensor, transducer or any other hardware component of the analysed system fails, the system reconfiguration (after repairing it) is often less than perfect. The notion of imperfection is called imperfect coverage, and it is defined as probability c that the system successfully reconfigures given that component fault occurs. The imperfect repair of a component implies that when the repair of the failed component is completed it is not "as good as new". A dependability model for diagnosing flexible information systems is presented. The meaning of dependability here is twofold:

- System diagnosis and availability;
- Dependence of the performance of the LCIS on the performance of its individual physical subsystems and components.

The model considers the task-based availability of an LCIS, where the system is considered operational as long as its task requirements are satisfied; respectively the system throughput exceeds a given lower bound. We model the LCIS with ADPN (Ciufudean & Filote, 2006). We decompose the LCIS in productions cells. In our assumption the availability of a cell j (j = 1, 2, ..., n, where n is the total number of part type cells in the LCIS) is calculated with a Markov chain which includes the failure rates, repair rates, and coverability of the respective devices in the production cell i. The colour domains of transitions that load cell i include colours that result in a value between 0 and 1, and the biggest value designates the cell (respectively the place in the ADPN model) which ensures the liveness of the net, respectively which will validate and burn its output transition. We assume that the reader is familiar with Petri nets theory and their applications to information systems or we refer the reader to (Kennedy & Eberhart, 2009), (Zurawski & Zhon 1994). Each part entering the system is represented by a token. The colour of the token associated with a part has two components. The first component is the part identification number and the second component represents the set of possible next operations determined by the process plan of the part. It is the second component that is recognized by the stochastic colours Petri net model, and the first component is used for part tracking and reference purposes. Let B_i be a (1 x m) binary vector representing all the operations needed for the complete processing of part type i. Let E_i be a (m x m) matrix representing the precedence relations among the operations of part type i, where m is the number of operations that are performed in the respective cell j (j = 1, 2, ..., n). For a part to be processed in the cell j it requires at least one operation that can be performed in the cell, that implies

$B_j > 0$. Also, for a part type where there is no precedent relationship between required operations, E_i is a matrix of zeros.

For a part with identification x and part type y, the initial colour of the corresponding token is:

$$V_{yx} = \left[yx, B_y - \left(B_y \cdot E_y \right) \right] \tag{9}$$

where $\left(B_y \cdot E_y \right)$ is a matrix of multiplication.

For example consider the process plan of part type L_1 and L_2 shown in Figure 2.

Our process plan first requires operation Op 1 and then operation Op 2 for complete processing. We assume that our LCIS can complete 5 different types of operations (e.g., for simplicity we consider only 5 different types of operations). For part type L_1, we have: $B_{L1} = [00011]$.

$$E_{L_1} = \begin{array}{c|ccccc} & op5 & op4 & op3 & op2 & op1 \\ \hline op5 & 0 & 0 & 0 & 0 & 0 \\ op4 & 0 & 0 & 0 & 0 & 0 \\ op3 & 0 & 0 & 0 & 0 & 0 \\ op2 & 0 & 0 & 0 & 0 & A_2 \\ op1 & 0 & 0 & 0 & A_1 & 0 \end{array}$$

Figure 2. Process plan of part type L_1 and L_2

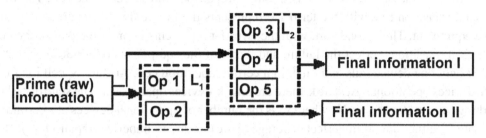

where A_1 is the availability of production cell 1 (which performs operation 1), and A_2 represents the availability of production cell 2 at time t. The availability A_i of cell i is calculated, as shown below, with Markov chains. We notice that A_i is re-evaluated at each major change in the process plan of LCIS (such as occurrence of events: damages of hardware equipments, changes of information process plan, etc.). Assuming that $A_1 > A_2$, then we assign to A_1 value 1 and to A_2 value 0, so that applying relation (9), the initial color of the token corresponding to a part that belongs to part type L_1 with identification mark 1, would be $V_{L1.1} = (L_{1.1}, 00001)$. Note that the information carried by the color of the tokens in the SCPN indicates the next operation to be performed by the LCIS. Generally, we may say that V is the set of colors that represent all the possible combinations of operations that can be performed in the LCIS. Each member of the set V is a vector with m components, where m is the maximum number of operations to be performed in the cells of the LCIS. For example, in an LCIS with 5 operations to be performed, we may have V = {00000, 00001, ... 11111}. For simplicity, we assume that operations in LCIS are maped to places in the SCPN model, places which are labeled with the operation identification number. The requirement for a production cell j (j=1, ..., n) which have N_i (i=1, ..., m) devices of type i, is that at least k_i of these devices must be operational for the LCIS to be operational. To determine the system availability which includes imperfect coverage and repair, a failure state due to imperfect coverage and repair was introduced (Campos, Chiola & Silva, 1991). To explain the impact of imperfect coverage, we consider the system given in Figure 3 which includes two identical information devices M_1 and M_2.

If the coverage of the system is perfect, i.e. c=1, then operation op1 is performed as long as one of the devices is operational. If the coverage is imperfect, then operation op 1 fails with probability 1-c, if one of the devices M_1 or M_2 fails. We may say that, if operation op 1 has been scheduled on device M_1 that has failed, then the system in Figure 3 fails with probability 1-c. The Markov chain for information cell j is shown in Figure 4. In Figure 4 the parameters λ, μ, c, r denote respectively the failure rate,

Figure 3. Example of operation performed by two identical devices

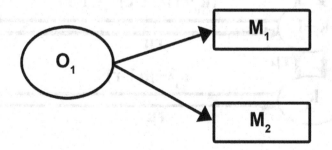

repair rate, coverage factor and the successful failure repair rate of devices in the cell. The first part of the horizontal transition rate with the term 1-c represents the failure due to imperfect coverage of an alternative equipment. The second part, with the term 1-r represents imprecise repair of the devices.

The vertical transitions reflect the failure and repair of the equipments (Ciufudean, s.a., 2006). We assume that only one device fails at a time, in a certain operation cell. At state N_i cell i is functioning with all N_i devices operational. At state k_i there are only k_i devices oparational. The state of cell i changes from working state w_i, for $k_i \leq w_i N_i$, where w_i is the number of operational devices at a certain moment, to failed state F_i, either due to imperfect coverage (1 - c) or due to imperfect repair (1 - r). If the fault coverage of the system and repair of the components are perfect, the Markov chain in Figure 4 reduces to one-dimension model. The solution of the Markov chain model given in Figure 4 is a probability that at least k_i devices are working at time t. The availability of cell i is given by the next relation (Ciufudean & Petrescu, 2005), (Ciufudean & Filote, 2006):

$$A_i(t) = \sum_{w_i=k_i}^{N_i} P_{k_i}(t), for\ i = 1, 2, \ldots, n \tag{10}$$

where $A_i(t)$ = the availability of cell i at moment t;

$P_{ki}(t)$ = probability of k_i devices being operational in cell I at time t;
N_i = total number of devices of type j in cell i;
K_i = required minimum number of operational devices in cell i.

Figure 4. Markov model for cell i

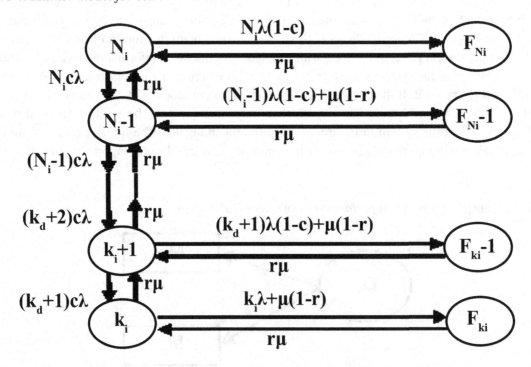

After a Markov chain for each cell of the measuring system is constructed and desired probabilities $A_i(t)$, i=1, 2, ..., n corresponding to each information cell are determined, the ant colony decision Petri net (ADPN) can be initialized and the simulation process of the LCIS begins. The status of this graph (e.g., the ADPN) at different moments t_k, gives us the diagnosis of the LCIS.

FRAMING THE MEAN CYCLE TIME

It has been proven (Ciufudean & Filote, 2005) that a marking belonging to the optimal solution under a periodic operational mode (POM) is an optimal solution under the earliest operational mode (EOM). So, we consider the earliest operational mode of the event graph, and we assume only non-pre-emptive transitions firings. We further assume that, when the transition fires, the related tokens remain in the input places until the firing process ends. They then disappear, and one new token appears in each output place of the transition. We use the following notations: M_i = the marking of the elementary circuits, $i \in N$; $X_t^k \in R^+$ = random variable generating the time required for the kth firing of the transition t, k $\in N$; $I_t(n)$ = instant of the nth firing initiation of the transition t; E = set of elementary circuits; s(e) = sum of the random variables generating the firing; $\sum_{t \in e} X_t^1$ = sum of times of the transitions belonging to e; E_t = set of elementary circuits containing the transition t. We assume that the sequences of transition firing times are independent sequences of integral random variables. It was proven in (Recalde, s.a., 1999) that there is a positive constant $s(M_0)$, so that:

$$\lim_{n \to \infty} \frac{I_t(n)}{n} = C_m \tag{11}$$

where: C_m = the average cycle time of the event graph.

Furthermore, we denote by m_t the mean value of X_t^k and by q_t the standard deviation of X_t^k, i.e.,

$$m_t = F\left[X_t^k\right] and \ q_t^2 = F\left[\left(X_t^k - m_t\right)^2\right].$$

The Lower Bound of the Mean Cycle Time

The cycle time of the deterministic problem obtained by replacing the random variables which generate the firing times by their mean values, is a lower bound of the mean cycle time (Ciufudean, s.a., 2007). The following relation proven in [9] provides a better lower bound for the value of the mean cycle time than the previous one:

$$C** \geq F\left[\max\left\{\frac{s\left[e \setminus \left\{t^*(e)\right\} + m_{t^*(e)}\right]}{M_0(e)}, m_{t^*(e)}\right\}\right] \tag{12}$$

where t*(e) is a transition, belonging to event e, with the greatest average firing time, i.e., $m_{t^*(e)} = \max_{t \in e} m_t$, and s[e\{t*(e)}] is the sum of the firing times of the transitions belonging to e except t*(e).

The Upper Bound of Mean Cycle Time

With M_0 being the initial marking, we derive a marking M_1 from M_0 by leaving the places which are empty in M_0, empty in M_1, and by reducing to one the number of tokens in the places containing more than one token in M_0. Thus, $M_1(p) \leq M_0(p)$ for any set of places of the strongly connected event graph. An earlier operation mode running with the initial marking M_1 leads to a greater mean cycle time than the one obtained when starting from M_0. Then, starting from M_1, we apply to the event graph the earliest operation mode, but we block the tokens as soon as they reach a place already marked in M_1. This operation mode is referred to as the constrained mode (Zurawski & Zhon, 1994). We denote by C* the mean cycle time obtained by using the constrained operation mode when M_1 is the initial marking. We know (Ciufudean, s.a., 2007). that C* is greater than the mean cycle time obtained by using the earliest operation mode starting from M_1 which, in turn, is greater than the mean cycle time obtained with the earliest operation mode when the initial marking is M_0. Thus, C* is an upper bound of the solution to our problem (i.e. the mean cycle time obtained starting from M_0 when using the earliest operation mode). The following relation defines this upper bound:

$$C^* = F\left[\max_{z \in Z} s(z)\right] \tag{13}$$

where, Z is the set of directed path verifying the following properties:

- The origin and the extremity of any path is a marked place;
- There is no marked place between the origin and the extremity of the path.

THE DIAGNOSIS OF THE EVENT GRAPH

In the reminder of the previous section, we compare the previous bounds with the existing ones. Under the assumption of non-preemptive transition firing, we have:

$$C^{**} \geq \max_{e \in E} F\left[\max\left\{\frac{s\left[e \setminus \{t^*(e)\} + m_{t^*(e)}\right]}{M_0(e)}, m_{t^*(e)}\right\}\right] = \text{old lower bound} \tag{14}$$

$$C^* \leq \sum_t m_t = \text{old upper bound} \tag{15}$$

The following relations show that the new bounds are better than the old ones. But how close are they to the optimal solution? In order to answer this question we give the next algorithm, inspired from the operational research area, for verifying the system performance:

1. Express the token loading in a (p x p) matrix P, where p is the number of places in the Petri net model of the system. Entry (A, B) in the matrix equals x if there are x tokens in place A and place A is connected directly to place B by a transition; otherwise (A, B) equals 0.

2. Express the transition time in a (p x p) matrix Q. The entry (A, B) in the matrix equals to the mean values of the random variables which generate the firing times (i.e. X_t^k) if A is an input place of the transition "i" and B is its output place. Entry (A, B) contains the symbol "w" if A and B are not connected directly as described above.

3. Compute matrix CP - Q (with p – w = ∞, and C = (C* + C**)/2, for p∈N), then use Floyd's algorithm to compute the shortest distance between every pair of nodes using matrix CP-Q as the distance matrix. The result is stored in matrix S. There are three cases.

    a. All diagonal entries of matrix S are positive (i.e., $CN_k - T_k > 0$ for all circuits - see relation (1)) and the system performance is higher than the given requirement;

    b. Few diagonal entries of matrix S are zero's and the rest are positive (i.e., $CN_k - T_k = 0$ for some circuits, and $CN_k - T_k > 0$ for the other circuits) - the system performance has just reached the given requirement;

    c. Few diagonal entries of matrix S are negative (i.e., $CN_k - T_k < 0$ for some circuits) - the system performance is lower than the given requirement.

In addition we may say that when a decision-free system runs at its highest speed, CN_k equals to T_k for the bottleneck circuit. This implies that the places in the bottleneck circuit will have zero diagonal entries in matrix S. The system performance can be improved by reducing the execution times of some transitions in the circuit, or introducing more concurrency in the circuit (by modifying the initial marking), or increasing the mean cycle time (by choosing another average value).

ESTIMATING THE AVAILABILITY OF INFORMATION SYSTEMS

When dealing with highly oscillating perturbations the inertia factor (see the relations (14) and (15)) the cognitive and social scaling factors may be used to control the swarm behaviour, especially the convergence and search diversity properties of the algorithm. Appropriate values guarantee convergence, but the optimal values regarding convergence rates and diversity are problem dependent. The inertia factor is usually decreased as the search progresses in order to improve the convergence rate (Yu, s.a., 2003), (Sedlaczek & Eberhard, 2007). The method described in the previous paragraphs can be easily extended to handle hybrid systems. Instead of satisfying the relations (14), and (15) in the whole system state space, these relations must be satisfied by the set of functions indexed by the system locations. Functions corresponding to different locations are linked via appropriate conditions that must be satisfied during discrete transitions between the locations. The idea is analogous to using multiple Lyapunov-like functions (Branicky, 1998), (Clerc, 2006) for stability of hybrid systems. One of the most important parameters that control the performance of a control system is its reliability. Reliability is measured by the fault exposure ratio (FER) (Stahl, 1989), (Bohoris & Yun, 1995), (Yang, 1995). It represents the average detection ability of the faults in the system. Other parameters that control FER are the size of the system and the execution speed of the control unit, which are both easily evaluated. In this section we illustrate our approach via the computation of pieces, damage probability in a system with 3 processing units and fed by a flow of two-state traffic streams. The processing units are P_1, P_2 and P_3, in which two type of parts are produced (T_{p1} and T_{p2}), and which are fed by independent traffic sources, T_1, T_2, ..., T_n. There are 4 different activities: a part of type T_{p1} is first processed on processing unit P_1 (activity 1) and then on processing unit P_3 (activity 3). A part of type T_{p2} is first processed on processing unit P_1 (activity 2) and then on processing unit P_4 (activity 4). The processing time for activity i is d_i.

In order to describe the information flow for the system depicted in Figure 5, we define: $u_i(k)$ = time instant at which raw information for a part of type T_i (T_{Pi}) is fed to the system for the $(k+1)^{st}$ time; $x_i(k)$ = time instant at which activity i starts for the kth time; $y_i(k)$ = time instant at which a finished product of type T_i (T_{Pi}) leaves the system. With these notations we have:

$$x_1(k+1) = \max (x_2(k) + d_2, u_1(k), x_3(k-N_3)) \tag{8}$$

$$x_2(k+1) = \max (x_1(k+1) + d_1, u_2(k), x_4(k-N_4)) \tag{9}$$

$$x_3(k+1) = \max (x_3(k) + d_3, x_1(k+1) + d_1) \tag{10}$$

$$x_4(k+1) = \max (x_4(k) + d_4, x_2(k+1) + d_2) \tag{11}$$

$$x_n(k+1) = \max (x_{n-1}(k) + d_{n-1}, u_n(k), x_{n-1}(k-N_{n-1})) \tag{12}$$

$$y_1(k) = x_3(k) + d_3 \tag{13}$$

$$y_2(k) = x_4(k) + d_4 \tag{14}$$

These equations help us to describe only the ideal behaviour of the information system, e.g. the behaviour without losses, without failures. But, of course, real behaviour implies a certain failure rate of the production mode within the system given in Figure 5. In order to describe the behaviour of our assembly system, we must consider that the system cells of the buffers B_i, i = 1, ..., n (where B_1, B_2, ...,B_5, B_6,..., B_n we assume that they have k cells for storage capacity, and B_3, B_4 have the storage capacity of N_3 cells, respectively N_4 cells, where $N_3+N_4= c$; c being the number of cells that can be served during one time unit) have the same length and, consequently, the same service time, which makes the discrete time Markov chain, and also discrete Stieltjes equations, natural modelling choice. Although the ap-

Figure 5. Example of an information system

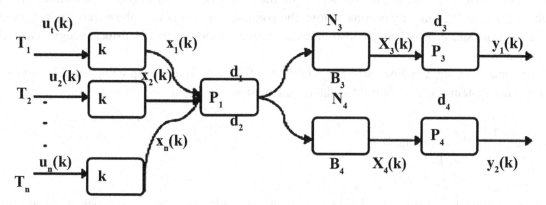

plications of our approach are not limited to the discrete time case, we focus our discussion on discrete event models in this paper (Gong, s.a., 1992), (Zhu & Li, 1993), (Gordon, 1994), (Nananukul & Gong, 1999). We model the production system given in Figure 5 as a discrete time server that can service c cells during one time unit. This server serves a queue with a capacity for k cells which is fed by an independent traffic source. The arrival process associated with a source has two states: active (ON) and idle (OFF), represented by 1 and 0 respectively. In the active state, an arrival can occur with probability α (in the experiments, α is assumed to be 1). No arrivals occur while the source is in the idle state. Each of these ON-OFF sources behave as follows: while an arrival process is in state 0, there is a probability $1-p_{00}$ that it will change to state 1 at the next time slot and a probability p_{00} that it will remain in state 0. While an arrival process is at state 1, there is a probability $1-p_{11}$ that it will transit to the idle state at the next time slot and a probability p_{11} that it will remain in state 1. When the server is busy, a maximum of c cells will deport in each time slot. The system can be modelled as a discrete time Markov chain with state (x_i, y_i), where x_i is the number of cells in the queue and y_i is the number of arrival sources in the active state at the ith time slot (Zhu & Li, 1993). We want to determine the steady state behaviour $(x, y) \overset{\Delta}{=} \lim_{i \to \infty} (x_i, y_i)$. Let S denote the state space. Let $T = [t_{n,m;k,l}]$ be the transition matrix for this Markov chain, where $t_{n,m;k,l} = \text{Prob}[x_{i+1}=k, y_{i+1}=l \mid x_i=n, y_i=m]$. Note that the dimension of this Markov chain is $(k+c) \cdot (n+1)$. The stationary probability distribution can be obtained by solving the equation:

$$\pi \cdot T = \pi \qquad (15)$$

In the literature the resources cell loss distribution P_L can then be calculated as in (Yang, s.a., 1995):

$$P_L = \frac{\sum_{(n,m) \in S} \max\left[(n + m - k), 0\right] \cdot P\left[x = n, y = m\right]}{\sum_{(n,m) \in S} m \cdot P\left[x = n, y = m\right]} \qquad (16)$$

We notice that when sizes of the model (either the buffers size, or the number of sources n, or the number of processing units P_i) becomes large, the computational cost is prohibitively high due to the size of the state space. We shall use also discrete Stieltjes equations and Henstock integrals to tackle this problem.

Numerical example: Let it be n=100, p_{00}=0.99, p_{11}=0.7, c=5. The corresponding traffic intensity of the assembly system given in Figure 5 is the following one:

$$\mu = \frac{n \cdot (1 - p_{00})}{c \cdot (2 - p_{00} - p_{11})} = 0,64 \; \langle 1 \tag{17}$$

The queue is stable. Applying techniques in (Yang, s.a., 1995) we obtain an exponential decay rate for the resources traffic loss probability $P_L(k)$ of μ=0.64 for this model. We consider that such an approach can not represent the resources cell loss distribution P_L among the k tasks of the considered system (e.g. for the information system given in Figure 5, it cannot represent the resources cell loss distribution between outputs $y_1(k)$ and $y_2(k)$). In order to achieve this goal we propose a new approach for determining P_L. We consider the next function:

$$P_{Lu}(t) = \sum_u \sum_n (-1)^u \cdot \left(n_u^\alpha + \frac{1}{n-1} \right) \cdot t^{\frac{n}{c \cdot v}} \tag{18}$$

where: u = the task accomplished by the system;

t = time required for accomplishing the process;

c_v = the coverage factor (e.g. the coverage probability) for the system fed by n sources through the input buffers;

n_u^α = is a function of probability α that the system receives in the uth task the resource number n.

We mention that the graphical representation of the function $P_{Lu}(t)$ given in the relation (18) allows to show the loss probability both for odd and even tasks according to the factor $(-1)^u$, respectively by positive and negative inflexions of the group.

We introduced in chapter 2 the coverage factor c_v and in order to explain it we consider that the equipment of a system (for instance a information system as in Figure 5) is divided into several cells, i = 1, 2, ..., k. The requirement for a cell i is that the cell including N_i equipment of type M_i ensures the functioning of at least k_i of the equipment, so that the system is operational. For the system in Figure 5 and for 10 sources as described before, we have:

$$P_{Lu}(t) = \sum_{u=1}^{2} \sum_{n=0}^{9} (-1)^u \cdot \left(n_u^\alpha + \frac{1}{n-1} \right) \cdot t^{\frac{n}{cv}} \tag{19}$$

As we notice that it is possible to occur in practice, and also as we perform the simulations in our laboratory of discrete event systems, we consider for α values between 0.4 and 0.7, by step of 0.1. Obviously, the rank of the $P_{Lu}(t)$ depends on the coverage factor c_v: as c_v decrease, the rank of the $P_{Lu}(t)$ grows, and that represents the case when the coverage of our system is poor and we can test it to the limit in order to simulate the functioning in the worst conditions. Such study allows us to determine the availability of the modeled information systems in real terms. Few results of the simulation we performed in Matlab for our information system (see Figure 5) for a typical work shift of 8 hours labor per day for one week, e.g. 40 hours, $\left(t \in \left[0, 40 \right] \right)$, are shown below. The graphic representations of $P_{Lu}(t)$ are drawn with discontinuous lines, while the blue continuous line represents the graphic of the real function $f(t) = 2t \sin \dfrac{1}{t^2} - \dfrac{2}{t} \cos \dfrac{1}{t^2}$ (see figures 6-10). In figures 6, 7, 8, 9, 10, we represent the simulation results for the coverage factors $c_v=1$, $c_v=0.66$, $c_v=0.5$, $c_v=0.33$, and $c_v=0.25$, respectively. We notice that as the coverage factor decreases, the performances of the information systems decrease too, and the stability of the system is diminished.

It can be seen that these functions converge (when the coverage factor tends to 0) to the real function defined on the interval [0, 40] hours by:

$$f(t) = \begin{cases} 2t \sin \dfrac{1}{t^2} - \dfrac{2}{t} \cos \dfrac{1}{t^2}, if\ t \neq 0 \\ 0, if\ t = 0. \end{cases} \tag{20}$$

Figure 6. P_{Lu} Matlab simulation of the information system given in Figure 5 with $c_v=1$

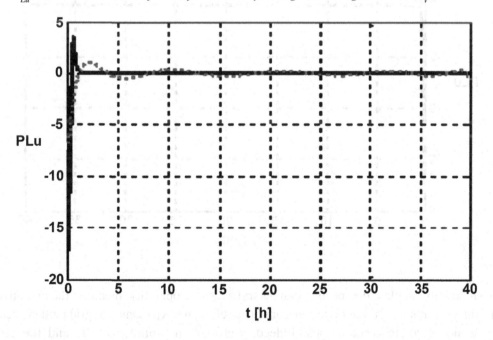

Figure 7. P_{Lu} Matlab simulation of the information system given in Figure 5 with $c_v=0,66$

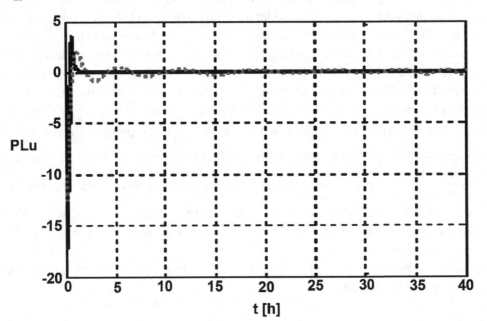

Figure 8. P_{Lu} Matlab simulation of the information system given in Figure 5 with $c_v=0,50$

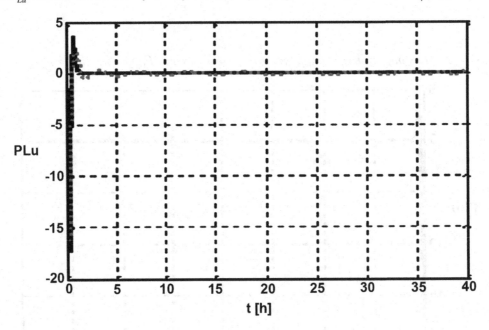

We wish now to calculate the mean resources traffic loss probability, meaning the primitive of our function. But $f(t)$ is not integral in Lebesgue sense (therefore, not Riemann-integral neither) because its primitive is not absolutely continuous. Indeed, consider an arbitrary $\delta > 0$, and the sequences $a_n = \dfrac{2}{\sqrt{(4n+1)\pi}}, b_n = \dfrac{2}{\sqrt{(4n)\pi}}$. Let $M < N$ be integers so that $\dfrac{1}{M} < \delta$ and $\displaystyle\sum_{n=M}^{N} a_n > 1$. This is

Figure 9. P_{Lu} Matlab simulation of the information system given in Figure 5 with c_v=0,33

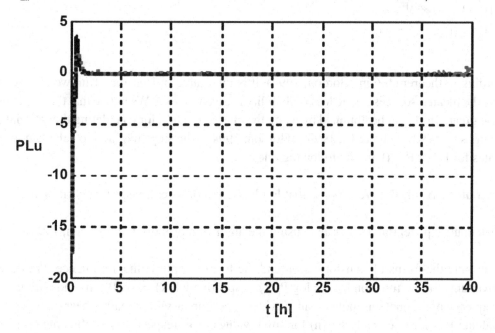

Figure 10. P_{Lu} Matlab simulation of the information system given in Figure 5 with c_v=0,25

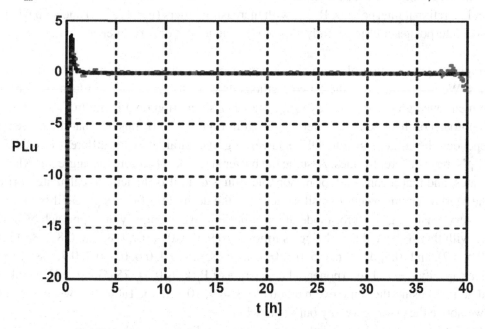

possible, since the series $\sum_{n=1}^{\infty}\frac{1}{\sqrt{n}}$ does not converge (it has an infinite sum). Then the finite collection

of intervals $\left\{\left[a_n, b_n\right], M \le n \le M\right\}$ satisfies $\sum_{n=M}^{N}\left(b_n - a_n\right) < \delta$, but $\sum_{n=M}^{N}\left|F(b_n) - F(a_n)\right| > 1$. So, F is not

absolutely continuous. On the other hand, as *f(t)* is the derivative of the real function:

$$F(t) = \begin{cases} t^2 \sin \dfrac{1}{t^2}, t \neq 0 \\ 0, t = 0 \end{cases} \qquad\qquad (21)$$

According to theorem 9.6 in (Gordon, 1994), it is integrable in Henstock-Kurzweil sense. Then its primitive, the mean resources traffic loss probability is given by *F(t)*. We notice that the same result, but with some great effort, can be obtained by using the classic method for calculating the rational interpolates (Gong, s.a., 1992), (Zhu & Li, 1993), (Gordon, 1994). The steps required to calculate the rational interpolates for P_L, or $P_{Lui}(t)$ are the following ones:

1. **Asymptotic Analysis:** We suppose that $\log P_L \approx \Theta*k$ ($k \to \infty$), respectively ($n \to \infty$).

    We calculate exponential decay rate Θ using the algorithm proposed in (Gordon, 1994).

2. Determine the forms of transformation and the form of approximant sequence. We develop approximants for the function $h(k) = \log P_L(k)$, respectively $h(k) = \log P_{Lui}(n)$ and will use an $R_{(n+1),}$ n sequence of rational interpolates since h(k) and h(n) are asymptotically linear (Gong, s.a., 1992).
3. Evaluate $P_L(k)$, respectively $P_{Lui}(n)$ for small values of k, respectively n (thus the corresponding values of h(k) and h(n) are known) by solving the Markov chain or using other available analytic methods (including Henstock integrals).
4. Calculate rational interpolants, $R_{(n+1),n}$ with increasing orders (n = 1, 2, …) and stop when the successive interpolates are sufficiently close to the range of k or u_i of interest.

The graphic representation of these classic approaches is drawn in figures 4-8, respectively, with dotted lines. We may notice that these two methods drive us to very close results as we observe from the above mentioned graph obtained by using Matlab 7 simulation on a Pentium IV computer. An interesting approach of the assembly system depicted in Figure 5 is to consider that the process P_1 is fed by n-independent heterogeneous ON-OFF sources, e.g. the parameters for different ON-OFF sources, P_{00}'s and P_{11}'s, have different values. Assume the buffer size is k. Therefore the aggregate Markov chain has $k \cdot 2^n$ states, and that means that for reasonable values of n, it is difficult to calculate traffic loss by solving the Markov chain, even for small values of k. the desired traffic loss probability in information systems is very small (e.g. 10^{-9}) for moderate-size buffers. We exemplify our approach by considering 15 sources with their P_{00}'s and P_{11}'s being as follows: (0.9, 0.9, 0.9, 0.9, 0.8, 0.8, 0.8, 0.8, 0.8, 0.7, 0.7, 0.7, 0.6, 0.5, 0.7), and (0.7, 0.8, 0.6, 0.5, 0.3, 0.6, 0.4, 0.8, 0.5, 0.9, 0.6, 0.5, 0.8, 0.7, 0.8), respectively. The service capacity c = 7. The exponential decay rate of $P_L(k)$ is 0.34375 (Gong & Nananukul, 1996). We calculate $P_L(k)$ using the same two methods for k = 9, 10, …, 15. Table 1 shows the results, in the brackets we notice the errors given by our approach.

As Table 1 shows, our results match the importance sampling results obtained in (Gong & Nananukul, 1996) to our approach.

Table 1. Traffic of resources cells loss probability for 15 independent information supply sources

K	$P_L(k)$	$P_{Lui}(t)$
9	$1.858327 \cdot 10^{-2}(\pm 0.018\%)$	$1.858438 \cdot 10^{-2}(\pm 0.018\%)$
10	$1.192634 \cdot 10^{-2}(\pm 0.013\%)$	$1.192567 \cdot 10^{-2}(\pm 0.013\%)$
11	$7.953624 \cdot 10^{-3}(\pm 0.021\%)$	$7.824341 \cdot 10^{-3}(\pm 0.021\%)$
12	$6.163792 \cdot 10^{-3}(\pm 0.029\%)$	$6.174915 \cdot 10^{-3}(\pm 0.029\%)$
13	$5.823462 \cdot 10^{-4}(\pm 0.033\%)$	$5.931628 \cdot 10^{-4}(\pm 0.033\%)$
14	$5.236143 \cdot 10^{-4}(\pm 0.042\%)$	$5.273810 \cdot 10^{-4}(\pm 0.042\%)$
15	$3.781632 \cdot 10^{-5}(\pm 0.053\%)$	$3.834193 \cdot 10^{-5}(\pm 0.053\%)$

CONCLUSION

This work establishes a necessary relation between diagnosis and performances of hybrid systems, such as information systems. We have proposed a new architecture for the diagnosis and performance evaluation of hybrid systems using swarm intelligence formalism. We assume that the necessary time to find the optimum solution is the cycle time of the Petri net that models synchronization, in the graph modelling the ACO algorithm. An important result in this paper is that it is always possible to reach a mean cycle time in an event graph modeled with Petri nets as close as possible to the greatest mean firing time using a finite marking, assuming that a transition cannot be fired by more than one token at each time. This result holds for any distribution of the transition firing time. An algorithm for verifying the distributed systems performance is introduced. An approach for computing upper and lower bounds of the performance of a conservative general system is mentioned. However, the bounds produced may be loose. Further research will focus on the conditions under which a mean cycle time can be reached with a finite marking, and we intend to deal with the Ant Colony Optimization theory in order to achieve this goal. A new and interesting approach is made by using the properties and applications of Henstock-Kurzweil integral to performance diagnosis of LCSI. We focus our research work on this frame and we have shown that it is a suitable manner to solve the LCIS's diagnosis. Our approach also allows an easy calculus for the availability refutation of uncertain (e.g. with low coverage factor c_v) LCIS with nonlinear dynamics. We use this method because some performance functions in HS's have been shown to belong to the class of functions to which rational interpolates converge geometrically fast, and this class of functions is characterized by the properties of functions on the complex plane (Gong, 1992). Therefore, we shall focus our future work on studying the convergence in capacity results for each concrete typical application in flexible information systems because it constitutes an illustrative and important field for LCIS applications. Considering that for real systems computer simulation is often the only means to get accurate performance values of down-sized systems, future research will provide the analysis of hybrid systems with nonlinear continuous dynamics.

REFERENCES

Bacelli, F., & Lin, Z. (1992). Compresion properties of stochastic decision free Petri nets. *IEEE Transactions on Automatic Control, 37*(12), 1905–1920. doi:10.1109/9.182477

Bohoris, G. A., & Yun, W. Y. (1995). Warranty costs for repairable products under hybrid warranty. *IMA Journal of Mathematics Applied in Business and Industry*, *6*, 13–24.

Branicky, M. (1998). Multiple Lyapunov functions and other analysis tools for switched and hybrid systems. *IEEE Transactions on Automatic Control*, *43*(4), 475–482. doi:10.1109/9.664150

Campos, J., Chiola, G., & Silva, M. (1991). Properties and performance bounds for closed free choice synchronized mono-class queuing networks. *IEEE Transactions on Automatic Control*, *36*(12), 1368–1382. doi:10.1109/9.106153

Ciufudean, C. (2005). Petri Net Based Diagnosis for Construction Design. *22nd International Symposium on Automation and Robotics in Construction, ISARC 2005*.

Ciufudean, C. (2006). Work-flows in Constructions Modelled with Stochastic Artificial Petri Nets. In *Proc. of The 23rd International Symposium on Automation and Robotics in Construction*.

Ciufudean, C. (2007). Reliability Markov Chains for Security Data Transmitter Analysis. In *Proc. of The Second International Conference on Availability, Reliability and Security*. Vienna University of Technology.

Ciufudean, C., & Filote, C. (2005). Performance Evaluation of Distributed Systems. *International Conference on Control and Automation, ICCA 2005*. IEEE.

Ciufudean, C., & Filote, C. (2006). Diagnosis of Complex Systems Using Ant Decision Petri Nets. *The First International Conference on Availability, Reliability and Security (ARES 2006)*, Vienna University of Technology. doi:10.1109/ARES.2006.52

Ciufudean, C., & Petrescu, C. (2005). Scheduling Diagnosis of Flexible Information Systems. In *Proc. Int. Conf. on Autom, Contr. And Syst. Eng. (ACSE'05)*.

Clerc, M. (1999). The Swarm and the Queen: Towards a Deterministic and Adaptive Particle Swarm Optimization. In *Proceedings of the IEEE Congress on Evolutionary Computation*. doi:10.1109/CEC.1999.785513

Clerc, M. (2006). *Particle Swarm Optimization*. London: ISTE. doi:10.1002/9780470612163

Dorigo, M. (1992). *Optimization, Learning and Natural Algorithms*. (PhD thesis). Polytechnic di Milano, Italy.

Gong, W. B. (1992). Rational representation for performance functions of queuing systems. In Princeton Conf. System & Information Science, (pp. 204-210).

Gong, W. B., & Nananukul, S. (1996). *Rational interpolation for rare event probabilities. In Stochastic Networks: Stability and Rare Events*. Springer – Verlag.

Gong, W. B., & Yang, H. (1995). Rational approximants for same performance analysis problems. *IEEE Transactions on Computers*, *44*(12), 1394–1404. doi:10.1109/12.477245

Gordon, R. A. (1994). The Integrals of Lebesgue, Denjoy, Perron and Henstock. Grad. Stud. In Math., 4, 234-245.

Kennedy, J., & Eberhart, R. (1995). Particle Swarm Optimization. In *Proceedings of the International Conference on Neural Networks*. doi:10.1109/ICNN.1995.488968

Kennedy, J., & Eberhart, R. (1999). Particle swarm optimization. In *Proceedings IEEE International Conference on Neural Networks*, (vol.4, pp. 1942-1948). doi:10.1109/ICNN.1995.488968

Kennedy, J., & Eberhart, R. (2009). *Swarm Intelligence*. San Francisco, CA: Kauffman publishers.

Laftit, S., Proth, J. M., & Xie, X. L. (1992). Optimization of invariant criteria for event graphs. *IEEE Transactions on Automatic Control, 37*(12), 547–555. doi:10.1109/9.135488

Michalewicz, Z., & Michalewicz, M. (1997). Evolutionary computation techniques and their applications. *IEEE International Conference on Intelligent Processing Systems, ICPIS*.

Nananukul, S., & Gong, W. B. (1999). Rational interpolation for stochastic DES's coverage issues. *IEEE Transactions on Automatic Control, 44*(5), 1070–1078. doi:10.1109/9.763231

Proth, J. M., & Xie, X. L. (1994). Cycle time of stochastic event graphs: Evaluation and marking optimization. *IEEE Transactions on Automatic Control, 39*(7), 1482–1486. doi:10.1109/9.299640

Recalde, L., s.a. (1999). Modeling and analysis of sequential processes that cooperate through buffers, *IEEE Trans. on Rob. and Autom., 14*(2), 267-277.

Sedlaczek, K. & Eberhard, P. (2007). Augmented Lagrangian Particle Swarm Optimization in Mechanism Design. *Journal of System Design and Dynamics, 1*(3).

Singh, J. (1979). *Operations Research*. Middlesex, UK: Penguin Books.

Stahl, H. (1989). On the convergence of generalized Padé approximants. *Constructive Approximation, 5*(1), 221–240. doi:10.1007/BF01889608

Yang, H. A. (1995). Efficient calculation of cell loss in ATM multiplexers. In *Proc. GLOBECOM*. doi:10.1109/GLOCOM.1995.502598

Yu, P. (2003). Virtual instrument parameter calibration with particle swarm optimization. *IEEE Swarm Intelligence Symposium*.

Zhu, Y., & Li, H. (1993). The Mac Laurin expansion for a GI/G/1 queue with Markov-modulated arrivals and services. *Queueing Systems, 14*(1-2), 125–134. doi:10.1007/BF01153530

Zurawski, R., & Zhon, M. C. (1994). Petri nets and industrial applications: A tutorial. *IEEE Transactions on Industrial Electronics, 41*(6), 567–583. doi:10.1109/41.334574

KEY TERMS AND DEFINITIONS

Ant Colony Optimization Algorithm (ACO): A probabilistic technique for solving computational problems which can be reduced to finding good paths through graphs.

Availability: The degree to which a system, subsystem is in a specified operable and committable state at the start of a mission, when the mission is called for at an unknown, *i.e.* a random, time.

Coverage: The proportion of the resources that the production system contains the true value of throughput.

Data Analysis: A process of inspecting, adapting, and modeling data with the goal of discovering useful information and decision making.

Diagnosis: The identification of the nature and cause of anything.

Discrete Event System: A discrete-state which contains solely discrete state spaces and event-driven state transition mechanisms.

Henstock–Kurzweil Integral: Also known as the narrow Denjoy integral, Luzin integral or Perron integral is a generalization of the Riemann integral, and in some situations is more general than the Lebesgue integral.

Information: Represents a form of knowledge and data which can be derived, as data represents values attributed to parameters, and knowledge signifies understanding of real things or abstract concepts.

Markov Chain: A mathematical system that undergoes transitions from one state to another, among a finite or countable number of possible states.

Subsequence: A sequence (set of elements) that can be derived from another sequence by deleting some elements without changing the order of the remaining elements.

Chapter 13
Software Process Paradigms and Crowdsourced Software Development:
An Overview

Nitasha Hasteer
Amity University, India

Abhay Bansal
Amity University, India

B. K. Murthy
Centre for Development of Advanced Computing, India

ABSTRACT

Production of quality software requires selecting the right development strategy. The process and development strategies for creating software have evolved over the years to cope with the changing paradigms. Cloud computing models have made provisioning of the computing capabilities and access to configurable pooled resources as convenient as having access to the common utilities. With the recent advancements in the use of social media and advent of software development through crowdsourcing, the need to comprehend and analyze the traditional process models of software development, with regard to the changed paradigm have become ever more necessary. The changes in the way software are being created and the continuous evolution in the processes of development and deployment has created a need to understand the development process models. This chapter provides an insight on the transition from the conventional process models of software development to the software development methodology being used to develop software through crowdsourcing.

DOI: 10.4018/978-1-4666-9688-4.ch013

INTRODUCTION

Software is a product and also a vehicle for delivering a product. Engineering of this Software is termed "Software Engineering". The term Software Engineering was coined at a NATO Conference in 1968 (Naur & Randell, 1968) and is defined as the systematic approach to the development, operation, maintenance and retirement of the software. The systematic approach enables the production of high quality software and focuses on the activities directly related to the production of the software and thus these activities form the core of a software project. The approach defines the tasks and the order in which they should be executed. The software development approach that we deploy plays a critical role in the quality of software being developed. A non-systematic development approach will produce software of low quality that may not meet user expectations and also might incur additional cost of development. Software engineers use software development models for creating fault free software that meets the requirements of the end users and is delivered within the pre specified time limit and budget. Following a non systematic approach and developing software without the use of development models, not only violates the budget and time deadlines but also results in creation of low-quality software that would be difficult to maintain.

Software development is a creative and ever evolving area. Organizations use various software development process models and methodologies for developing software. A Software Process Model defines the stages in which a project should be divided, order of execution of these stages, and other constraints and conditions on the execution of these stages (Pankaj, 2010). The process models are based on the paradigms of software development and have undergone evolution from their inception to the present times. From the first published water fall model in 1970 to the Agile Software Development Methodologies until the recent times, the software industry has seen the evolution over years to help the industry cope up with the changing scenarios. There exists some generic process models and many variations of these generic process models have been proposed and in practice a combination of different models may be used. A generic process model defines five key activities: communication, planning, modeling, construction and deployment (Pressman, 2014). Each generic model describes a process flow that describes the activities and tasks along with their sequence of occurrence.

The computing paradigm has seen a revolution with the advent and popularity of distributed computing. Gone are the days when organizations used to buy additional hardware to increase the computing capabilities of the systems. Cloud Computing has much changed the way organizations operate and do business. According to the definition given by the US National Institute of Standards and Technology (NIST), Cloud computing is a model for enabling ubiquitous, convenient, on-demand network access to a shared pool of configurable computing resources (for e.g., networks, servers, storage, applications and services) that can be quickly provisioned and released with minimal management effort or interaction of the service provider (Mell & Grance, 2011). This framework provides three service models and four deployment models. The service models are Infrastructure as a Service (IaaS), Platform as a Service (PaaS) and Software as a Service (SaaS). The deployment of the distributed network can be as a Public, Private, Hybrid or a Community cloud. The software that is created for the cloud is designed and developed to cater to the multi tenancy architecture. The task of creating such software requires process models and frameworks that could incorporate all the security requirements of the multi tenancy environment (Hasteer, Bansal, & Murthy, 2013). The development methods, tools and models for engineering cloud services needs attention and must be addressed to realize the potential of the technology completely as the cloud drives new needs.

Regardless of the model being used for development, it has always been important for all the team members to have a precise understanding of the software product developed and have coordination among them. In today's scenario, the social media has drastically changed the landscape of software engineering (Storey, Singer, Cleary, Figueira Filho, & Zagalsky, 2014). Social networking has made the creation of "virtual communities" possible through various web based hosting services. Social networking and social software are becoming prominent in every aspect of our life (Begel, Bosch, & Storey, 2013). People from all around the world have started demonstrating an unprecedented social behavior, wherein they are coming together and joining hands to perform tasks, usually for little or no money, that were once the sole province of employees. This phenomenon commonly known as 'Crowdsourcing', is sweeping through industries ranging from professional photography to journalism to sciences, to software development (Howe, 2006). The objective of crowdsourcing software development is to produce high quality and low cost software products by harnessing the power of the crowd. To meet this objective, the crowd workers who agree to work on the task are given some financial or social incentives. Large numbers of people coordinate and compete to solve complex problems. Through participation and closer coordination among large number of people and use of new computing and communication technologies it is possible to address a wide range of problems through crowdsourcing. Wikipedia is considered to be one such Technology Mediated Social Participation System(TMSP) (Kraut, Maher, Science, Olson, & Thomas, 2010).

BACKGROUND

Several studies in literature focus on different types of process models and their evolution. The choice of methodology for a software project has varied largely due to the diversity in nature and type of software applications. Technological advancements too have led to new paradigms of software development. The development of software by outsourcing it to different organizations has been in practice for quite some time now. But the use of cloud to outsource the development of software to a crowd is new and is relatively in its nascent stage.

The term Crowdsourcing was coined by J Howe in 2006 through his article 'The Rise of Crowdsourcing' in the wired magazine (Howe, 2006). Researchers have described Crowdsourcing as a problem solving model that makes use of collective intelligence and wisdom of the crowd. It has been termed as a flexible model that is applicable to wide range of activities from producing consumer goods and media contents to creating software. Many crowdsourced platforms are coming up to solve complex problems and facilitate the development of software. Organizations like RentACoder, oDesk, Elance, Topcoder, uTest adopt different approaches for crowdsourcing (R., A, & Eric, 2010). Most of these organizations serve as a liaison between the free lance software developers and the clients. They provide workforce to the clients and in return take payment of the same. Some organizations on the other hand provide the solution and get payment from the client for developing complete software from crowd workers. To understand the

role that social networking plays in software development, Andrew Begel et al. (2013) conducted interviews with leaders from GitHub, MSDN, Stack Exchange and TopCoder. The focus of these interviews was to understand these social networks, their features and to know the incentives given to encourage participation. Their work reveals the evolution, hurdles faced and the future of crowdsourcing software development.

PROCESS MODELS

A 'Process' is defined as the collection of activities and tasks to be performed for creation of a Product. Activities describe broad objectives and tasks focuses on small and well defined objectives. The activities and tasks reside within a framework that describes their relationship with one another. These frameworks are termed 'Models'. A Process Model provides a useful framework and a roadmap for software engineering work to all concerned with the development of a product. It is defined as an abstract representation of a design process, which in turn is a structured set of activities required to conceive a full software system or parts of it.

Changes in technology and improvement in the hardware affects the process of software development. Any organization involved in development of the software requires choosing the right kind of process model for executing the project. Depending on the complexity of the software, nature and scope of the project and tools used for developing the software different process models are chosen. Due to the diversity in the nature of the projects, designers and developers many a times find it difficult to stick to one process model and may change the implementation of the model during the development process. The literature reveals many generic process models and also shows how these models have changed with time to cater to the changing development paradigms.

Classic Life Cycle Model

The classic lifecycle model, popularly known as the waterfall model or a linear model is one of the oldest and the simplest process models. The model follows a sequential approach to the development of software. The model was originally proposed by Royce in 1970 (Royce, 1970) and later has been modified depending on the nature of the activities. The phases of this model moves linearly from the specification of the requirements to design, implementation, testing and operation. The end of one phase and the beginning of the other is clearly marked in the model. The fundamental activities in the model are well defined and are as listed below:

- **Requirement Analysis:** The purpose of this activity is to define in detail the system specifications. The goals and constraints of the system are established in discussion with the customer during this phase. It focuses on analysis of the requirements and planning for them.
- **System and Software Design:** The purpose of this activity is to establish the overall system architecture. The phase focuses on identifying and describing the software system abstraction and their relationship.
- **Implementation and Unit Testing:** The implementation phase deals with converting the design to a set of programs and testing at a unit level ensures that each unit meets its specifications.
- **Integration and System Testing:** During this phase the individually tested components are integrated and it is ensured that the software meets the requirements.
- **Operation and Maintenance:** During the operations phase, the software system is installed and put to use. Maintenance phase involves correction of any errors that went unnoticed during the earlier phases of the life cycle.

The model is illustrated in Figure 1 and is a good choice for the projects where the requirements are well understood and the project is a routine type of project. Although very popular, this model has some

Figure 1. Water fall model

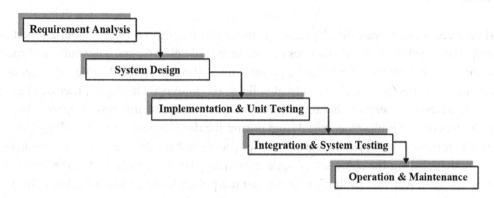

constraints. It is a rare possibility to have a smooth sequential flow of activities in case of real projects. Additionally, the model requires all the system requirements to be well defined at the start of the project that may not be feasible in most of the projects. Since the technology changes at a rapid rate and in case of a project that extends to few years, freezing of the requirements at an early stage might result in developing a product that would work on an obsolete technology. This model requires formal documents at the end of each phase. Further, since the developed product is delivered at the end, the working version of the software is not seen until late in the process. In today's world, development of software is fast paced and changes are frequent. This makes the use of waterfall model inappropriate in many cases.

Incremental Process Model

The incremental model combines the elements of linear and parallel process flows. The model applies the linear sequences in a staggered fashion. At the end of each linear sequence a deliverable increment of the software is produced. In this process model, the customers identify the services to be provided by the system. The requirements for the services that are to be delivered in the first increment are then defined in detail. The increment is completed and delivered to the customer and then further requirement analysis for the later increments is done. The model is illustrated in Figure 2. Though these models have lower risk of project failure because of a working version seen early, it sometimes can be difficult to map the customer requirements to a small sized increment.

Figure 2. Incremental process model

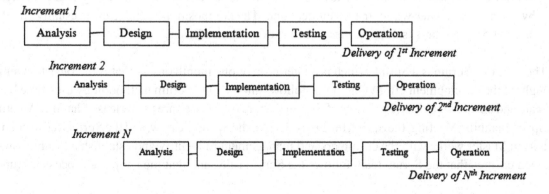

AGILE MODELS

In the earlier decades of software development, software teams used to do work under defined roles and with the help of defined models and practices to execute a well defined plan. The traditional models that were used for development were rigid. In the present times, the resource, schedule, cost and other constraints are managed effectively with the help of agile methodologies. The Agile processes have a group of software development methods that focus on iterative and incremental development. These models replace high-level design with frequent redesign during the development process. All agile models are based on the incremental development and delivery concept and have different processes to achieve their objectives, though sharing the same set of agile principles. These models enable the development of high-quality, adaptive software by small teams using the principles of continuous design improvement and testing based on rapid feedback and change. Some of the popular agile models are Scrum, Extreme programming (XP), Crystal, Adaptive Software Development (ASD) and Feature Driven Development (FDD). The scope of this chapter focuses on detailed discussion on Scrum Based Development.

Scrum Development

For decades organizations have relentlessly pursued to create optimized and repeatable processes. Out of the many models of agile, scrum based development is one lightweight methodology that has been used to develop complex products since 1990 (Tuli, Hasteer, Sharma, & Bansal, 2014). Scrum based development uses an iterative and incremental approach to optimize predictability and control risk while developing software. The Scrum framework is transparent, is inspection based and encourages adaptability. It includes a set of Scrum teams and their associated roles. The process is time boxed and has a set of defined artifacts and rules (Schwaber, 2009).

Each Scrum Team has the following three important roles in a project:

1. **Scrum Master:** The Scrum Master acts as a facilitator whose job is to resolve obstructions and also to help in creating a self-organizing environment. He is responsible for ensuring that the process is thoroughly understood and rigorously followed.
2. **Product Owner:** The product owner is responsible for product vision, maximizing return on investments, prioritizing the product backlog, and also deciding whether to continue development or not under certain conditions. Product backlog is a list of customer-centric features. Product owner is responsible for maximizing the value of the work that the scrum team does and ensuring the completion of tasks as specified in the product backlog.
3. **Scrum Development Team:** The team is responsible for doing each and every task as assigned by the product owner within the sprint deadline. The composition of the team is usually small. A team of 5-7 members is made to carry out the development tasks.

The core of a Scrum is a Sprint, which is an iteration of one month or less that is of consistent length throughout the development effort. One sprint follows another and Scrum methodology uses time boxes to create and ensure regularity. Processes of Scrum that are time boxed are the Release Planning Meeting, the Sprint Planning Meeting, the Sprint, the Daily Scrum, the Sprint Review, and the Sprint Retrospective.

Scrum Master works with the customers and management to identify and establish a Product owner and then ensures that the Scrum team follows the Scrum rules and that the adoption process is smooth and continuous. The Scrum teams are self organizing teams and are cross functional. The Product Owner has the responsibility of managing the Product Backlog. His responsibility is to ensure the value of the work that the 'Team' is contributing along with the prioritization of the Product Backlog. The members of the Scrum Development team are responsible for translating items of the Product Backlog into increments of deliverables at every Sprint. The members possess all required skills to create and deliver an increment. The optimal size of the team as specified earlier is generally from 5 to 7 members. Teams do not have sub teams and the composition of a team may change at the end of a Sprint.

Sprints are time boxed iterations. Sprints consist of the Sprint Planning meeting, the development work, the Sprint Review, and the Sprint Retrospective. There is no gap between successive Sprints and each Sprint has a goal which is an objective that covers a defined scope of a Product Backlog to be accomplished. The goal is met by successful implementation of the same. Product owner has the authority to cancel a Sprint, in case of a Sprint goal getting obsolete. During a Sprint planning meeting, the Product owner presents the top priority Product Backlog to the Scrum Team. Team figures out what functionality is to be developed during the next Sprint and how it can be developed. A list of tasks required to convert the Product Backlog into working software is formulated. It is called a Sprint Backlog. At the end of a Sprint, a Sprint Review meeting is held wherein the Product Owner identifies what has been accomplished so far. The Team discusses the issues encountered during Sprints and how they managed to solve them. The Team then demonstrates the work that has been accomplished. The Product Owner then discusses the Product Backlog as it stands, in the presence of the Scrum Master. The Sprint Review provides valuable inputs to subsequent Sprint Planning meeting. After the Sprint Review and before the next Sprint planning meeting, the Sprint Retrospective meeting is held. The purpose of the Retrospective is to inspect how the last Sprint went with regard to people, relationships, process and tools. The focus is to inspect Scrum team composition, methods, processes. Each Scrum development team meets daily for 15 minutes for a status check and to see the progress towards attaining the Sprint's goal. These Daily Scrum meetings helps to identify gaps, improve communication and promote quick decision making. The flow of various Scrum meetings in Scrum based development of software is illustrated in Figure 3.

Following are the artifacts of the model:

1. **Product Backlog:** It is a list of prioritized requirements that are needed for the development of the end product.

Figure 3. Scrum meeting flow in Scrum based development

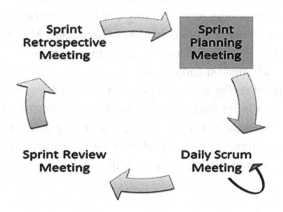

2. **Sprint Backlog:** It is a list of tasks to turn the Product Backlog for one Sprint into an increment of potentially shippable product.
3. **Release Burndown:** It is instrumental in measuring remaining Product Backlog across the time of a release plan.
4. **Sprint Burndown:** It measures remaining Sprint Backlog items across the time of a Sprint.

The Product Backlog usually has User Stories and represents everything required to develop and launch a successful product. It contains list of all features, functionalities, enhancements, technologies, and bug fixes that consist of the changes that will be made to the product for its future releases. The items contained have attributes of description and priority. The Team modifies Sprint Backlog throughout the Sprint, as well as Sprint Backlog that occurs during the Sprint. The Sprint Backlog has a list of tasks the Team performs to turn Product Backlog items into an increment and it is a real picture of the work that the team plans to accomplish. A Burndown measures the remaining backlog over time. Sprint Backlog Burndown is a graph of the amount of Sprint Backlog work remaining in a Sprint across time in the Sprint. The Release Burndown graph records the sum of remaining Product Backlog estimated effort across time.

CROWDSOURCING

A revolution that marks the rise of online community made up of like minded individuals who work together or compete with each other to deliver solutions to problems is termed Crowdsourcing (Howe, 2008). One of the oldest known commercial crowdsourcing applications is Amazon's Mechanical Turk (AMT) that was launched in 2006. It is a market place for small tasks that cannot be easily automated (Ipeirotis, 2010). Crowd workers called 'Turkers' select 'Human Intelligence Tasks' (HITs) from this market place of work that matches their interests and abilities. This gives them freedom to work as per their convenience and also gain financial and social incentives. Some work for fun and others for gaining extra income or to earn a reputation. HITs on AMT platform are generally of the kind like characterizing data, transcribing spoken languages, or creating data visualizations. What makes crowdsourcing interesting is that virtually everyone and anyone can contribute. The proliferation of crowdsourcing practices indicates a new peer production paradigm (Doan, Ramakrishnan, & Halevy, 2011). It has been classified into various typologies based on its aim of practice. Typologies vary depending on whether the power of crowd is being used for solving complex problems (crowd wisdom), or for seeking a creative input (crowd creation), or for opinion polling (crowd voting), or for the purpose of outsourcing tasks (crowd production) and or for raising money (crowd funding) (Kleemann & Voß, 2008). Crowdsourcing is efficient with the right community at place and for gathering information quickly and effectively. It is inexpensive and can fill the information gap quickly.

Crowdsourcing as described earlier is also an emerging form of outsourcing software development. Software development process is revolutionizing from the usual outsourcing to an organization or a client to the outsourcing to the masses. The idea of crowdsourcing software development is to take the services of voluntary online community to build software rather than taking the services of traditionally employed workers. Enterprises can outsource the task of developing software to the general crowd in either collaborative or competitive manner and can have access to scalable workforce. In a collaborative crowdsourcing environment, people collaborate to produce software products. Competitive crowdsourcing on the other hand is reward based.

Organizations can harness the power of crowd and take benefit of Crowdsourcing in order to produce quality software, and have reduction in the total cost incurred on the development process. Apart from these direct benefits, there are many other reasons why organizations would like to get software developed through crowdsourcing. Some of the benefits include rapid acquisition of the solution, identification of talent, and promoting diversity of ideas, broadening participation, better marketing and encouraging the use of some specific tools.

Software Application Development Methodology

Many platforms for facilitating software development through crowdsourcing are coming up. One such successful TMSP platform is TopCoder. TopCoder founded by Jack Hughes is one of the largest competition based software development portal that posts software development tasks as contests. The business model at TopCoder has been refined since its inception to accommodate the changes in the software industry. However its USP has always been the competition based software development approach (R. et al., 2010). The platform has an online community of digital creators who compete to develop and refine technology, various web assets, perform extreme value analytics, and develop mobile applications for customers. This commercial crowdsourced platform has seen significant growth in its community members in the past few years, which itself speaks volumes about the popularity of the platform. Many organizations are using the services of this platform, to develop low cost software. The platform has also been successfully used to develop spaceflight software for small satellites among many other projects. SPHERES Zero Robotics is an initiative of DARPA to develop spaceflight software through crowdsourcing (Nag, Heffan, Alvar, & Lydon, 2012). It is a project between MIT, Aurora Flight Sciences and TopCoder that is supported by NASA and DARPA. It is a robotics programming competition in which students learn to a write a program for controlling a satellite in space with the help of a graphics or a C editor. The robots are miniature satellites called SPHERES. The participants compete to win a technically challenging game by programming their strategies into these miniature satellites. Thousands of developers compete during the contests hosted by TopCoder to creatively design a web interface for these students and assemble the software components to build a robust framework to allow satellite control.

TopCoder provides mechanisms and infrastructure to manage and facilitate the creation of problem statements and their solutions. A platform manager is assigned to each project who closely works with the client to formulate the problem and host it onto the platform in the form of competitions. This model of software development is known as a 'Competition Based Development Model'. The software application development methodology being used by the TopCoder platform is illustrated in Figure 4.

Software Development Phases of Competition Based Development Model

The application development process of a competition based development model progresses in phases. Each phase is executed through a competition or a series of competitions and the winning entry serves as an input to the subsequent phases. The client of a crowdsourced platform may use an existing com-

Figure 4. Software application development methodology at crowdsourced platform
Source: Company Docs.

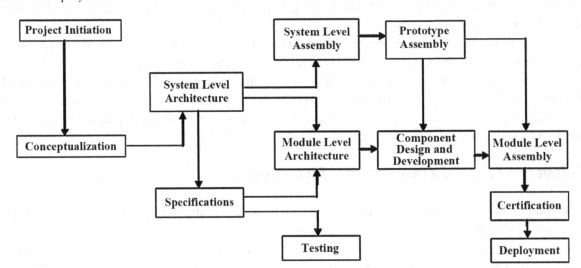

ponent from the platform catalogue or request for creation of a new component. There are six broad phases namely Conceptualization, Specification, Architecture, Component Design and Development, Testing and Assembly.

- **Conceptualization Phase:** The development begins with the Conceptualization phase. The competitions under this phase are conducted to identify and document the needs and ideas of all the project stakeholders. The competitions under this phase may commence by either running a series of studio competitions to create graphical conceptualization artifacts like Storyboards, Wireframes and Prototypes, or a series of Conceptualization contests to create a Business Requirement Document and High-Level Use Cases.
- **Specification Phase:** The Specification phase deals with the formulation of application requirements in as much detail as necessary in order to accomplish the goals for the creation of application module. The high level use cases that are identified during Conceptualization contests are assigned to modules during System Architecture phase, and during the Specification phase all the individual scenarios that make up those use cases are broken up in text and graphical form with the help of UML Activity Diagrams.
- **Architecture Phase:** There is a System Level and a Module Level Architecture phase of software development. The competitions at a System Level Architecture phase accepts the business requirements and prototype defined in Conceptualization phase as input to define the overall technical approach that will be employed to meet those requirements. Module Level Architecture Phase defines the lower level technical design of an independent module of a larger application. This phase is responsible for defining the components and their interactions that will implement the requirements for a particular module.
- **Design and Development Phase:** During the component design phase, competitors get an opportunity to clarify any unclear requirements and define technical details for implementation. Component design competitions take the component requirements developed during the architec-

ture phase as input and produce a detailed component design specifications from the same. After the component design competition is completed, the detailed component design specification acts an input into Component development competition phase. During the component development competition phase, the component is actually implemented.

- **Testing Phase:** The competitions in the testing phase provide a mechanism for verifying that all of the requirements that were identified during the initial phases of the project were properly implemented and that the system performs as expected. The test scenarios developed through these competitions ensures that the requirements are met end-to-end.
- **Assembly Phase:** The System Assembly phase hosts competitions to create the foundation for the application. This includes creating the build scripts that will be used throughout the application as well as incorporating all identified components into the shell that implements the application's cross-cutting concerns.

Most Crowdsourced platforms in general effectively handle the security and Intellectual property issues that might resist the potential clients from using their services. Company sources reveal that customers who are not familiar with the security and privacy policy of the platform may have apprehensions in choosing the platform initially (R. et al., 2010). Crowdsourced platforms produce white papers on their security and privacy policies to attract new customers and communicate their processes to them. These papers reveal in detail the confidentiality policy and the approach to software development being used at the platform. The platforms also allow the clients to generate test data sets to be used during the competitions and keep anonymous names for their organizations to ensure privacy. The competitors or the contestants too are allowed to set their privacy policies at the platform.

Benefits of Competition Based Development Model

The competition based development model has successfully created software for the use of individuals and organizations. Some of the benefits of the competition based development model with regard to a project in an organization are as listed below:

- The time and cost needed to hire, train and fire people are lowered.
- The cost of networking, communication and infrastructure is reduced.
- The participants possess diverse skills and experience there by creating innovative solutions.
- The individual's interest and choice of working on a particular problem increases the chances of submitting solutions as per the deadline.
- The solutions to the problems are not dependent on individuals.
- The intensive review process ensures the selection of the best and quality work as a winning solution.
- Winning solution is rewarded with a fixed pre decided amount only if the solution meets the specifications and is delivered on time there by reducing cost of development.

There are numerous benefits that crowd workers realize so as to be active participants in the competition based development model. Individuals are keen to participate as competitors either to spend quality time on the internet for fun or to earn extra income. The social and financial incentives gained by competitors are often a driving factor for continuous participation in the competitions at a crowdsourced platform. The flexibility of working as per their convenience and having no requirement of reporting to their bosses is an attraction for many.

Cloud Based Software Crowdsourcing Architecture

The distributed development of software through crowdsourcing needs reference architecture to facilitate the development process. The cloud based software crowdsourcing reference architecture proposed by Wei-Tek Tsai et al (Tsai, Wu, & Huhns, 2014) has the following components or elements:

- Cloud Service Management dash board to help system administrators and software crowdsourcing organizers.
- Collaboration and communication tools, such as a distributed black board system to enable participants to participate in discussions.
- Software development tools for modeling, simulation, code editing and compilation, design notation and documentation and testing.
- Tools for raking participants and displaying scoreboards.
- Cloud Payment tools and credit management tools.
- Repository of software modules, specifications, architectures, and design patterns for possible reuse.

The reference architecture for a Cloud Based Software Crowdsourcing system that encapsulates the above stated elements is as shown in Figure 5. The use of architecture can enable the achievement of the ultimate goal of developing high-quality and low-cost software products.

With the use of a properly specified Platform-as-a-Service (PaaS), software project managers can establish a customized cloud software environment to facilitate software crowdsourcing process. A manager can make use of software networking and collaboration tools to create a reward mechanism to motivate the crowd workers. Depending on the scale of development and value of the software product, a project manager can specify the appropriate budget to attract as many talented developers as possible and provision computing resources to sustain their activities.

The manager uses project management tools to coordinate development tasks among the crowd workforce by ranking individual's expertise and matching their skills with the different task levels and types. For a specific software project, manager sets up a virtual system platform with all the necessary software development gears to assist crowd workers with their tasks. Social Network and Collaboration tools can prove to be of great use for creating virtual teams and generating skilled crowd workforce.

ISSUES AND CHALLENGES

Crowdsourcing offers both advantages and disadvantages. With regard to Crowdsourcing in general, one of the challenges is development of trusted sources and platforms. It is not possible to examine all the contents and restricting who posts data online defeats the entire purpose of crowdsourcing. Participants are humans and as a common saying goes 'To err is human'. It might be false to assume that there is no error in the information provided. Unfortunately, the hard fact is that the accuracy of information is an

Figure 5. Reference architecture Cloud Based Software crowdsourcing
Source: Internet Computing, (Tsai et al. 2014).

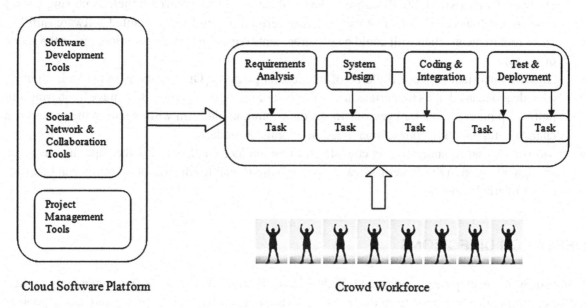

assumption most of the times. Another big challenge is publicizing a crowdsourcing platform and establishing a network of volunteers. The publicizing task requires resources that might be a scarcity. Higher participation rates translate into a greater volume of data and managing this process may become tedious.

Stol and Fitzgerald (Stol & Fitzgerald, 2014) have identified six key concerns for crowdsourcing software development. According to them decompositions of tasks, coordination and communication, planning and scheduling, knowledge and intellectual property, quality assurance, motivation and remuneration are important and relevant in the context of software crowdsourcing. The challenges of a competition based Development Model are as listed below:

- Formulating a proper work flow design by decomposition is an area of concern. The appropriate decomposition of the tasks into smaller units and establishing a workflow design is a challenge specifically with regard to software. Understanding complex interfaces and interdependencies to decompose the problem into smaller tasks by the platform mangers in order to host the competition could be challenging.
- Establishing proper Coordination and Communication among the participants is another challenge. The competitors who compete and submit their solutions are not a part of the crowdsourced platform. Coordination between the platform manager and all the competitors is an issue that needs to be handled effectively as interdependent tasks are performed by different competitors.
- Planning and scheduling of the competitions need to be handled effectively for timely delivery of the product. Once the competition and deadline has been scheduled, it is the difficult for the platform manager to intervene and expedite the development for meeting any demands of the customer for early delivery of the product.
- Quality of the end product might be of concern. Findings in the literature (Peter, Laura, Sara, & Haakon, 2013) suggest that more the number of contestants, poor is the quality of the work delivered. It is sometimes considered that crowd mostly consists of amateurs; hence the quality of

work is their least concern. Also, there could be disagreement about a solution for any published problem. Kittur (Kittur, 2010) distinguished 'subjective' tasks for which there is no single right answer, and 'objective' tasks that can be easily verified. Even if the submitted software fulfills a set of requirements, there still could be disagreements regarding certain functionality or the scope of a task.

- Another important concern could be of intellectual property. Organizations might be hesitant to provide too many details for certain tasks. The risk of losing competitive advantage by sharing and putting up many details for everyone to see on a common platform with regard to the project is a challenge for adopting competition based development model.

- Motivation and remuneration is considered to be another challenge and the topic has received enough attention in the field of software engineering research. Motivational factors can be external or internal to the tasks.

RESEARCH DIRECTIONS

Many software development models and methods have been proposed to assist organizations in their software development process: from the traditional waterfall approach to the recent and highly popular agile methods. Crowdsourcing can be incorporated into conventional software development processes and can contribute to any software development phase (Tsai et al., 2014).

Crowdsourcing continues to advance and solve increasingly complex issues (Greengard, 2011). It has generated interest amongst software engineering researchers worldwide and scientists are attempting to solve complex problems through crowdsourcing. Researchers across the globe are working on various aspects of crowdsourcing software development.

Techniques to formulate reward mechanisms in order to minimize cost have been an area of active research. The work of Dominic and Milan (Dipalantino & Vojnovic, 2009) demonstrated the relationship between incentives and crowd participation at a crowdsourced platform. Many researchers have compared features offered by various crowdsourced platforms. The work of M Vukovic (Vukovic, 2009) proposes a taxonomy for classification of crowdsourcing platforms, and evaluates a number of existing systems against the set of identified features.

One of the prime challenges for crowdsourcing software development as discussed in the earlier section is the decomposition of the tasks into microtasks. Thomas D Latoza et al (Latoza, Towne, Adriano, & Hoek, 2014) developed an approach to decompose programming work into microtasks. They proposed the coordination of work through tracking the changes to a graph of artifacts, generating appropriate microtasks and propagating change notifications to artifacts with dependencies. They implemented the approach in a Cloud based IDE CrowdCode.

Scientists are investigating the age, skill set and demographics of crowd workers to propose models best suited for development of the software at crowdsourced platforms. The work of Wenjun Wu (Wu, Tsai, & Li, 2013) examines software crowdsourcing processes including TopCoder and AppStori. The work also identifies the min max nature among contestants as an important design element with regard to creativity and quality. The strategic behavior of contestants at a crowdsourced platform has been analyzed by Archak (Archak, 2010). The research found that individual specific traits along with project requirements and project payment are significant predictors of the final project quality. He found significant evidence of strategic behavior of contestants and the study reveals that high rated contestants face tougher competition. The research conjectured that the reputation and cheap talk mechanism employed by TopCoder platform positively effects the efficiency of simultaneous all-pay contests and this approach should be considered for adoption in other crowdsourcing platforms as well. In their work Zhenghui and Wenjun (Hu & Wu, 2014) applied the famous game theory to model two player algorithm challenges at TopCoder to study the competitive behavior of contestants.

It is changing the way organizations handle complex problems and issues. Research in future can focus on solving issues with regard to multiple winners at a crowdsourced platform. Another area that needs attention is how cloud infrastructure can be well utilized for hosting scalable application development environment for dynamic, task specific crowd teaming.

CONCLUSION

Software development requires a disciplined engineering approach to create cost effective and quality software. Development of quality software requires creative skills along with sound technical expertise. The emerging crowdsourcing practices would inspire new research directions and give new dimensions to stimulate while ensuring high quality of the developed software products. With social networking getting so predominant in our daily life, future would see more integration of the social networks with the creator networks and crowdsourced software development platforms. While our traditional models aim at providing a systematic way to develop software, one cannot overlook and deny the emergence of successful open source projects where there is no existence of a predefined process. Platforms that promote crowdsourcing software development would have to incorporate flexibility in terms of scaling up the resources and building a trust among stakeholders for the wide acceptance of this development methodology.

REFERENCES

Archak, N. (2010). Money, Glory and Cheap Talk: Analyzing Strategic Behavior of Contestants in Simultaneous Crowdsourcing Contests on TopCoder.com. In *Proceedings of WWW 2010*. Raleigh, NC: ACM. doi:10.1145/1772690.1772694

Begel, A., Bosch, J., & Storey, M.-A. (2013). Social Networking Meets Software Development: Perspectives from GitHub, MSDN, Stack Exchange, and TopCoder. *IEEE Software*, *30*(1), 52–66. doi:10.1109/MS.2013.13

Dipalantino, D., & Vojnovic, M. (2009). Crowdsourcing and All-Pay Auctions. In *Proceedings of EC'09*. Stanford, CA: ACM.

Doan, A., Ramakrishnan, R., & Halevy, A. Y. (2011). Crowdsourcing systems on the World Wide Web. *Communications of the ACM*, *54*(4), 86. doi:10.1145/1924421.1924442

Greengard, S. (2011). Following the Crowd. *Communications of the ACM*, *54*(2), 20–22. doi:10.1145/1897816.1897824

Hasteer, N., Bansal, A., & Murthy, B. K. (2013). Pragmatic assessment of research intensive areas in cloud: A systematic review. *Software Engineering Notes*, *38*(3), 1. doi:10.1145/2464526.2464533

Howe, J. (2006). The rise of crowdsourcing. *Wired Magazine*. Retrieved November, 2014, from http://www.wired.com/wired/archive/14.06/crowds_pr.html

Howe, J. (2008). *Crowdsourcing*. New York: Crown Business.

Hu, Z., & Wu, W. (2014). A Game Theoretic Model of Software Crowdsourcing. In *Proceedings of 8th International Symposium on Service Oriented System Engineering (SOSE)*. Oxford, UK: IEEE. doi:10.1109/SOSE.2014.79

Ipeirotis, P. G. (2010). Analyzing the Amazon Mechanical Turk Marketplace. *ACM XRDS*, *17*(2), 16–21. doi:10.1145/1869086.1869094

Kittur, A. (2010). Crowdsourcing, collaboration and creativity. *ACM XRDS*, *17*(2), 22–26. doi:10.1145/1869086.1869096

Kleemann, F., & Voß, G. G. (2008). Un(der)paid Innovators: The Commercial Utilization of Consumer Work through Crowdsourcing. *Science. Technology & Innovation Studies*, *4*(1), 6–26.

Kraut, R., Maher, M., Lou, U. S. N., Olson, J., & Thomas, J. C. (2010). Scientific Foundations: A Case For Technology Mediated Social Participation Theory. *IEEE Computer*, *43*(11), 22-28.

Latoza, T. D., Ben Towne, W., Adriano, C. M., & Van Der Hoek, A. (2014). Microtask Programming: Building Software with a Crowd. In *Proceedings of User Interface Software and Technology Symposium (UIST)*. Honolulu, HI: ACM. doi:10.1145/2642918.2647349

Mell, P., & Grance, T. (2011). *The NIST Definition of Cloud Computing Recommendations of the National Institute of Standards and Technology*. NIST Special publication 800-145.

Nag, S., Heffan, I., Alvar, S.-O., & Lydon, M. (2012). SPHERES Zero Robotics Software Development: Lessons on Crowdsourcing and Collaborative Competition. In *Proceedings of Aerospace Conference*. IEEE. doi:10.1109/AERO.2012.6187452

Naur, P., & Randell, B. (1968). Software Engineering: Report of a Conference Sponsored by the NATO Science Committee. Garmisch, Germany: NATO.

Pankaj, J. (2010). *Software Engineering: A Precise Approach*. Wiley India.

Peter, K., Laura, D., Sara, K., & Haakon, F. (2013). Co-worker transparency in microtask marketplace. In *CSCW'13 Proceedings of 2013 conference on computer supported cooperative work*. San Antonio, TX: ACM.

Pressman, R. S. (2014). *Software Engineering A Practitioner's Approach (7th ed.)*. Mc Graw Hill Education.

Royce, W. W. (1970). Managing the development of large software systems. In *Proceedings of International Conference on Software Enginerring (ICSE-9)*. IEEE.

Schwaber, K. (2009). *Scrum Guide*. ScrumAlliance.

Stol, K., & Fitzgerald, B. (2014). Two's Company, Three's a Crowd: A Case Study of Crowdsourcing Software Development. In *Proceedings of ICSE 2014*. Hyderabad, India: IEEE. doi:10.1145/2568225.2568249

Storey, M.-A., Singer, L., Cleary, B., Figueira Filho, F., & Zagalsky, A. (2014). The (R) Evolution of social media in software engineering. In *Proceedings of the Future of Software Engineering - FOSE 2014*. Hyderabad, India: IEEE.

TopCoder. (n.d.). Retrieved November, 2014 from https://www.topcoder.com

Tsai, W.-T., Wu, W., & Huhns, M. N. (2014). Cloud-Based Software Crowdsourcing. *Internet Computing*, *18*(3), 78–83. doi:10.1109/MIC.2014.46

Tuli, A., Hasteer, N., Sharma, M., & Bansal, A. (2014). Empirical Investigation of agile software development: A Cloud Perspective. *Software Engineering Notes*, *39*(4), 1–6. doi:10.1145/2632434.2632447

Vukovic, M. (2009). Crowdsourcing for Enterprises Maja Vukovi. In *Congress on Services-I*. IEEE. doi:10.1109/SERVICES-I.2009.56

Wu, W., Tsai, W., & Li, W. (2013). Creative software crowdsourcing: From components and algorithm development to project concept formations. *International Journal of Creative Computing*, *1*(1), 57–91. doi:10.1504/IJCRC.2013.056925

KEY TERMS AND DEFINITIONS

Human Intelligence Tasks: Human intelligence tasks are a variety of simple tasks that humans can accomplish better than computers. One of the common examples of such tasks is the identification of some attributes in multiple images.

IaaS: Infrastructure as a Service (IaaS) model of Cloud provides the capability to the consumer to provision processing, storage, networks, and other fundamental computing resources so as to be able to deploy and run software that can include operating systems and applications. The consumer does not manage or control the underlying cloud infrastructure but has control over operating systems, storage, and deployed applications and some limited control of select networking components.

Multi-Tenancy: A term used for a software architecture where a single instance of an application serves multiple customers. Each customer or a group of customers is called a tenant. Multi tenancy applies to all the three layers of a cloud.

PaaS: Another service model of Cloud computing. In Platform as a Service (PaaS) model the capability provided to the consumer is to deploy applications onto the cloud infrastructure. These applications can be either *consumer created* or *acquired* applications created through programming languages, tools and services supported by the provider. The consumer does not manage or control the underlying cloud infrastructure including network, servers, operating systems, or storage.

SaaS: One of the service models of Cloud Computing. In Software as a Service (SaaS) model of Cloud Computing the service or the capability provided to the consumer is to use the provider's applications running on a cloud infrastructure. The applications are accessible from various client devices through either a thin client interface, such as a web browser (e.g., web-based email), or a program interface. The consumer does not manage or control the underlying infrastructure or individual application capabilities.

Technology Mediated Social Participation Systems: Systems that involve the participation of people, communication tools and actions to transform and build new systems. These systems are centered toward better engagement of people with diverse experiences, perspectives, skills, motivation and knowledge and to create new systems.

USP: Unique Selling Proposition (USP) is a concept of marketing that is used to explain a successful pattern of an organization.

Chapter 14
Using Soft Systems Ideas within Virtual Teams

Frank Stowell
University of Portsmouth, UK

Shavindrie Cooray
Curry College, USA

ABSTRACT

Recent research adds support to the view that the way that individuals act as part of a virtual group is different from behavior in face-to-face meetings. Specifically researchers have discovered that conflicts are more prevalent within virtual teams as opposed to face to face teams. This is because research has shown that participants are more likely to change their initial points of view (shaped by personal values, biases and experience) when discussions are held in a face to face environment rather than through virtual means. This insight raises doubts upon the effectiveness of CMCs as an instrument of organizational cohesion. In this paper we reflect upon this position and attempt to discover if these concerns can be overcome through the employment of Systems methods used in organizational inquiry. We do this through an evaluation of the results of a preliminary study between Curry College in Boston, Massachusetts, USA and Richmond University in London, UK.

1. INTRODUCTION AND PURPOSE

In this chapter we explore the use of a method of inquiry based on soft systems thinking as a means of ameliorating the effects of conflict in synchronous virtual teams (Stowell & Welch, 2012 pp116-117; Champion & Stowell, 2001). We describe the results of an Action Research (AR) project undertaken between two educational institutions geographically separated by several thousand miles who share a common language and intent, as a means of gaining understanding of synchronous virtual communication. The authors add to synchronous virtual team research and systems literature through the lessons learnt from an investigation of the feasibility of using a soft method of inquiry within a synchronous virtual environment. The approach was adopted as a means of creating shared understanding and managing conflicts. The lessons learnt from the study are relevant in the light of the rising use of virtual teams in various sectors including IT project development, education and business management.

DOI: 10.4018/978-1-4666-9688-4.ch014

2. BACKGROUND AND LITERATURE REVIEW

2.1 Seeking Understanding of a Situation of Interest

In order to explore the communal impact of CMC's compared to face-to-face communication it is prudent to reflect upon the setting in which it exists; namely an organization.

Arriving at a definition of an organization is not an easy task. From an observer's point of view a social grouping of any kind will be recognised by something that gives it a meaning for *them,* which they could describe as an organization. For example, supporters of a soccer team are affiliates of both the soccer club itself and to a wider social grouping. For the observer an organisation is a recognisable entity. In short, everyone has a general, uninspected, idea of what an organization means; consequently it is exceptionally difficult to pin down a sharp definition. Yet if we are to intervene into an organization we need a model that allows us to gain understanding. Lessons learnt from extensive research into organizations undertaken at the University of Lancaster over 30 years (See Checkland, 1999) showed that an ontological notion of organization was flawed. The research demonstrated that each organization is unique which led to the conclusion that such positivist notions should be abandoned in favour of an epistemological approach. That is to say that each organizational intervention had to be understood afresh.

We argue that each organization is unique and those that make up the organization may not share the view of it as espoused by its executive. Checkland and Holwell (1998, p8) point out an organization is not "...simply a rational machine whose members willingly combine together to pursue organizational goals". Gudykunst, (1997) and Hofstede, (1980) refer to individualistic and collectivist cultures, a view that has much in common with the sociologist Tönnies whose ideas about social groups, or organizations, he conceptualized through the notions of Gemeinschaft and Gesellschaft. Gemeinschaft, he advised, is achieved through morals, conformism and social control whereas Gesellschaft achieves stability through police and laws; Rules within Gemeinschaft are implicit whilst Gesellschaft are explicit written laws, but even so Tönnies argued that nothing happens in Gesellschaft that is more important for the individuals wider group than it is for themselves..." (Truzzi,1971). Harris (2009, p52) is more explicit and suggests that everyone is out for themselves and "...living in a state of tension against everyone else".

If we accept the view that an organisational member's first concern is for their own security and position then we must also question what affect such attitudes might have upon the way the organization operates, particularly when parts are linked through virtual communication technologies. As a means of understanding this challenge we incorporated the notion of commodity as a means of 'managing' potential discord. Stowell [1989] suggested that the metaphor of 'a commodity' embodying power could be helpful in gaining understanding of these tensions. He suggested that the expression of power through the idea of a metaphor could provide a practical means of addressing its management. This idea has proved useful because of the difficulty in defining power and, as a consequence, when intervening in an organisation we can be left with impressions of power rather than something that can be explained to others. 'Commodity' was chosen in an attempt to provide a neutral account of perceived inputs of individuals and groups in which attempts are made to translate a situation into something else.

The notion of commodity relates to something that is considered of value by organisational members within their culture, or the value systems of a given organisation; it will vary between and within each social grouping. The use of the word commodity should, through discussion, lead to a more precise definition or explanation of what is meant. The notion of commodity is intended to open up free discussion between participants in which participants work to exchange knowledge rather than seek to impose a particular opinion. Stowell felt that by surfacing these "commodities of power" it would enable a stream of analysis to be created. He suggests that in any organisational setting we can ask: what are these commodities? How are they obtained, exercised, defended, suffused or limited, relinquished? In previous studies [Stowell, 2012] researchers have used the metaphor as a way of thinking and surfacing the kinds of 'commodities' that individuals use. In this study we attempt to apply this idea even though the engagement of participants was to be through the use of virtual synchronous communication technology rather than in a face-to-face situation.

The growing use of remote communications, sometimes between groups that have never met, raises uncertainties about the way decisions are made arising from discussions in virtual meetings. Our research task was to investigate how organizational interaction is affected by CMC and to do this we needed a method of expression that was suited to the management of the complexities described above. We return to the ideas developed at the University of Lancaster from which the research program into organizational intervention led to ideas of inquiry suited to the objective of this project.

2.2 Soft Systems Thinking and Its Value to Organizational Intervention

One outcome of the research programme at the University of Lancaster was the development of soft systems thinking (Checkland, 1981; 2006). These ideas, with their intellectual grounding in the work of Schutz and Husserl (Checkland, 1999, pA20) emerged as a result of dissatisfaction with the limitations of traditional hard systems thinking. Hard systems thinking approaches assume an objective reality where the client already knows his actual needs. Those needs are then considered to be the goals to be achieved by the system; after which the focus is on "how" to attain those objectives. In soft systems thinking emphasis is put on "what" the current state of the situation is, "what" stakeholders would ideally prefer to happen and "how' to attain those requirements. Checkland and Poulter (2006) argue that many "problematic" situations do not have a clear problem to solve and that what exists is a sense that all is not well; the kind of situation that the management of an enterprise might face. We challenge the notion of an objective reality since the views of participants in any situation are subjective and it makes sense that by allowing multiple views to surface a more holistic (systemic) interpretation of the situation can emerge. This view has implications for developing relationships between organizational members if the various parts of the organization are geographically separated.

Importantly the cornerstone of soft systems thinking is its learning component. Checkland and Poulter (2006) argue that stakeholders should learn about the situation they are a part of in order to better understand "what" they hope to achieve promoting the view that each stakeholder has a unique perspective of the situation. This idea requires suitable tools to facilitate the expression of all views within a collaborative atmosphere allowing a more holistic, systemic picture of the situation to emerge.

Before investigating the usefulness of soft systems thinking in the development and operation of synchronous virtual teams, we examine how virtual teams and their dynamics are described in the literature.

2.3 The Organization and the Management of Virtual Teams

Teams are the typical elementary units of an organization (Furst et al, 1999). They can be defined as " a small number of people with complementary skills who are equally committed to a common purpose, goals, and working approach for which they hold themselves mutually accountable" (Zenun et al, 2007;p 701). Teams are formed in order to combine the diverse views and talents of individuals to achieve objectives and in the past team members were co-located because of the high levels of interdependencies that are inherent in group work. An examination of the literature suggests that organizations are beginning to implement projects over distance, with teams consisting of people who are based in dispersed geographical locations. Mason (2014) point out that three quarters of organizations want to introduce a mobile video-conferencing solution, and 61 per cent want to integrate video-conferencing.

Such teams (commonly known as virtual teams) use computer mediated communicating (CMC) in order to collaborate on tasks. The definition of a virtual team is unclear. For example, Jaarvanpaa, et al (1998) limits the term "virtual team" to teams that never meet face to face whilst others, such as Kirkman et al, (2004) and Cohen and Gibson (2003) argue that virtuality lies on a continuum ranging from highly to the minimally virtual.

It is obvious that the formation of a virtual team requires new methods of supervision, which is supported in the literature e.g. Geislar, (2014) and Jarvenpaa and Leidner (1998). When interacting face-to-face people are able to use verbal and nonverbal cues to regulate the flow of conversation; provide immediate feedback, and convey subtle meanings. Establishing the identity of and between team members is more difficult where the teams are remote, perhaps not sharing language or culture, particularly as there is rarely, if any, physical contact and visual contact is artificial Geisler (2014). It is important to establish trust between team members whether collocated or remote, since there are interdependencies between members (Griffith et al, 2003). Experience tells us that it is more difficult to establish trust in groups who never meet and only share a specific area of interest.

Kimble, Li and Barlow, (2000, p5) argue that without trust and identity the management of a virtual organization '...cannot be conceived'. Geislar (2014) points out, "...that unlike the physical world that consists of matter, the virtual world is composed of information that is diffused over time and space."

Geisler (2014) found that "...trust and identity are two significant issues for efficient creation and operation of virtual teams. Identity plays a critical role in communication and yet, when spatial borders separate team members, identity is ambiguous. Basic indicators of personality traits and social roles are harder to identify". Interdependency comes from trust in each team member and is a basic component of any team, virtual or real. "The extent to which trust is needed varies also with the relationship of the communicating parties ....for trust is both the fruit of communication and its necessary pre-condition" (Vickers, 1983, p.212). In virtual teams the conveyance of cues is hindered, feedback is delayed, and interruptions or long pauses in communication often occur (Montoya Weiss et al, 2001).

These difficulties in forming relationships are especially apparent in teams that interact asynchronously. Typically, in an asynchronous discussion, members do not meet at the same time, but instead use channels such as email and discussion boards to make contributions at different times, usually on multiple topics/threads. This pattern can reduce the synergy of team members if there are no links among the responses. Long delays between communications can lead to discontinuous and disjointed discussions. These limitations make interaction and consensus building difficult (Montoya Weiss et al, 2001; Dennis, 1996). As a result, effective communication in an asynchronous CMC environment tends to require a great deal of effort (Levin, He & Robbins, 2004; Rourke & Anderson, 2002).

Notwithstanding these difficulties organizational management are increasingly exploring the use of *synchronous* communication environments as a means of overcoming the challenges of asynchronous environments whilst still maintaining the advantages of virtual team-work. Synchronous virtual teams suggest the means of overcoming problems associated with asynchronous virtual teams; although they are not devoid of challenges (e.g. conflict, trust). (See Sarkar and Valacich 2010; Kirkman et al 2004; Kuo and Yu, 2009;Jarvanpaa et al, 2004).

2.5 Synchronous Virtual Teams

Whilst synchronous environments provide location-independent opportunities for conversation they are not time independent since team members must be logged in at the same time. Synchronous CMC's range from text-only chat tools to more advanced tools that feature audio, video and shared whiteboard capabilities. Studies have reported more equal participation among team members in synchronous environments due to the higher levels of social presence created by the synchronous use of audio and video (e.g. Paulus and Phipps, 2008; Pena-Shaef et al., 2001) Social presence "is the extent to which one feels the presence of a person with whom one is interacting" (Burke and Chidambaram 1999, p. 559) or as Sproull and Keisler (1986) argue the extent to which one becomes aware of others. Social presence is highest when a communication channel is able to effectively transmit facial, proximal and other non-verbal cues in addition to what is actually spoken. As a result social presence is highest in face to face communication while as Sproull and Keisler (1986) argue the social presence of CMCs is limited due to difficulties in transmitting visual and non verbal cues. Additionally different types of CMC environments have higher or lower social presence. Text-only CMC environments restrict "socio-emotional communication" (Burke and Chidambaram 1999, p. 559) and are believed to have lower levels of social presence since it is harder to transmit non textual cues. In contrast audio, video- based synchronous CMC environments have a comparatively higher social presence since some facial, proximal and other non- verbal cues can be communicated albeit not as much as with face to face communication.

The social presence of the communication media used influences the degree to which a team can address issues such as conflict resolution (see Sarkar and Valacich, 2010; Klein, 1991), empowerment (Kirkman et al, 2004) and trust (Kuo and Yu, 2009; Jarvanpaa et al, 2004). As Sarkar and Valacich (2010) suggest the lack of a social presence in CMC environments leads to members in virtual teams refusing to change their a priori (initial biases, preferences) attitudes toward a group level view hence delaying conflict resolution. Rahim (1992) and Thomas and Kilmann (1974) delineate five conflict-handling modes to describe conflict management in teams: avoidance, accommodation, competition, collaboration, and compromise. Researchers investigating the effects of these conflict management modes on virtual teams provide suggestions such as introducing a temporal coordination mechanism (Montoya Weiss et al, 2001) and trust mechanisms (Griffith et al (in Gibson and Cohen (2003), as a means of dealing with conflict management.

Unfortunately, conflict theory that is applicable to asynchronous virtual teams may not be wholly transferable to synchronous virtual teams because of the fundamental differences in the communication environments (Montoya Weiss et al, 2001). It is difficult to identify trends in the area as there is a dearth of literature relating to *synchronous* virtual teams (Paulus and Phipps, 2008). Although some examples

Figure 1. Relationship between communication media and social presence

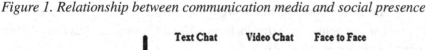
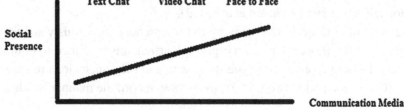

conducted in synchronous virtual environments exist (for examples see Im and Lee, 2004; Paul et al, 2004; Pena-Shaef et al, 2001) to our knowledge no previous study has examined soft systems inquiry as a viable means of managing synchronous virtual teams.

3. THE FIELD RESEARCH

The study was conducted with participants from Curry College in Boston, Massachusetts, USA and Richmond University in London, UK. Founded in 1879, Curry College is a private, 4 year liberal arts-based college while Richmond University is a private, liberal arts and professional studies university established in 1972. At the time of the study Curry College and Richmond University were in negotiations to establish a 'study abroad' program where each institution could send students for a semester to the other institution. It is in this context that the second author was invited by the Chief Academic Officer (CAO) of Curry College to organize a field study with participants from both Curry College and Richmond University to explore issues pertaining to the implementation of the study abroad program. The field study sought to gain an in-depth understanding of how to set up the study abroad program and how to create a long term sustainable relationship between the two institutions. Since key stakeholders were located in two different continents it was difficult to facilitate face to face meetings, which called for the setting up of either an asynchronous (e.g. email, discussion boards) or synchronous (Skype, Gotomeeting, Wiggio) virtual environment for collaboration. This is similar to the situation faced by modern organizations where geographically dispersed employees are expected to collaborate on projects using virtual teams.

4. RESEARCH APPROACH: ACTION RESEARCH

The objective of the study was two-fold. The organizational goal was to enhance understanding of practical issues associated with setting up the study abroad program while the research goal was to add to theory about the viability of using soft methods of inquiry within synchronous virtual teams. The dual goals of enhancing practice and theory in the study are conducive to the tenets of Action Research (AR), which combines theory generation with researcher intervention to solve immediate organizational problems (Sein et al, 2011; Baskerville and Wood-Harper 1998). AR is typically an iterative process (Susman, 1978) that helps stakeholders in a specific situation and creates general knowledge that is useful in similar situations (West and Stansfield, 1999; Iverson et al, 2004).

Although AR's linking of theory with practice is aligned with the dual goals of this study, it had its challenges. According to See Pui-Ng and Gable (2010) the personal biases of the researcher can threaten the validity of qualitative AR which corresponds to arguments made that AR researchers are active participants in an inquiry, both influencing and being influenced by it (Susman and Evered, 1978; Checkland and Holwell, 1998 and Davidson et al, 2012).

Researcher bias was minimized in the study by the researchers consciously striving to let the participants be in *control* of the process of inquiry, giving participants a chance to edit and confirm the study data, the researcher's interpretation of the study data and the conclusions reached; and using the mnemonic PEArL (Champion and Stowell, 2001) as a tool to record the manner in which the researcher participated in each session. [for PEArL see section 5] External parties can use the information when interpreting results. Also the researcher can use the PEArL records to reflect on her own interactions with participants.

Rigor is another challenge faced by qualitative AR researchers especially since it is difficult to reproduce the exact results of social inquiries due to the dynamic and unpredictable nature of participants (Checkland,1981; Cooray, 2010). Checkland and Holwell (1998) suggested the notion of 'recoverability' as a way of addressing the issue and argue that external parties should be allowed to see the intellectual thought process that led the inquirers to reach a conclusion even if they might not be able to reach the same conclusion themselves. Checkland (1991) argues that AR researchers should document the potential application area (A), framework (F) for understanding (A) and a methodology (M) for problem solving within (A) based on (F) before the study takes place. This is to ensure that external parties can see the process that led to conclusions being reached and understand how the conclusions were arrived at even if they might not be able to reproduce the exact same results.

Based on Iverson et al.'s (2004) interpretation of Checkland's (1991) action research cycle and FMA model this study combines theory and practice as follows.

- **Research Theme:** The general area of interest was how soft methods of inquiry could be used within synchronous virtual teams
- **Research Framework (F):** Theory and concepts about organizations, soft methods of inquiry and virtual teams framed the study as outlined in the literature review
- **Research Methodology (M):** Action Research and AIM (see below)
- **Real World Problem Situation (A):** The research addressed the challenges associated with setting up and maintaining a study abroad program between Curry College and Richmond University
- **Reflection Based on (F) and (M):** While working on (A), the researchers continuously made sense of the accumulating experience based on (F) and (M).
- **Findings:** The researchers eventually exited from the situation and critically reviewed the results and experiences to identify research contributions and to document the research.

Klein and Myers (1999) argued that the epistemological foundations of AR can be positivist, interpretivist or critical in nature and the (M) adopted, in this study is interpretive i.e. that knowledge is subjective and socially constructed. The epistemological nature of the method of research gave the researchers a dilemma. They wished to remain faithful to the intellectual roots of the approach but at the same time provide a third party with enough signposts to 'recover' the process. This means that the research should be undertaken in such a way that someone unconnected with the research can follow the logic of their conclusions. In order to achieve this the researchers should make explicit at the start of the research the intellectual frameworks and the process of applying the said frameworks [see Checkland, 1999, pA40]. To this end the method of research chosen was the Appreciative Inquiry Method because the approach satisfies the underlying philosophy of soft systems, and provides a record of what takes place in the form of the various diagrams used as part of the inquiry process.

In addition to the diagrams produce by AIM and to add to the authenticity [Champion and Stowell, 2001] of the study the researchers produced PEArL records for each session to document the manner or atmosphere in which the inquiry took place. These measures helped to reduce potential bias and also provided the means of recording the *manner* or atmosphere in which the inquiry took place. This meant that a third party could follow the way in which the conclusions had been reached.

5. THE APPRECIATIVE INQUIRY METHOD (AIM) AND PEArL

As our interest centres on investigating the use of soft methods of inquiry within synchronous virtual teams the Appreciative Inquiry Method (AIM), as indicated above, was chosen as the 'soft' method of inquiry to be used in the field study (see Stowell, 2012). AIM was chosen because it has proven to be useful in situations where a central issue or question to explore can be identified (as was the case with the Curry College/Richmond University partnership) and it is aligned with the dual goals of this study. Action researchers such as Davison et al, (2004) and Susman and Evered, (1978) argue for an iterative process of problem diagnosis, action and reflection. This is reflected in AIM through a repeated set of activities that use simple sense making devices to 'appreciate' a situation of interest, take action based on that understanding and learn from the results of actions.

AIM has shown itself to be valuable over a number of field studies and workshops (e.g. Stowell, 2012, West and Stansfield, 1999, West and Thomas, 2005), but as it stood AIM, like SSM, still needed to find ways of appreciating and managing the effects of organizational power. To this end the notion of PEArL evolved from the original conception of Champion and Stowell (2001). Champion and Stowell conceived the mnemonic to help the analyst/researcher assess the relationships within the participant groups. Subsequently, we have found it useful in clarifying participants thinking about the situation of interest and helping them to manage the use of power by members of the participant group (see Stowell, (2014); Stowell and Welch, (2012).

PEArL stands for the following: P = Participants; E = Environment; A = Authority; r= relationships; L== Learning. PEArL enables the researcher and client to be explicit about who is included within and excluded from the situation of interest. We have found that using PEArL allows the practitioner to ask about A and about the lower case 'r' (which relates to the commodity of power- see Stowell 2014) which helps the researcher, and often the client, to explain the influences within the boundary of the situation of interest. For example, by asking the A and r in PEArL helps surface the power relationships between the participants that are thought to exist (Stowell, 1989, 2014; Checkland, 1999; Checkland and Poulter, 2006, pp36-38).

The purpose of the first stage of AIM is to enable participants to express their thoughts on a central question or issue with minimal influence from the researcher. The tool used for this purpose is the systems map. Participants are expected to write down their thoughts on the central issue as bubbles around the central element. Research has demonstrated that organizing one's thoughts in the form of a systems map is useful in enabling a more holistic (systemic) understanding of the central element to be generated (Stowell, 2012).

The second phase of AIM is dedicated to engaging in a discussion about the combined systems map, which is created by amalgamating the individual maps that were produced in phase one. The objective is for participants to be exposed to each other's views thereby helping them to expand on their original perceptions of the central question and generating a more holistic view of the situation. Once participants debate the combined map and agree on *a* version of it the next step is for participants to select an element/s to explore further. Questions related to PEArL and CATWOE are then used to help participants to explore the selected element/s. Questions corresponding to the five elements of PEArL are first used to reflect on how the selected element/s is currently being implemented (what *is* the case) and on how the element could ideally be implemented (what *ought* to be the case).

In the final stage a definition for *a* system that could implement the selected element is produced. The definition is produced based on the answers participants gave to the PEArL and CATWOE questions in the previous stage. The definition encapsulates information about what the system will do, who will do it, why it will be done, who owns the system and environmental constraints.

Once participants edit and agree on the definition, it is then used to identify activities that should be actioned to operationalize the system defined by the definition.

Although researchers have examined the feasibility of AIM within a face-to-face environment (West, 1995; Stowell, 2012; Hart, 2013) and to an extent within an asynchronous virtual environment (Hart, 2013; Stansfield 1997), to our knowledge AIM and other 'soft' methods of inquiry have not been tested within a synchronous virtual team.

6. PREPARING FOR THE STUDY

Since the key stakeholders that were involved in implementing "the study abroad program" were located in two continents a virtual communication environment was created to engender discussion and collaboration. Although asynchronous communication environments (i.e. email) were used by participants initially it became apparent to them that they needed to meet virtually at the same time to facilitate productive live discussions. As a result a tool that could create a synchronous communication environment was sought. The free, web based software tool Wiggio was chosen because: First, it provided, audio, video and texting capabilities for participants to communicate synchronously, which assisted in replicating a face to face environment as much as possible. Second, Wiggio provides a whiteboard feature that participants could use to create/edit a diagram together in real time. Since AIM requires the collaborative production and edition of systems maps the whiteboard feature in Wiggio helped to ensure that participants could take control of the inquiry process and create maps themselves with minimal input from the researcher.

Typically a group consists of participants with varying levels of knowledge in technology and soft methods of inquiry meaning that provisions had to be made to account for those differences. Prior to the study the participants were sent "how to" videos and reading material to learn about Wiggio in order

Figure 2. Schematic diagram of AIM

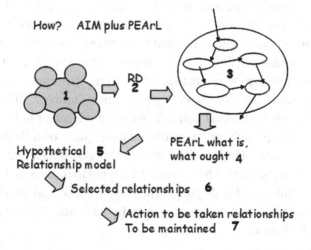

to familiarize them with the technology used. Videos demonstrating the use of AIM and systems maps were also sent to participants through email. In the case of our study participants were located in two places- Boston, USA and London, UK. Before the study the researcher used a survey to identify one participant from each location who had good technological skills and held a separate online session with them to familiarize them with the technology to be employed and requirements of the study. The two participants were then considered to be "power participants" in that they could help others at their location with any immediate technology problems. Thereafter an online training session was held with all participants so that they could try out the various features of Wiggio prior to the study.

During the online training session participants were also introduced to AIM as a method of inquiry. Since AIM consists of the simplest of tools that require minimal preparation participants appeared to quickly comprehend what was expected from them. This ready acceptance of the method employed gives credence to the importance of selecting a simple method of inquiry (such as AIM); participants should not be distracted by the technology or the method of inquiry itself. The participants' priority should always be focussed on the main task which is to analyze the agreed problem or issue.

As the participants were in different locations it was not possible to meet face to face prior to the study, which is a common scenario in most modern and geographically dispersed organizations. Although steps were taken to help participants to understand the technology and objectives of the study it cannot be overlooked that each group has its unique dynamics that may include individuals who do not want to use technology. In such cases if individuals are more comfortable with asynchronous methods, such as email, then they could still participate, using AIM, by preparing a systems map of their perceptions on the problem theme using a text editor such as Microsoft Word. They could then email the document to the facilitator who could add their ideas to the combined map that will be discussed by participants during the synchronous virtual sessions (see Hart 2014). This method is less effective since the participant will not be able to join in the live discussions with other participants.

7. CONDUCTING THE STUDY

7.1 Phase One

Two separate sessions were held during the first stage of AIM to create two systems maps that represented the views of Curry College and Richmond University respectively. The first session was held with Richmond University. The researcher sent emails to the invitees who then clicked on a link to enter the meeting. Participants were able to see and hear other participants using the audio and video features of Wiggio. The participants were asked to focus on the central question, agreed in advance, "How can we make the exchange program between Curry and Richmond more attractive to students so that a sustainable relationship between the colleges becomes feasible?". The whiteboard on Wiggio was opened and the central question added within a bubble at the center of the board. All participants could see this in real time. The participants were invited to add their own thoughts to the central question by adding bubbles to the central bubble. The shared whiteboard meant that participants were able to modify the systems map in real time. They were then invited to discuss and clarify any components of the map that they didn't understand or disputed. The session ended with participants from Richmond University agreeing on their version of the systems map. In the second session participants from Curry College repeated a similar exercise.

7.1.1 Data Generated

The systems map shown in Figure 3 was collectively produced by participants from Richmond University.

After the session the researcher created two types of PEArL records. The first was to record the researcher's interpretation of the manner or atmosphere in which the session took place (i.e.: interactions between participants) so that an external party could understand how the results were reached even though they may not be able to reproduce the exact same results. This type of record was created to address the challenge in qualitative social research of being unable to replicate social situations in exactly the same way due to the dynamic and unpredictable nature of people. The second type of PEArL record was to document the researcher's own interactions during the session. This type of PEArL record was used to assist both an external party and the researcher to trace how the researcher may have influenced the sessions.The researcher can use these records to improve upon the way in which future sessions are conducted. Both types of records can be used to reflect on how each session was conducted and by external parties to interpret and authenticate the results of the study.

The systems map shown in Figure 4 was collectively produced by participants from Curry College during phase one.

PEArL records similar to those in Tables 1 and 2 was produced after the session with Curry College to record the researcher's interpretation of the manner or atmosphere in which the session took place (i.e.: interactions between participants) and to document the researcher's own interactions during the session. These records can be used to reflect on how the session was conducted and by external parties to authenticate the results of the study and identify how much influence the researcher had.

Figure 3. Systems map produced by members of Richmond University

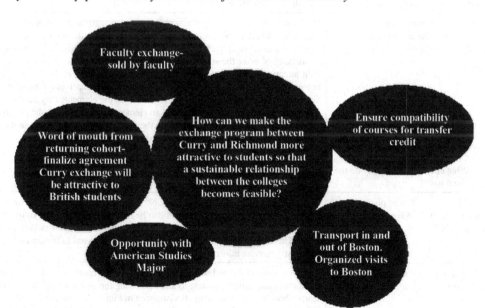

Figure 4. Systems map produced by particpants from Curry College

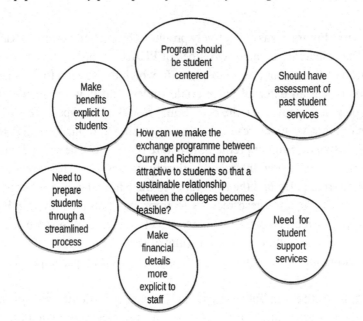

Table 1. PEArL record to document the manner or atmosphere in which the session took place (for the meeting with representatives from Richmond University

PEArL Elements	Observations
Participants- Who is involved? Why are they invloved? What is their role in the study?	Representatives from Richmond University including the Director of Admissions (North America), Researcher
Engagement- How will P engage? What methods will be used? What environmental issues influence the engagment process?	Participants seemed comfortable with the use of Wiggio as demonstrated by their willingness to draw systems maps and engage in the live discussions. Since AIM is simple and intuitive all participants appeared to grasp it easily. This demonstrated the importance of using the simplest methods of inquiry so that participants can focus their attention on the most important task- which is to analyze the problem issue at hand. Most participants made changes to the map themselves on the whiteboard although at times the researcher was asked to add the elements they described. When the researcher was asked to add an element by a participant she did so and then immediately sought feedback from the participant to ensure that the map truly reflected the participant's perceptions and not the researcher's.
Authority- What formal authority is associated with the roles? What embedded authority do the tools for engagement have?	Director of Admissions North America had the highest formal rank among those present
relationships-What commodities of power have people used in the situation? What control strategies were present?	An atmosphere of cordiality prevailed. No explicit power commodities were detected although one participant did mention that he had spearheaded many study abroad partnerships in the past and frequently drew on his past experiences to articulate a point. This may have led to that participant being perceived as having more credibility and hence exercise informal power over the others.
Lessons-What has been learnt about the situation and the approach used	Participants were able to learn more about the central question as demonstrated by them constantly editing, adding to and discussing the systems map. What participants learnt about the problem theme is visualized in Figure 3. Participants stated that being able to see the map displayed visually on the whiteboard helped them to take a step back and think clearly which helped them to see the 'big picture'. Learned more about who had more knowledge about the study abroad program

Table 2. PEArL record to demonstrate researcher's involvement in the session with Richmond University

PEArL Elements	Observations
Participants	Representatives from Richmond University, Researcher
Engagement	The researcher explained the tasks to the participants and then only spoke when participants wanted some clarification. Some of the participants were not comfortable editing the whiteboard themselves and requested the researcher to add their ideas to the map. In such cases the researcher immediately sought the participant's feedback. Since the researcher is also a faculty member of a participant institution the other participants constantly tried to draw out the researcher's opinion. Care had to be taken to not influence the participants so as to ensure a participant driven process.
Authority	Participants from Richmond- mainly the Director of Admissions North America Researcher had some power since she knew the technology and method of inquiry the best
relationships	An atmosphere of cordiality prevailed. The researcher was both the facilitator and a faculty member of Curry College. In the study the researcher tried to just play the role of facilitator and tried to ensure that she did not influence the proceedings. However in such a situation where the facilitator also has a stake in the problem situation if proper care is not taken the facilitator can possibly yield informal power using her extra knowledge in the method of inquiry and technology used. The approach adopted by the researcher was to explain what was required to participants and then only speak when participants had questions or needed clarification
Lessons	Learned that some participants knew much more about the study abroad program than others. Made a note to figure out how to get the others more involved. Also in a situation such as this where the facilitator is also an active stakeholder in the problem situation extra care should be taken to ensure that the facilitator does not use her knowledge of the communication technology and method of inquiry to influence proceedings in her favor.

7.2 Phase Two

The researcher combined the two maps (produced by Richmond and Curry participants) during phase one into one combined systems map on Wiggio's whiteboard and presented it during phase two to a group that included participants from both Curry and Richmond. Participants were invited to review the combined map, discuss any elements they disagreed with and make changes to the map. As participants were exposed to each other's' views they began to better understand the 'big picture' surrounding the central question and learn their way to a more systemic shared view of the situation. They were also able to edit and agree on a version of the combined systems map. Thereafter participants were asked to select the most important element in the map to explore further. The researcher then asked questions, based on PEArL and CATWOE, to facilitate a thorough investigation of the element chosen.

7.2.1 Data Generated

From the beginning of the meeting it became apparent that there were many issues that some of the Richmond participants didn't know (e.g. that there was a delay in getting the J1 visa for UK students) and many issues that were surprising to some participants from Curry College (e.g. that Richmond University could only realistically market to 15% of their students). The participants said that the combined map helped them to expand on their understanding of the central question and to learn about new issues of which they had been previously unaware of. After some discussion participants agreed that they wanted to explore the marketing aspect further. PEArL and CATWOE (Checkland and Poulter, 2006) questions were used to help participants to reflect on the marketing element, the results of which follow. Participants were first asked to reflect on the marketing component from the systems map in terms of the 6 CATWOE elements in order to get a rich understanding of the transformation (T) that participants hoped to achieve through the implementation of the marketing component, who would implement the marketing component (A), who would benefit or be affected by it (C), who would have the ownership of marketing (O), what environmental constraints could have an effect and the worldview (W) that made the participants choose that particular component as the most important.

Results of CATWOE inquiry into selected element is shown in Table 3.

Participants were then asked to think about the five PEArL elements in terms of "what is the case" concerning the marketing component at present which helped them to reflect on how marketing of the program is currently actioned. The answers given by participants are as shown in Table 4.

Thereafter participants were asked to think about the five PEArL elements in terms of "what ought to be the case" concerning the marketing component which helped them to reflect on how marketing of

Table 3. Results of CAWOE inquiry into the selected element

CATWOE Elements	Observations
C-Customer	Students
A-Actors	Director of Admissions in North America (Richmond) and Director of Communications(Curry)
T-Transformation	Students who don't know about the exchange program become aware and excited about the program
O-Owner	Director of Admissions in North America (Richmond) and Director of Communications(Curry)
W-Worldview	Increasing revenue and providing students with international exposure that will make them more marketable to employers
E-Environment	English students are averse to studying abroad, most Richmond students are international students and they get an international experience just by coming to Richmond so may not be interested in coming to the US, J1 visa issues from the US end, problems with housing availability

Table 4. Results of PEArL inquiry into how the selected element is currently implemented (what is the case)

PEArL Elements	Observations
Participants	Currently registrar's office and Director of Admission's office, Director of Communications
Engagement	There is lots of confusion about who does what and who is in charge. Currently the Marketing department is under the authority of the Director of Communications and publishes brochures and limited web content to publicize the study abroad program. There does not appear to be a holistic marketing policy yet.
Authority	Director of Admissions in North America (Richmond) and Director of Communications(Curry)
relationships	The role of the marketing department is to oversee the way the college is marketed but there appeared to be some competition between one participant at Curry College and the Marketing Department. The tension was demonstrated when the participant made several comments on how the marketing department did not know enough about the study abroad program to market it and that the marketing should be overseen by the people directly involved with getting the program implemented (aka 'the insiders'). This was in contrary to the usual scenario where the marketing department spearheads all the marketing campaigns for the college. The participant appeared to use their extensive knowledge of the study abroad program to bring the marketing of the study abroad program under the overview of his department rather than the marketing department of the college.There were hints that the marketing department did not have adequate information to market the program effectively since they had not been involved in all the meetings and previously. This influence may have led to the conclusion that a separate study abroad office had to be created which among other duties would lead the marketing efforts as well. The use of 'insider knowledge' and wider access to information by the said participant can be considered to be a 'commodity' of power used by him as an informal or 'soft' means of influencing others. This was not a formal use of power since all participants had a similar formal rank but some had participated in the meetings on the study abroad program (outside of this study) since the start while others had not.
Lessons	Some suspicions about who is in charge and about financial details being held back from certain key players.

the program should ideally be actioned. The intention was to get participants to think 'outside the box' and identify what they would ideally like to happen if they had unlimted resources and no constraints. The answers given by participants are as shown in Table 5.

The reflection on "what is the case" and "what ought to be the case" was operationalized as a way of assisting participants to consider how things are done now with regards to the marketing aspect of the program and how things ought to be done. The reflection also enabled participants to discuss the gap between "is" and "ought" and the economic and organizational feasibility of implementing the ideas generated during the "what ought to be the case" reflection using PEArL.

7.3 Phase Three

Prior to Phase Three of AIM the researcher produced a root definition for a system that could implement the selected element (i.e. marketing). The definition was created using the answers that participants gave to the PEArL and CATWOE elements in the previous stage. Some of the components of the root definition were left blank since they were ambiguous and the intention was to get the participants to review and edit it. The incomplete root definition (RD) was represented on the whiteboard on Wiggio and participants both supplied the information to fill in the blanks and proceeded to edit it directly on the whiteboard in real time. The edited RD is as follows and participants both supplied the information to fill in the blanks and proceeded to edit it directly on the whiteboard in real time. The edited RD is as follows

"A system owned by the Director of Communication and Study abroad office at Richmond and staffed by the study abroad office at Curry and Study abroad office at Richmond to promote Curry to Richmond students and promote Richmond to Curry students by cherry picking and promoting programs that are especially attractive to students at the other institution and using social media, brochures and web references in order to provide students with international exposure to make them more employable and in order to make the institutions more attractive to potential new students in light of Curry being marketed to only 13% of Richmond students, limited housing availability and the threat of long J1 visa delays."

Once the root definition was agreed by all participants they then reflected on activities that would need to be actioned in order to implement the system defined by the root definition. The activities defined could be traced back to the root definition and are as follows.

Table 5. Results of PEArL inquiry into how the selected element should be ideally implemented (what ought to be the case)

PEArL Elements	Observations
Participants	Should have a separate study abroad office at Curry College that is the main point of contact and does the coordination.
Engagement	Marketing Richmond to Curry students is easy since it sells itself. Marketing Curry to Richmond students is hard to do due to the constraints discussed in E in CATWOE. Here Curry needs to cherry pick certain programs that would be especially attractive to Richmond students such as TV and radio in the COM department and market those aggressively. Curry should especially be marketed to students taking the new American studies major at Richmond. Use brochures, web references and social media.
Authority	Head of study abroad office
relationships	none that they could foresee
Lessons	How to market the program effectively to targeted students

- *Create a study abroad office at Curry College*
- *Create a marketing plan to promote Curry College to Richmond students*
- *Promote Curry College to Richmond students*
- *Create a marketing plan to promote Richmond to Curry College students*
- *Promote Richmond to Curry College students*
- *Identify current programs of study that are especially attractive to the student population at the other institution*
- *Promote selected programs to students of the other institution*
- *Train employees in the use of new media such as social media to promote programs*
- *Determine the best way to use the study abroad program as a way of marketing the institutions to potential new students (not students from the other institution)*
- *Periodically assess the percentage of Richmond students, Curry College is being marketed to*
- *Monitor J1 visa wait times and changes*

8. RESULTS AND DISCUSSION

The following discussion analyzes the Action Research detailed above and its contribution to literature on synchronous virtual teams and soft systems thinking. Iverson et al (2004) in their seminal paper argue that certain criteria should be articulated in order to maintain relevance and rigor in an qualitative AR study.

The discussion validates the research process by detailing the following criteria specified by Iverson et al in relation to the field study detailed here:

- *Clarifying the roles played by the actors involved,*
- *Reviewing the approach to data collection,*
- *Explicating how the process was controlled,*
- *Examining the usefulness of the implemented ideas*
- *Considering the theoretical contribution made by the study*
- *Considering the conditions under which the results can be transferred to or adapted to other contexts. .*

Each of these issues played a key role in ensuring a satisfactory level of relevance and rigor.

- **Roles:** Although AR researchers cannot be objective observers (Checkland, 1981) clarifying roles can help establish the impartiality of researchers. An AR researcher is expected to reflect on his own practice (Iverson et al, 2004). This criterion was met in the study when the researcher documented her interactions in each session in the form of PEArL records (see Table 2), which helped the researcher to reflect on her role in the study, learn from it and change her approach. Furthermore, all models and documentation produced were validated and edited by participants. The practitioner-participants in the study included the Director for Admissions in North America (Richmond), Associate Dean of Postgraduate, Business and Economics at Richmond, Director of Academic Enrichment Center at Curry College, Dean of Academic Affairs and Registrar of Curry College, professors and staff. The role of participants was to provide context relating to the "study abroad program" and provide a social test bed for the exploration of soft methods in virtual teams.

- **Documentation:** Describing the data collection approach in detail is a key discipline that distinguishes research from consulting (Baskerville and Wood-Harper 1996). All three phases of the study produced data which were documented in several ways.
  - ◦ Collaboratively designed models (i.e. systems maps, root definitions, activities) helped document what participants learnt and agreed upon during the study
  - ◦ In PEArL records where researchers documented the manner or atmosphere in which the sessions took place
  - ◦ In PEArL records where the researchers documented their interactions with other participants
  - ◦ In email correspondence between participants
  - ◦ In minutes of sessions

The data was collected throughout all phases and represent a time span of eight months. The information contained in data models in one phase could be traced back to the models produced in the previous phase leading to a high level of transparency and traceability of the data. When participants were able to see how the information they shared was mapped throughout the process in several models their sense of ownership in the process and results increased. Furthurmore, PEArL records detailing the manner in which the study took place (see Table 1) and describing the researcher's interactions (see Table 2) were kept to help external parties to interpret how results were reached.The FMA (Checkland, 2006) model was also used to articulate the framework of ideas, methodology and area of interest assocated with the study (see section 4).

- **Control:** Avison et al. (2001, p.38) propose that action researchers should be aware of and report on three control structures: control over initiation, determination of authority, and degree of formalization.

The study was commissioned by the CAO of Curry College in light of the upcoming partnership with Richmond University. The research was governed by a formal contract between the CAO of Curry College, the Dean of International Programs at Richmond University and the researcher. This contract was complemented by informal commitments between the participants and the researcher. Updates were regularly reported back to CAO of Curry College and the Dean of International Programs at Richmond University.

- **Usefulness (and Areas for Further Research):** Establishing usefulness of results in the problem situation supports the impartiality of the research and creates a baseline upon which the results might be transferred (Baskerville and Wood-Harper 1996). Checkland (1981, p. 253) states that "…the criterion by which the research was judged internally was its practical success as measured by the readiness of actors to acknowledge that learning had occurred, either explicitly or through implementation of changes". The usefulness of the study was determined by looking for traces in the documentation of the participants perceptions of the usefulness of soft methods of inquiry within a synchronous virtual environment.

In all three phases participants changed their perceptions of how they viewed the partnership between the two institutions; they realized that they had not understood the whole picture and as their views were exposed to others they-elaborated upon their perceptions. This was demonstrated when participants engaged in debate and iteratively modified the models they produced in each phase; some participants even changed their own bubbles in the systems maps following the discussions. The "Appreciation" (Vickers 1983) gained about the situation contributed to a formal contract being signed by the institutions and the launch of the "study abroad program": The first batch of US students were sent to the UK in the fall of 2014 while the first batch of UK students were sent to the US in Spring 2015.

The study has implications for the success of outsourcing and IS development (i.e: systems analysis) in general, which is increasingly conducted by globally dispersed project members using virtual collaborative environments. The study showed the usefulness of AIM in two ways. First, as a soft method of inquiry conducive to the attainment of a shared group level understanding of a situation which systems thinkers argue should take place within systems analysis. Second, it demonstrates the viability of AIM as a way of creating shared understanding and reducing conflict within virtual synchronous teams.

Although participants were trained in the use of Wiggio before the study began some participants lacked confidence in the use of technology and constantly requested the researcher's assistance. One repercussion was that the researcher was compelled to edit the models based on what the participants said, which may have increased the researcher's influence on the proceedings. The researcher's interactions in each instance were documented in the PEArL records to reflect her influence in the study and also to enable third parties to understand how the data was interpreted. The study highlighted the challenge of ensuring that researcher influence is kept to a minimum when dealing with less technically competent participants.

The r in PEArL refers to informal power relationships (Stowell, 2014) where participants use various *commodities* such as aggressiveness, experience, implications of rank and vocal prowess to exert power within groups. The researcher recorded several informal power relationships that were observed during the sessions. The researcher was interested in studying if any new *commodities* would emerge within a virtual synchronous team, in particular whether more technically able participants would use their technical confidence to impose their will. Interestingly in this study that did not appear to be the case indeed the reverse occurred. The participants who were less technical able publicly announced their lack of confidence several times and seemed to be reluctant to engage with the technology (which itself could be the use of soft power; the implication being a form of ready made excuse if things went wrong). They asked the researcher and other participants to add/edit the models. In this case it appeared that the less able participants attempted to *give* control of the sessions to others in the group rather than the more technically able participants taking control themselves. Further research could investigate if certain factors (such as prowess in technology) can prompt participants to *give* power away and if so what researchers could do to avoid this pitfall to encourage participation.

Although most of the discussions were devoid of conflict, in phase two participants from Curry College were surprised to discover that Richmond University could only market the study abroad program to 15% of their students due to the high levels of international students at Richmond and the low interest among English students. The issue caused some conflict when the participants from Richmond were questioned on why this fact had not been divulged earlier. In order to reduce tension the researcher asked questions corresponding to PEArL and CATWOE. Questions related to PEArL and CATWOE were used to first understand Richmond participants' perceptions of the situation and then what ideally they would like to happen. Participants from Curry College stated that the deeper exploration of the issue using PEArL and CATWOE enabled them to understand Richmond University's future plans and assisted in diffusing the situation. Participants from Richmond University stated that the questions corresponding to the elements from PEArL and CATWOE helped them to organize their thoughts and explain their future plans in a more structured and comprehensible fashion. A limitation of the research is that decision-making tasks (such as the ones reported in the study) are characterized as moderately conflict-provoking. Further research could uncover insights on the application of soft methods is areas that are highly conflictual in nature such as negotiation and judgmental tasks both within a virtual and face to face setting.

Another challenge faced by the authors and by qualitative AR researchers in general is the difficulty of reproducing the exact same results again due to the dynamic and unpredictable nature of participants. In this study in addition to the documentation produced by participants (systems maps, CATWOE,root defintion, activities etc) the researchers also documented the manner or atmosphere in which the study took place with the use of PEArL. Checkland's (1999) FMA model was also used to elucidate the framework of ideas, methodology and area of interest *before* the study started so that the intellectual framework surrounding the study is defined early. By doing the above the researchers strived to provide third parties with the ability to understand the process through which conclusions were reached even though they might not be able to attain the same results Furthur research is needed to explore other means of increasing the rigor of qualitative AR so as that external parties can interpret the results better.

9. CONCLUSION

The contribution of the research is three-fold. First, it demonstrated the usefulness of soft methods as a viable mode of creating shared understanding within virtual synchronous teams that can be useful to all types of virtual teams including IS project teams conducting IS development remotely. Second, it offered a structured understanding of some of the challenges of using soft methods in virtual synchronous teams and insight on how to address them. Third, it provided an example of a study where soft methods assisted in diffusing a conflictual situation. The outcome of the AR study suggested that soft methods had a practical application within virtual synchronous teams. For example, the participants implemented the ideas generated from the study in the 'study abroad' program which was launched at the end of 2014. The study also showed the importance of the facilitator having had experience with 'soft' inquiries. Key elements of the facilitators skills include; minimal intrusion; knowledge of suitable record keeping; action research and be able to address group dynamics (such as power) that emerge. The ideas generated from the study provide the foundation for further research into the use of soft methods of inquiry in virtual synchronous groups.

REFERENCES

Avison, D. E., Baskerville, R., & Myers, M. (2001). Controlling action research projects. *Technology & People*, *14*(1), 28–45. doi:10.1108/09593840110384762

Baskerville, R.L., & Wood-Harper, A.T. (1996) A Critical Perspective on Action esearch as a Method for Information Systems Research. *Journal of Information Technology, 11*, 235-246.

Burke, K., & Chidambaram, L. (1999). How Much Bandwidth is Enough? A Longitudinal Examination of Media Characteristics and Group Outcomes. *Management Information Systems Quarterly*, *3*(4), 557–580. doi:10.2307/249489

Champion, D., & Stowell, F. A. (2001). PEArL: A Systems Approach to Demonstrating Authenticity in Information System Design. *Journal of Information Technology*, *16*(1), 3–12. doi:10.1080/02683960010028438

Checkland, P. (1981). *Systems Thinking, Systems Practice*. Chichester: John Wiley.

Checkland, P. B. (1999). *Systems Thinking, Systems Practice, a Thirty Year etrospective*. Chichester: Wiley.

Checkland, P. B., & Holwell, S. (1998). *Information, Systems and Information Systems*. Chichester: Wiley.

Checkland, P. B., & Poulter, J. (2006). *Learning for Action*. Chichester: Wiley.

Cooray, S. F. (2010). *End-User driven Development of Information Systems-evisiting Vickers notion of 'Appreciation'*. (Thesis). University of Portsmouth.

Davison, R., Martinsons, M., & Ou, C. (2012). The Roles of Theory in Canonical Action Research. *Management Information Systems Quarterly*, *36*(3), 763–786.

Dennis, A. (1996). Information exchange and use in group decision making: You can lead a group to information, but you can't make it think. *Management Information Systems Quarterly*, *20*(4), 433–458. doi:10.2307/249563

Furst, S., Blackburn, R., & Rosen, B. (1999). Virtual Team Effectiveness: A proposed Research Agenda. *Information Systems Journal*, *9*(4), 249–259. doi:10.1046/j.1365-2575.1999.00064.x

Geisler, B. (2014). *Virtual Teams*. Retrieved from http://www.newfoundations.com/OrgTheory/Geisler721.html

Gibson, C., & Cohen, G. (2003). *Virtual Teams that work: Creating conditions for virtual team effectiveness*. San Francisco: Jossey Bass.

Griffith, T., Mannix, E., & Neale, T. (2003). *Conflict and virtual teams in Virtual teams that work: creating conditions for virtual team effectiveness* (S. Cohen, Ed.). San Francisco: Jossey Bass.

Gudykunst, W. B. (1997). Cultural variability in communication. *Communication Research*, *24*(4), 327–348. doi:10.1177/009365097024004001

Harris, J. (2009). *Tönnies, Community and Civil Society*. Cambridge, UK: Cambridge Univerity Press.

Hart, P. J. (2014). *Investigating Issues Influencing Knowledge Sharing in a Research Organization Using AIM*. (Thesis). University of Portsmouth.

Im, Y., & Lee, O. (2004). Pedagogical implications of online discussion for preservice teacher training. *Journal of Research on Technology in Education*, *36*(2), 155–170. doi:10.1080/15391523.2003.10782410

Iverson, J. H., Mathiassen, L., & Nielsen, P. A. (n.d.). Managing Risk in Software Process Improvement: An Action Research Approach. *Management Information Systems Quarterly*, *28*(3), 395–433.

Jarvenpaa, S. L., & Leidner, D. E. (1998). Communication and trust in Global virtual teams. *Journal of Computer-Mediated Communication*, *3*(4). Available at http://hyperion.math.upatras.gr/commorg/jarvenpaa/

Jarvenpaa, S. L., Shaw, T., & Staples, S. (2004). The Role of Trust in Global Virtual Teams. *Information Systems Research, 15*(3), 250–267. doi:10.1287/isre.1040.0028

Kimble, C., Li, F., & Barlow, A. (2000) Effective virtual teams trough communities of Practice. Unpublished manuscript, Strathclyde Business School, University of Strathclyde, Glasglow, Scotland.

Kirkman, B., Rosen, B., Tesluk, B., & Gibson, B. (2004). The impact of team empowerment in virtual team performance: The moderating role of face to face interaction. *Academy of Management Journal, 47*(2), 175–192. doi:10.2307/20159571

Klein, H., & Myers, M. (1999). A set of principles for conducting and evaluating interpretive field studies in information systems. *Management Information Systems Quarterly, 23*(1), 67–93. doi:10.2307/249410

Klein, M. (1991). Supporting conflict resolution in cooperative design systems. *IEEE Transactions on Systems, Man, and Cybernetics, 21*(6), 1379–1390. doi:10.1109/21.135683

Kuo, F.-y., & Yu, C. (2009). An Exploratory Study of Trust Dynamics in Work-Oriented Virtual Teams. *Journal of Computer-Mediated Communication, 14*(4), 823–854. doi:10.1111/j.1083-6101.2009.01472.x

Levin, B., He, Y., & Robbins, H. (2004) Comparative study of synchronous and asynchronous online case discussions. In C. Crawford et al. (Eds.), *Proceedings of society for InformationTechnology and Teacher Education International Conference* (pp. 551–558). Chesapeake, VA: AACE.

Mason, W. (2014). *How videoconferencing in the cloud can transform business collaboration.* Retrieved from http://www.techradar.com/us/news/world-of-tech/how-video-conferencing-in-the-cloud-can-transform-business-collaboration-1250490

Montoya-Weiss, M., Massey, A., & Song, M. (2001). Getting it together: Temporal oordination and conflict management in global virtual teams. *Academy of Management Journal, 44*(6), 1251–1262. doi:10.2307/3069399

Paulus, T., & Phipps, G. (2008). Approaches to case analyses in synchronous and asynchronous environments. *Journal of Computer-Mediated Communication, 13*(2), 459–484. doi:10.1111/j.1083-6101.2008.00405.x

Pena-Shaef, J., Martin, W., & Gray, G. (2001). An epistemological framework for analyzing student interactions in computer-mediated communication environments. *Journal of Interactive Learning Research, 12*(1), 41–68.

Pui, S. (2010). Maintaining packaged software: A revelatory study. *Journal of Information Technology, 25*(1), 65–90. doi:10.1057/jit.2009.8

Rahim, M. A. (1992). *Managing Conflict in Organizations* (2nd ed.). New York: Praeger.

Rourke, L., & Anderson, T. (2002). Exploring social presence in computer onferencing. *Journal of Interactive Learning Research, 13*(3), 259–275.

Sarker, S., & Valacich, J. (2010). An Alternative To Methodological Individualism: A Non-Reductionist Approach To Studying technology Adoption By Group. *Management Information Systems Quarterly, 34*(4), 779–808.

Sein, M. K., Henfridsson, O., Puraro, S., Rossi, M., & Lindgren, R. (2011). Action Design Research. *Management Information Systems Quarterly*, *35*(1), 37–56.

Sproull, L., & Keisler, S. (1986). Reducing social context cues: Electronic mail in organizational communication. *Management Science*, *32*(11), 1492–1512. doi:10.1287/mnsc.32.11.1492

Stansfield, M. H. (1997), *The Effect Of Computer Based Technology In Attempting To Enhance A Subjective Method Of Knowledge Elicitation*. (Thesis) University of Paisley.

Stowell, F. A. (1989). *Change, organizational power and the metaphor 'commodity'*. (Unpublished PhD Thesis). Department of Systems, University of Lancaster.

Stowell, F. A. (2012). The Appreciative Inquiry Method – A Suitable Framework for Action Research? *Systems Research and Behavioral Science*, *30*(1), 15–30. doi:10.1002/sres.2117

Stowell, F. A. (2014). Organizational Power and the Metaphor Commodity. *International Journal of Systems and Society*, *1*(1), 12–20. doi:10.4018/ijss.2014010102

Stowell, F. A., & Welch, C. (2012). *The Managers Guide to Systems Practice, Making Sense of Complex Problems*. Chichester: Wiley. doi:10.1002/9781119208327

Susman, G., & Evered, R. (1978). An Assessment of the merits of scientific action research. *Administrative Science Quarterly*, *23*(December), 583–603.

Thomas, K. W., & Kilmann, R. H. (1974). *Thomas-Kilmann Conflict Mode Instrument*. Mountain View, CA: Xicom.

Truzzi, M. (1971). *Sociology: The Classic Statements*. New York: Oxford University Press.

Vickers, G. (1983). *The Art of Judgement*. London: Harper and Rowe.

West, D. (1995). The Appreciative Inquiry Method: A Systemic Approach To information Systems Requirements Analysis. In F. A. Stowell (Ed.), *Information systems Provision: The Contribution of Soft Systems Methodology* (pp. 140–158). Maidenhead, UK: McGraw-Hill.

West & Stansfield. (1999). Systems maps for interpretive inquiry: Some comments and experiences. *Computing and Information Systems*, *6*, 64–82.

West & Thomas, L. (2005). Looking for the Bigger Picture: An Application of the Appreciative Inquiry Method in RCUS. In *Conference proceedings 'Information Systems Unplugged'*. Northumbria University.

Zenun, M. M. N., Loureiro, G., & Araujo, C. S. (2007). The Effects of Teams' Colocation on Project Performance. In G. Loureiso & R. Curran (Eds.), *Complex Systems Concurrent Engineering Collaboration, Technology Innovation and ustainability*. London: Springer. doi:10.1007/978-1-84628-976-7_79

Section 3
Innovative Solutions and Games for Crises, Virtual Leading, and Effective Remote Teams

Chapter 15
Empowering Crisis Response–Led Citizen Communities:
Lessons Learned from JKFloodRelief.org Initiative

Hemant Purohit
George Mason University, USA

Arun Vemuri
Google, USA

Mamta Dalal
InCrisisRelief.org, India

Vidya Krishnan
InCrisisRelief.org, India

Parminder Singh
Twitter, USA

Raheel Khursheed
Twitter, USA

Bhavana Nissima
InCrisisRelief.org, India

Surendran Balachandran
InCrisisRelief.org, India

Vijaya Moorthy
InCrisisRelief.org, India

Harsh Kushwah
InCrisisRelief.org, India

Aashish Rajgaria
InCrisisRelief.org, India

ABSTRACT

Crisis times are characterized by a dynamically changing and evolving need set that should be evaluated and acted upon with the least amount of latency. Though the established practice of response to rescue and relief operations is largely institutionalized in norms and localized; there is a vast sea of surging goodwill and voluntary involvement that is available globally to be tapped into and channelized for maximum benefit in the initial hours and days of the crisis. This is made possible with the availability of real-time, collaborative communication platforms such as those facilitated by Facebook, Google and Twitter. They enable building and harnessing real-time communities as an amorphous force multiplier to collate, structure, disseminate, follow-through, and close the loop between on-ground and off-ground coordination on information, which aids both rescue as well relief operations of ground response organizations. At times of emergencies, amorphous online communities of citizens come into existence on their own, sharing a variety of skill sets to assist response, and contribute immensely to relief efforts during

DOI: 10.4018/978-1-4666-9688-4.ch015

earthquakes, epidemics, floods, snow-storms and typhoons. Since the Haiti earthquake in 2010 to the most recent Ebola epidemic, online citizen communities have participated enthusiastically in the relief and rehabilitation process. This chapter draws from real world experience, as authors joined forces to set up JKFloodRelief.org initiative, to help the government machinery during floods in the state of Jammu & Kashmir (JK) in India in September 2014. The authors discuss the structure and nature of shared leadership in virtual teams, and benefits of channelizing global goodwill into a purposeful, and sustained effort to tide over the initial hours when continued flow of reliable information will help in designing a better response to the crisis. The authors discuss the lessons learned into 5 actionable dimensions: first, setting up response-led citizen communities with distributed leadership structure, in coordination with the on-ground teams. Second, communicating clearly and consistently about sourcing, structuring, and disseminating information for both internal team challenges, solutions, and plans with shared goal-preserving policies, as well as external public awareness. Third, developing partner ecosystem, where identifying, opening communication lines, and involving key stakeholders in community ecosystem - corporates, nonprofits, and government provide a thrust for large-scale timely response. Fourth, complementing and catalyzing offline efforts by providing a public outlet for accountability of the efforts, which recognizes actions in both off-ground and on-ground environments for volunteers, key stakeholders and citizens. Lastly, the fifth dimension is about follow-up & closure, with regrouping for assessing role, next steps, and proper acknowledgement of various stakeholders for a sustainable partnership model, in addition to communicating outcome of the efforts transparently with every stakeholder including citizen donors to ensure accountability. With the extensive description of each of these dimensions via narrative of experiences from the JKFloodRelief.org initiative, the authors aim to provide a structure of lessons learned that can help replicate such collaborative initiatives of citizens and organizations during crises across the world.

INTRODUCTION

According to Seeger, Sellnow, and Ulmer (1998), crises have defining characteristics that are specific, unexpected, and non-routine events or series of events, creating high levels of uncertainty and threat or perceived threat to an organization's high priority goals. More loosely, it is an escalated "emergency event", and in this chapter's context, for natural hazards. An emergency is defined as an exceptional event that exceeds the capacity of normal resources and organizations to cope with it (Alexander, 2002).

Considering the online community knowledge source - *Wikipedia*, a crisis is defined as any event that is, or is expected to lead to, an unstable and dangerous situation affecting an individual, group, community, or whole society. Crises lead to negative changes in the economic, societal, or environmental affairs of a region, and therefore, require a systematic approach to management. Crisis management is the process by which organizations deal with the crisis. It may include activities such as identifying the nature of the crisis, intervening to reduce or nullify damage and streamline rehabilitation efforts.

As of today, there are proven practices and guidelines adopted to respond to a crisis and aid rescue and relief operations. These are often localized for each country or region, and often highly institutionalized under a variety of complex norms and processes of responding organizations. However, every large-scale crisis event brings unique and unexpected situations, which challenge the response management to engage with citizens and respond efficiently.

To aid the potential solution, as a consequence of growing internet penetration and social media adoption, real-time platforms facilitated by Facebook, Google, and Twitter have made it much quicker and simpler to connect and coordinate the huge human goodwill available worldwide. It can be optimized to bring great benefit during the critical early response days post a crisis event. Volunteering citizens observe surroundings and situations, share updates on events, engage in discussions and spread awareness of priorities as they bring on board a variety of skillsets while teaming up organically to form highly effective and dynamic online communities that end up playing pivotal roles during crisis response.

Through shared experiences and learning, these communities have even enabled crisis readiness, averting many a death or catastrophe. Communication, invariably, is the first casualty of any crisis, and online communities play a critical role in plugging the big communication gap - as authors learnt while coordinating the relief efforts.

In the following sections, authors share their experiences, observations, and learning from the JKFloodRelief.org initiative formed during the 2014 floods in the state of Jammu & Kashmir (JK) in India. Beginning with introduction of how an online volunteering team first came into existence during the Uttarakhand floods in June 2013, briefly described are also the efforts undertaken towards assisting rescue and relief operations in the aftermath of these floods. The description then throws light upon the efforts by the team during the time cyclone Phailin struck Indian coast in the state of Odisha. In this case, authors picked up alerts about the impending cyclone before it made landfall and therefore, the team was able to sound out evacuation calls and reach out in advance.

By the time, the JK floods occurred, equipped with the past learnings from these two natural crises, authors plunged headlong into coordinating and aiding rescue and relief efforts by assisting the Indian Army and the JK state government. The cumulative energies of existing and new team members were invested into assisting with rescue and relief that were supported by collaborative technologies oriented for an impactful response. From identifying the on-ground needs to setting up countrywide collection centers for relief materials, and dispatching them to the flood hit areas, this team was actively involved in every sphere.

Important lessons from the challenges and employed practices in virtual teams are discussed along five dimensions in the following - team structure and leadership, communication, collaborative partner eco-system, complementary offline-online coordination, and constructive closure. Authors conclude this case study and observations with an acknowledgement of all those who contributed to the JKFloodRelief. org initiative in their own way.

BACKGROUND IN DIGITAL CRISIS RESPONSE

Citizens as a complementary crisis response-led community on the ground for assisting rescue, response and relief beyond the formal (institutionalized) response organizations has been in existence through ages. For instance, during Hurricane Katrina in 2005, people in the local communities were helping each other for relief as well as locating missing people (Palen et al., 2007). However, the lack of organizational coordination (Quarantelli, 1998) and chaos among the on-ground communities, as they track affected people's needs and respond to the prioritized needs is a critical challenge. The authors emphasize the role of technology assistance (Palen et al., 2010; Meier, 2015; Purohit et al., 2014) in building a collaborative system between citizens and responding organizations for efficient response and relief management. For instance, the Asian Tsunami in 2004 that was a large-scale humanitarian crisis, led to Information and Communication Technology (ICT) based solutions to manage relief and response efforts for the response organizations namely the *Sahana* system for information management (Careem et al., 2006). However, a big challenge to assist such ICT based solution for better information management of the response organizations is access to the information. Interestingly, a new phenomenon of online virtual communities of on and off-ground citizen volunteers to assist crisis response information collation has come into existence in recent years, and is recognized as '*digital humanitarians*' (Meier, 2015).

One such widely known digital volunteering initiative was after the Haiti earthquake in January 2010, which created a crowdsourced *Ushahidi* crisis map of resource needs based on sourcing SMS information from Haiti (Meier et al., 2010). Ushahidi is well known for its crisis map tool for situational awareness based on crowdsourced inputs, which was deployed first time during Kenya election to report any incidences of atrocities using SMS (Okolloh, 2009). Such crowdsourcing information based tools and initiatives in addition to surge of social media brought a new light towards participatory and collaborative crisis response of volunteering citizens and formal response communities, which is not limited by geographical boundaries of affected regions (Starbird, 2011). However, there are obvious challenges in incorporating the virtual volunteers in the workflows of the formal crisis response, especially articulation of organizational tasks and situational awareness (Purohit et al., 2014). State-of-the-art research on social media to organize information for efforts during crisis coordination and management have reflected upon these challenges and potential solutions such as identifying needs and resource availability, categorizing massive data streams into resource categories, etc. (Purohit et al., 2013; Imran et al., 2013; Reuters et al., 2013; Olteanu et al., 2014). On the other hand in the context of organizational coordination, it has led to creation of a Digital Humanitarian Network[1] (DHN), which facilitates a platform for collaboration between volunteer-technical community of citizens and the formal response organizations. The members of volunteer-technical community include various volunteering organizations with dedicated skillsets, such as statistics, data analysis, mapping, etc., and the majority of them contribute online. However, a variety of coordination concerns between online and offline (on-ground) responses continue to arise in the course of a rapid response execution; and the organizational adoption of the volunteered work with the formal governmental (GOs) and non-governmental organizations (NGOs) further challenges the integration of citizen-led responses.

In this spirit of a committed digital volunteer-technical community, JKFloodRelief.org initiative swung into action immediately after devastating floods of J&K state of India relying on the experience of some of the volunteer coordinators working together earlier. JKFloodRelief.org acted as an information collector, an amplifier, a mediating bridge between online and offline response coordination while collaborating with both on-ground formal response organizations (GOs and NGOs) and off-ground online volunteers worldwide, using a variety of technology platforms. The authors first describe the emergence of such an online volunteer team in the next section.

CASE OF PRE-ERA OF *JKFloodRelief.org*: UTTARAKHAND AND PHAILIN CRISES

In June 2013, the northern state of Uttarakhand in India witnessed flash floods due to a cloudburst and melting of glaciers that led to landslides. The scale of the crisis was such that thousands died and hundreds went missing in the aftermath. The Indian Army was responding to this crisis by arranging several sorties each day to rescue those alive in the flooded areas. However, the massive scale of the tragedy needed more volunteers to compliment the government's efforts. It was then that a non-formal team that included the authors forged connections through online links, and sprung into action.

The team members started coordinating with each other to define tasks, take roles, and execute various tasks that could bring awareness about and help rescue of missing persons and reunite them with their families. The role of a clear structure for tasks, and communication among team members was paramount to avoid duplication of efforts. When various Governmental and Army units released rescue lists (including several that were only scanned images) on various Web resources, the team converted and uploaded these lists into searchable databases. Some team members also translated lists from Hindi, the local language into English for easy accessibility and searching. The team also scanned *Save Our Soul* (*SOS*) rescue calls on the Web including Twitter social network messages (*tweets*) and made use of mobile numbers for fetching the last call received information of missing persons to locate their probable coordinates while redirecting all this data to the Indian Army. This constant and cohesive flow of pertinent information helped the Army in their efforts to locate missing persons.

Along with Google, the team put in efforts in working with the *Person Finder* tool from Google that listed crucial information about each rescued/missing person (refer Figure 1). At final count, the Google Person Finder database listed 14,967 records, all of which had been manually inserted.

This new team also joined forces with CrisisMappers[2] (network of digital humanitarians) in collaboration with Google Crisis Response, and found support from Humanity Road (http://humanityroad.org/). Along with several volunteers from universities around the world and prestigious organizations (especially Standby Task Force, Info4, OpenCrisis, Humanitarian OpenStreetMap Team, and Crisis-MappersUK), this team developed a district wise situational awareness map—*crisis map*[3] of status of area, rescue work, relief camp details, hospitals, roads, control rooms information and donation centers in states across India. The volunteer team not only accomplished tasks at hand, but also harmoniously worked to build trust among members for creation of a response-led community; the key learning and takeaway being clear and consistent communication.

In October 2013 when information about Cyclone Phailin striking coastal India came in, the team once again came together without hesitation to respond to the arising needs.

Figure 1. Snapshot of Google Person Finder for Uttarakhand flash floods in 2013

Figure 2. Snapshot of a collaborative crowdsourced Google Spreadsheet that was regularly updated to maintain situational awareness map as shown in Figure 3

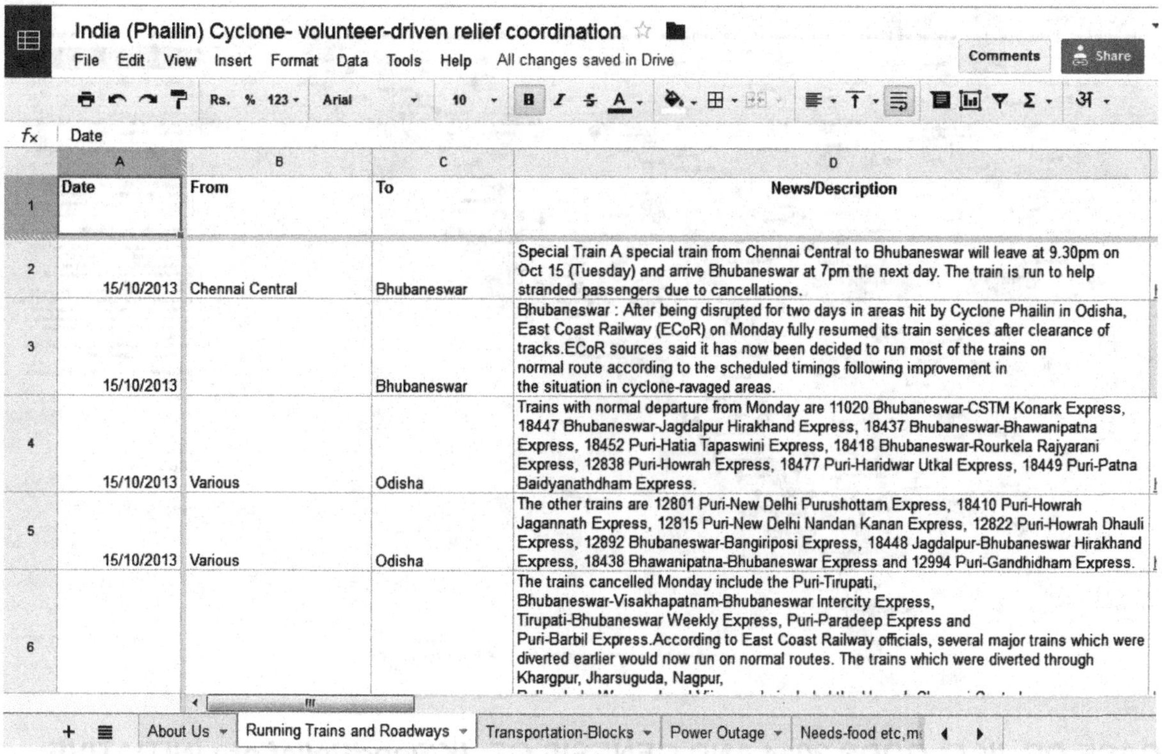

	Date	From	To	News/Description
2	15/10/2013	Chennai Central	Bhubaneswar	Special Train A special train from Chennai Central to Bhubaneswar will leave at 9.30pm on Oct 15 (Tuesday) and arrive Bhubaneswar at 7pm the next day. The train is run to help stranded passengers due to cancellations.
3	15/10/2013		Bhubaneswar	Bhubaneswar : After being disrupted for two days in areas hit by Cyclone Phailin in Odisha, East Coast Railway (ECoR) on Monday fully resumed its train services after clearance of tracks.ECoR sources said it has now been decided to run most of the trains on normal route according to the scheduled timings following improvement in the situation in cyclone-ravaged areas.
4	15/10/2013	Various	Odisha	Trains with normal departure from Monday are 11020 Bhubaneswar-CSTM Konark Express, 18447 Bhubaneswar-Jagdalpur Hirakhand Express, 18437 Bhubaneswar-Bhawanipatna Express, 18452 Puri-Hatia Tapaswini Express, 18418 Bhubaneswar-Rourkela Rajyarani Express, 12838 Puri-Howrah Express, 18477 Puri-Haridwar Utkal Express, 18449 Puri-Patna Baidyanathdham Express.
5	15/10/2013	Various	Odisha	The other trains are 12801 Puri-New Delhi Purushottam Express, 18410 Puri-Howrah Jagannath Express, 12815 Puri-New Delhi Nandan Kanan Express, 12822 Puri-Howrah Dhauli Express, 12892 Bhubaneswar-Bangiriposi Express, 18448 Jagdalpur-Bhubaneswar Hirakhand Express, 18438 Bhawanipatna-Bhubaneswar Express and 12994 Puri-Gandhidham Express.
6				The trains cancelled Monday include the Puri-Tirupati, Bhubaneswar-Visakhapatnam-Bhubaneswar Intercity Express, Tirupati-Bhubaneswar Weekly Express, Puri-Paradeep Express and Puri-Barbil Express.According to East Coast Railway officials, several major trains which were diverted earlier would now run on normal routes. The trains which were diverted through Khargpur, Jharsuguda, Nagpur,

For Cyclone Phailin, in the days before the cyclone made landfall, the team focused on bridging communication flow of key information from credible organizations to citizens. The scope of the work was digital amplification for critical information such as cyclone path status, evacuation plan, and shelter information that was collated from various government sites, from tweets of journalists as well as from phone calls to key reporters on ground. After the cyclone struck, the focus shifted to identifying and mapping phone line cutoffs, road and infrastructure damage, flooding status, and relief distribution status for remote areas (refer Figure 2). The volunteers from CrisisMappers and Humanity Road again collaborated with the team. Humanity Road conducted statewide meetings of governmental and non-governmental groups and reported relevant information. The team promoted relief needs based on ground inputs. Status of blocked roads, running/halted trains was regularly updated in a centralized location. Team members led by *Twitris* research group at Kno.e.sis Center provided complementary support for a crisis map (refer Figure 3) in collaboration with Google Crisis Response, and joined by volunteers from universities and organizations.

From an entirely management perspective, the team was focusing on structure of dynamic roles of volunteers for specific micro tasks regarding information collation, filtering, mapping, and amplification in the online world. The takeaway was efficiency in roles management for maximum impact.

Figure 3. Crisis Map for cyclone Phailin in Oct 2013, which used crowdsourced data curated by CrisisMappers volunteers, led by Twitris group at Kno.e.sis Center

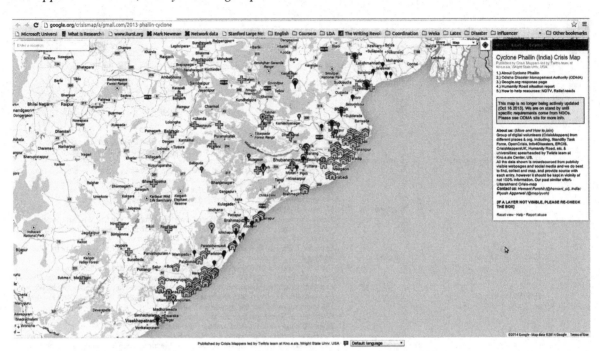

CASE OF *JK FLOODS 2014* AND GENESIS OF *JKFloodRelief.org* INITIATIVE

In early September 2014, when unprecedented floods hit the J&K state of India, the state capital, Srinagar went down in the deluge. With phone lines under water, during the first four days of the flood, there was a lack of communication between Chief Minister's office, police, hospitals and other service providers, creating a dire situation for rescue and relief operational coordination. JKFloodRelief.org came into existence to address these challenges (summary in Figure 4.)

#JKFloodRelief Inception

It was the first week of September, when personal Twitter timelines started to show up SOS messages from flood-affected people in JK. These were tough to ignore and the compelling need to help was an emotion that connected all of the authors (examples of interactions in Figure 5 to launch a volunteer effort), some of whom were already connected through volunteering in previous initiatives during the Uttarakhand floods and Cyclone Phailin[4]. Some of these members of the aforementioned online team again took the initiative to form a small group of volunteers spread across countries - connected with a common desire to reach out and help towards ensuring efficient rescue and relief reaching the affected.

As the team was joined in by new volunteers, most of them didn't know each other yet one connection led to another with the common desire to help. In hardly the course of a day, there was a core coordination group in place with defined roles for specific tasks. The team had learned two important lessons from prior experience - communication between structured roles for efficient team coordination, and a need

Figure 4. Summary of JKFloodRelief.org work. First (left) figure shows a media report on the team's function, the second figure shows impact of the citizen-led response, and the third figure shows a banner the team put up to propagate offline action items in the online world. Updates went up late night daily in India with the help of a global volunteer team.

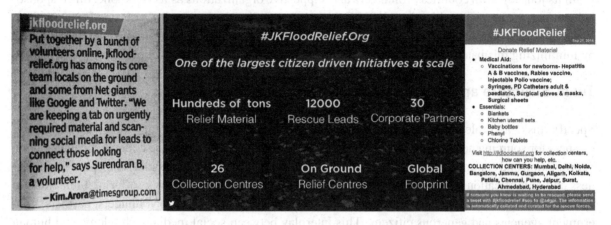

Figure 5. Live anywhere, help anywhere! Examples of how VOICE team members self organized via Twitter globally, India, Singapore & US. The age of new communication platforms helped VOICE team's distributed coordination.

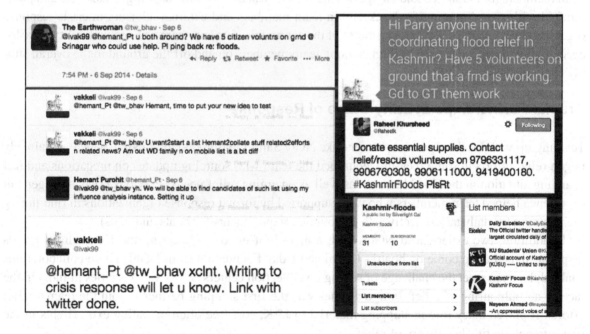

for building a collaborative effort for maximum impact of the relief initiative. Therefore, specific task teams were introduced this time to smoothen team communication for distributed task coordination, and for building an ecosystem of strategic partnerships to collaborate with. The emergent team had a strong offline component based on partnering local on-ground organization in the affected region. Volunteers dedicated time and effort and in quick time a well-coordinated citizen-led online and on-ground relief and rescue effort was underway. This was how #JKFloodRelief or JKFloodRelief.org was born in the digital world.

In its journey, with countless volunteers and supportive organizations as its backbone, this response-led volunteer team took an identity - *VOICE* (Volunteers Online for Impact in Crisis and Emergencies). The team had a primary goal - to transform public goodwill into positive action for relief coordination by minimizing mismatch between what is needed versus what is offered by people.

Team Structure and Communication

Specific tasks and roles were designed in the team structure, such as communication with on-ground partners for sourcing needs, new corporate and NGO tie-ups, setup of city-wise collection centers, team coordination and communication, creating digital and traditional media (e.g., radio) campaign for awareness and amplification besides being connected to ground volunteers and those offering their skills in cyberspace. The team leveraged on the power of social media to reach out to volunteers, partners, government agencies and generous citizens. This interplay between social media, technology, and human effort paid off in expansion of the team, online and offline, thus augmenting impact.

Owing to the distributed leadership driven coordination, the team was able to effectively support relief efforts by parallel, non-duplicated execution of required tasks, supported by communication across the team members for the shared knowledge of what-where-when-who-how mapping, challenges and plans (crafted with crowdsourcing feedback within team members and external advisors in crisis response organizations). Overall, a key responsibility of the communication sub-team was accountability of tasks, execution, reporting, and trust building among team members on and off the ground in the overall supply chain discussed next.

Crucial Partnerships at Early Stage of Response

Teaming up with local on-ground partners like NGO Sajid Iqbal Foundation (SIF), which had already begun relief efforts in some measure, equipped the team with sourcing updates on precarious and fast changing situation in the flood hit areas as well as emerging relief needs. Previous crisis management experience helped team members who were supported by formal response organizations in fine-tuning the planning for daily task objectives (e.g., sourcing requirements for boats and tents).

In addition, two experienced NGOs with a history of good work during the Uttarakhand floods response signed up to come on board - Goonj and Uday Foundation from Delhi. These collaborations enabled the team meet the challenge of setting up collection centers across various cities in India in the face of an outpouring of relief. Indigo airlines was the first shipping partner to join the team's relief efforts. It played a key role in shipping material to J&K, cost free often operating extra flights in the sector to help in timely delivery of relief.

Owing to such partners and many others (mentioned later) who came on board, the Team VOICE was able to establish a supply chain for sourcing needs from on-ground in affected regions, collecting the relief materials at centers across India, shipping them to J&K, and having them received for systematic distribution in the affected regions.

Rescue and Relief Coordination in the Information Age

Given the dynamic situation on ground zero, Team *VOICE* was aware that the relief initiative alone was inadequate as innumerable people were stranded with little or no supplies, houses were on the verge of collapse, and patients needed medical intervention. With incessant rains continuing to worsen the situation days after the first flood hit, the team had to brace up to help those with SOS rescue needs. The focus had to also shift to people posting on social media messages that sought urgent rescue of lives. A major coordination effort on data and technology sharing by the *VOICE* team partners Twitter India, Google, Kno.e.sis Center, Facebook and the Indian Army's public information branch ADGPI provided the solution. Technology assistance to sharpen the efforts played a significant role, since availability of data on social media already existed – for example in Figure 6 and 7, volunteers were following messages relaying SOS calls.

Technology solutions that helped widen relief needs information dissemination through influencers, as well as filtering, organizing and following up on the rescue related information for both need-to-rescue & rescued/evacuated categories were used. The use of technology helped redirect SOS calls to the army's rescue information monitoring teams, and curate evacuated/rescued information for sharing with the affected people, via *Smart-Feed* from Twitter, *Twitris* from Kno.e.sis Center, *Person Finder* from Google, and also through a crowdsourced *Google Form* and *Spreadsheet*. The authors list these technologies and the purpose for how they were utilized in the relief efforts:

- **Twitter Handle, Facebook Page, & Website:** To reach citizens effectively and faster towards spreading critical updates. Using WordPress, http://jkfloodrelief.org website was created for curating, hosting and disseminating detailed information. The *VOICE* team regularly updated content and with high frequency in the initial phase of the initiative due to the dynamically changing situation of the needs and priorities. Twitter handle of *@JKFloodRelief* (now *@InCrisisRelief*), a hashtag *#JKFloodRelief*, and a Facebook page with a similar name identity were created overnight. Uniformity of identity including the name was crucial to ensure accessibility, identity and collaboration towards a sustained effort.
- **Smart-Feed:** To collate all the messages related with SOS calls on Twitter, based on prior defined filters for search. It was added on the team's website to provide dedicated search results of SOS messages, updated periodically.
- **Twitris Rescue Stream & Map:** (http://twitris2.knoesis.org/dev/app/#JKFloodRelief) To automatically filter the social media streams in real-time from Twitter for Need-to-rescue (Red stream on the left) & Evacuated/Rescued (Green stream on the right) messages and map them for situational awareness. Upon clicking on the map pin or the message in the stream, an editable message-box allowed volunteers to redirect SOS calls to appropriate agencies, in this case ADGPI (refer Figure 7.)
- **Twitris User & Network Analyzer:** (http://twitris.knoesis.org/kashmirfloods/network) To identify users in online communities for spreading information quickly, to engage with influential informants of various rescue/relief need types, as well as to redirect or follow up on messages with SOS calls (which in turn helped data collection for *Google Person Finder/ Rescue Form*). The tool also allowed analyzing effectiveness of strong versus weak coordination between influential users by connectedness of who-talks-to-whom communication network, and helped the Team *VOICE* assess the daily engagement level for its initiative on social media (refer Figure 8).

Figure 6. Exemplary follow-ups on SOS call by volunteers

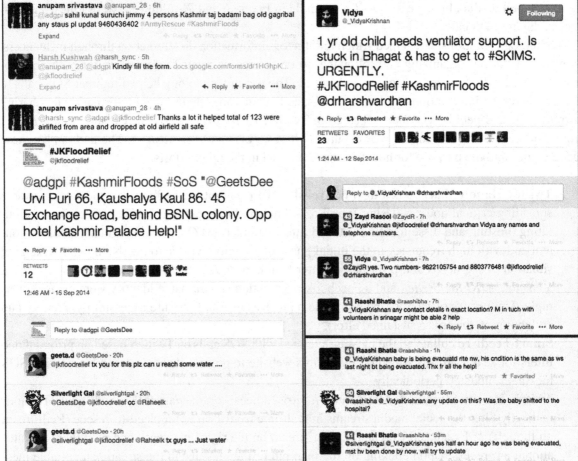

- **Google Form:** For rescue requests of missing people, and also crowdsourcing who were evacuated. This information was fed into in the Google *Person Finder* database. The team spread the *Google Form* link via influencers in the social media.
- **Google Spreadsheet:** To regularly update information for specific resource needs, and priorities received from information via the offline, on-ground partners. This allowed collaborative work between sub-teams of the relief initiative (refer Figure 9.) Spreadsheets were also used for crowdsourcing rescue requests.

Figure 7. Rescue Stream & Map tool for automatically identifying and enabling sharing of Need-to-rescue, the SOS calls (first figure) from the left/red stream, and situational awareness using geolocation of messages on the map (second figure)

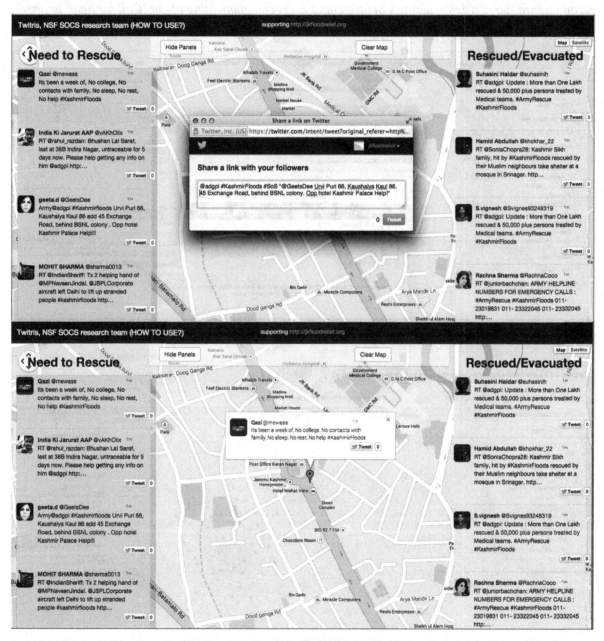

- **Google Person Finder:** To create a searchable database of rescued and missing people (refer Figure 1). This tool was helpful for connecting missing family members. This tool also facilitates searching records via a simple SMS based structured query.
- **Online-Offline Information Dissemination Tools:** For maximizing public awareness about both priority rescue and relief needs. Team *VOICE* informed citizens about the prioritized relief needs

Figure 8. Influential User & Network Analyzer tool. First figure shows identifying influential users under specific need (here 'Call for help'), and engaging with them by category (here 'Medical') for effectively spreading critical information (e.g., daily needs) or redirecting/following up SOS calls. The second figure shows a dense network of influential users communicating about 'Call for help' implying higher degree of engagement amongst them, which was useful for coordination.

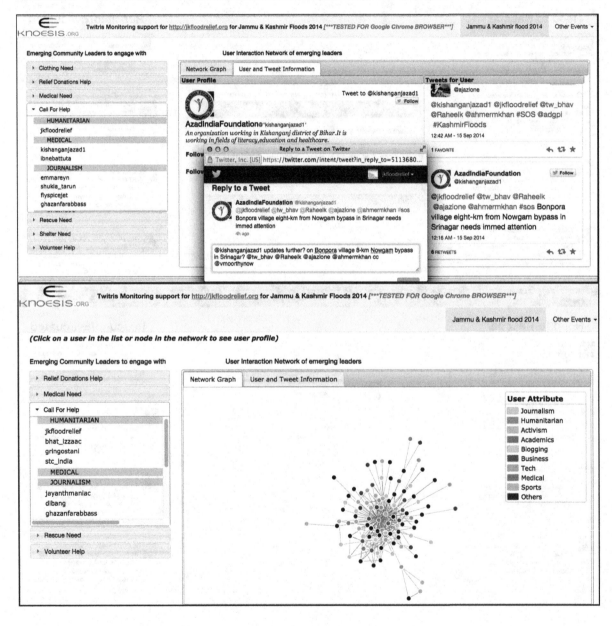

and how-to-help via regular online updates across social media platforms using posters, images, text posts and updated website links. For targeted information dissemination of resource priorities, the team utilized influencer analysis for specific city-wise focus. Offline updates via support from traditional communication partners also helped extensively, such as the Fever 104 FM & Big FM

Figure 9. Snapshot of online collaborative spreadsheet for information management, containing priorities sourced by coordination with the offline and on-ground teams

Category	ITEMS	ITEM-DETAILS	LAST UPDATED
Medicines	***CHECK EXPIRY FIRST***		
Immediate Priority	Disinfectants	Dettol, savlon, hand sanitizers	Sept 13 9:00 pm IST
Low Priority	First aid	bandages, band aid, betadine, scissors, hydrogen peroxide, soframicine powder/tube	Sept 10 11:30 am IST
Low Priority	Syringes	Syringes	Sept 10 11:30 am IST
Low Priority	Needles, ORS masks	Needles, ORS masks	Sept 10 11:30 am IST
Essentials			
Immediate Priority	Lifejackets	Lifejackets for places where boats cannot go	Sept 13 3:00 am IST
Immediate Priority	Rafts to help rescue victims	Rafts, boats	Sept 13 3:00 am IST
Immediate Priority	Disinfectants/Bleach Powder	Bleach Powder	Sept 13 3:00 am IST
Immediate Priority	Ropes, Rubber boots		Sept 13 3:00 am IST
Medium Priority	Ready to eat food	which doesnt take much to pack	Sept 11 11:00 am IST
Medium Priority	Undergarments (new) for men, women and children	New	Sept 11 11:00 am IST
Medium Priority	Power extension	Power banks/ electrical extensions and socket hubs	Sept 11 11:00 am IST

(radio channels) helped boost the public awareness campaign. The team shared a daily blogpost for creating awareness as also building transparency and trust in the process with stakeholders. A key lesson from this work was the accountability of the team to the citizens (refer Figure 10.)

Invaluable support via Partner Ecosystem

The scale of the crisis being so large, citizen provided donation supplies needed to be augmented through corporate support. It was at this point that Team *VOICE* volunteer coordinators reached out to business

Figure 10. Blog update for daily work's accountability: It helped build credibility for the voluntary initiative, and gained support to deliver impactful response

RELIEF EFFORTS UPDATE: SEP 9, 2014

SEPTEMBER 9, 2014   #JKFLOODRELIEF

The morning saw 1300kg shipment from Goonj and 126 cartons of medicines from Cipla leave for Srinagar via an Indigo flight. In the meanwhile, all SOS tweets were tagged appropriately, streamlined into our site and directed to Indian Army website.

Relief needs assessment, donation offer curation and redirection, collection centre development across country continued to grow at a furious pace. As a result, we now have several collection centres in Bangalore, Mumbai, Delhi and Delhi NCR and Jammu. For latest updates on our essential needs and where to drop off, please see here: http://jkfloodrelief.org/?page_id=29 and here http://jkfloodrelief.org/?page_id=13

Further, the Website was synched through online monitoring tools to provide real time info on relief work. Please see our Live Twitter feed which pools out SOS calls. The popular application by Google, Google Person Finder was launched and is being rapidly populated

RELIEF EFFORTS UPDATE: SEPT 23-26, 2014

SEPTEMBER 27, 2014   #JKFLOODRELIEF

Good day/night/morning the great power, citizens! We have been working in the backend for few days, and it is time to update our progress–a big thanks to our coordinators, volunteers and the generous donors for ensuring priority relief supplies. Among the most interesting day of our on-ground coordination was today, given that we were informed yesterday about emerging priority need for shrinking food and other supplies, in addition to blankets to deal with cold weather in many relief camps. Although some of relief supplies were stocked in specific relief warehouses, the relief distribution on ground was not well-coordinated due to tactical approach to provide immediate relief. Also, some items like rice and blankets were indeed unavailable locally, and emergency supplies had to be raised across the nation.

organizations. The *VOICE* team received generous support from several organizations—Cipla, Biocon (medical aid, sanitation & doctors); Emami, Essar (personal hygiene needs & chlorine tablets); Max India, Helios Eco Vidyut (essential supplies); Indigo, Air India, SpiceJet, Air Asia (shipments through airlifting & sponsoring); and Uber (for in-city transportation). To consolidate the relief collection efforts in several cities by Goonj and Uday Foundation, the Sikhi Awareness Foundation and Surat Children Welfare Organization came on board, helping the team build a comprehensive set of more than 25 collection centers across India.

Well-known, experienced response organizational actors, including AmeriCares, JK Red Cross, Oxfam, and Save the Children, also actively supported the team, which was crucial for a collaborative relief effort minimizing duplication and achieving greater impact. In creating public awareness, the *VOICE* team received support from several media organizations namely ABP News, NDTV, BGR, Zee News, Hindustan Times, FirstPost, LightHouse Insights, Dailyo, NY Times, NPR and OneIndia.

At the end of Week 1, the team had delivered a total of 15 tons of food, 4 tons of medical aid, 3.5 tons of clothing and blankets (refer Figure 11), all a reflection of the generosity of donors and support of selfless volunteers including online and offline partners (refer Figure 12 and 13). During the course of such initiatives where engagement of known organizations takes the spotlight, it is easy to lose the heroic efforts of local heroes who work relentlessly on the ground and contribute hugely to the effort. To address this concern the communications team within *VOICE* frequently shared details to highlight the work done by local participants in offline coordination efforts.

Figure 11. Milestones from the first week of #JKFloodRelief initiative

work done by www.jkfloodrelief.org 6 sept -14 sept

1. Deployment of Smart Feed resulting in 12000 Rescue leads for @adgpi, helped deploy people finder.
2. Network of 26 collection centres set up across the country.
3. 15 tons of food shipped.
4. 4 tons of life saving drugs shipped.
5. 3.5 tons of blankets & clothes shipped.
6. 1 ton of Sanitary Napkins shipped.
7. 200 Kgs of Baby food shipped.
8. Additional 3.5 ton relief plane full of meds & food shipped.
9. Close to 150 tonnes of material ready to be shipped.
10. 2 Lakh Chlorine tablets shipping by 15th Sept.
11. Establishment of on ground volunteer team, doctor team. warehouse & dispatch of relief to Rajori, Poonch, Anantnag & other inaccessible areas.

#ReliefPartners
@Bioconlimited
@IndiGo6E
@Essar
@goonj
@udayfoundation
@AmeriCares
Emami
Cipla
@OxfamIndia
Sajid Iqbal Foundation
Sikhi Awareness Foundation

#DataEffort
@adgpi
@TwitterIndia
@google
@facebook
@Aircel
@knoesis

Tag all your SOS tweets #KashmirFloods & mark them to @adgpi & @jkfloodrelief to enable auto-curation & action.

Figure 12. Online and offline coordination results: on-ground relief work in the affected region
More pictures at http://incrisisrelief.org/?page_id=623

Figure 13. Support poured in for Team VOICE's efforts from all sections

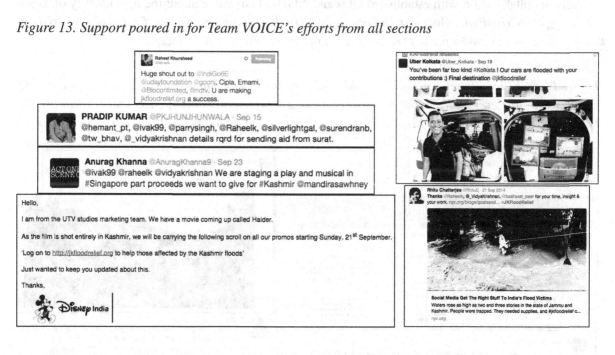

Continuation and Expansion of Campaign

By the end of the third week, rehabilitation and relief had been streamlined by the government and other agencies of the state. As such Team *VOICE* began winding down its efforts in JK but just then another calamity struck. This time in the North East of India, where the states of Assam and Meghalaya were ravaged by floods in late September. How could the team quickly gear up to respond was an important question but relying on the valuable learning from experience and availability of members onboard since JK efforts were ongoing, Team *VOICE* got its act together and prepared a plan which focused on adding energy through new online volunteers and local on-ground associations.

The *VOICE* team launched the *#NEFloodRelief* initiative to help affected people in the two North-eastern states of Assam and Meghalaya. In Assam, the team worked closely with the AIDRC team, who provided an in-depth need assessment. This enabled sourcing inputs on prioritized relief supplies and within 3 days, 635 families in 3 affected villages who had lost everything in the floods were provided much needed basic help (refer Figure 14). The last engagement of the year 2014 was in Meghalaya where the *VOICE* team worked with strong local organizations, VHAM (Voluntary Health Association of Meghalaya) and Bakdil for providing relief support in the most affected remote Garo Hills districts while also syncing in with the West Garo district collector, a governmental organizational actor.

While *VOICE* planned its response in Assam and Meghalaya, team members realized the unpredictable nature of a crisis event that could happen anytime, anywhere. This is when the team decided to reaffirm its objectives - which was to continue doing citizen-led response work - *planning*, *organizing* and *coordinating* towards providing rescue and relief to victims of crises and emergencies anywhere in India in collaboration with established GOs and NGOs. Thus came about the new identity of Team *VOICE* - *@InCrisisRelief* (Incrisisrelief.org), and *VOICE – Volunteers Online for Impact in Crisis & Emergencies*. The transition from JKFloodRelief.org had happened at this stage.

Figure 14. #NEFloodRelief distribution by AIDRC team in the Kamrup district of Assam, impacted 635 families who lost everything in the floods. The second figure shows citizen engagement crediting the team's aim to transform public goodwill into actual on-ground action via a strong partner ecosystem.

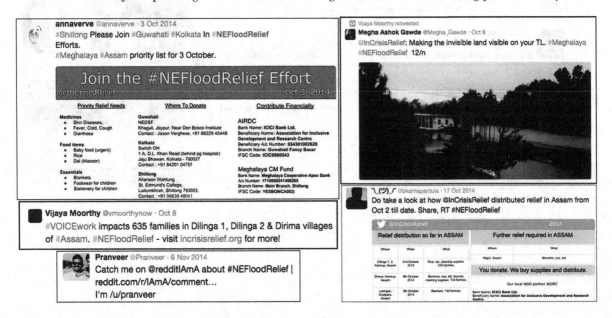

CHALLENGES OF RESPONSE-LED COMMUNITIES

The narration of the story of JKFloodRelief.org would be incomplete without this section. While the latest digital tools and technologies vastly ease creation of response-led citizen communities, several challenges arise and need to be addressed. Here is a list of some of these:

1. **Identifying Signal from Noise:** During times of crisis while social media is a genuine conduit for relief related information, it also carries significant noise that may throw the efforts off-track. Some of it is malicious though most of it is well meaning. People often tweet messages of hope, well wishes, etc. using the same hashtags used for relief measures. On the other hand during the MH370 crisis, some inauthentic sources cropped up providing misleading information. Technological advancement is required to mine goal-oriented information and measure its credibility.
2. **Being a Safekeeper of Public Goodwill is a Huge Responsibility:** Such communities often find themselves at the epicenter of outpouring of public goodwill that gets channelized in the form of donations and funds. This is often overwhelming and in extreme cases unmanageably destructive since such communities are neither resourceful nor structured for this purpose given their emerging nature.
3. **Sustenance is Tougher than Initiation:** Initiating such communities could be the output of a transient burst of inspiration or empathy. However sustaining them over a long period, or at least, till its objectives are met, requires genuine commitment to the cause. It is important to assess clarity of purpose and motivation of all members at the outset.

PROCESS AND LEARNING

The authors list various lessons from the processes and structure of the virtual team leadership experience during *JKFloodRelief.org* initiative (summarized in Figure 15).

1. **Setup and Structure:**
    ◦   **Distributed Leadership:** A basic comfort with the idea of 'Distributed' or 'Diffused' leadership is needed. Response-led communities may not follow the hierarchical structures of traditional communities and the members need to be open to this. To a traditionalist, this might even appear chaotic but that's just how response-led communities operate. Distributed leadership requires commitment for collaboration to enable flow of shared knowledge among team members for challenges, processes, planning, and accountability.
    ◦   **Amorphous Structure:** Related to the above, such communities usually have amorphous structures with regular role shifts based on needs, time, availability, etc.
2. **Communication:** The central pillar of response-led communities is the power of communication, and hence, it is paramount to emphasize its following functions.
    ◦   **Listen:** Members need to be good listeners and observers. An obvious point but often overlooked. The needs and expectations from such communities continuously evolve requiring members to regularly listen, internalize and adapt.

Figure 15. Summary of lessons for virtual team leadership[5]

- ○ **Reinforce:** Limited attention spans necessitate message reinforcement and repetition that might appear redundant but is often needed. If one has to err, err on the side of over-sharing. In the case of *JKFloodrelief.org*, each call for donation had to be reinforced several times for it to start spreading and having a focused impact.
- ○ **Share:** As the community starts making an impact, sharing the work done and its results become important. This is not 'chest thumping', or creating a buzz of own work. Sharing is important since these communities are accountable to those who have put their trust in them, and to garner further support for future efforts.
3. **Ecosystem Development:** Creating a partner network is almost mandatory. Communities have to be realistic about their capacity and capability, and hence it is important to identify and empower relevant partners.
   - ○ **Express Specific Needs:** Instead of saying 'Help needed', communication has to be specific about what's needed. For instance, '*Urgent antibiotics required from pharmaceutical companies*' is likely to get a better response than '*urgent help required from corporates*'.
   - ○ **Recognize:** Make it a point to recognize the partners. It is gratifying and encourages others to step forward. One needs to remember '*goodwill is contagious*'.
   - ○ **Trust & Empower:** Ensure the members fully trust partners and empower them as much as possible. Three important fundamentals of partner management are: Remove friction from the system; Let experts do their job; Get out of the way.

4. **Complement Offline:** Online communities complement offline efforts instead of aiming to replace them. Understand and recognize the work being done on the ground and be a 'force multiplier' for them. Whether it is government agencies, NGOs or regular on-ground volunteers, they are the real unsung heroes and possibly putting in much more efforts than online activities.

5. **Closure:** As an initiative reaches completion, ensure all efforts and processes are documented, donations accounted for and all stakeholders are duly recognized. There should be a clear articulation of a phase or initiative coming to its logical conclusion. The response-led communities owe it to themselves, their partners and supporters and very importantly for future efforts.

DISCUSSION AND FUTURE INITIATIVES

Online response-led communities are often born at times of crippling human tragedy, led by an overwhelming compassion from volunteers, who may otherwise be silent spectators. The need for any such community is to outlive the tragedy and live through the relief and rebuild phase, in both an impactful and systematic manner. It channels the goodwill in the society and gives it purpose and direction so that everyone in the community can contribute to the rescue and rehabilitation efforts. This chapter has attempted to identify lessons from the real world experiences to inform future crisis response initiatives connecting citizens and organizations.

Documenting Lessons Learned: A Necessity

While the energy and contribution of many amorphous platforms emerging during crises and then vanishing, cannot be depreciated; what is lost upon their withering away is an entire valuable knowledge base. It is, thus, imperative for online communities to realize the importance of endurance, and to value their legacy. As in any other domain or perhaps greater in the area of online relief coordination efforts, the seamless networks; contacts; workflows; specific event learning; challenge management; outcomes and impact assessment - all of these must be preserved and not lost in time. The *"Why-What-When-Where-How"* must be documented for another day. Crises are never easy to deal with. Starting from scratch after a crisis requires colossal effort and immense understanding of what can help and what could be a waste. In a situation as such, the more that is known in terms of learning from past experiences, the better enabled an initiative would be towards responding to an emerging crisis.

JKFloodRelief.org, now *InCrisisRelief.org* is the product of this thought process. While the structure of the team is flexible and members may float in or out in the course of mapping an arising need to skills required and their availability; it is the work, the synergy, the massive goodwill of people and their trust that must be retained and cherished.

The authors hope that all this information from experiences, and analysis put logically in one place could contribute to understanding of response-led communities driven by online and offline coordination. They undertake the responsibility of sharing these learnings with other online and offline groups who might find the experiences and lessons learned useful. To summarize, for a response-led community like the *VOICE*, the end of active engagement during a crisis relief or rebuild phase is actually the beginning of the inevitable task to put together all facts and experiences of work done while dealing with the crisis. The goal is to make cogent dossiers that can effectively inform peers and the next generation of digital volunteering communities about the challenges, solutions, and best practices.

NGO 2.0: The 'Uber'ization of Goodwill

Just like 'Uber' and 'Air BnB' business models have revolutionized transport and rental accommodation, authors anticipate a potential in 'Uber'-ization of NGOs. While there are only a few who opt for a full time career in philanthropy, there are plenty more who are keen to make a difference and willing to allocate a fraction of their time to humanitarian causes. Modern digital tools enable channelizing public goodwill into positive action in ways that were not possible previously. They allow rapid formation of real time communities allocated to a specific goal. These communities could be spread across geographies, and include participants from various walks of life. In fact the more diverse such communities, the better it is, as that would allow for wider perspectives and skill sets. The authors think the time has come for the NGO model to be reimagined and anticipate the birth of NGO 2.0.

CONCLUSION

The surge of online response-led citizen communities enabled by social media platforms has revolutionized the reach of humanitarian response. This story is an amalgamation of experiences of challenges, processes and milestones during the journey of *JKFloodRelief.org*, with an anticipation of reproducing such relief initiatives globally based on the learnings shared. To effectively connect this outreach and the volunteer workflow with formal crisis response during relief and rehabilitation, more attention to such integration of a cooperative system between citizens and organizations is required. The authors have described understanding of a few of such challenges from the technical, and organizational standpoints and have also discussed various dimensions of structure and nature of roles that assume leadership in virtual teams. Channelizing global goodwill into a purposeful and sustained effort to tide over the initial hours of shock when a crisis strikes is a critical aspect, while forging partnerships and empowering partners during the relief phase is another key to providing effective and timely relief.

There is unfortunately no escape from crises and the uncertainty around it, but the narration of *JKFloodRelief.org* effort seeks to create a cohesive and logical guide through experiences on how digitally connected citizen communities can set themselves up for engaging constructively in crisis response coordination.

ACKNOWLEDGMENT

The authors and Team *VOICE* acknowledge the invaluable contribution via a *Thank You* page (http://incrisisrelief.org/?page_id=1380) for all online and offline volunteers - the true unsung heroes of the real on-ground action; organizations and corporate partners some of who were mentioned in the chapter earlier, without whom *JKFloodRelief.org* and other initiatives would not have been possible.

REFERENCES

Alexander, D. E. (2002). *Principles of emergency planning and management*. Oxford University Press.

Careem, M., De Silva, C., De Silva, R., Raschid, L., & Weerawarana, S. (2006, December). Sahana: Overview of a disaster management system. In *Information and Automation, 2006. ICIA 2006. International Conference on* (pp. 361-366). IEEE.

Meier, P. (2015). Digital Humanitarians: How Big Data is changing the face of humanitarian response. Taylor & Francis Press. Spring 2015.

Meier, P., & Munro, R. (2010). The unprecedented role of SMS in disaster response: Learning from Haiti. *SAIS Review of International Affairs, 30*(2), 91–103.

Okolloh, O. (2009). Ushahidi, or 'testimony': Web 2.0 tools for crowdsourcing crisis information. *Participatory Learning and Action, 59*(1), 65-70.

Olteanu, A., Castillo, C., Diaz, F., & Vieweg, S. (2014, May). CrisisLex: A lexicon for collecting and filtering microblogged communications in crises. In *Proceedings of the 8th International AAAI Conference on Weblogs and Social Media (ICWSM'14)*.

Palen, L., Anderson, K. M., Mark, G., Martin, J., Sicker, D., Palmer, M., & Grunwald, D. (2010, April). A vision for technology-mediated support for public participation & assistance in mass emergencies & disasters. In *Proceedings of the 2010 ACM-BCS visions of computer science conference* (p. 8). British Computer Society.

Palen, L., Hiltz, S. R., & Liu, S. B. (2007). Online forums supporting grassroots participation in emergency preparedness and response. *Communications of the ACM, 50*(3), 54–58. doi:10.1145/1226736.1226766

Purohit, H., Castillo, C., Diaz, F., Sheth, A., & Meier, P. (2013). Emergency-relief coordination on social media: Automatically matching resource requests and offers. *First Monday, 19*(1). doi:10.5210/fm.v19i1.4848

Purohit, H., Hampton, A., Bhatt, S., Shalin, V. L., Sheth, A. P., & Flach, J. M. (2014). Identifying Seekers and Suppliers in Social Media Communities to Support Crisis Coordination. *Computer Supported Cooperative Work, 23*(4-6), 513–545. doi:10.1007/s10606-014-9209-y

Quarantelli, E. L. (1988). Disaster Crisis Management: A Summary of Research Findings. *Journal of Management Studies, 25*(4), 373–385. doi:10.1111/j.1467-6486.1988.tb00043.x

Reuter, C., Heger, O., & Pipek, V. (2013). Combining real and virtual volunteers through social media. In *Proceedings of the 10th International ISCRAM Conference*. KIT.

Seeger, M. W., Sellnow, T. L., & Ulmer, R. R. (1998). Communication, organization, and crisis. *Communication Yearbook, 21*, 231–275.

Starbird, K. (2011). Digital volunteerism during disaster: Crowdsourcing information processing. In *Conference on Human Factors in Computing Systems*.

ENDNOTES

[1] Digital Humanitarian Network website: http://digitalhumanitarians.com/content/guidance-collab-orating-volunteer-technical-communities

[2] CrisisMappers, a Digital Humanitarian Network: http://crisismappers.net

[3] Uttarakhand Crisis Map by CrisisMappers volunteers: http://www.thehindu.com/sci-tech/technol-ogy/gadgets/using-crisis-mapping-to-aid-uttarakhand/article4854027.ece

[4] Blog on details of pre-era of JK Floods relief initiative is available at: https://goo.gl/a3TMeJ

[5] Presentations from the *VOICE* team members on JKFloodRelief.org lessons: http://www.slideshare.net/InCrisisRelief/the-jkfloodrelieforg-story-39683653, http://www.slideshare.net/InCrisisRelief/crisisresponselearningsjkfloodrelieforgefforts

Chapter 16
Virtual Shipping Entrepreneurial Leadership Styles in Maritime and Shipping Industry

Tarek Taha Mohamed Kandil
Helwan University, Egypt

Shereen Hassan Nassar
Sadat Academy for Management Sciences (SAMS), Egypt

ABSTRACT

The concept of entrepreneurial leadership in any challenging industry involves fusing the concepts of a strategic approach of the management change, enhancing capabilities for continuously, creating and appropriating support, and development of value and competitive advantages in the company and technological growth. The challenge for shipping and transportation industry is to build a compatibility to continuously explore and reduce the threats of secure and safety and enhance new successful and competitive opportunities. Shipping Entrepreneurial Leadership style that scanned The twenty one attributes of outstanding leadership dimensions in the GLOBE project scales to identify nineteen attributes of their extraordinary performance, The chapter categorizes ten shipping behavior likely to be expected in loading on the two roles that impede the shipping entrepreneurial leaders' performance enactment scenario and Eight leadership behavior likely to be facilitated in their roles towered encouraging the team performance.

INTRODUCTION

Economical Features of Shipping Marketplace Forces

From the economical view point, the demand for shipping services is, basically, affected by the GDP of one country. As the country's GDP fluctuates, that will drive the forces of demand in the shipping industry (Davis, Pitts, & Cormier, 2000; W. Talley, 2013). The shipping services are also having possibility of double-dip influences on the market demand side(W. Talley, 2013). The global demand for

DOI: 10.4018/978-1-4666-9688-4.ch016

the shipping services is directly and immediately impacted by an economic depression For instance, the last financial crises happened between 2007 and 2009 in the UK and European was negatively and doubled impacting the European financial services that were dominant in ship liquidity and shipping firms' shares in the world financial markets (Bhandarker, 1990; Davis et al., 2000; W. Talley, 2013).

Stakeholders of the shipping companies are not a hidden anymore .They are all realized the influences of the potential changes that the industry hold (Davis et al., 2000; Leiss, 2013; Niman, 2014). Moreover, shipping market place is as much a part of the economic changes forcing challenges as it represents a force on its own (Davis et al., 2000; Leiss, 2013; Niman, 2014). As key factors to the shipping industry, it is very important to highlight the main contributions and value added that major making policies processes in shipping industry will be seeking from shipping and container sea-lanes:

1. Reduced the operating stemming costs to increase enable the shipping companies to compete the market price of the service (Dizard & Gadlin, 2014; Halavais, 2004; Leiss, 2013).
2. Raise sustainability competitive advantages and shipping competencies from narrow scope of customer shipping security to the supply chain for everything related to the human side of shipping industry (La Monte & Woytowitz, 2015)
3. Increased efficiency and reliability through incorporating a strategic entrepreneurial mind-set as a key element of strategic management in the new shipping environment (La Monte & Woytowitz, 2015)
4. Creating a sustainability innovation for the company team work and its network of shipping stakeholders by managing the existing and potential resources to achieve the target goals of the shipping companies (La Monte & Woytowitz, 2015)

Those economic features may add even more challenges for shipping decision makers. The sustainability innovation scenario will be the only challenge for shipping leaders in order to create valuable relationship followerswith and then will be considered as the shipping company's social responsibility performance as well (Dizard & Gadlin, 2014; Halavais, 2004; Leiss, 2013).

Since that moment, a growing stream of research in management science and Shipping Companies theory has begun to acknowledge the magnitude of virtual Shipping Companies to business (W. Talley, 2013). The concept of virtual Shipping Companies is relatively new and there is no single clear-cut definition of what a VOC is. The present chapter has conducted intensive literatures review to address the various conceptual framework and the various aspects of the virtual Shipping Companies discussed in the literature of the past 20 years(Corbin & Musante, 2015; Davis et al., 2000; Leiss, 2013; Niman, 2014; Wild, Wild, & Han, 2014). The chapter presents in table Bhandarker various definitions of Virtual Shipping Companies VSCs.

Political Shipping Industry Changes: The Verticality of Sea Lanes for The Middle-East Shipping Services

The shipping industry in some port of the world faces some demographics shifts of the human effects Succar, 2013 #95. The social-economical and demographics farces raise new paradigms in the shipping environment. That population of the middle-income consumers in some big cities; like Delhi, Beijing, Shanghai Singapore and Dubai, moves the shipping and transportation complicity toward South-Asian gravity centres (Kvapil, 2012). Those Socio-economic changes in such a part of the world have some significant changes for the international container shipping industry in the Gulf Area. These changes are (Corbin & Musante, 2015; Halavais, 2004; Kvapil, 2012; La Monte & Woytowitz, 2015):

1. Raising the need for trigger more regulation in order to stand against more pollution and human safety and security around major Arabian Peninsula port cities (Corbin & Musante, 2015; Halavais, 2004; Kvapil, 2012; La Monte & Woytowitz, 2015).
2. Raising the opportunities to serve new south-south markets (W. Talley, 2013).
3. Increasing the new entry mode of domestic owners-enterprise size of shipping in companies to the international container shipping industry (Corbin & Musante, 2015; Halavais, 2004; Kvapil, 2012; La Monte & Woytowitz, 2015).

In order to ensure sustainable comparative advantages from the local socio-economic development shift to the international shipping market place (Corbin & Musante, 2015; Halavais, 2004; Kvapil, 2012; La Monte & Woytowitz, 2015). The new Arabian Peninsula shipping environment has been raised the volume of safety and security involvement in shipping industry – that of the economic support concern like the increasing number of enterprising shipping and maritime companies', owners especially in the Gulf area (Cooke, 2013); the shipping industry witnessed an increasing these type of shipping companies, in the last two decades. That leads to an increase of the customer protection regulations in the new shipping competitive environment. That is directed into the creation of shipping marketing and employment in such an area, and the legality environment and protection international port laws from monopoly tendencies, otherwise such tenderness will increased in the shipping market place in the area (Cooke, 2013; Kuratko, 2007). In addition to that, and like the European ports, public safety and security in shipping and ports maintenance in the Gulf ports have been organised in the gulf area in order to protect shipping and port environment from the air, sea water and noise pollution (Cooke, 2013; Corbin & Musante, 2015; Halavais, 2004; Kvapil, 2012; La Monte & Woytowitz, 2015).

On the other hand, The Arabian Peninsula port cities have special political features. The location is one of the main sources and exporters of oil and natural gas around the world (Herbert, 2011). The unique geographical feature and accessibility of the location have made it one of the most important and dynamic transport and logistics structures as a growing international trade area in volume. However, shipping industry and sea ports in such an area made the gulf countries' shipping leaders express their worries about the Gulf's shipping and maritime industries(Herbert, 2011). The Strait of Hormuz regional worries include economic and politic crises threats and political conflict risks of some sea lanes; in Yamane and Iran, violent piracy, gulf's unpredictable oil prices and recently close shipping lanes and ports (Dizard & Gadlin, 2014; Goulielmos & Plomaritou, 2014; Herbert, 2011).

Maritime transport is one of the main elements of logistics system that does the job of carrying and handling cargoes across the oceans. It includes activities such as contracting, shipping, moving shipment, loading and unloading activities. The maritime chain is a network that connects widely dispersed transportation linkages between consigners, ports and consignees (W. K. Talley & Ng, 2013). Shipping and ports have important direct and indirect impacts on the development of oceans and coasts.

Maritime transport is responsible for 90% of the world international trade (Rushton, Croucher, & Baker, 2010). This mode of transport offers many advantages over other modes of transportation. It is adequate for cargos with high volume and long delivery lead times. Maritime transport mode is slow and prone to delay that justifies its low cost compared to other modes. Global outsourcing, advances in technology, industrial and trade regulations and sustainability pressure have affected the way maritime transport is managed and resulted in higher levels of in-transit shipments leading to crucial economic and environmental consequences (Rushton et al., 2010) .

In their recent review, United nations conference on trade and development UNCTAD (2014) has presented the recent development in international maritime transport. In 2013, world merchandise trade volumes has slightly increased by 2.2% that influenced a modest growth in sea freight by 3.8% constituting a total volume of 9.6 billion tonnes (UNCTAD, 2014). Dry cargo especially bulk commodities (e.g. coal, grain, iron ore and forest products, etc.) has formed the highest percentage of this expansion by 5.5%. The largest share (70.2%) of sea freight is accounted for containerized trade and general cargo (break-bulk), whereas tanker trade (petroleum products, crude oil and gas) has formed 29.8%. In 2014, the world fleet including bulk carriers (42.9%), oil tankers (28.5%) and container ships (12.8%) constituted 1.69 billion dwt representing 4.1% annual growth. 42.9% of the total tonnage is accounted for bulk carriers, 28.5% represents oil tankers an d container ships is allocated 12.8% (UNCTAD, 2014). There is an increase in the throughput of world containers port by 5.1% constituting 651.1 million 20-foot equivalent units (TEUs) (UNCTAD, 2014).

Maritime Fleet Ownership

The recent review of Maritime Transport has provided a pioneer classification of fleet ownership pattern. They differentiate between beneficial ownership location and the nationality of ultimate owner. The former refers to the location in which a company that have the main commercial responsibility for the vessel is positioned, whereas the latter indicates the nationality of the owner of the ship regardless the location. In today's maritime environment, the predominant fleet ownership style is beneficial ownership location rather than owner's nationality. Ship-owners tend to locate their companies in their countries.

Environmental Regulations

In April 2014, International Maritime Organization has developed guidelines for implementing a number of technical and operational measures for international shipping operations to enhance energy efficiency and diminish greenhouse gas emissions that are still exempted from Kyoto Protocol liabilities (Fitzgerald, Howitt, & Smith, 2011; UNCTAD, 2014). The main dimensions to reduce the environmental effect of international maritime transport comprise technological options and market-based measures for regional and global policy actions (Miola, Marra, & Ciuffo, 2011). Research informed by policy proposals into measuring these emissions are running to decrease greenhouse gas emissions as well as other toxic substances that dramatically increase air pollution. McConnell, (2002) argues that maritime transport is the main factor in the development and protection of effective integrated marine ecosystem. Training and education associated with integrated management systems and sustainable development need to be established especially in the early phases of maritime education and training.

Security

Maritime transport security aspects are crucial for the whole maritime transport chain. Recently, maritime transport security has gained significant interest at the international maritime level. Although international maritime transport has flourished after the first half of 20th century due to major advances in technology and the globalization of international trade, terrorist attack dramatically affects maritime transport industry at all levels. The nature of this mode of transportation makes it highly vulnerable to terrorist attack considering the massive volume, extremely diversified routes along with the wide variety of cargoes that are sometimes very difficult to trace their origins, ownership and description. According to UNCTAD (2014) Piracy as key dimension of security concerns is continued to threaten sea freight specially in the West African Gulf of Guinea that is still very severe, whereas in the Gulf of Aden, the coast of Somalia and the Western Indian Ocean it is continued off. Global economy is greatly affected by any sort of disruption of maritime transport chain that may lead to international serious actions that justifies why maritime security regulations are at the top of police agenda. At the organizational, national and international levels, many security initiatives and measures have been implemented such as container security initiative (CSI), IMO's international ship and port facilities (ISPS) code, International Labour Organizations and World Custom Organizations, and the United States Custom-Trade partnership Against Terrorism. Literature has referred to different approaches for managing security concerns of maritime transport and ports. In US, number of initiatives have concentrated on maximisation of internal security, whereas EU has implemented a strategy that its main focus is to balance security requirements with other objectives such as trade defence and privacy safeguard (Papa, 2013).

The requirements of maritime security have directed many security improvements that are able to improve service quality and maritime transport performance. Security improvements may improve service quality through achieving more reliable service, more efficient operations and management, more awareness of social responsibility and better firm image in the market (Thai, 2007).

Maritime Safety

According to International Maritime Organization (IMO), maritime accidents are primarily caused by human errors. The seriousness of this topic has motivated IMO to develop guidelines for examining of human factors in marine accidents (O'Neil, 2003). The investigation of marine accidents should include not only the actions at the time of accident occurrence, but also the conditions that form the accident actions and determining hidden catalysts that facilitated unsafe circumstances to exist (Chen et al., 2013). The factors that may lead to the occurrence of accidents include internal and external. Driven by the IMO guidelines, the SHEL model was developed as a useful tool to identify the discrepancies between the work environment and the sharp-end personnel to interpret the failure action (Chen et al., 2013). Here are the causal factors that may influence accident occurrence

- External failure factors combine legislations gap, administration oversight and design flaws,
- Internal failure elements include:
  - Organizational influences, comprising resource management, organizational process and organizational climate;
  - Unsafe supervision in terms of inadequate supervision, planned inappropriate operations, failure to correct known problem and supervisory violations;
  - Preconditions such as software, hardware, physical and technological environment condition of operator(s) and life ware;
- Failure factors results in unsafe act that can be classified as error (skill-based errors, rule-based mistakes and knowledge-based mistakes) or violation (routine violations and exceptional violations).

In considering Swiss Cheese Model, disaster and accidents are very seldom the outcome of a single factor; usually they are caused by a mix of human errors committed by one or more than one person (Reason, 1997).

Port Traffic

This section explains how and why specific kinds of port traffic are located in specific kinds of regional economies. Extant research on port regions has revealed relatively contradictory outcomes related to the functional and physical disconnection between port traffic and port city (e.g. Hoyle, 1989).

Literature has indicated that development in technological aspects in the ports and maritime sectors resulted in the disconnection between the level of port throughputs and port cities demographic size (Ducruet & Lee, 2006). Hub-and-spokes network configuration has been implemented by container shipping companies in the 1980s that allow them to tolerate huge traffic at specific hub ports to perceive economies of scale as well as eliminating a number of ports from their services to minimise time and cost (Cullinane & Khanna, 2000). Therefore, the volumes of port throughput were directly associated with the port nodes that are centralized in the international shipping network regardless local economic activity simply because large number of these central nodes were located in less urbanized places (Ducruet & Itoh, 2015). This result supports previous research that confirmed the disconnection between urban growth and port growth (e.g. Jacobs, Ducruet, & De Langen, 2010).

In relation to the use of local port facilities, literature has illustrated that shippers are less likely to use nearby terminal for their cargoes. In some cases, shippers might value the position of freight handling facilities conditioned by if they are near the production factory, on-dock or somewhere else (Itoh, Tiwari, & Doi, 2002). The concentration of port and maritime organizations tend to be in large non-port cities, however there is no relationship between port traffic and the positioning of maritime and port firms in large cities. In contrast, material flows are strongly affected by certain localities.

Research has shown that advanced producer services have an impact on network formation and worldwide urban hierarchies between cities (e.g. Derudder, Taylor, & Ni, 2010; Taylor, 2004). Furthermore, the literature illustrated that the location of advanced producer services is associated with the economies of maritime localisation that considers ship owners, advanced producer services and port related industry, however no relationship has been confirmed related to the flows of throughput of ports (Jacobs, Koster, & Hall, 2011).

New Approaches of Security Dimensions in Global Shipping Industry Environment

Since 9/11, shipping industry security and safety are major shipping and transportation issues that need internationally agreed security adjustments (Ingram, 2011). The boundaries between authorities, institutions and localities are not addressed for security issues in the container and port services (W. Talley, 2013). The new approaches of security in shipping should to take in to considerations sustainability and the environment which is closely related to globalisation of market place (Ingram, 2011). Security and safety principles in the shipping industry environment are very crucial issues central to the shipping and maritime policy-making(Ingram, 2011). These security principles in shipping polices-making processes intervene in the shipping global market place by related governmental institutions of all jurisdictions to prevent considerable social and human harm like; security violations inadvertent transport of terrorist materials around the globe and pollution from taking place (Ingram, 2011).

Efforts to increase the scope of shipping security have been accompanied by attempts to challenge the shipping industry of the traditional security and safety paradigms, which on related to the shipping enterprise-customer or consumer relationships prospective to a more essential concept through the "concept of human security", which refocuses security on the human effect. The concept of human security affects the human welfare in any industry by numerous ways such as environmental degradation, poor governance, and organized terrorist crime (Curtis, 2014; Forsgren & Johanson, 2014; Goulielmos & Plomaritou, 2014; Halavais, 2004; Herbert, 2011).That means that the fundamental concept of security is addressing an integrating "Ocean business community" that can create opportunities for extraordinary performance of shipping and maritime industries (Curtis, 2014; Forsgren & Johanson, 2014; Goulielmos & Plomaritou, 2014; Halavais, 2004; Herbert, 2011) . In order to contribute to a more comprehensive approach to human security in shipping industry, in this chapter, the author proposes classifying security approach into four levels: human, regional, national, and international. It is suggested, in this chapter, that the problem of human effect security dimensions in shipping and maritime industries does not lie primarily with the penalties who fail to respond to international shipping standards of security, or even mechanisms that shape the framework for shipping policy-making process but the problem lies with lacking of shipping sustainability leaders which case the failure to provide clearly vision to understand the relationships between jurisdictions operating at human, international, national and regional levels (Sharma, 2014). Then, in case that a turbulent security challenge might happen, the shipping companies fail to achieve a successful ordinary performance (Curtis, 2014; Forsgren & Johanson, 2014; Goulielmos & Plomaritou, 2014; Halavais, 2004; Herbert, 2011). Author suggests that the new approach of shipping industry security dimensions start primarily with a new paradigm of shipping leadership style (see Figure 1).

In the increasing of complex and volatile contexts of shipping environment, effective and sensitive security developments of each all the above turbulent threats and opportunities concerns are based; in the first place, on a style of leadership and unique behavioural attributes form the shipping and transportation industries' leaders in the Gulf area. In the recent leadership and entrepreneurial behaviours literatures observe that the new approaches of effective behavioural style of leaders, in turbulent and competitive

Figure 1. Human effect security dimensions rainbow and shipping leadership style

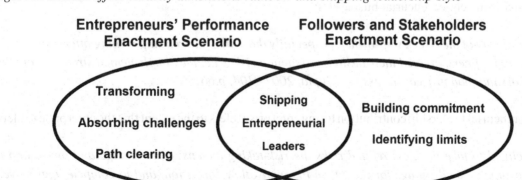

organisations, must be more entrepreneurial in order to enhance their organisational contributions and performance. "Entrepreneurial mind-set" has been strongly recommended as a key element of successful companies in a high-velocity industry of change and uncertainty environments. Entrepreneurial leadership involvement as a new concept in shipping industry environment is very important implementations of a new paradigm for effective and successful development of shipping industry (Curtis, 2014; Forsgren & Johanson, 2014; Goulielmos & Plomaritou, 2014; Halavais, 2004; Herbert, 2011).

Virtual Shipping Entrepreneurial Leadership Style (VSEL) Literatures

The concept of entrepreneurial leadership in any challenging industry involves fusing the concepts of a strategic approach of the management change, enhancing capabilities for continuously, creating and appropriating support, and development of value and competitive advantages in the company and technological growth (Bhandarker, 1990; Davis et al., 2000). The challenge for shipping and transportation industry is to build a compatibility to continuously explore and reduce the threats of secure and safety and enhance new successful and competitive opportunities (Sharma, 2014). The entrepreneurial leadership literature suggests that such a leadership style provide adequate making-policy processes to secure resources through its organisational efforts. The literature revealed the need for such processes to transform the existing allocation of the public resources security and safety in the challengeable industry in fundamental ways (W. Talley, 2013). The entrepreneurial leaders are capable of making their firms even more responsive to change taking place in the environment of their industry through sensitive and participative policy-making processes and effective communication skills (W. Talley, 2013).

In addition, fostering the followers autonomous initiatives for any unforeseen circumstances may happen in their work which is very important to organisation's team work to enhance their capability and competency in "autonomous strategic decision-making process" (W. Talley, 2013). The shipping entrepreneurial leaders persevere in the face of shipping environmental change, through an effective development of meaningful and defined processes of security and safety policy-making of within all political and economic institutions (W. Talley, 2013). These processes allow the most effective determination of different jurisdictional hierarchy policy-levels for safety, security in shipping environment for example international policy level (Gulf area, national policy level Kuwait and Qatar), regional policy level Doha (Kirk, 2014).

The shipping entrepreneurial companies promise their staff team efforts will lead to extraordinary performance in challengeable shipping environment. Gupta et al., 2007, p.66 emphasis the role of shipping and maritime entrepreneurial businesses:

The vast majority of the Greek ship-owners, especially those that entered ship owning after a successful career as ship officers, created their business assuming risk... i.e. they implemented strategies of their predecessors that had proved successful. (Gupta, 2007 #104, p.66)

Entrepreneurial leadership contribution to any enterprise refers to the extent that the enterprise leaders:

... are inclined to take business-related risks the risk-taking dimension, to change and innovation to obtain a competitive advantage for their firm the innovation dimension, and to compete aggressively with other firms the proactiveness dimension. (Covin and Slevin, 1988, p. 218.)

Promoting the entrepreneurial leadership style in the shipping industry encourages the shipping companies, which have a diversification of their shipping and port activities, to emphases on shared values among shipping company's team members and gain access for required technological information through multiple channels in shipping and transportation industries (Kirk, 2014).

The entrepreneurial shipping companies are generating new techniques and ideas through which idea, if pressures, are continuously creating new opportunities from existing shipping traditional works and activities that are institutionalized that were dominant in mobilizing the capacity of the shipping firms and its stakeholders for more competitive advantages in their turbulent environment . That led the literatures in entrepreneurs and shipping industry to assert that the entrepreneur leading men is the most important and decisive factor in the growth of modern shipping and transportation industry undertaking (Bhandarker, 1990; Davis et al., 2000; Goulielmos & Plomaritou, 2014). The Entrepreneurial leadership style, would then add values to the shipping undertaking into two main areas – that of the shipping industry environment security environment and policy making (Brüggemann, 2014). Entrepreneurial leadership literatures state that this style of leadership evokes super-ordinate outcomes from the followers through emerging environment contingencies of the industry context. The similarity between the entrepreneurial leadership approach of leaders behaviours and team-oriented leadership is that the both approaches emphasis the participation and the involvement by the company group (Brüggemann, 2014). Moreover, the entrepreneurial leadership scholars reveal that those styles of leaders deal effectively with uncertain Shipping Companies context and they are capable of orchestrate constantly the diversifying role driven by the dynamic environment of the organisation. A previous research that summarised the role of the entrepreneur in shipping industry states that the entrepreneurial companies' role in shipping industry is very important comparing other industries (Berson, Halevy, Shamir, & Erez, 2014). The value added of the entrepreneurial leaders in shipping industry is focusing on the leaders' capabilities to create a great vision and transform confidence to the group to achieve that vision. The shipping entrepreneurial leaders are those who style of leadership who enable the shipping owned companies to mobilize the capacity to meet the shipping industry challenge. They are, precisely, capable of articulating a compelling shipping and transportations vision through encouraging their followers that they can achieve their target goals (Abbas & El, 2015; Adzei & Atinga, 2012; Akinci, Aksoy, & Atilgan, 2004; Anderson, 1998; Asariotis et al., 2011; Atkins, 2001).

The entrepreneurial shipping companies offer a new paradigm of shipping and maritime entrepreneurial leadership style that enters the international shipping market places has the capability to be successful in the majority of cases in delivering a sustainable competitive advantage (Berson et al., 2014; Borchert, Gootiiz, & Mattoo, 2014).

The Scenarios Facing Entrepreneurial Leaders in the Shipping Security Environment

Drawing on the idea of promoting the entrepreneurial leadership style in the shipping industry, we suggest that the shipping companies' leaders are ready to face two interrelated enactment scenarios of action in order to deal with unforeseen circumstances of the shipping security issues (Tarling, 2013).

First scenario is envisaging a compatibility to continuously explore security and safety enhancements for super-ordinary performance. The entrepreneurial leadership provides adequate making-policy processes to secure shipping resources through sensitive and participative policy-making processes effective communication. In additional to that the leadership style can create innovative techniques from the challengeable shipping environment constrains to raise the ability for sustainable competitive advantages. In this chapter, the author named the first scenario entrepreneurs' performance enactment. The second scenario for shipping entrepreneurial leaders is to evoke the shipping super-ordinate outcomes through persuading their followers and the company's network of stakeholders that the making-policy processes are to meet the shipping challengeable environment is possible by securing resources to achieve the target vision underlying the scenario. We call this role shipping sub-ordinates and stakeholders' enactment scenario (Wong, 2012). The two scenarios are integrated to each other. Through entrepreneurs' performance enactment scenario cannot be conceived without an appropriate followers and stakeholders' enactment scenario, also the followers and stakeholders' enactment scenario cannot be assembled until a persuading scenario is activated. Both scenarios act much like the parallel evolution of cognitive integrating shipping participation of a team work or a concrete and abstract learning (Tarling, 2013).

In addition, shipping entrepreneurial leadership goes beyond Shipping Companies adaptation of the shipping and maritime industry, which is the focus of the new approach of security dimensions or human side effect. The roles that shipping entrepreneurial leaders participate in convince of their vision form proactive "enactment scenarios" of new sustainable competiveness in the shipping industry reconfiguration (Renko, El Tarabishy, Carsrud, & Brännback, 2015). Base of these roles of shipping entrepreneurial leaders, we suggest four specific roles of shipping leadership modified from two of which are related to entrepreneurs' performance enactment scenario and two are related to followers and stakeholders' enactment scenario (Renko et al., 2015).

First, with the awareness of the role of "transforming and path-clearing," shipping entrepreneurial leaders "transform a challenge that will encourage and motivate their followers to their limits of their accomplishments not over their limits" Transforming of shipping team is considered as a vision "worthy of persistence"(Brüggemann, 2014; Cooke, 2013). The shipping entrepreneurial leadership, in this track always absorbs uncertainty of shipping industry through negotiating the shipping strategies of the internal and external environment and framing the innovation challenges as "path-clearing," role of the leaders (Brüggemann, 2014; Cooke, 2013).

In the shipping entrepreneurial leaders' role of "absorbing security challenges of shipping industry," the entrepreneurial leaders formulates an inspire imagination vision challenge for the future of the company's super-ordinary performance in the new shipping environment and, then, the leaders will take the responsibility for any unexpected result about the future (Sharma, 2014).

In order for being sure about the best the shipping leaders can get from followers, they are responsible of building their confidence, trust and innovation and creativity competencies, which will enable them to act as much as it could be possible to realize the vision. That will solve lot of problems related to potential resistance of the new organisational changes and gaining support from shipping companies' stakeholders (Sharma, 2014). The role of "transforming, absorbing security challenges of shipping industry and path-clearing, help to achieve the entrepreneurs' performance enactment scenario—the following two roles are related to the followers and stakeholders' enactment scenario. The inspiration skills of entrepreneurial leaders will help them "building high organisational commitment," that push follower toward extraordinary effort to accomplish the vision and make their responsibilities even more clear as much as they described by their leader (Sharma, 2014).

The role of identifying the followers' limits is what the shipping entrepreneurial leaders need to accomplish in order to achieve the sustainable organisational commitment in the face of the scenario constraints. The shipping entrepreneurial leader frames the followers' capabilities by stop self-imposed ideas of limitation. That will result in new innovation competences. In Figure 2 we illustrates these four

Figure 2. the two scenarios facing shipping entrepreneurial leaders

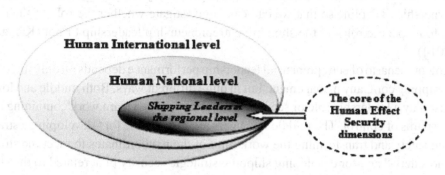

specific roles of the shipping entrepreneurial leaders subsumed under the two dimensions of enactment scenarios, provide the theoretical framework for the construct of shipping entrepreneurial leadership. In the balance of the article, we empirically test this construct in the following section (Sharma, 2014).

An Empirical Measurement Model for the Shipping Entrepreneurial Leadership (SEL)

We identify the measurement of the shipping entrepreneurial leadership dimensions in this Chapter by:

1. Categorising the shipping leaders' behavioural attributes mentioned in the above before. These attribute that underlying of the four specific roles of shipping leaders lead to entrepreneurs' performance enactment scenario and followers and stakeholders' enactment scenario (Davids, 2012; Kilgour, 1992; Walumbwa, Orwa, Wang, & Lawler, 2005).
2. Develop the measured scale of Entrepreneurial leadership developed by Gupta et al., 2004 who provide an evidence of the universal appeal of the entrepreneurial leaders' dimensions in different managerial styles across cultures (Dimovski, Penger, Peterlin, & Uhan, 2013; Kuratko & Hornsby, 1999; Mustafa & Lines, 2012; Zaccaro & Banks, 2001).
3. Using the project of GLOBE Scale that design leadership effectiveness survey. The project used a sample collected from 62 societies worldwide of around 15,000 middle managers during 1995–1997 (Zaccaro & Banks, 2001). The survey includes more than 900 different firms of 10 major cultural clusters of the world: Latin America, Latin Europe, Anglo, Nordic, Germanic, Eastern Europe, Southern Asia, Confucian Asia, Middle East, and Sub-Sahara Africa (Mustafa & Lines, 2012).

In this chapter, the author attempts to use GLOBE project data scales from managerial subordinate respondents who rated 22 attributes of effective leadership attributes. These behavioural attributes measured on a five-point Likert scale, ranging from 1 strongly disagree to 5strongly agree (Choi, 2009; Hale & Fields, 2007; Kilgour, 1992; Svendsen, 1981).

The performance of the shipping entrepreneurial leadership depends on the subordinates all the managerial department of the shipping selected companies in Qatar (Abdalla & Al-Homoud, 2001). These subordinates who respond to the paper questionnaire must be the shipping company leaders' team work and must perceive tasks that related to safety and security of shipping and transportation transactions during the last three years, so that they had experienced clear perception of the style of leadership that they dealt with. The selected companies of the paper survey were based on the criteria of private enterprise ownership discipline so that we can easily investigate whether the role of their managers is matching the four specific roles of the shipping entrepreneurship leadership or not (Kantabutra, 2014; Muniapan, 2014)

The enactment scenario of entrepreneurial leadership performance depends on only middle, and lower levels of the shipping company management, but in quite different ways. Both middle and lower levels of managers must play a leadership role of security and safety within a team work containing a group of at least 10 to 15 spams of control. The middle managers are responsible for "developing a strategic vision of security and safety and transforming the work limit of their subordinates to meet the strategic vision of security and safety " and for developing shipping strategic security plan related to the challengeable environment in operational levels of shipping companies(Carton, Murphy, & Clark, 2014)

THE RESEARCH METHODS AND FINDINGS

Validity and Reliability of Shipping Entrepreneurial Leadership Model

Based on Gupta et al., (2004) Shipping Entrepreneurial Leadership style that scanned The 21 attributes of outstanding leadership dimensions in the GLOBE project scales to identify 19 attributes of their extraordinary performance, we categorized 10 shipping behaviours likely to be expected in loading on the two roles that impede the shipping entrepreneurial leaders' performance enactment scenario and 8 leadership behaviours likely to be facilitated in their roles towered encouraging the team performance in dealing with safety and security missions in shipping environment (Gupta, MacMillan, & Surie, 2004). We listed 22 leadership behavioural roles in the present paper, 12 of them have been short listed from Gupta et al., (2004) scale of entrepreneurial behaviours and the rest of these attributes have been impeded based on the GLOBE projects scales (Gupta et al., 2004; Kuratko, 2007)

Of these, we included five shipping behavioural attributes of the shipping companies 'leaders- performance oriented, ambitious, diplomatic, inspirational, intellectually Stimulating, visionary, forethought, trust builder, persuasive and accommodating- in the shipping entrepreneurial leadership constructs of Entrepreneurs' performance enactment scenario to test the shipping leaders' roles of transforming & path-clearing and absorbing security challenges. Others, for example, - has extra insight, and negotiate- have been excluded from the shipping entrepreneurial leadership constructs of entrepreneurs' performance enactment scenario because these items had low positive or negative less than +0.50 or -0.5 correlation with the remaining items of the leaders attributes of the questionnaire. Conducting an exploratory factor analysis EFA of the shipping leaders' roles of transforming & path-clearing and absorbing security challenges' dimensions, in which all of the above included items is above 0.70, only 4 of the 22 shipping leadership attributes did not load over 0.60 onto their predicted role factors.

CONCLUSION AND DISCUSSION

In the present chapter, the author attempted to the same way in the present chapter, the author used to embed items in the first two enactment scenarios of entrepreneurial performance in the last two dimensions of the Shipping followers in the present chapter, the author addressing and stakeholders enactment scenarios. Evidently, enthusiastic, improvement-oriented, team builder inspirational integrator intellectually stimulating positive critical are entrepreneurial leaders' behaviours and attributes that are deemed efficient only in shipping companies level. The final shipping leaders' behaviours and dimensions underlying the four enactment scenarios of entrepreneurial leadership constructs are shown in Table 5.

Composite Reliability and Confirmatory Factor Analysis CFA

Thereafter, with using a confirmatory factor analysis CFA on the same sample supported our research findings. Scales measuring each of the four enactment scenarios of shipping leadership had a composite reliability scores ranging from 0.59 to 0.68, n=306 at the leaders level, which is acceptable for an individual-level of shipping construct. In addition, the four dimensions scales shin the present chapter, the author d sufficient and acceptable composite reliability scores at shipping companies level range .64–.75, n=89. Further, the scales for the shipping entrepreneurial leaders subsumed under the two dimensions of enactment scenarios had more than 0.70 composite reliability scores at both of the individual and shipping company levels, as they are given in Table 2.

Construct validity used in the present paper is consisted of two parts: internal validity and external validity. Internal validity is of two types of validities: convergent and discriminate. Convergent validity of the scales is indicated if each factor loading for the scales is statistically significant. This condition was satisfied for all the leadership roles except 4 leaders' roles that In the present chapter, the author have mentioned earlier, two sub-dimensions of entrepreneurial leadership performance and shipping followers.(Kuratko, 2007) In the present chapter, the author addresses enactment scenarios. The overall entrepreneurial leadership scale at individual and shipping company levels are also satisfied. Using Average variance extracted analysis AVE in this paper, In the present chapter, the author tests to see if the square root of every AVE value belonging to each construct of shipping entrepreneurial leaders is much larger than any correlation among any pair of these constructs. AVE is calculated as:

$$AVE = \frac{\sum[\lambda 2]}{\sum[\lambda_I^2] + \sum[\mathrm{Var}\varepsilon i]},$$

Fornell and Larcker, 1981 suggest that the value of AVE for each latent construct should be at least 0.50. In the present paper, the value of AVE for each shipping leadership latent construct is more than 0.50 which refers to a satisfied discriminate validity for the research model.

THE CHAPTER INSTRUMENT

A self-organized questionnaire has distributed for 180 Shipping Leaders working in two Arabian Gulf Port in United Arabe Emirates and Kuwait mobile banking users in the UAE banking customers. The research sample of the present study has been designed to be two sample group for Mobile banking leaders and customers. The chapter analyses the data based on using data coding system using Nvivo 10 software. And structural equational modeling using Amos IBM V.22. The two questionnairesare used in the present chapter use Likert Scale of Five main scales started by cosidered (1) for totally disagree to (5) for totally agree. The present research received 288 valid applications of a response rate of about 75.6%. In Table 1 shows the demographic data for the chapter sample.

THE CHAPTER FINDINGS

The results of confirmatory factor analysis (Second Order) for both formative and reflective measurements show that the calculated the composite reliability values range from 0.9341 to 0.8113. The variance extracted scored from 0.564 to 0.833 which indicates an acceptable level range. The Cronbach's Alpha shows an acceptable rang of loading which all above 0.7 (Table 2).

Table 1. The demographic data for the chapter mobile banking customer sample

Demographic Data	Items	Numbers of Mobile Payment Users	Percentage (%)
Gender	Male	193	67%
	Female	95	33%
Age	Below 20	66	11%
	20-30	153	53%
	30-40	49	18%
	Above 40	20	18%
Job	student	124	45%
	professional	146	51%
	Others	10	4%
Frequency of shipping trips	At times	144	50%
	often	98	34%
	Very often	46	16%

Table 2. The Cronbach's Alpha shows an acceptable range of loading

Measures	Construct	Items	Cronbach'sAlpha	Composite Reliability	Variance Extracted
Formative	Transforming & Path-Clearing	4	0.8120	0.8113	0.632
	Absorbing Security Challenges of Shipping performance	3	0.8223	0.8799	0.564
	Building Commitment	4	0.9132	0.9341	0.833
	Identifying Limits	4	0.7899	.80221	0.726
Reflective	Transforming & Path-Clearing	4	0.8101	0.8113	0.632
	Absorbing Security Challenges of Shipping performance	3	0.8211	0.8778	0.564
	Building Commitment	4	0.9132	0.9341	0.833
	Identifying Limits	4	0.7878	.80210	0.723

In the Table 3 the researcher attempt to calculate the discriminant validity.

The discriminant validity measured by Average Variance extracted (AVE) of each constructs.(AVE) "should be larger than all the cross-correlations between the construct and items should load more strongly on their related construct than on other constructs."(VELMURUGAN, Nadu, & VELMURU-GAN, 2014). The composite reliability was evaluated and all constructs were greater than 0.70 (Fornell & Larcker, 1981).

The Sample Group Structural Equation Modelling

The two sample-group SEM analysis uses confirmatory factor analysis (CFA) shows that the two self-organized questionnaires had an acceptable reliability and validity tests. As shown in Table 4, the good-ness of fitting indices summary for the measurement and structural models used in the chapter.

Table 5 explains that the overall goodness of fitting summary and statistics show that the structural model used in the present chapter fits the data collected well.

Structure Equation Model Approach

This chapter proposed an applied and descriptive research using confirmatory approach based on using structural equations modeling (AMOS software V.21). Modelling the causal relationship between the external factors and customer retention of mobile payments mediating the influence of the leadership style on the mobile customer retention, is such a structural model needs to be fitted by using values of normalized fitness index which has used the following fitted indexes: (NFI), relative fitness index (RFI), increasing fitness index (IFI) and comparative fitness index (CFI). Using the factors adapted from litera-ture, the present research improves content and discriminant validity a long with composite reliability.

Table 3. The correlation matrix of the research variables

Construct	Cronbach's Alpha	(EV)	(EELS)	(CT)	(M-PA)
(EV)	0.8120	1,00			
(WEELS)	0.8223	0.334	1.00		
(CT)	0.9132	0.455	0.343	1.00	
(M-PA)	0.7899	0.643	0.533	0.343	1,00

Table 4. Goodness of fitting indices summary of the chapter models

Fitting Summary	X2	X2/df	GFI	CFI	TLI	IFI	RMSEA
Values for the Measurement Model	202.165	1.671	0.921	0.935	0.957	0.965	0.058
Values for the Structural Model	242.133	2.90	0.908	0.915	0.932	0.945	0.065
Standard accepted values		>0.05 <3	>0.80	>0.90	>0.90	>0.90	<0.08

Table 5. Measures of shipping entrepreneurial leadership

Shipping Enactment Scenarios of Leaders	Shipping Leaders Dimensions	Shipping Leadership Behaviours Items in the Questionnaire	Explanation of the Items
Entrepreneurs' performance enactment scenario	Transforming & Path-Clearing	Performance oriented	1-Transform a challenge that will encourage and motivate their followers to their limits of their accomplishments not over their limits.
		Ambitious	2-Sets high goals and works hard to achieve them
		Has extra insight	3- Instinctive
		Negotiate	4-Is able to bargain effectively
		Diplomatic	5-have the ability to persuade others
Entrepreneurs' performance enactment scenario	Absorbing Security Challenges of Shipping performance	Visionary	1- The entrepreneurial leaders formulates an inspire imagination vision challenge for the future of the company's super-ordinary performance in the new shipping environment.
		Forethought	2-Anticipates possible future challengeable event
		Trust builder	3-Instills the followers with trust and confidence
		Persuasive	4- at interpersonal Skilful and tactful
		Accommodating	5-Gives advices accommodating and motivate.
Shipping followers and stakeholders enactment scenario	Building Commitment	Enthusiastic	1-Demonstrates a strong positive emotions for work
		Improvement-oriented	2-Development-oriented of the team group
		Team builder	3-Able to induce followers members to be a team work
		Inspirational	4-Motivate values of the followers to work hard
		Integrator	5-Integrates followers into cohesive
Shipping followers and stakeholders enactment scenario	Identifying Limits	Intellectually Stimulating	1-"*Encourages others to use their mind—challenges beliefs, stereotypes, and attitudes of others*"
		Positive	2-Bright, confident optimistic
		Critical	3-Makes decisions securely and quickly

Also the research uses a These items were first translated into Persian by a researcher. Then techniques of 'Translation-back translation' by using the research items to be translated into Arabic from English by Arabic English speakers and from back to English from Arabic by another English–Arabic speaker to ensure the same shared meaning and understanding to keep consistency for the research instrument.

Thereafter, with using a confirmatory factor analysis (CFA) on the same sample supported our research findings. Scales measuring each of the four enactment scenarios of electronic entrepreneurial leadership had a composite reliability scores ranging between (0.59 to 0.68) at the leaders level, which is acceptable for an individual-level of electronic entrepreneurial construct. In addition, the four dimensions scales showed sufficient and acceptable composite reliability scores at electronic entrepreneurial banking level (range between 0.64–0.75). Further, the scales for the electronic entrepreneurial leaders subsumed under the two dimensions of enactment scenarios had more than 0.70 composite reliability scores at both of the individual and electronic entrepreneurial Mobile Banking services (DiMaggio & Hargittai, 2001; Eddy et al., 2014; Gladwin et al., 1995; van Riel & Pura, 2005; Warschauer, 2007).

REFERENCES

Abbas, Hussein H, & El, Abd El Halim Omar Abd. (2015). The Role of Suez Canal Development in Logistics Chain. In *Global Supply Chain Security* (pp. 163-180). Springer.

Abdalla, I. A., & Al-Homoud, M. A. (2001). Exploring the implicit leadership theory in the Arabian Gulf states. *Applied Psychology*, *50*(4), 506–531. doi:10.1111/1464-0597.00071

Adzei, F. A., & Atinga, R. A. (2012). Motivation and retention of health workers in Ghana's district hospitals: Addressing the critical issues. *Journal of Health Organization and Management*, *26*(4), 467–485. doi:10.1108/14777261211251535 PMID:23115900

Akinci, S., Aksoy, S., & Atilgan, E. (2004). Adoption of internet banking among sophisticated consumer segments in an advanced developing country. *International Journal of Bank Marketing*, *22*(3), 212–232. doi:10.1108/02652320410530322

Anderson, G. L. (1998). Toward authentic participation: Deconstructing the discourses of participatory reforms in education. *American Educational Research Journal*, *35*(4), 571–603. doi:10.3102/00028312035004571

Asariotis, R., Benamara, H., Finkenbrink, H., Hoffmann, J., Lavelle, J., Misovicova, M., . . . Youssef, F. (2011). *Review of Maritime Transport, 2011*. Academic Press.

Atkins, E. T. (2001). Blue Nippon: authenticating jazz in Japan. Duke University Press.

Berson, Y., Halevy, N., Shamir, B., & Erez, M. (2014). Leading from different psychological distances: A construal-level perspective on vision communication, goal setting, and follower motivation. *The Leadership Quarterly*.

Bhandarker, A. (1990). *Corporate success and transformational leadership*. New Age International.

Borchert, I., Gootiiz, B., & Mattoo, A. (2014). Policy barriers to international trade in services: Evidence from a new database. *The World Bank Economic Review*, *28*(1), 162–188. doi:10.1093/wber/lht017

Brüggemann, H. (2014). *Entrepreneurial leadership styles: a comparative study between Startups and mature firms*. Academic Press.

Carton, A. M., Murphy, C., & Clark, J. R. (2014). A (blurry) vision of the future: How leader rhetoric about ultimate goals influences performance. *Academy of Management Journal*, *57*(6), 1544–1570. doi:10.5465/amj.2012.0101

Chen, S.-T., Wall, A., Davies, P., Yang, Z., Wang, J., & Chou, Y.-H. (2013). A Human and Organisational Factors (HOFs) analysis method for marine casualties using HFACS-Maritime Accidents (HFACS-MA). *safety. Science*, *60*, 105–114. doi:10.1016/j.ssci.2013.06.009

Choi, E. K. (2009). Entrepreneurial leadership in the Meiji cotton spinners' early conceptualisation of global competition. *Business History*, *51*(6), 927–958. doi:10.1080/00076790903266877

Cooke, P. (2013). *Complex adaptive innovation systems: Relatedness and transversality in the evolving region*. Routledge.

Corbin, D. L., & Musante, C. F. (2015). *Package assembly for thin wafer shipping.* US Patent 20,150,076,029.

Cullinane, K., & Khanna, M. (2000). Economies of scale in large containerships: Optimal size and geographical implications. Journal of Transport Geography. *Journal of Transport Geography*, *8*(3), 181–195. doi:10.1016/S0966-6923(00)00010-7

Curtis, S. (2014). *Global Cities and International Relations.* Routledge.

Davids, J. P. (2012). *Entrepreneurial leadership in dynamic markets.* University of Johannesburg.

Davis, J. A., Pitts, E. L., & Cormier, K. (2000). Challenges facing family companies in the Gulf Region. *Family Business Review*, *13*(3), 217–238. doi:10.1111/j.1741-6248.2000.00217.x

Derudder, B., Taylor, P. J., & Ni, P. (2010). Pathways of change: Shifting connectivities in the world city network, 2000-2008. *Urban Studies (Edinburgh, Scotland)*, *47*(9), 1861–1877. doi:10.1177/0042098010372682

Dimovski, V., Penger, S., Peterlin, J., & Uhan, M. (2013). Entrepreneurial Leadership In The Daoist Framework. *Journal of Enterprising Culture*, *21*(04), 383–419. doi:10.1142/S0218495813500167

Dizard, J. E., & Gadlin, H. (2014). Family Life and the Marketplace: Diversity and Change in the American Family. *Historical Social Psychology*, 281.

Donald, F. (2007). Entrepreneurial leadership in the 21st century. *Journal of Leadership & Organizational Studies*, *13*(4).

Ducruet, C., & Itoh, H. (2015). Regions and material flows: Investigating the regional branching and industry relatedness of port traffic in a global perspective. *Journal of Economic Geography*, 1–26. doi:10.1093/jeg/lbv010

Ducruet, C., & Lee, S. W. (2006). Frontline soldiers of globalisation: Port-city evolution and regional competition. *GeoJournal*, *67*(2), 107–122. doi:10.1007/s10708-006-9037-9

Fitzgerald, W. B., Howitt, O. J. A., & Smith, I. J. (2011). Greenhouse gas emissions from the international maritime transport of New Zealand's imports and exports. *Energy Policy*, *39*(3), 1521–1531. doi:10.1016/j.enpol.2010.12.026

Forsgren, M., & Johanson, J. (2014). *Managing networks in international business.* Routledge.

Goulielmos, A. M., & Plomaritou, E. (2014). The Shipping Marketing Strategies within the Framework of Complexity Theory. *British Journal of Economics, Management & Trade*, *4*(7), 1128–1142.

Gupta, V., MacMillan, I. C., & Surie, G. (2004). Entrepreneurial leadership: Developing and measuring a cross-cultural construct. *Journal of Business Venturing*, *19*(2), 241–260. doi:10.1016/S0883-9026(03)00040-5

Halavais, R. A. (2004). *Load bearing structure for composite ecological shipping pallet.* Google Patents.

Hale, J. R., & Fields, D. L. (2007). Exploring servant leadership across cultures: A study of followers in Ghana and the USA. *Leadership*, *3*(4), 397–417. doi:10.1177/1742715007082964

Herbert, D. E. J. (2011). Theorizing religion and media in contemporary societies: An account of religious 'publicization'. *European Journal of Cultural Studies, 14*(6), 626-648.

Hoyle, B. S. (1989). The port-city interface: Trends, problems, and examples. *Geoforum, 20*(4), 429–435. doi:10.1016/0016-7185(89)90026-2

Ingram, J. (2011). A food systems approach to researching food security and its interactions with global environmental change. *Food Security, 3*(4), 417–431. doi:10.1007/s12571-011-0149-9

Itoh, H., Tiwari, P., & Doi, M. (2002). An analysis of cargo transportation behaviour in Kita Kanto (Japan). *International Journal of Transport Economics, 29*, 319–335.

Jacobs, W., Ducruet, C., & De Langen, P. W. (2010). Integrating world cities into production networks: The case of port cities. *Global Networks, 10*(1), 92–113. doi:10.1111/j.1471-0374.2010.00276.x

Jacobs, W., Koster, H. R. A., & Hall, P. V. (2011). The Location and Global Network Structure of Maritime Advanced Producer Services. *Urban Studies (Edinburgh, Scotland), 48*(13), 2749–2769. doi:10.1177/0042098010391294

Kantabutra, S. (2014). Visionary leadership at a Thai apparel manufacturer: Surprising evidence? *International Journal of Business Excellence, 7*(2), 168–187. doi:10.1504/IJBEX.2014.059547

Kilgour, F. G. (1992). Entrepreneurial leadership. *Library Trends, 40*(3), 457–474.

Kirk, D. J. (2014). The "Singapore of the middle east": The role and attractiveness of the Singapore model and TIMSS on education policy and borrowing in the Kingdom of Bahrain. In *Education for a Knowledge Society in Arabian Gulf Countries*. Emerald Group Publishing Limited.

Kuratko, D. F. (2007). Entrepreneurial leadership in the 21st century. *Journal of Leadership & Organizational Studies, 13*(4), 1–11. doi:10.1177/10717919070130040201

Kuratko, D. F., & Hornsby, J. S. (1999). Corporate entrepreneurial leadership for the 21st Century. *Journal of Leadership & Organizational Studies, 5*(2), 27–39. doi:10.1177/107179199900500204

Kvapil, L. A. (2012). *The Agricultural Terraces of Korphos-Kalamianos: A Case Study of the Dynamic Relationship Between Land Use and Socio-Political Organization in Prehistoric Greece*. University of Cincinnati.

La Monte, D. P., & Woytowitz, P. J. (2015). *Landscape controller with feature module*. Google Patents.

Leiss, W. (2013). *Social communication in advertising: Consumption in the mediated marketplace*. Routledge.

McConnell, M. (2002). Capacity building for a sustainable shipping industry: a key ingredient in improving coastal and ocean and management. *Ocean & Coastal Management, 45*(10), 617-632. doi:10.1016/S0964-5691(02)00089-3

Miola, A., Marra, M., & Ciuffo, B. (2011). Designing a climate change policy for the international maritime transport sector: Market-based measures and technological options for global and regional policy actions. *Energy Policy, 39*(9), 5490–5498. doi:10.1016/j.enpol.2011.05.013

Muniapan, B. (2014). The Bhagavad-Gita and Business Ethics: A Leadership Perspective. *Asian Business and Management Practices: Trends and Global Considerations: Trends and Global Considerations*, 232.

Mustafa, G., & Lines, R. (2012). The triple role of values in culturally adapted leadership styles. *International Journal of Cross Cultural Management*, 1470595812452636.

Niman, N. B. (2014). *The Gamification of Higher Education: Developing a Game-based Business Strategy in a Disrupted Marketplace*. Palgrave Macmillan. doi:10.1057/9781137331465

O'Neil, W. A. (2003). The human element in shipping. *WMU Journal of Maritime Affairs*, 2(2), 95–97. doi:10.1007/BF03195037

Papa, P. (2013). US and EU strategies for maritime transport security: A comparative perspective. *Transport Policy*, 28(0), 75–85. doi:10.1016/j.tranpol.2012.08.008

Reason, J. (1997). Managing the Risks of Organizational Accidents. Ashgate, UK: Aldershot.

Renko, M., El Tarabishy, A., Carsrud, A. L., & Brännback, M. (2015). Understanding and measuring entrepreneurial leadership style. *Journal of Small Business Management*, 53(1), 54–74. doi:10.1111/jsbm.12086

Rushton, A., Croucher, P., & Baker, P. (2010). The Handbook of Logistics & Distribution Management (4th ed.). Kogan Page.

Sharma, K. L. (2014). *Entrepreneurial performance in role perspective*. Abhinav Publications.

Svendsen, A. S. (1981). *The role of the entrepreneur in the shipping industry*. Academic Press.

Talley, W. (2013). *Maritime safety, security and piracy*. CRC Press.

Talley, W. K., & Ng, M. W. (2013). Maritime transport chain choice by carriers, ports and shippers. *International Journal of Production Economics*, 142(2), 311–316. doi:10.1016/j.ijpe.2012.11.013

Tarling, N. (2013). *Status and Security in Southeast Asian States*. Routledge.

Taylor, P. J. (2004). *World City Network: A Global Urban Analysis*. London: Routledge.

Thai, V. V. (2007). Impacts of Security Improvements on Service Quality in Maritime Transport: An Empirical Study of Vietnam. *Maritime Econ Logistics, 9*(4), 335-356.

UNCTAD. (2014). *Review of Maritime Transport*. New York, Geneva.

Walumbwa, F. O., Orwa, B., Wang, P., & Lawler, J. J. (2005). Transformational leadership, organizational commitment, and job satisfaction: A comparative study of Kenyan and US financial firms. *Human Resource Development Quarterly*, 16(2), 235–256. doi:10.1002/hrdq.1135

Wild, J., Wild, K. L., & Han, J. C. Y. (2014). *International business*. Pearson Education Limited.

Wong, W. Y. C. (2012). *Cognitive metaphor in the West and the East: A comparison of metaphors in the speeches of Barack Obama and Wen Jiabao*. Academic Press.

Zaccaro, S. J., & Banks, D. J. (2001). Leadership, vision, and organizational effectiveness. In *The nature of organizational leadership: Understanding the performance imperatives confronting today's leaders* (pp. 181–218). Uratko.

Chapter 17
Virtual Strangers No More:
Serious Games and Creativity
for Effective Remote Teams

Howard Bennett Esbin
Heliotrope, Canada

ABSTRACT

In this chapter we examine how virtual team trust and effectiveness may be improved through the transformative power of serious games and creative process. To start we explore the pervasive lack of emotional intelligence within the workplace at an individual level and which we call 'the EQ Gap'. This is followed by an examination of challenges faced by both traditional and virtual teams. We then consider how the same EQ Gap also manifests in both traditional and virtual teams as well. Indeed, it's worse for the latter. This leads to a review of the kinds of EQ training needed for both team types. A discussion then follows as to how serious games, play, and creativity can help virtual teams in particular to become more emotionally intelligent, trusting, and ultimately more collaborative. A brief case study of a serious game called Prelude is shared to illustrate these findings in a practical context.

INTRODUCTION

Globally linked virtual teams will transform every government and company in the world. Any of our peers who don't do it won't survive. – John Chambers, CEO of Cisco (DeRosa & Lepsinger, 2010)

In this chapter we examine how virtual team formation and trust may be improved through the transformative power of serious games and creative process. The phrase 'virtual team' first surfaced in business literature during the late 1980s and early 90s. "A virtual team is a group of individuals who work across time, space and organizational boundaries with links strengthened by webs of communication technology" (Lipnack & Stamps, 2000). Empirical research since then has largely focused on virtual team trust, communication, leadership, and performance (Bodiya, 2013). Practically every sector and field of endeavour now uses virtual teaming including, business, healthcare, education, volunteerism, and the military, for example. In 2007, "Cisco CEO John Chambers said the first phase of Internet productivity gains for business is over, but a second phase, based on virtual teams able to capitalize on virtualized corporate resources, is beginning" (Babcock, 2007).

DOI: 10.4018/978-1-4666-9688-4.ch017

Between 2005 and 2013, the number of employees who worked virtually grew by 80%. In the USA alone "almost half the adult labour force, 64 million individuals, will be involved in telecommuting and remote working at least part of the time" (Burt, 2013). By 2015, 1.3 billion people are expected to soon be working on virtual teams (Johns & Gratton, 2013). This is expected to grow another 21% in 2016 (Equichord, 2014).

This accelerating global development has been facilitated by a convergence of factors – business (massive mergers, take-overs, and partnerships), economic (global recession), technological (inexpensive cloud-based collaboration platforms), and environmental (global warming). Moreover, virtual teams offer very attractive benefits including: reduced travel costs, reduced carbon footprint, access to a more diverse global talent pool, and greater potential for innovation.

However, there's a significant gap between management expectations and actual virtual team performance outcomes. In one global study, 27% of virtual teams were found to be not fully performing (Onpoint Consulting, 2010). Another study revealed that only 18% of seventy global business virtual teams were found to be highly successful (Siebdrat, Hoegl & Ernst, 2009). Essentially, 80% of virtual teams are performing significantly below capacity. Unsurprisingly, 19 out of 20 "executives say they have experienced difficulty in managing virtual teams" (Dachis Group, 2013). "The financial cost of this gap is enormous due to lost productivity, missed deadlines, declining morale, and failure to innovate" (Lepsinger, 2014).

One challenge stems from technology constraints, for example, sporadic Internet availability. Global time zone differences are another challenge, for example the 15 hours between Tokyo and Winnipeg. However, the greatest challenge stems from lack of trust between employees who are essentially 'virtual strangers'. This is despite the fact they may work for the same organization, share common goals, and be members of the same virtual team.

In this chapter we begin by examining the significant lack of emotional intelligence within the workplace generally at an individual level. We call this the "EQ Gap". This is followed by an examination of challenges faced by both traditional and virtual teams. We then consider how the same EQ Gap also manifests in both traditional and virtual teams as well. Indeed, it's worse for the latter. This leads to a review of the kinds of EQ training needed for both team types. A discussion then follows as to how serious games, play, and creativity can help virtual teams in particular to become more emotionally intelligent, trusting, and ultimately more collaborative. A brief case study of a serious game called Prelude is shared to illustrate these findings in a practical context.

THE EQ GAP

When executives from the Big Three auto makers hopped their individual jets in November 2008 to fly from Detroit to Washington and throw themselves at the feet of Congress to beg for federal financial support, we all marveled at the disconnect … They never thought to find that emotional quotient, their EQ. Asking…for money while traveling in imperious style. And then they were taken aback with the public outcry. Remember, it wasn't their IQs people were questioning as much as their EQs … therein lies the EQ gap. – Dr. Nancy Snyderman, NBC Chief Medical Editor (Snyderman, 2009)

The 'soft skills' of an organization's managers and staff are crucial to its success. These skills are also known as Emotional Intelligence or EQ. EQ has four broad dimensions – self-awareness, self-management, social awareness, and relationship management. It's a natural complement to Cognitive Intelligence, or IQ (Intellectual Quotient). Like IQ, EQ is also needed at all life stages.

People who accurately perceive others' emotions are better able to handle changes and build stronger social networks (Cherniss, 2000). Research has shown that 80% of adult "success" is due to emotional intelligence (Goleman, 1995). The best-run firms promote emotional ties and wellbeing. People who accurately perceive others' emotions are better able to handle changes and build stronger social networks (Cherniss, 2000). Senior management understand that employee attitudes and wellbeing are linked to financial results (Mental Health Roundtable, 2011).

The EQ Gap is a phrase coined in 2009 by Dr. Nancy Snyderman, then the NBC Chief Medical Officer. She has used this phrase to describe the lack of emotional intelligence on the part of Detroit automakers that flew to Washington on their private jets to ask for a government bailout during the worst of the global economic crash.

I use this phrase broadly in this chapter to describe the pervasive lack of emotional intelligence that we see manifesting throughout the business world more generally. For example, only 36% of employees are able to identify emotions as they occur (Bradberry & Greaves, 2009). Research also indicates 90% of executives derail professionally due to: the lack of emotional and social competencies including inability to handle interpersonal problems; unsatisfactory team leadership during times of difficulty or conflict; and the inability to adapt to change or elicit trust (Center for Creative Leadership, 2012).

To extend analysis of the EQ Gap further, the core causes of employee workplace conflict are:

- **94%:** Have worked or currently work with a toxic person;
- **86%:** Warring egos & personality clashes;
- **73%:** Poor leadership;
- **67%:** Lack of honesty;
- **64%:** Stress;
- **59%:** Clashing values;
- **40%:** Verbal confrontations;
- **40%:** Scapegoating;
- **32%:** Spreading Rumors. (Psychometrics Canada, 2009; Immen, 2009; Marketwired, 2011; University of Phoenix, 2013)

In a dire related context, over 54 million employees experience bullying, which is over a third of the US workforce (Dye, 2009). The consequences of the workplace EQ Gap are:

- Declining Productivity;
- Declining Morale;
- Rising Absenteeism and Turnover;
- Rising Depression and Anxiety;
- Rising Health & Benefits Costs;
- Increasing Disability Premiums.

TEAMING CHALLENGES: TRADITIONAL AND VIRTUAL

We cannot avoid teaming. We can only team well or badly. – Jessica Lipnack & Jeffrey Stamps (Lipnack & Stamps, 2000)

Virtual teaming is a 21st-century survival skill. – Jamie Feild Baker, Executive Director, Martin Institute (Lipnack & Stamps, 2000)

Regardless of their specific tasks, all virtual teams can increase their human, social, and knowledge capital. – Jessica Lipnack & Jeffrey Stamps (Lipnack & Stamps, 2000)

Contemporary Business Teams

A simple definition of "team" is: "Two or more people working together" (http://www.oxforddictionaries.com/definition/english/team) Harvard Professor Amy C. Edmondson has called teaming "the engine of organizational learning" (Starvish, 2012).

This observation is borne out empirically using Google Search. The phrase "importance of teams in business" produced 344,000,000 results in 49 seconds. Yet for all this emphasis, both traditional co-located teams and the more recent virtual teams experience the same fundamental challenges. The antidote is training and more training. We seem to have this conceit that being human we are naturally social. However our entire education system is largely predicated on developing, evaluating, and rewarding students on an individual level. Few youth within the overall student population get to work on teams of any kind, sport, music, volunteer, etc. By they time they graduate into the workplace most young people have had little experience as productive team members. It is therefore no wonder that both traditional and virtual team challenges abound. The antidote is training and even more training.

Traditional Team Challenges

Despite the growing importance of teamwork for organizational learning and success, as the research shows, too many workers are largely out of touch with their emotions and those of others. It stands to reason that this "EQ Gap" will also manifest within teams. This is borne out with the research literature. Indeed, the state of emotional intelligence in co-located project teams is dismal. Some 68% of employees have experienced dysfunctional teams and 81% employees attribute poor communication as the cause for cross-team failure. No wonder that only 24% of employees prefer to work on teams. Moreover, less than a third of employees see their organizations as highly collaborative. This is largely attributed to a lack of time for familiarity or the sharing of personal history (Marketwired, 2011; University of Phoenix, 2013).

Traditional teams do have ways to overcome these persistent challenges including naturally occurring opportunities for members to get to know each other informally. This is generally referred to as 'water cooler' relationship building. In many companies today, physical workspaces are deliberately designed to encourage this kind of casual interaction between employees. For example, Apple's new headquarters as described by the late Steve Jobs: "If a building doesn't encourage [collaboration], you'll lose a lot of innovation and the magic that's sparked by serendipity. So we designed the building to make people get out of their offices and mingle in the central atrium with people they might not otherwise see" (Stepper, 2013).

Virtual Team Challenges

Virtual teams have all the aforementioned challenges of traditional teams but practically none of the advantages. In virtual teaming, there are no readily apparent equivalents to these informal 'water cooler' exchanges. Consequently, "many organizations recycle the same guidelines and best practices they use for co-located teams and hope for the best. Frankly, that just doesn't work. Virtual teams and face-to-face teams are the proverbial 'apples and oranges' (Onpoint Consulting, 2010). Moreover, virtual team experience is more concentrated and accelerated given the short duration of most projects. For example, the classic team development stages of forming and storming tend to now occur simultaneously as well as loop back onto previous stages.

Team EQ Training: Traditional and Virtual

Research and practical experience show that with training, workers can be more emotionally intelligent. Moreover, EQ training can yield positive results with only modest investments in time and capital (https://hbr.org/2013/05/can-you-really-improve-your-em, Tomas Chamorro-Premuzic, May 29 2013). Problematically, less than 33% of companies provide a proper training framework for co-located employees. As well, less than 15% evaluate their training program effectiveness.

This training challenge is even more pronounced regarding virtual teams. 65% of virtual work teams report that they've never had a team building session (Cragan, Wright & Casch, 2009). A recent study found that many virtual teams need special leadership, that trust is essential, that teambuilding exercises pay off, and that, unless a combination of high-tech and high-touch is maintained, performance peaks are often followed by declines in the productivity (Right Management, 2006).

Indeed, those virtual teams who have participated in team building exercises scored significantly higher in leadership, decision-making and team performance. Problematically, usually there is not much time to build (trust) little by little because often the teams are short-lived in projects. Nemiro (2004) identified eleven competencies for effective virtual teamwork. Most are related to social and emotional intelligence.

Table 1.

Virtual Team Challenges
1. Inability to read non-verbal cues
2. Insufficient time to build relationships
3. Absence of collegiality
4. Time zones challenges
5. Difficulty establishing rapport and trust
6. Difficulty seeing the whole picture
7. Difficulty managing conflict
8. Making decisions
9. Sense of isolation
10. Language problems
11. Fear of expressing opinions
12. Poor understanding of other local factors - holidays, laws, customs

Table 2.

Virtual Team Competencies
1. Developing an awareness of yourself and how you interact with others;
2. Developing and practicing supportive communication skills;
3. Building the ability to communicate effectively across cultures;
4. Resolving conflict effectively;
5. Problem solving and decision making skills;
6. Managing stress because virtual work schedules are often 24/7;
7. Time management and personal productivity skills;
8. Developing and motivating others – coaching and empowering;
9. Utilizing positive political skills to push ideas forward;
10. Knowledge management, data gathering, and information access skills; and
11. Developing ways to advance one's career in the virtual workplace.

Ultimately, "virtual teams whose members spend time at the onset of their work getting to know each other, experience greater trust among members down the road, which facilitates the overall effectiveness of their working together" (Martins & Shalley, 2009).

To amplify this observation further, "for every day their organization spends in the storming stage, inefficiencies abound, project deadlines are missed, and financial resources are drained without moving the organization forward...By spending a bit more time in the forming stage (where positive relationships begin), it is possible to actually shorten the amount of time required for the group to pass through the storming stage and move on to the norming and performing stages where real productivity and effort reside".

There's a growing need for more responsive types of online training expressly for virtual team formation and deployment. Our original research has identified twenty best practices developed on an ad hoc basis and drawn from several different fields using virtual teams including business, the military, and healthcare.

Team Mental Models

A team mental model is a representation of the world that individuals working and/or playing together share in common. All teams share mental models, which are generally tacit. Team mental models help anticipate, predict, and coordinate what team members are going to do. This shared cognition also enables team members to respond to change on a cooperative basis. Research indicates that the greater the convergence between the mental models of team members, the greater their effectiveness will be as a team. The ability to perceive the needs of other team members is the result of the overlapping of their mental models. This "produces a mutual awareness, with which team members can reason not only about their own situation, but the status and activities of the other members of the team and progress of the team toward its goal" (Implementing Shared Mental Models for Collaborative Teamwork/ http://agentlab.psu.edu/lab/publications/cast_iat03wkshp.pdf). Conversely, the larger a team is, the lower the likelihood of a shared mental model altogether.

Table 3.

Virtual Team Best Practices
1. Pre-project online activity 2. Engage entire team
3. Use multiplayer games 4. Sub-teams
5. Develop shared language 6. Find or create common symbols 7. Use visuals in communications 8. Practice storytelling 9. Make the invisible visible
10. Pro-active exchange 11. Encourage personal connections 12. Build empathy
13. Leverage "swift trust" (expertise-based) 14. Showcase team member expertise/competence 15. Co-create knowledge for collective repository and team memory 16. Feature annotated and author identified documents and drawings
17. Introduce distributed leadership
18. Provide Intranet site 19. Social-networking features 20. Establish synchronous / asynchronous rhythms

We know the greater convergence of mental models, the better that team's performance. Because virtual teams lack interpersonal clues, like body language, found in traditional live experience, alternative online means are needed. Indeed, virtual teams have an "especially urgent need for visible, explicit, shared models to give meaning to the online world" (Jessica Lipnack & Jeffrey Stamps / NetAge 2008). Visual representations help catalyze and crystalize the development of team mental models and shared understanding. These visual resources facilitate communication within teams as well as contribute to the strengthening of their shared mental models and ultimately to enhancing creative output.

SERIOUS GAMES AT WORK

I think of play as training for the unexpected. – Marc Bekoff, evolutionary biologist (Beckoff, 2012)

To be playful and serious at the same time is ... the ideal mental condition. – John Dewey, How We Think (Dewey, 1910)

Research has shown that play helps people of all ages learn new skills. This has become increasingly important given the 21st Century knowledge economy's focus on life long learning and ongoing professional development (Kücklich, 2014). Moreover, given that the 21st Century has been called the Age of Imagination, play is now enshrined as "the essence of inventive activity. Invention begins in the joyful, free association of the mind" (Root-Bernstein & Root-Bernstein, 2013).

Clark Abt coined the term "serious game" in a book by that name in 1970 (wikipedia.org/wiki/Serious_game#Definition_and_scope). A "serious game" may be defined as any game like activity that has a purpose beyond the traditional play principal. That is, it transcends play for play's sake. However games have been used for serious end goals much longer. War games, for example, have been practiced by militaries around the world for centuries. The original form of chess, developed in India during the Sixth Century C.E, is a case in point.

Since the advent of personal computers in the 1980s, the use of serious games has continued to grow (Wikipedia, 2014). They are seen as vital way to foster team development via emotional engagement, heightened communication, and social interactivity (Ellis, Luther, Bessiere & Kelogg, 2008). Serious online games are now offered in many fields. Simulation games, for example, are extensively used for training in medicine, disaster relief, the military, and other high-risk, high stake work. Commercial pilots, for example, are strapped into a box that replicates the inside of a cockpit and every conceivable flying condition. This also now includes 3D virtual world related games such as the Sims (www.thesims.com).

"Gamification" is a recent term that refers to the use of game dynamics to engage players – employees, customers, and/or students – to help realize their varied objectives. (https://badgeville.com/wiki/Gamification). Another example of this trend's ubiquity is that mobile handheld devices now enable players to access virtual games practically anywhere as well as 24 hours a day. Elearning in gamification form is also highly economical. For example, according to Deloitte's Leadership Academy, face-to face instruction is about $50 US per employee but only $5 US per employee online (Virtual College, 2014).

CREATIVITY AT WORK

Play, Games, and Creativity

As more emphasis in business and industry is put on the need for creativity and innovation, the boundary between work, play, and games has become increasingly fine. Play is at the root of all games. It's also indispensible for creative thought and ultimately invention and innovation. "When rule-bound work does not yield the insights or results we want to achieve, when conventional thought, behavior, and disciplinary knowledge become barriers to our goals, play provides a fun and risk-free means of seeing from a fresh perspective, learning without constraint, exploring without fear" (Root-Bernstein & Root-Bernstein, 2013). Of course, many adults think play is a child's pastime. Therefore they are often at a disadvantage when creativity and innovation are called for. The subtext is that one cannot play, have fun, or be inventive if one is fearful, stressed, or without feeling trust in oneself, others, and one's surroundings. Research shows that as we close down emotionally our reptilian brain comes to the fore manifesting in either a flight or flight response. Notwithstanding typical adult self-consciousness about play and creativity, a 2010 IBM Global Study, 60% of CEOs said that creativity would be a top leadership trait in the ensuing 5 years.

Traditional Teams and Creativity

Dr. Edward de Bono is ne of the foremost pioneers in fostering creative thinking in the workplace. His classic 1970 book "Lateral Thinking: Creativity Step by Step " set the stage for a new level of serious consideration by senior management of this historically undervalued cognitive faculty. In 1985 he wrote that: "Creative thinking…is not a mystical talent. It is a skill that can be practised and nurtured" (Six

Thinking Hats, p. 99). The growth of creativity training for teams has been staggering. Another Google search yielded 137,000,000 results in 54 seconds for the phrase "creativity training for teams". Typical is Stanford University's "Crash Course on Creativity". "This crash course is designed to explore several factors that stimulate and inhibit creativity in individuals, teams, and organizations. In each session we will focus on a different variable related to creativity, such as framing problems, challenging assumptions, and creative teams" (http://online.stanford.edu/creativity-fa12).

Virtual Teams and Creativity

Dr. Jill Nemiro first examined the role of creativity in virtual teams in her 2004 seminal text "Creativity in Virtual Teams: Key Components for Success". In the eleven years since then, the literature has grown somewhat. It consists largely of company white papers and case studies as well as academic research. This literature generally examines how creativity may be used to help virtual teams develop trust. The findings tend to generally agree that emotional intelligence, or EQ, promotes team trust. Moreover, without trust team creativity is constrained (Barczak, Lassk & Mulki, 2010). Trust and creativity, in turn, contribute to engagement, wellbeing and self-actualization for team m (Barczak, Lassk & Mulki, 2010).

However, there is still little research specifically on how virtual teams can perform more effectively through the use of creativity (Kisielnicki, 2008). Indeed, very little is still really known about creativity and collaboration within virtual teams, (Jonti, 2013). There is a significant need for more studies in this area altogether (Letaief, Favier & Le Coat, 2006).

There is also a second research literature theme looks at how creativity is used by virtual teams to innovate new methods, processes, products and services. However, there is much less research in this area. We do know that visual representations help catalyze and crystalize the development of team mental models and shared understanding. Such visual representations, now often digital, consist of appropriated and/or self-generated images, diagrams, photographs, and sketches. These visual resources facilitate communication within teams as well as contribute to the strengthening of their shared mental models and ultimately to enhancing creative output. This is particularly evident in creative fields such as design, film, advertising, and architecture where team-based visual thinking and expression is paramount (https://www.academia.edu/8147519/The_Psychology_of_Creativity_Mental_Models_in_Design_Teams). Ultimately, "a design team sharing a mental model from the beginning has better chances to perform better, and reach a larger number of innovative solutions" (ibid).

Assessing Virtual Team Creative Climate

There is an instrument to assess the creative climate of a virtual team. The Virtual Team Creative Climate is a survey developed by Jill Nemiro in 2004. This includes five components of creativity: design, climate, resources, norms and protocols, and continual assessment and learning. There are also eleven dimensions that may affect the creativity of a virtual team: "acceptance of ideas and constructive tension; challenge; collaboration; dedication commitment; freedom; goal clarity; information sharing; management encouragement; personal bond; sufficient resources and time; and trust" (Nemiro, 2004). These eleven dimensions are then aggregated into three broad categories: connection, raw materials, and management and team member skills, VTCC survey findings are depicted in a graphic feedback display. The instrument is helpful for suggesting areas of improvement for virtual team creativity.

Virtual Team Creative Exercises

Over the course of the past decade a variety of creative activities have been developed to help virtual teams become more trusting, cohesive, and productive. Many of these activities are listed as virtual team best practices in the Table featured in the previous section of this chapter.

These activities are designed to help teams to develop a shared language and common identity, or conscious mental model, impossible with only text or voice based communication. The creative artifacts produced enable virtual teams to make visible the invisible ((Kimball, 1999). This reflects the old saying that a picture is worth a thousand words. "Because intuition emerges out of images and symbols better than it emerges out of words, drawing is an effective way to pull out intuitive creativity" (Nemiro, 2004).

Lisa Kimball of Groupjazz.com also suggests several generic strategies for artifact creation. This includes creating a "signifier of the team" such as a co-created, shared image of "what team success looks and feels like." The goal is to make this visible to everyone in that environment for shared understanding. This will convey the sense that ideas are coming from the whole team rather than a single member. This exercise also helps put the process in a team context.

The 2011 "Big Book of Virtual Teambuilding Games: Quick, Effective Activities to Build Communication, Trust and Collaboration from Anywhere" by Mary Scannell also offers a range of exercises for virtual teams at each stage of development, forming, storming, norming, and performing. For more examples of virtual team exercises using creativity to foster trust please see the Appendix.

PRELUDE VIRTUAL: A CASE STUDY

The following introduces a serious game called Prelude that fosters trust prior to the start of a new project. Virtual teams are guided through five sequential 30-60 minute activities, synchronous and asynchronous, depending on the number of participants. The game provides virtual teams with a unique opportunity for self and interpersonal exploration through the first activity called iStar, which is proprietary psychometric. In the creative process that follows, players create and co-create a series of artifacts on an individual and team basis called iTags, weTags, and allTags respectively. Prelude helps 'draw out' the best from individuals and teams, literally and metaphorically. Participants better understand each other. They learn to identify and harness their diverse assets in a cohesive, purposeful way. This occurs during the forming and storming phases of the team's development. Not only does this experiential process reduce the tensions inherent in these initial phases, it also helps enhance the team's mental model explicitly and positively. Prelude is a good practical example of the concepts and themes introduced throughout the chapter.

Figure 1. Prelude game process by author H. B. Esbin PhD

Best Practices

Twenty emerging best practices have been identified drawn from several sectors using virtual teams. They involve revealing and engaging the "whole" person and team emotionally in an integral way at the very start of a project. These also share common characteristics involving the transformative power of serious play, games, and creativity. These are outlined below along with an explanation as to how Prelude incorporates each within a seamless integrated learning experience.

1. Prior to the main project, offer non-project related online activity.
2. Engage entire team in online activity.

Prelude is an online interactive game for virtual teams played prior to a new project. The entire team is guided through a series of five short activities, synchronous and asynchronous. This helps naturally accelerate team rapport and trust.

3. Use multiplayer games; breaking teams into sub-teams.

Figure 2. Prelude game process by author H. B. Esbin PhD

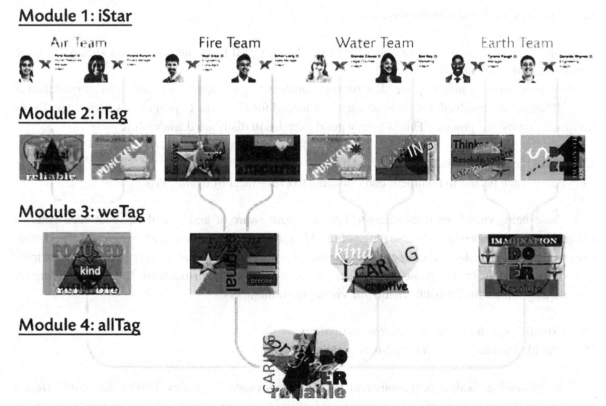

323

Prelude is a multiplayer game involving individual, sub team, and whole group activities.

4.   Teams need to develop a shared language.
5.   Find or create common symbols and rituals to co-create a new culture.
6.   Use visuals in communications.
7.   Practice art of storytelling.

"Virtual teams have an especially urgent need for visible, explicit, shared models to give meaning to the online world" (Lipnack & Stamps, 2008). Prelude helps virtual teams draw together, literally and metaphorically. This involves players using visual communication and storytelling. The digital symbols they generate - iStars, iTags, weTags, and allTags – become powerful markers of identity and cohesion. This contributes to developing a positive team mental model. As Rao observes: "Symbols, team names and logos help give teams a tangible identity to which they "belong"" (2014).

8.   Encourage personal relations.
9.   pro-actively build interpersonal connections.
10. Enable staff to learn about each other.
11. Help build empathy.

Prelude is a simple, powerful, cost effective way to pro-actively build interpersonal relations, learn about each other, encourage personal connections, and develop empathy.

12. Introduce distributed leadership.

Distributed leadership is defined as something shared by all team members, "not simply from the appointed leader" (Pearce & Sims, 2001).
Prelude provides a safe way for all virtual teammates to participate as equals. A key guideline is that 'all hands' are involved and that consensus is needed for the game to progress at each stage. Every teammate 'owns' the process. This is a very good exercise in distributed leadership.

13. Make the invisible visible to team itself.
14. Allow each person to comment on/contribute to every facet of overall task.

Prelude helps virtual team members to become more aware of and articulate about each other as integral human beings with a diverse range of thinking styles, skills, talents, and experience. This sensitizes the team as a whole to its collective strengths. The entire process allows every person to participate and contribute to the overall game result. These are shared online in virtual team galleries. In doing so, Prelude helps make the invisible visible for virtual team members.

15. Establish synchronous / asynchronous rhythms.
16. Provide Intranet site with social-networking features.

Prelude involves both synchronous and asynchronous player activities. Playprelude.com offers a safe game portal with range of social networking features to ensure a positive team learning experience.

17. Leverage "swift trust" (expertise-based).
18. Showcase team member expertise/competence.
19. Co-create knowledge for collective repository and team memory.
20. Feature annotated and author identified documents and drawings.

All sports teams and orchestras practice and tune up prior to the main event, be it soccer or symphony. However, the research literature shows virtual teams are not usually afforded this opportunity. Virtual strangers are simply expected to become high performing teams cold. Playing Prelude is like a project rehearsal in game form. It provides a simple, safe, and positive way for team members to learn about each other's character and expertise through a unique combination of psychometrics and creative process. Each virtual team has its own private game space and galleries for sharing. Post game this online space continues to be a rich area for virtual team identity, memory, learning, and online game co-created artifacts.

Background

Prelude's interdisciplinary design is informed by my doctoral research combined with over 25 years of senior management experience in the private sector, international development and philanthropy (Esbin, 1998). Its universality of application is due to an underlying holistic philosophy and pedagogy, which is perennial. Prelude paradoxically illustrates how, "the newest digital technologies are returning us to the most ancient form of media — one in which a natural order is restored" (Cohen, 2007).

Prelude was originally developed as a blended learning resource for in traditional classrooms and workshops. To date over 25000 players, from age 12 to 60 plus, have played it in very diverse cultures and contexts including middle schools, high schools, alternative schools, independent programs such as drop out prevention, life-work skills, and pro-social skills, aggression replacement training, prison youth gang reclamation, youth entrepreneurism, and colleges. This also includes Aboriginal youth programs and Arab education in the Persian Gulf. In some institutions the entire staff uses Prelude for their own professional development prior to using it with students.

In 2012, the Software Information Industry Association [SIIA] chose Prelude as one of that year's most "relevant and innovative education tech products". This same year, Prelude was also awarded the Seal of Quality from Curriculum Services Canada, a national quality assurance organization. In 2012 the blended learning game was further adapted for use in business training and professional development.

Prelude Virtual was piloted from May to December 2014 with a dozen companies around the world. In February 2015 a new game version was launched incorporating the invaluable feedback from these company pilots.

CONCLUSION

Over the course of two decades, the use of virtual teams has continued to grow, as has the literature on theory and practice. Throughout we've seen two consistent refrains from thought leaders and practitioners. First, virtual teams have unique interpersonal challenges. Second, technology alone is not a solution. High tech must be complemented by "high touch". What's called for therefore is a purposeful, committed focus on enhancing the social and emotional skills of virtual team leaders and members. This is a crucial prerequisite for better trust, which has been called 'human glue' (Nemiro, 2004).

"Allowing time to play games, share humor, or respond to one another's personal issues and crises, perhaps thought of as inappropriate behavior in conventional teams, may not be inappropriate at all in the virtual world of work" (Nemiro, 2004). Accordingly, in this chapter we examined how virtual team trust and effectiveness may be improved through the transformative power of serious games and creative process. To start we explored the pervasive lack of emotional intelligence within the workplace at an individual level and which we call 'the EQ Gap'. This was followed by an examination of challenges faced by both traditional and virtual teams. We then considered how the same EQ Gap also manifests in both traditional and virtual teams as well. This led to a review of the kinds of EQ training needed for both team types. A discussion then followed as to how serious games, play, and creativity can help virtual teams in particular to become more emotionally intelligent, trusting, and ultimately more collaborative. A brief case study of a serious game called Prelude was shared to illustrate these findings in a practical context.

In conclusion, with over one and a half billion people now working remotely, demand will certainly continue to grow for innovative resources to support more effective virtual teaming. Gaming and creativity are proven methods and especially powerful when used in tandem.

REFERENCES

Appelo, J. (2014). *How identifying symbols can enhance team culture*. Retrieved from http://switchand-shift.com/how-identifying-symbols-can-enhance-team-culture

Babcock, C. (2007). *Chambers sees virtual teams competing with virtualized resources*. Retrieved from http://www.informationweek.com/chambers-sees-virtual-teams-competing-with-virtualized-resources/d/d-id/1059100

Barczak, G., Lassk, F., & Mulki, J. (2010). Antecedents of team creativity: An examination of team emotional intelligence, team trust and collaborative culture. *Creativity and Innovation Management, 19*(4), 332–345. doi:10.1111/j.1467-8691.2010.00574.x

Bateson, G. (1967). *A theory of play and fantasy*. Bobbs-Merrill.

Beckoff, M. (2012). *Article*. Retrieved from http://www.psychologytoday.com/blog/animal-emotions/201202/the-need-wild-play-let-children-be-the-animals-they-need-be

Bhagwatwar, A., Massey, A., & Dennis, A. R. (2013). *Creative virtual environments: Effect of supraliminal priming on team brainstorming*. Paper presented at the 46th Hawai International Conference on System Sciences. Retrieved from http://scholar.google.com/citations?view_op=view_citation&hl=es&user=qh2OnA4AAAAJ&citation_for_view=qh2OnA4AAAAJ:W7OEmFMy1HYC

Bodiya, A. (2013). *Virtual reality: the impact of task interdependence and task structure on virtual team productivity and creativity*. (Doctoral Dissertation). Alliant International University, Los Angeles, CA. Retrieved from ProQuest (UMI 3452402).

Bradberry, T., & Greaves, J. (2009). *Emotional Intelligence 2.0*. Talentsmart. Retrieved from http://www.talentsmart.com/products/emotional-intelligence-2.0/

Burt, J. (2013). *Dell wants half of employees working remotely by 2020.* Retrieved from http://www. eweek.com/mobile/dell-wants-half-of-employees-working-remotely-by-2020.html

Center for Creative Leadership. (2012). *White paper: Prelude at work.* Retrieved from http://www. heliotrope.ca/wp-content/uploads/2012/10/White-Paper-Prelude-At-Work.pdf

Chamarro-Premuzic, T. (2013). *Can you really improve your team?* https://hbr.org/2013/05/can-you-really-improve-your-em

Cherniss, C. (2000). *Emotional intelligence: What it is and why it matters.* Retrieved from http://www. eiconsortium.org/reports/what_is_emotional_intelligence.html

Cohen, J. (2007). *Article.* Retrieved from http://blog.ted.com/2007/03/09/ted2007_day_two/

Cragan, J. F., Wright, D. W., & Casch, C. R. (2009). *Communication in small groups.* Bradley University / Cengage Learning.

Dachis Group. (2013). *Enabling virtual teams.* Retrieved from http://cdn2.hubspot.net/hub/178806/file-41102699-pdf/Enabling_Virtual_Teams.pdf

DeRosa, D., & Lepsinger, R. (2010). *Virtual team success: A practical guide for working and leading from a distance.* San Francisco: Jossey-Bass/A Wiley Imprint.

Dewey, J. (1910). *How we think.* Lexington, Mass: D.C. Heath. doi:10.1037/10903-000

Dragusha, C. (2012). *Managing virtual teams: guidelines to effective leadership.* Retrieved from https://www.theseus.fi/bitstream/handle/10024/50321/dragusha_cajup%20thesis.pdf?sequence=1

Dye, L. (2009, October 21). Why is your boss a bully? *ABC News.* Retrieved from http://abcnews.go.com/Technology/DyeHard/boss-bully/story?id=8872662

Ellis, J. B., Luther, K., Bessiere, K., & Kellogg, W. A. (2008). *Games for virtual team building.* Retrieved from http://pdf.aminer.org/000/248/298/vicious_and_virtuous_cycles_in_global_virtual_team_role_coordination.pdf

Equichord. (2014). *Merging the ability to learn with a virtual world.* Retrieved from http://equichord.com/merging-the-ability-to-learn-with-a-virtual-world/

Esbin, H. B. (1998). *Carving lives from stone.* (Doctoral Thesis). McGill University, Canada. Retrieved from http://www.mosaichub.com/member/p/howard-b-esbin-phd

Goleman, D. (1995). *Emotional intelligence.* Bantam Books.

Gratton, L., & Erickson, T. J. (2007). *Eight ways to build collaborative teams.* Retrieved from https://hbr.org/2007/11/eight-ways-to-build-collaborative-teams/

Henricks, T. S. (2010). Callois´ Man, Play, and Games. An appreciation and evaluation. *American Journal of Play*, 157–185.

Huizinga, J. (1949). Homo ludens: A study of the play-element in culture. London: Routledge & Kegan Paul.

Immen, W. (2009, August 23). Power has its own price. *The Globe and Mail*. Retrieved from http://www.theglobeandmail.com/report-on-business/power-has-its-own-price/article4215325/

Johns, T., & Gratton, L. (2013). *The third wave of virtual work*. Retrieved from https://hbr.org/2013/01/the-third-wave-of-virtual-work

Jones, T. L. (2009). *Virtual team communication and collaboration in army and corporate applications*. (Master's thesis). Fort Leavenworth, Kansas. Retrieved from file:///C:/Users/m/Downloads/ADA502091.pdf

Jonti, D. (2013). *How does creativity work in a highly virtual environment?* Retrieved from https://www.linkedin.com/groups/How-does-creativity-work-in-3980060.S.260012070

Kimball, L. (1999). *The virtual team: Strategies to optimize performance*. Retrieved from http://www.groupjazz.com/pdf/opt-perf.pdf

Kisielnicki, J. (2008). *Virtual technologies: Concepts, methodologies, tools, and applications*. Hershey, PA: Information Science Reference / IGI Global. doi:10.4018/978-1-59904-955-7

Kratzer, J., Leenders, R., & Van Engelen, J. (2006). Managing creative team performance in virtual environments: An empirical study in 44 R&D teams. *Technovation*, *26*(1), 42–49. doi:10.1016/j.technovation.2004.07.016

Kücklich, J. (2014). *Play and playability as key concepts in new media studies*. Retrieved from http://es.scribd.com/doc/205038181/Play-and-Playability-as-Key-Concepts-in-New-Media-Studies#scribd

Ladimeji, K. (2014). *Why you should hire candidates for EQ rather than IQ*. Retrieved from https://www.recruiter.com/i/why-you-should-hire-candidates-for-eq-rather-than-iq/

Larsson, A., Larsson, T., Bylund, N., & Isaksson, O. (2007). Rethinking virtual teams for streamlined development. In S. P. MacGregor & T. Torres-Coronas (Eds.), *Higher creativity for virtual teams: Developing platforms for co-creation* (pp. 138–156). Hershey, PA: Information Science Reference. doi:10.4018/978-1-59904-129-2.ch007

Lepsinger, R. (2014). Virtual team failure: Six common reason why virtual teams do not succeed. *Business Know-How*. Retrieved from http://www.businessknowhow.com/manage/virtualteam.htm

Letaief, R., Favier, M., & Le Coat, F. (2006). *Creativity and the creation process in global virtual teams: Case study of the intercultural virtual project*. DBLP.

Lipnack, J., & Stamps, J. (2000). *Virtual teams: People working across boundaries with technology*. New York: John Wiley & Co.

Mahajan, N. (2014). *The importance of creativity in business*. Retrieved from http://knowledge.ckgsb.edu.cn/2014/08/13/marketing/the-importance-of-creativity-in-business/

Marantz Henig, R. (2008, February 17). *The New York Times*. Retrieved from http://www.nytimes.com/2008/02/17/magazine/17play.html?pagewanted=all&_r=0

Marketwired. (2011). *Canadian study reveals lack of team collaboration despite its positive business impact*. Retrieved from http://www.marketwired.com/press-release/canadian-study-reveals-lack-team-collaboration-despite-its-positive-business-impact-lse-inf-1581172.htm

Martins, L. L., & Shalley, C. E. (2009). *Creativity in virtual work: Effects of demographic differences*. Retrieved from http://icos.umich.edu/sites/icos6.cms.si.umich.edu/files/lectures/MartinsShalley.pdf

Martins, L. L., Shalley, C. E., & Gilson, L. L. (2009). Virtual teams and creative performance. *Proceedings of the (42th) Annual Hawaii International Conference on Systems Science*.

Mental Health Roundtable. (2011). Retrieved from http://www.mentalhealthroundtable.ca/

Nemiro, J. E. (2001). Assessing the climate for creativity in virtual teams. In M. M. Beyerlein, D. A. Johnson, & S. T. Beyerlein (Eds.), *Virtual teams. Advances in interdisciplinary studies of work teams* (Vol. 8, pp. 59–84). Emerald Group Publishing Limited. doi:10.1016/S1572-0977(01)08019-0

Nemiro, J. E. (2004). *Creativity in virtual teams: Key components for success*. New York: John Wiley & Co.

Online Etymology Dictionary. (2014). Retrieved from http://www.etymonline.com/

Onpoint Consulting. (2010). *Six lessons for effective virtual teams: A recipe for success*. Retrieved from http://www.onpointconsultingllc.com/2010/05/six-lessons-for-effective-virtual-teams-a-recipe-for-success/

Oxford Dictionary Online. (2015). Retrieved from http://www.oxforddictionaries.com/definition/english/team

Pearce, C. L., & Sims, H. P. (2001). Shared leadership: Toward a multi-level theory of leadership. *Advances in Interdisciplinary Studies of Work Teams*, 7, 115–139. doi:10.1016/S1572-0977(00)07008-4

ECB Project. (2011). *What we know about joint evaluations*. Retrieved from file:///C:/Users/m/Downloads/What%20We%20Know%20About%20Joint%20Evaluations%20April%202011%20Section3.pdf

Psychometrics Canada. (2009). *Warring egos, toxic individuals, feeble leadership. A study of conflict in the Canadian workplace*. Retrieved from http://www.psychometrics.com/docs/conflict%20study%20news%20release%20feb%202009.pdf

Rao, A. (2014). *Constructing a team identity*. Retrieved from http://www.selfgrowth.com/articles/constructing-a-team-identity

Right Management. (2006). *Virtual teaming study*. Retrieved from http://www.right.com/global/includes/pdfs/Virtual_Teaming_Study.pdf

Root-Bernstein, R. S., & Root-Bernstein, M. M. (2013). *Sparks of genius: The thirteen thinking tools of the world's most creative people*. Houghton Mifflin Harcourt.

Siebdrat, F., Hoegl, M., & Ernst, H. (2009). *How to manage virtual teams*. Retrieved from http://sloanreview.mit.edu/article/how-to-manage-virtual-teams/

Snyderman, N. (2009). *Article*. Retrieved from http://www.huffingtonpost.com/living/the-blog/featured-posts/2009/11/11/

Stahl, G., & Hesse, F. (2007). Welcome to the future. *International Journal of Computer-Supported Collaborative Learning, 2*(1). Retrieved from http://ijcscl.org/?go=contents&article=24

Starvish, M. (2012). *Why leaders need to rethink teamwork*. Retrieved from http://www.forbes.com/sites/hbsworkingknowledge/2012/12/28/why-leaders-need-to-rethink-teamwork/

Stepper, J. (2013). *The best office design for collaboration is also the cheapest*. Retrieved from http://johnstepper.com/2013/02/23/the-best-office-design-for-collaboration-is-also-the-cheapest/

The Virtual Leader. (2013). *Using telework to lean out the enterprise*. Retrieved from https://thevirtual-leader.wordpress.com/2013/07/02/using-telework-to-lean-out-the-enterprise/

University of Phoenix. (2011). *University of Phoenix survey reveals nearly seven-in-ten workers have been part of dysfunctional teams*. Retrieved from http://www.phoenix.edu/news/releases/2013/01/university-of-phoenix-survey-reveals-nearly-seven-in-ten-workers-have-been-part-of-dysfunctional-teams.html

Virtual College. (2014). *Most global firms to use gamification in e-learning by 2014*. Retrieved from http://www.virtual-college.co.uk/news/Most-global-firms-to-use-gamification-in-elearning-by-2014-newsitems-801528054.aspx

Watkins, M. (2013). *Making virtual teams work: Ten basic principles*. Retrieved from https://hbr.org/2013/06/making-virtual-teams-work-ten/

Wikipedia. (2015). *Serious Games*. Retrieved from wikipedia.org/wiki/Serious_game#Definition_and_scope

Wood, & Associates. (2003). *Why emotional intelligence matters on the job*. Retrieved from http://www.woodassociates.net/toi/Search/PDF/Why%20Emotional%20Intelligence%20Matters%20on%20the%20Job.pdf

Yen, J., Fan, X., Sun, S., Want, R., Chen, C., Kamali, K., & Volz, R. A. (2003). *Implementing Shared Mental Models for Collaborative Teamwork*. Retrieved from http://agentlab.psu.edu/lab/publications/cast_iat03wkshp.pdf

APPENDIX: EXERCISES USING CREATIVITY TO FOSTER VIRTUAL TEAM TRUST

Mathematics Virtual Team

The integration of graphical, narrative, and symbolic semiotic modalities facilitates joint problem solving. It allows group members to invoke and operate with multiple realizations through the creation of their artifacts. (Stahl & Hesse, 2007)

Draw A Picture

Draw a picture that represents your view of your team's current reality (or your current life situation). Use whatever images come to mind to describe the current situation Draw a picture that represents what you feel the future holds for your team or what you would ideally like to see for the team or for yourself. Use whatever images come to mind to describe your future or ideal situation. 3. Examine the two sets of drawings. What have you learned about your current and ideal team reality (or life situation)? Record the insights that emerge from your reflections. 4. Now consider what you might do to make that second drawing more of a reality in the future (Nemiro, 2004).

Caricature Match Game

Procedure: Have team members draw a picture to depict themselves without using any words—a picture that describes them in any creative manner. It can be a sketch of themselves, their hobbies, interests, family, or any and all of the above. Have everyone scan their drawing and email it to the facilitator or game leader. After collecting the pictures, the leader can compile them and post them on the team website creating a "Team Self-Portrait Gallery." The goal is to see who can guess which drawing goes with which team member (Big Book of Virtual Teambuilding Games, 2011),

Totem Game

The IVP Project conducted among twenty-six universities distributed throughout fifteen countries. The teamwork consists in the elaboration of a 'totem' and a descriptive draft. The 'totem' is a complex symbol that is created to represent the team members' identity. The descriptive draft is a two-page document compiled by each GVT in order to describe its totem and to facilitate its comprehension. The total duration of the project is eight weeks. Four weeks are devoted to the elaboration of a totem and its description. Each GVT is made up of students belonging to at least three different universities. A calendar is proposed to the students in order to help them get organized amongst themselves and to respect the deadlines. Each GVT use a discussion forum to communicate and share ideas and files. Although the members are free to post their messages, the GVT members belonging to the same university must post at least one message twice a week. Two different sessions are organized annually, one in the Fall and one in the Spring. Our particular study sample is composed of twenty-five teams from a given session. About fifteen individual members compose each team. Ten universities from 7 countries participated in this particular IVP session (Letaief, Favier & Le Coat, 2006).

The Artist Game

Participants have a few minutes to go to the Web and find an image or screenshot that best conveys who they are and then emails it to the facilitator. The facilitator can then flash each image on the screen one at a time and participants try to guess who it relates to. Each person then has a chance to explain why they chose that image (The Virtual Leader, 2013, http://thevirtualleader.wordpress.com).

Compilation of References

"Ted" Weston, F.C., Jr. (2002). *A vision for the future of extended enterprise systems.* Presentation, J.D. Edwards FOCUS Users Conference, Denver, CO.

"Ted" Weston, F.C., Jr. (2003, November/December). ERPII: The extended enterprise system. *Business Horizons,* 49-55.

Aart, C. J., Wielinga, B., & Schreiber, G. (2004). Organizational building blocks for design of distributed intelligent system. *International Journal of Human-Computer Studies, 61*(5), 567–599. doi:10.1016/j.ijhcs.2004.03.001

Abbas, Hussein H, & El, Abd El Halim Omar Abd. (2015). The Role of Suez Canal Development in Logistics Chain. In *Global Supply Chain Security* (pp. 163-180). Springer.

Abdalla, I. A., & Al-Homoud, M. A. (2001). Exploring the implicit leadership theory in the Arabian Gulf states. *Applied Psychology, 50*(4), 506–531. doi:10.1111/1464-0597.00071

Achrol, R., & Kotler, P. (1999). Marketing in the network economy. *Journal of Marketing, 63*(Special Issue), 146–163. doi:10.2307/1252108

Adzei, F. A., & Atinga, R. A. (2012). Motivation and retention of health workers in Ghana's district hospitals: Addressing the critical issues. *Journal of Health Organization and Management, 26*(4), 467–485. doi:10.1108/14777261211251535 PMID:23115900

Ahuja, J. (2010). A study of virtuality impact on team performance. *The IUP Journal of Management Research, 9*(5), 27–56.

Ahuja, M., & Galvin, J. (2003). Socialization in virtual groups. *Journal of Management, 29*(2), 161–185. doi:10.1177/014920630302900203

Akinci, S., Aksoy, S., & Atilgan, E. (2004). Adoption of internet banking among sophisticated consumer segments in an advanced developing country. *International Journal of Bank Marketing, 22*(3), 212–232. doi:10.1108/02652320410530322

Akkermans, H., Bogerd, P., Yucesan, E., & Van Wassenhove, L. (2003). The impact of ERP on supply chain management: Exploratory findings from a European Delphi Study. *European Journal of Operational Research, 146*(2), 284–294. doi:10.1016/S0377-2217(02)00550-7

Aldea, C. C., Popescu, A. D., Draghici, A., & Draghici, G. (2012). ICT tools functionalities analysis for the decision making process of their implementation in virtual engineering teams. *Procedia Technology, 5,* 649–658. doi:10.1016/j.protcy.2012.09.072

Alexander, D. E. (2002). *Principles of emergency planning and management.* Oxford University Press.

Allameh, S. M., Momeni, Z. M., Esfahani, Z. S., & Bardeh, M. K. (2011). An assessment of the effect of information communication technology on human resource productivity of Mobarekeh Steel Complex in Isfahan (IRAN). *Procedia Computer Science, 3,* 1321–1326. doi:10.1016/j.procs.2011.01.010

Al-Mashari, M., Al-Mudimigh, A., & Zairi, M. (2003). Enterprise resource planning: A taxonomy of critical factors. *European Journal of Operational Research*, *146*(2), 352–364. doi:10.1016/S0377-2217(02)00554-4

Al-Mudimigh, A., Zairi, M., & Al-Mashari, M. (2001). ERP software implementation: An integrative framework. *European Journal of Information Systems*, *10*(4), 216–226. doi:10.1057/palgrave.ejis.3000406

alternativeTo. (2014). *Unfuddle.* Retrieved November 20, 2014 from: http://alternativeto.net/software/unfuddle/

Anderson, A. H., McEwan, R., Bal, J., & Carletta, J. (2007). Virtual team meetings: An analysis of communication and context. *Computers in Human Behavior*, *23*(5), 2558–2580. doi:10.1016/j.chb.2007.01.001

Anderson, E., & Weitz, B. (1986). spring). Make-or-buy decisions: Vertical integration and marketing productivity. *Sloan Management Review*, *27*(3), 3–19.

Anderson, G. L. (1998). Toward authentic participation: Deconstructing the discourses of participatory reforms in education. *American Educational Research Journal*, *35*(4), 571–603. doi:10.3102/00028312035004571

Andrews, D., Conway, J., Dawson, M., Lewis, M., McMaster, J., Morgan, A., & Starr, H. (2004). *School revitalization: The IDEAS way (ACEA Monograph Series No. 34).* Winmalee, Australia: Australian Council for Educational Leaders.

Ann & Lee. (2005). Utilizing knowledge context in virtual collaborative works. *DSS Journal*, 35.

Anna et al. (2004). Building a project ontology with extreme collaboration and virtual design &construction. *Advanced Engineering Informatics*, 71-83.

Ann, D., Lenore, N., & Chris, L. (2010). Facilitating transdisciplinary sustainable development research teams through online collaboration. *International Journal of Sustainability in Higher Education*, *11*(1), 36–48. doi:10.1108/14676371011010039

Anthony, T., DeMarie, S., & Anthony, H. (1998). Virtual teams: Technology and the workplace of the future. *The Academy of Management Perspectives*, *12*(3), 17–29. doi:10.5465/AME.1998.1109047

Anussornnitisam, P., & Nof, S. Y. (2003). E-work: The challenge for next generation ERP systems. *Production Planning and Control*, *14*(8), 753–765. doi:10.1080/09537280310001647931

Appelo, J. (2014). *How identifying symbols can enhance team culture.* Retrieved from http://switchandshift.com/how-identifying-symbols-can-enhance-team-culture

Archak, N. (2010). Money, Glory and Cheap Talk: Analyzing Strategic Behavior of Contestants in Simultaneous Crowdsourcing Contests on TopCoder.com. In *Proceedings of WWW 2010.* Raleigh, NC: ACM. doi:10.1145/1772690.1772694

Ardichvili, A., Page, V., & Wentling, T. (2003). Motivation and barriers to participation in virtual knowledge-sharing communities of practice. *Journal of Knowledge Management*, *7*(1), 64–77. doi:10.1108/13673270310463626

Argyres, N. S. (1996). Capabilities technological diversification and divisionalization. *Strategic Management Journal*, *17*(5), 395–410. doi:10.1002/(SICI)1097-0266(199605)17:5<395::AID-SMJ826>3.0.CO;2-E

Arnison, L., & Miller, P. (2002). Virtual teams: A virtue for the conventional team. *Journal of Workplace Learning*, *14*(4), 210427294. doi:10.1108/13665620210427294

Arvey, R. (2009). *Why face-to-face business meetings matter.* Retrieved November 20, 2014, from: http://www.iacconline.org/content/files/WhyFace-to-FaceBusinessMeetingsMatter.pdf

Arya, A., & Mittendorf, B. (2008). Pricing internal trade to get a leg up on external rivals. *Journal of Economics & Management Strategy*, *17*(3), 709–731. doi:10.1111/j.1530-9134.2008.00192.x

Asariotis, R., Benamara, H., Finkenbrink, H., Hoffmann, J., Lavelle, J., Misovicova, M., . . . Youssef, F. (2011). *Review of Maritime Transport, 2011*. Academic Press.

Atkins, E. T. (2001). Blue Nippon: authenticating jazz in Japan. Duke University Press.

Avison, D. E., Baskerville, R., & Myers, M. (2001). Controlling action research projects. *Technology & People, 14*(1), 28–45. doi:10.1108/09593840110384762

Avolio, B. J., & Kahai, S. (2003). Adding the "e" to e-leadership: How it may impact your leadership. *Organizational Dynamics, 31*(4).

Avolio, B. J., Kahai, S., & Dodge, G. E. (2001). E-Leadership: Implications for theory, research, and practice. *The Leadership Quarterly, 11*(4), 615–668. doi:10.1016/S1048-9843(00)00062-X

Avolio, B., Walumbwa, F. O., & Weber, T. J. (2009). Leadership: Current theories, research, and future directions. *Annual Review of Psychology, 60*(1), 421–449. doi:10.1146/annurev.psych.60.110707.163621 PMID:18651820

Babcock, C. (2007). *Chambers sees virtual teams competing with virtualized resources*. Retrieved from http://www.informationweek.com/chambers-sees-virtual-teams-competing-with-virtualized-resources/d/d-id/1059100

Bacelli, F., & Lin, Z. (1992). Compresion properties of stochastic decision free Petri nets. *IEEE Transactions on Automatic Control, 37*(12), 1905–1920. doi:10.1109/9.182477

Bagchi, S., Kanungo, S., & Dasgupta, S. (2003). Modeling use of enterprise resource planning systems: A path analytic study. *European Journal of Information Systems, 12*(2), 142–158. doi:10.1057/palgrave.ejis.3000453

Bailenson, J. N., Yee, N., Merget, D., & Schroeder, R. (2006). The effect of behavioral realism and form realism of real-time avatar faces on verbal disclosure, nonverbal disclosure, emotion recognition, and copresence in dyadic interaction. *Presence (Cambridge, Mass.), 15*(4), 359–372. doi:10.1162/pres.15.4.359

Banker, R. D., Chang, H., & Kao, Y. (2010). Evaluating cross-organizational impacts of information technology – an empirical analysis. *European Journal of Information Systems, 19*(2), 153–167. doi:10.1057/ejis.2010.9

Barabasi, A. L., & Oltvai, Z. N. (2004). Network biology: Understanding the cell's functional organization. *Nature Reviews. Genetics, 5*(2), 101–113. doi:10.1038/nrg1272 PMID:14735121

Barczak, G., Lassk, F., & Mulki, J. (2010). Antecedents of team creativity: An examination of team emotional intelligence, team trust and collaborative culture. *Creativity and Innovation Management, 19*(4), 332–345. doi:10.1111/j.1467-8691.2010.00574.x

Baruch, Y., & Lin, C. P. (2012). All for one, one for all: Coopetition and virtual team performance. *Technological Forecasting & Social Change, 79*(6), 1155–1168. doi: 10.1016/j.techfore.2012.01.008

Baskerville, R.L., & Wood-Harper, A.T. (1996) A Critical Perspective on Action esearch as a Method for Information Systems Research. *Journal of Information Technology, 11*, 235-246.

Bass, B. M. (1985). *Leadership and performance beyond expectations*. New York: Free Press.

Bass, T., & Mabry, R. (2004). Enterprise architecture reference models: A shared vision for Service-Oriented Architectures. In *Proceedings of the IEEE MILCOM* (pp. 1-8).

Bateson, G. (1967). *A theory of play and fantasy*. Bobbs-Merrill.

Beckoff, M. (2012). *Article*. Retrieved from http://www.psychologytoday.com/blog/animal-emotions/201202/the-need-wild-play-let-children-be-the-animals-they-need-be

Begel, A., Bosch, J., & Storey, M.-A. (2013). Social Networking Meets Software Development: Perspectives from GitHub, MSDN, Stack Exchange, and TopCoder. *IEEE Software, 30*(1), 52–66. doi:10.1109/MS.2013.13

Beheshti, H. M. (2006). What managers should know about ERP/ERPII. *Management Research News, 29*(4), 184–193. doi:10.1108/01409170610665040

Bélanger, F., & Watson-Manheim, M. B. (2006). Virtual teams and multiple media: Structuring media use to attain strategic goals. *Group Decision and Negotiation, 15*(4), 299–321. doi:10.1007/s10726-006-9044-8

Bell, B. S., & Kozlowski, S. W. J. (2002). A typology of virtual teams: Implications for effective leadership. *Group & Organization Management, 27*(1), 14–49. doi:10.1177/1059601102027001003

Bendoly, E., Soni, A., & Venkataramanan, M. A. (2004). *Value Chain Resource Planning (VCRP): Adding value with systems beyond the enterprise.* Retrieved January 17, 2010, from www.fc.bus.emory.edu/~elliot_bendoly/VCRP_BH.pdf

Benetytė, D., & Jatuliavičienė, G. (2014). Building and sustaining trust in virtual teams within organizational context. *Regional Formation and Development Studies, 10*(2), 18–30. doi:10.15181/rfds.v10i2.138

Bengtsson, M., & Kock, S. (2000). "Coopetition" in business networks – to cooperate and compete simultaneously. *Industrial Marketing Management, 29*(5), 411–426. doi:10.1016/S0019-8501(99)00067-X

Benlian, A., & Hess, T. (2011). Comparing the relative importance of evaluation criteria in proprietary and open-source enterprise application software selection – a conjoint study of ERP and Office systems. *Information Systems Journal, 21*(6), 503–525. doi:10.1111/j.1365-2575.2010.00357.x

Beranek, P. M., Broder, J., Reinig, B. A., Romano, N. C., & Sump, S. (2005). Management of virtual project teams: Guidelines for team leaders. *Communications of the AIS, 16*(10), 247–259.

Berry, G. (2011). Enhancing Effectiveness on Virtual Teams: Understanding Why Traditional Team Skills Are Insufficient. *Journal of Business Communication, 48*(2), 186–206. doi:10.1177/0021943610397270

Berry, G. R. (2011). Enhancing effectiveness on virtual teams: Understanding why traditional team skills are insufficient. *Journal of Business Communication,* 0021943610397270.

Berry, M. W., Drmac, Z., & Jessup, E. R. (1999). Matrices, vector spaces, and information retrieval. *SIAM Review, 41*(2), 335–362. doi:10.1137/S0036144598347035

Berson, Y., Halevy, N., Shamir, B., & Erez, M. (2014). Leading from different psychological distances: A construal-level perspective on vision communication, goal setting, and follower motivation. *The Leadership Quarterly.*

Bertels, H., Kleinschmidt, E., & Koen, P. (2011). Communities of Practice versus Organizational Climate: Which One Matters More to Dispersed Collaboration in the Front End of Innovation? *Journal of Product Development and Management, 28,* 757–772.

Bhagwatwar, A., Massey, A., & Dennis, A. R. (2013). *Creative virtual environments: Effect of supraliminal priming on team brainstorming.* Paper presented at the 46th Hawai International Conference on System Sciences. Retrieved from http://scholar.google.com/citations?view_op=view_citation&hl=es&user=qh2OnA4AAAAJ&citation_for_view=qh2OnA4AAAAJ:W7OEmFMy1HYC

Bhandarker, A. (1990). *Corporate success and transformational leadership.* New Age International.

Bilgihan, A., Okumus, F., Nusair, K., & Kwun, D. (2011). Information technology applications and competitive advantage in hotel companies. *Journal of Hospitality and Tourism Technology, 2*(2), 139–153. doi:10.1108/17579881111154245

Binder, M., & Clegg, B. T. (2005b). Partial evolutionary multiplicity: An approach to managing the dynamics of supply structures. In *Proceedings of the 18th International Conference on Production Research*. Universita di Salerno.

Binder, M., & Clegg, B. T. (2006). A conceptual framework for enterprise management. *International Journal of Production Research*, *44*(18/19), 3813–3829. doi:10.1080/00207540600786673

Birnholtz, J., & Finholt, T. (2007). Cultural challenges to leadership in cyberinfrastructure development. In S. Weisband (Ed.), *Leadership at a distance: Research in technology-supported work* (pp. 195–207). New York, NY: Lawrence Erlbaum Associates.

Blackburn, R. S., Furst, S. A., & Rosen, B. (2003). Building a winning virtual team. In C. Gibson & S. Cohen (Eds.), *Virtual teams that work: Creating conditions for effective virtual teams* (pp. 95–120). San Francisco, CA: Jossey–Bass.

Blackburn, R., Furst, S., & Rosen, B. (2003). Building a winning virtual team. In *Virtual teams that work* (pp. 95–120). Academic Press.

Blackstone, J. H., Jr., & Cox, J. F. (2005). APICS Dictionary (11th ed.). APICS: The association for Operations Management.

Bodiya, A. (2013). *Virtual reality: the impact of task interdependence and task structure on virtual team productivity and creativity.* (Doctoral Dissertation). Alliant International University, Los Angeles, CA. Retrieved from ProQuest (UMI 3452402).

Boell, S. K., Campbell, J., Cecez-Kecmanovic, D., & Cheng, J. E. (2013). *The Transformative Nature of Telework: A Review of the Literature*. Academic Press.

Bohoris, G. A., & Yun, W. Y. (1995). Warranty costs for repairable products under hybrid warranty. *IMA Journal of Mathematics Applied in Business and Industry*, *6*, 13–24.

Boiney, L. G. (2001). Gender impacts virtual work teams. *Graziadio Business Report*, *4*(4), 5.

Bolman, L. E., & Deal, T. E. (2003). *Reframing organizations: Artistry, choice, and leadership*. San Francisco, CA: Jossey-Bass.

Bond, B., Genovese, Y., Miklovic, D., Wood, N., Zrimsek, B., & Rayner, N. (2000). *ERP is dead - long live ERPII*. Retrieved November 8, 2009, from www.pentaprise.de/cms_showpdf.php?pdfname=infoc_report

Bond, C. F., & Titus, L. J. (1983). Social facilitation: A meta-analysis of 241 studies. *Psychological Bulletin*, *94*(2), 265–292. doi:10.1037/0033-2909.94.2.265 PMID:6356198

Borchert, I., Gootiiz, B., & Mattoo, A. (2014). Policy barriers to international trade in services: Evidence from a new database. *The World Bank Economic Review*, *28*(1), 162–188. doi:10.1093/wber/lht017

Bosch-Sijtsema, P. (2007). The impact of individual expectations and expectation conflicts on virtual teams. *Group & Organization Management*, *32*(3), 358–388. doi:10.1177/1059601106286881

Boston Business Journal. (2015). *How Boston-area companies and hospitals are dealing with the snow and their employees*. Retrieved February 28, 2015 from http://www.bizjournals.com/boston/blog/startups/2015/02/how-boston-area-tech-companies-are-dealing-with.html?page=all

Boudreau, M.-C., Loch, K. D., Robey, D., & Straud, D. (1998). Going global: Using information technology to advance the competitiveness of the virtual transnational organization. *The Academy of Management Executive*, *12*(4), 120–128.

Bourgault, M., Drouin, N., Daoudi, J., & Hamel, E. (2008). Understanding decision making within distributed project teams: An exploration of formalization and autonomy as determinants of success. *Project Management Journal*, *39*(S1Suppl.), S97–S110. doi:10.1002/pmj.20063

Boutellier, R., Gassmann, O., Macho, H., & Roux, M. (1998). Management of dispersed product development teams: The role of information technologies. *R & D Management*, *28*(1), 13–25. doi:10.1111/1467-9310.00077

Bradberry, T., & Greaves, J. (2009). *Emotional Intelligence 2.0*. Talentsmart. Retrieved from http://www.talentsmart.com/products/emotional-intelligence-2.0/

Brady, M., Fellenz, M., & Brookes, R. (2008). Researching the role of information communication technology (ICT) in contemporary marketing practices. *Journal of Business and Industrial Marketing*, *23*(2), 108–114. doi:10.1108/08858620810850227

Brandt, V., England, W., & Ward, S. (2011). Virtual teams. *Research & Technology Management*, *54*(6), 62–63.

Brandt, V., England, W., & Ward, S. (2011). Virtual teams. *Research Technology Management*, *54*(6), 62–63.

Branicky, M. (1998). Multiple Lyapunov functions and other analysis tools for switched and hybrid systems. *IEEE Transactions on Automatic Control*, *43*(4), 475–482. doi:10.1109/9.664150

Brown, J. (2013, February 12). *Virtual Team Management Trends and Telecommuting*. Retrieved on February 24, 2015 from http://blog.hubstaff.com/virtual-team-management-trends/

Browne, J., & Zhang, J. (1999). Extended and virtual enterprises-Similarities and differences. *International Journal of Agile Management Systems*, *1*(1), 30–36. doi:10.1108/14654659910266691

Brown, S. L., & Eisenhardt, K. M. (1995). Product development: Past research, present findings, and future directions. *Academy of Management Review*, *20*(2), 343.

Brown, W. C. (2006). IT governance, architectural competency, and the Vasa. *Information Management & Computer Security*, *14*(2), 140–154. doi:10.1108/09685220610655889

Bruce, J. et al. (2003). Adding the "E" to leadership: How it may impact your leadership. *Organizational Dynamics*, *31*(4).

Bruce, J., & Arolio, G. (2002). E-Leadership:implications for theory, research and practices. *Leadership Quarterly*, *11*(4).

Brüggemann, H. (2014). *Entrepreneurial leadership styles: a comparative study between Startups and mature firms*. Academic Press.

Brynjolfsson, E., & Hitt, L. (2004). Computing productivity: Firm-level evidence. *The Review of Economics and Statistics*, *85*(4), 793–808. doi:10.1162/003465303772815736

Buco, M. J., Chang, R. N., Luan, L. Z., Ward, C., Wolf, J. L., & Yu, P. S. (2004). Utility computing SLA management based upon business objectives. *IBM Systems Journal*, *43*(1), 159–178. doi:10.1147/sj.431.0159

Burke, K., & Chidambaram, L. (1999). How Much Bandwidth is Enough? A Longitudinal Examination of Media Characteristics and Group Outcomes. *Management Information Systems Quarterly*, *3*(4), 557–580. doi:10.2307/249489

Burns, J. M. (1978). *Leadership*. New York: Harper & Row.

Burt, J. (2013). *Dell wants half of employees working remotely by 2020*. Retrieved from http://www.eweek.com/mobile/dell-wants-half-of-employees-working-remotely-by-2020.html

Business Dictionary. (2014). *Electronic collaboration*. Retrieved November 20, 2014, from: http://www.businessdictionary.com/definition/electronic-collaboration.html

Byrne, J. A., & Brandt, R. (1993, February 8). The virtual corporation. *Business Week,* 36-41.

Cabrera, J. (2013, March 15). *Redarquía. Change management in the era of collaboration* [Slideshare]. Retrieved from http://es.slideshare.net/fullscreen/jcabrera/redarqua-gestin-del-cambio-en-la-era-de-la-colaboracin/1

Cabrera, J. (2013a, July 19). *Redarquía* [Slideshare]. Retrieved from http://es.slideshare.net/jcabrera/redarquia-el-orden-emrgente-en-la-era-de-la-colaboracin

Cabrera, J. (2013b). *Redarquía and organizational change* [Web log post]. Retrieved from http://cabreramc.files.wordpress.com/2013/02/11_redarquc3ada-y-cambio-organizacional.pdf

Callaway, E. (2000). *ERP – the next generation: ERP is Web Enabled for E-business*. Charleston: Computer Technology Research Corporation.

Campbell, B., Kay, R., & Avison, D. (2005). Strategic alignment: A practitioner's perspective. *Journal of Enterprise Information Management, 18*(6), 653–664. doi:10.1108/17410390510628364

Campos, J., Chiola, G., & Silva, M. (1991). Properties and performance bounds for closed free choice synchronized mono-class queuing networks. *IEEE Transactions on Automatic Control, 36*(12), 1368–1382. doi:10.1109/9.106153

Candido, G., Barata, J., Colombo, A. W., & Jammes, F. (2009). SOA in reconfigurable supply chain: A research roadmap. *Engineering Applications of Artificial Intelligence, 22*(6), 939–949. doi:10.1016/j.engappai.2008.10.020

Canie, L. et al. (2004). Would the real project management language please stand up? *International Journal of Project Management*, 22-43.

Capra, F. (1999). *The web of life*. Barcelona, Spain: Anagram.

Careem, M., De Silva, C., De Silva, R., Raschid, L., & Weerawarana, S. (2006, December). Sahana: Overview of a disaster management system. In *Information and Automation, 2006. ICIA 2006. International Conference on* (pp. 361-366). IEEE.

Carmel, E. (1999). *Global software teams: Collaborating across borders and time zones*. Upper Saddle River, NJ: Prentice-Hall.

Carson, J. B., Tesluk, P. E., & Marrone, J. A. (2007). Shared leadership in teams: An investigation of antecedent conditions and performance. *Academy of Management Journal, 50*(5), 1217–1234. doi:10.2307/20159921

Carte, T. A., Chidambaram, L., & Becker, A. (2006). Emergent leadership in self-managed virtual teams. *Group Decision and Negotiation, 15*(4), 323–343. doi:10.1007/s10726-006-9045-7

Carton, A. M., Murphy, C., & Clark, J. R. (2014). A (blurry) vision of the future: How leader rhetoric about ultimate goals influences performance. *Academy of Management Journal, 57*(6), 1544–1570. doi:10.5465/amj.2012.0101

Cascio, F. W. (2015). *The Virtual Workplace: A Reality Now*. SIOP Retrieved from: http://www.siop.org/tip/backissues/tipapril98/cascio.aspx

Cascio, W. F. (2000). Managing a Virtual Workplace. *The Academy of Management Executive, 14*(3), 81-90.

Center for Creative Leadership. (2012). *White paper: Prelude at work*. Retrieved from http://www.heliotrope.ca/wp-content/uploads/2012/10/White-Paper-Prelude-At-Work.pdf

Chamarro-Premuzic, T. (2013). *Can you really improve your team?* https://hbr.org/2013/05/can-you-really-improve-your-em

Champion, D., & Stowell, F. A. (2001). PEArL: A Systems Approach to Demonstrating Authenticity in Information System Design. *Journal of Information Technology, 16*(1), 3–12. doi:10.1080/02683960010028438

Chang, C. M. (2011). New organizational designs for promoting creativity: A case study of virtual teams with anonymity and structured interactions. *Journal of Engineering and Technology Management, 28*(4), 268–282. doi:10.1016/j.jengtecman.2011.06.004

Chapman, C. S., & Kihn, L. A. (2009). Information systems integration, enabling control and performance. *Accounting, Organizations and Society, 34*(2), 151–169. doi:10.1016/j.aos.2008.07.003

Chaur, Paul, & Chillarege. (1996). Virtual project management for software. *NSF Workshop on Workflow &Process Automation.*

Checkland, P. (1981). *Systems Thinking, Systems Practice.* Chichester: John Wiley.

Checkland, P. B. (1999). *Systems Thinking, Systems Practice, a Thirty Year etrospective.* Chichester: Wiley.

Checkland, P. B., & Holwell, S. (1998). *Information, Systems and Information Systems.* Chichester: Wiley.

Checkland, P. B., & Poulter, J. (2006). *Learning for Action.* Chichester: Wiley.

Chen, I. J. (2001). Planning for ERP systems: Analysis and future trend. *Business Process Management Journal, 7*(5), 374–386. doi:10.1108/14637150110406768

Chen, S.-T., Wall, A., Davies, P., Yang, Z., Wang, J., & Chou, Y.-H. (2013). A Human and Organisational Factors (HOFs) analysis method for marine casualties using HFACS-Maritime Accidents (HFACS-MA). *safety. Science, 60*, 105–114. doi:10.1016/j.ssci.2013.06.009

Cherniss, C. (2000). *Emotional intelligence: What it is and why it matters.* Retrieved from http://www.eiconsortium.org/reports/what_is_emotional_intelligence.html

Chidambaram, I., & Tung, I. (2005). Is out of sight out of mind? An empirical study of social loafing in technology supported groups. *Information Systems Research, 16*(2), 27–39. doi:10.1287/isre.1050.0051

Childe, S. J. (1998). The extended enterprise – a concept of co-operation. *Production Planning and Control, 9*(4), 320–327. doi:10.1080/095372898234046

Childs, R., Gingrich, G., & Piller, M. (2009). The future workforce: Gen Y has arrived. *Engineering Management Review, 38*(3), 32–34.

Chiu, C. M., Hsu, M. H., & Wang, E. T. G. (2006). Understanding knowledge sharing in virtual communities: An integration of social capital and social cognitive theories. *Decision Support Systems, 42*(3), 1872–1888. doi:10.1016/j.dss.2006.04.001

Cho, H., Jung, M., & Kim, M. (1996). Enabling technologies of agile manufacturing and its related activities in Korea. *Computers & Industrial Engineering, 30*(3), 323–334. doi:10.1016/0360-8352(96)00001-0

Choi, C., & Hoon Yi, M. (2009). The effect of the Internet on economic growth: Evidence cross-country panel data. *Economics Letters, 105*(1), 39–41. doi:10.1016/j.econlet.2009.03.028

Choi, E. K. (2009). Entrepreneurial leadership in the Meiji cotton spinners' early conceptualisation of global competition. *Business History, 51*(6), 927–958. doi:10.1080/00076790903266877

Chon, F. (2003). A collaborative project management architecture. In *Proceeding of the 36th Hawaii International Conference on System Science.*

Chorafas, D.N. (2001). *Integrating ERP, CRM, Supply Chain Management, and Smart Materials.* New York, NY: CRC Press LLC and Auerbach Publications.

Chudoba, K. M., Wynn, E., Lu, M., & Watson-Manheim, M. B. (2005). How virtual are we? Measuring virtuality and understanding its impact in a global organization. *Information Systems Journal, 15*(4), 279–306. doi:10.1111/j.1365-2575.2005.00200.x

Chung, A. A. C., Yam, A. Y. K., & Chan, M. F. S. (2004). Networked enterprise: A new business model for global sourcing. *International Journal of Production Economics, 87*(3), 267–280. doi:10.1016/S0925-5273(03)00222-6

Chung, K. S. K., & Hossain, L. (2010). Towards a social network model for understanding information and communication technology use for general practitioners in rural Australia. *Computers in Human Behavior, 26*(4), 562–571. doi:10.1016/j.chb.2009.12.008

Chu, S. K., & Kennedy, D. M. (2011). Using online collaborative tools for groups to co-construct knowledge. *Online Information Review, 35*(4), 581–597. doi:10.1108/14684521111161945

Cicei, C. C. (2012). Assessing members' satisfaction in virtual and face-to-face learning teams. *Procedia: Social and Behavioral Sciences, 46*, 4466–4470. doi:10.1016/j.sbspro.2012.06.278

Ciotti, G. (2013). *Why Remote Teams Are the Future (and How to Make Them Work)*. Retrieved February 28, 2015 from: http://www.helpscout.net/blog/virtual-teams/

Cisco. (2009). *Creating a Collaborative Enterprise*. Retrieved from http://www.cisco.com/c/dam/en/us/solutions/collateral/enterprise/collaboration-strategies/C11-533734-00_collab_exec_guide.pdf

Ciufudean, C. (2006). Work-flows in Constructions Modelled with Stochastic Artificial Petri Nets. In *Proc. of The 23rd International Symposium on Automation and Robotics in Construction*.

Ciufudean, C. (2007). Reliability Markov Chains for Security Data Transmitter Analysis. In *Proc. of The Second International Conference on Availability, Reliability and Security*. Vienna University of Technology.

Ciufudean, C. (2005). Petri Net Based Diagnosis for Construction Design. *22nd International Symposium on Automation and Robotics in Construction, ISARC 2005*.

Ciufudean, C., & Filote, C. (2005). Performance Evaluation of Distributed Systems. *International Conference on Control and Automation, ICCA 2005*. IEEE.

Ciufudean, C., & Filote, C. (2006). Diagnosis of Complex Systems Using Ant Decision Petri Nets. *The First International Conference on Availability, Reliability and Security (ARES 2006)*, Vienna University of Technology. doi:10.1109/ARES.2006.52

Ciufudean, C., & Petrescu, C. (2005). Scheduling Diagnosis of Flexible Information Systems. In *Proc. Int. Conf. on Autom, Contr. And Syst. Eng. (ACSE'05)*.

Clark, H. H., & Brennan, S. E. (1991). Grounding in communication. In L. B. Resnick, J. M. Levine, & S. D. Teasley (Eds.), *Perspectives on socially shared cognition* (pp. 127–149). Washington, DC: American Psychological Association. doi:10.1037/10096-006

Clegg, B., & Wan, Y. (2013). ERP systems and enterprise management trends: A contingency model for the enterprization of operations. *International Journal of Operations & Production Management, 33*(11/12), 1458–1489. doi:10.1108/IJOPM-07-2010-0201

Clerc, M. (1999). The Swarm and the Queen: Towards a Deterministic and Adaptive Particle Swarm Optimization. In *Proceedings of the IEEE Congress on Evolutionary Computation*. doi:10.1109/CEC.1999.785513

Clerc, M. (2006). *Particle Swarm Optimization*. London: ISTE. doi:10.1002/9780470612163

Coates, J. (2007). *Generational learning styles.* LERN Books.

Cohen, J. (2007). *Article.* Retrieved from http://blog.ted.com/2007/03/09/ted2007_day_two/

Coleman, D. (2012, June 20). *The remains of social enterprises* [Web log post]. Retrieved from http://www.cmswire.com/cms/social-business/the-challenges-of-the-social-enterprise-016147.php#null

Collaborative Work. (2014, December 15). In *Wikipedia, the free encyclopedia* [Electronic version]. Retrieved from http://es.wikipedia.org/wiki/Trabajo_colaborativo

Collines. (2002). Virtual & networked organizations. *Express Exec.*

Connaughton, S. L., & Daly, J. (2004). Leading from afar: Strategies for effectively leading virtual teams. In S. H. Godar & S. P. Ferris (Eds.), *Virtual and Collaborative Teams: Process, Technologies and Practice* (pp. 49–75). Hershey, PA: Idea Group Publishing. doi:10.4018/978-1-59140-204-6.ch004

Connaughton, S. L., & Daly, J. (2005). Leadership in the new millennium: Communicating beyond temporal, spacial, and geographical boundaries. In P. Kalbfleisch (Ed.), *Communication Yearbook 29* (pp. 187–213). Mahwah, NJ: Psychology Press.

Cooke, P. (2013). *Complex adaptive innovation systems: Relatedness and transversality in the evolving region.* Routledge.

Cooray, S. F. (2010). *End-User driven Development of Information Systems-evisiting Vickers notion of 'Appreciation'.* (Thesis). University of Portsmouth.

Corbin, D. L., & Musante, C. F. (2015). *Package assembly for thin wafer shipping.* US Patent 20,150,076,029.

Cordery, J. L., & Soo, C. (2008). Overcoming impediments to virtual team effectiveness. *Human Factors, 18*(5), 487–500. doi:10.1002/hfm.20119

Coronas, T. T., Oliva, M. A., Luna, J. C. Y., & Palma, A. M. L. (2015). Virtual Teams in Higher Education: A Review of Factors Affecting Creative Performance. In *International Joint Conference* (pp. 629–637). Springer. doi:10.1007/978-3-319-19713-5_55

Could, D. (2005). *Virtual team.* Retrieved from www.seanet.com/~daveg/vrteams.html

Covey, S. M. R. (2009). *How the Best Leaders Build Trust.* Retrieved February 1, 2015 from http://www.leadershipnow.com/CoveyOnTrust.html

Cox, J., Pearce, C. L., & Perry, M. (2003). Toward a model of shared leadership and distributed influence in the innovation process. In C. L. Pearce & J. A. Conger (Eds.), *Shared leadership: Reframing the hows and whys of leadership* (p. 48). Thousand Oaks, CA: Sage Publications. doi:10.4135/9781452229539.n3

Cragan, J. F., Wright, D. W., & Casch, C. R. (2009). *Communication in small groups.* Bradley University / Cengage Learning.

Crampton & Sheila. (2005). Relationships among geoghraphical dispersion, team processes and effectiveness in software development work team. *Journal of Business Research.*

Cramton, C. D., & Orvis, K. L. (2003). Overcoming barriers to information sharing in virtual teams. In C. B. Gibson & S. G. Cohen (Eds.), *Virtual teams that work: Creating conditions for virtual team effectiveness* (pp. 21–36). San Francisco, CA: Jossey-Bass.

Crandall, R. (1968). Vertical integration and the market for repair parts in the United States automobile industry. *The Journal of Industrial Economics, 16*(3), 212–234. doi:10.2307/2097561

Cremers. A., Kahler. H., & Rittenbruch. M. (2005). Supporting cooperation in a virtual organization. *Proceeding pf ICIS* (pp. 3-38).

Crisp, C. B., & Jarvenpaa, S. L. (2015). Swift trust in global virtual teams. *Journal of Personnel Psychology.*

Cross, R., & Rieley, J. (1999). Team learning: Best practices and tools for an elusive concept. *National Productivity Review, 18*(3), 9–18. doi:10.1002/npr.4040180303

Crowdsourcing. (2014, November 24). In *Wikipedia, the free encyclopedia* [Electronic version]. Retrieved from http://es.wikipedia.org/wiki/Crowdsourcing

Crowdvoting. (2014, July 3). In *Wikipedia, Die freie Enzyklopädie* [Electronic version]. Retrieved from http://de.wikipedia.org/wiki/Crowdvoting

Cullinane, K., & Khanna, M. (2000). Economies of scale in large containerships: Optimal size and geographical implications. Journal of Transport Geography. *Journal of Transport Geography, 8*(3), 181–195. doi:10.1016/S0966-6923(00)00010-7

Cummins, F. A. (2009). *Building the Agile Enterprise with SOA, BPM and MBM*. Burlington, VT: Morgan Kaufmann Publishers and Elsevier Inc.

Curtis, S. (2014). *Global Cities and International Relations*. Routledge.

D'Innocenzo, L., Mathieu, J. E., & Kukenberger, M. R. (2014). A meta-analysis of different forms of shared leadership-team performance relations. *Journal of Management.* doi:10.1177/0149206314525205

Dachis Group. (2013). *Enabling virtual teams*. Retrieved from http://cdn2.hubspot.net/hub/178806/file-41102699-pdf/Enabling_Virtual_Teams.pdf

Daim, T., Ha, A., Reutiman, S., Hughes, B., Pathak, U., Bynum, W., & Bhatla, A. (2012). Exploring the communication breakdown in global virtual teams. *International Journal of Project Management, 30*(2), 199–212. doi:10.1016/j.ijproman.2011.06.004

Damm & Schindler. (2002). Security issue of a knowledge medium for distributed project work. *International Journal of Project Management*, 37-48.

Daniel, E. M., & White, A. (2005). The future of inter-organizational system linkages: Findings of an international delphi study. *European Journal of Information Systems, 14*(2), 188–203. doi:10.1057/palgrave.ejis.3000529

Daniels, S. (1998). The virtual corporation. *Work Study, 47*(1), 20–22. doi:10.1108/00438029810196685

Davenport, T. H. (1998). Putting the enterprise into the enterprise system. *Harvard Business Review*, 121–131. PMID:10181586

Davenport, T. H., & Brooks, J. D. (2004). Enterprise systems and the supply chain. *Journal of Enterprise Information Management, 17*(1), 8–19. doi:10.1108/09576050410510917

Davenport, T., & Pearlson, K. (1998). Two cheers for the virtual office. *Sloan Management Review, 39*(4), 51–65.

David, K. (2004). Examining effective technology project leadership traits &behavior. *Computers in Human Behavior.*

Davids, J. P. (2012). *Entrepreneurial leadership in dynamic markets*. University of Johannesburg.

Davis, E. W., & Spekman, R. E. (2004). *Extended enterprise: Gaining competitive advantage through collaborative supply chains*. New York, NY: Financial Times Prentice-Hall.

Davis, J. A., Pitts, E. L., & Cormier, K. (2000). Challenges facing family companies in the Gulf Region. *Family Business Review*, *13*(3), 217–238. doi:10.1111/j.1741-6248.2000.00217.x

Davison, R., Martinsons, M., & Ou, C. (2012). The Roles of Theory in Canonical Action Research. *Management Information Systems Quarterly*, *36*(3), 763–786.

Day, D. V., Fleenor, J. W., Atwater, L. E., Sturm, R. E., & McKee, R. (2014). Advances in leader and leadership development: A review of 25 years of research and theory. *The Leadership Quarterly*, *25*(1), 63–82. doi:10.1016/j.leaqua.2013.11.004

Day, D. V., Gronn, P., & Salas, E. (2004). Leadership capacity in teams. *The Leadership Quarterly*, *15*(6), 857–880. doi:10.1016/j.leaqua.2004.09.001

De Maria, F., Briano, C., Brandolini, M., Briano, E., & Revetria, R. (2011). Market-leader ERPs and cloud computing: A proposed architecture for an efficient and effective synergy. In *Proc. of the 10th WSEAS Conference on Applied Computer and Applied Computational Science*. Madison, WI: WSEAS.

de Pillis, E., & Furumo, K. (2007). Counting the cost of virtual teams. *Communications of the ACM*, *50*(12), 93–95. doi:10.1145/1323688.1323714

Deeter-Schmelz, D., & Kennedy, K. (2004). Buyer-seller relationships and information sources in an e-commerce world. *Journal of Business and Industrial Marketing*, *19*(3), 188–196. doi:10.1108/08858620410531324

Dekker, D., Rutte, C., & Van den Berg, P. (2008). Cultural differences in the perception of critical interaction behaviors in global virtual teams. *International Journal of Intercultural Relations*, *32*(5), 441–452. doi:10.1016/j.ijintrel.2008.06.003

Denison, D. R., Hooijberg, R., & Quinn, R. E. (1995). Paradox and performance: Toward a theory of behavioral complexity in managerial leadership. *Organization Science*, *6*(5), 524–540. doi:10.1287/orsc.6.5.524

Dennis, A. (1996). Information exchange and use in group decision making: You can lead a group to information, but you can't make it think. *Management Information Systems Quarterly*, *20*(4), 433–458. doi:10.2307/249563

DeRosa, D., & Lepsinger, R. (2010). *Virtual team success: A practical guide for working and leading from a distance*. San Francisco: Jossey-Bass/A Wiley Imprint.

Derudder, B., Taylor, P. J., & Ni, P. (2010). Pathways of change: Shifting connectivities in the world city network, 2000-2008. *Urban Studies (Edinburgh, Scotland)*, *47*(9), 1861–1877. doi:10.1177/0042098010372682

Desanctis, G., & Monge, P. (1999). Communication processes for virtual organizations. *Organization Science*, *10*(6), 693–703. doi:10.1287/orsc.10.6.693

Dewey, J. (1910). *How we think*. Lexington, Mass: D.C. Heath. doi:10.1037/10903-000

Diffin, J., Chirombo, F., & Nangle, D. (2010). *Cloud Collaboration: Using Microsoft SharePoint as a Tool to Enhance Access Services*. Retrieved February 24, 2015, from http://contentdm.umuc.edu/cdm/ref/collection/p15434coll5/id/1046

Dimovski, V., Penger, S., Peterlin, J., & Uhan, M. (2013). Entrepreneurial Leadership In The Daoist Framework. *Journal of Enterprising Culture*, *21*(04), 383–419. doi:10.1142/S0218495813500167

Dipalantino, D., & Vojnovic, M. (2009). Crowdsourcing and All-Pay Auctions. In *Proceedings of EC'09*. Stanford, CA: ACM.

Dizard, J. E., & Gadlin, H. (2014). Family Life and the Marketplace: Diversity and Change in the American Family. *Historical Social Psychology*, 281.

Doan, A., Ramakrishnan, R., & Halevy, A. Y. (2011). Crowdsourcing systems on the World Wide Web. *Communications of the ACM*, *54*(4), 86. doi:10.1145/1924421.1924442

Donald, F. (2007). Entrepreneurial leadership in the 21st century. *Journal of Leadership & Organizational Studies*, *13*(4).

Doolin, B., & McLeod, L. (2012). Sociomateriality and boundary objects in information systems development. *European Journal of Information Systems*, *21*(5), 570–586.

Dorigo, M. (1992). *Optimization, Learning and Natural Algorithms*. (PhD thesis). Polytechnic di Milano, Italy.

Dorogovtsev, S. N., & Mendes, J. F. (2013). *Evolution of networks: From biological nets to the Internet and WWW*. Oxford University Press.

Dorr, M. (2011). *Developing Real Skills for Virtual Teams*. Retrieved from http://onlinemba.unc.edu/wp-content/uploads/developing-real-skills.pdf

Dougherty, D. (1992). Interpretive barriers to successful product innovation in large firms. *Organization Science*, *3*(2), 179–202. doi:10.1287/orsc.3.2.179

Dragusha, C. (2012). *Managing virtual teams: guidelines to effective leadership*. Retrieved from https://www.theseus.fi/bitstream/handle/10024/50321/dragusha_cajup%20thesis.pdf?sequence=1

Driskell, J. E., Radtke, P. H., & Salas, E. (2003). Virtual teams: Effects of technological mediation on team performance. *Group Dynamics*, *7*(4), 297–323. doi:10.1037/1089-2699.7.4.297

Ducruet, C., & Itoh, H. (2015). Regions and material flows: Investigating the regional branching and industry relatedness of port traffic in a global perspective. *Journal of Economic Geography*, 1–26. doi:10.1093/jeg/lbv010

Ducruet, C., & Lee, S. W. (2006). Frontline soldiers of globalisation: Port-city evolution and regional competition. *GeoJournal*, *67*(2), 107–122. doi:10.1007/s10708-006-9037-9

Dulin, L. (2005). *Leadership preferences of a Generation Y cohort: A mixed methods investigation*. (Ph.D. dissertation). University of North Texas. Retrieved August 15, 2014, from ABI/INFORM Global. (Publication No. AAT 3181040).

Durate. (2001). *Mastering virtual team: Strategies, tools, and techniques the succeed*. Academic Press.

Dye, L. (2009, October 21). Why is your boss a bully? *ABC News*. Retrieved from http://abcnews.go.com/Technology/DyeHard/boss-bully/story?id=8872662

Earley, C. P., & Gibson, C. (2002). *Multinational work teams - a new perspective*. Mahwah, NJ: Lawrence Erlbaum Associates, Inc.

Ebrahim, N., Ahmed, S., & Taha, Z. (2011). Virtual Teams and Management Challenges. *Academic Leadership Journal*, *9*(3), 1-7. Retrieved on February 24, 2015 from https://www.academia.edu/302000/Virtual_Teams_and_Management_Challenges

Ebrahim, A., Ahmed, S., & Taha, Z. (2009). Virtual teams: A literature review. *Australian Journal of Basic and Applied Sciences*, *3*(3), 1–9.

ECB Project. (2011). *What we know about joint evaluations*. Retrieved from file:///C:/Users/m/Downloads/What%20We%20Know%20About%20Joint%20Evaluations%20April%202011%20Section3.pdf

Eckartz, S., Daneva, M., Wieringa, R., & Hillegersberg, J. V. (2009). Cross-organizational ERP management: How to create a successful business case? In *SAC'09 Proceedings of the 2009 ACM Symposium on Applied Computing*. Honolulu, HI: ACM.

Efendi, J., Mulig, E., & Smith, L. (2006). Information technology and systems research published in major accounting academic and professional journals. *Journal of Emerging Technologies in Accounting*, *3*(1), 117–128. doi:10.2308/jeta.2006.3.1.117

El-Kassrawy, Y. A. (2014). The impact of trust on virtual team effectiveness. *International Journal of Online Marketing*, *4*(1), 11–18. doi:10.4018/ijom.2014010102

Ellemers, N., Gilder, D., & Haslam, S. A. (2004). Motivating individuals and groups at work: A social identity perspective on leadership and group performance. *Academy of Management Review*, *29*(3), 459–478.

Ellis, J. B., Luther, K., Bessiere, K., & Kellogg, W. A. (2008). *Games for virtual team building*. Retrieved from http://pdf.aminer.org/000/248/298/vicious_and_virtuous_cycles_in_global_virtual_team_role_coordination.pdf

Elmore, B. (2006). It's a SMALL world after all. *Baylor Business Review, 25*(1), 8-9. Retrieved May 14, 2015 from: http://web.a.ebscohost.com.ezproxy.hpu.edu/ehost/pdfviewer/pdfviewer?vid=6&sid=6c324351-f088-43c9-ab84-c188741ab7b0%40sessionmgr4003&hid=4101

El-Sofany, H., Alwadani, H., & Alwadani, A. (2014). Managing Virtual Teamwork in IT Projects: Survey. *Journal of Advanced Corporate Learning*, *7*(4), 28–33. doi:10.3991/ijac.v7i4.4018

Elvekrog, J. (2015). *5 Ways to Ensure Remote Employees Feel Part of the Team*. Entrepreneur. Retrieved May 11, 2015 from: http://www.entrepreneur.com/article/243795

Emprendelandia.es. (2015). *Crowdfunding reward (reward-based crowdfunding): Collective funding for job creation*. Retrieved from http://www.emprendelandia.es/que-es-el-crowdfunding

Ensley, M. D., Hmieleski, K. M., & Pearce, C. L. (2006). The importance of vertical and shared leadership within new venture top management teams: Implications for the performance of startups. *The Leadership Quarterly*, *17*(3), 217–231. doi:10.1016/j.leaqua.2006.02.002

Eoi. (2012). *Crowdcreating: conceptualization and production of the product or service (Campus EOI Sevilla)*. Retrieved from http://www.eoi.es/portal/guest/evento/1994/crowdcreating-conceptualizacion-y-produccion-del-producto-o-servicio-campus-eoi-sevilla

Equichord. (2014). *Merging the ability to learn with a virtual world*. Retrieved from http://equichord.com/merging-the-ability-to-learn-with-a-virtual-world/

Erhardt, N. (2011). Is it all about teamwork? Understanding processes in team-based knowledge work. *Management Learning*, *42*(1), 87–112. doi:10.1177/1350507610382490

Erhardt, N. L., Martin-Rios, C., & Way, S. A. (2009). From bureaucratic forms towards team-based knowledge work systems: Implications for human resource management. *International Journal of Collaborative Enterprise*, *1*(2), 160–179. doi:10.1504/IJCENT.2009.029287

Erhardt, N. L., Werbel, J. D., & Shrader, C. B. (2003). Board of director diversity and firm financial performance. *Corporate Governance: An International Review*, *11*(2), 102–111. doi:10.1111/1467-8683.00011

Erhardt, N., & Gibbs, J. L. (2014). The Dialectical Nature of Impression Management in Knowledge Work: Unpacking Tensions in Media Use Between Managers and Subordinates. *Management Communication Quarterly*, *28*(May issue), 155–186. doi:10.1177/0893318913520508

Ericson, J. (2001). *What the heck is ERPII?* Retrieved May 27, 2012, from http://www.line56.com/articles/default.asp?ArticleID=2851

Esbin, H. B. (1998). *Carving lives from stone*. (Doctoral Thesis). McGill University, Canada. Retrieved from http://www.mosaichub.com/member/p/howard-b-esbin-phd

Espinosa, J., Cummings, J., Wilson, J., & Pearce, B. (2003). Team boundary issues across multiple global firms. *Journal of Management Information Systems, 19*(4), 157–190.

European Commission. (2003). Commission recommendation of 6 May 2003 concerning the definition of micro, small and medium sized enterprises. *Official Journal of the European Union, L, 124*(1422), 36–41.

Evaristo, R., & Muntvold, E. (2002). Collaborative infrastructure formation in virtual projects. *Journal of Global Information Technology Management, 5*(2), 29–47. doi:10.1080/1097198X.2002.10856324

Eveland, J. D., & Bikson, T. K. (1988). *Work group structures and computer support: A field experiment*. Retrieved at: http://tuiu.academia.edu/JDEveland/Papers/1207189/Work_group_structures_and_computer_support_A_field_experiment

Fan, Z. P., Suo, W. L., Feng, B., & Liu, Y. (2011). Trust estimation in a virtual team: A decision support method. *Expert Systems with Applications, 38*(8), 10240–10251. doi:10.1016/j.eswa.2011.02.060

Faraj, S., & Sambamurthy, V. (2006). Leadership of information systems development projects. *IEEE Transactions on Engineering Management, 53*(2), 238–249. doi:10.1109/TEM.2006.872245

Ferreira, P. G. S., de Lima, E. P., & da Costa, S. E. G. (2012). Perception of virtual team's performance: A multinational exercise. *International Journal of Production Economics, 140*(1), 416–430. doi:10.1016/j.ijpe.2012.06.025

Feynman, R. P., Leighton, R. B., & Sands, M. (2013). *The Feynman Lectures on Physics, Desktop Edition* (Vol. 1). Basic Books.

Fitzgerald, W. B., Howitt, O. J. A., & Smith, I. J. (2011). Greenhouse gas emissions from the international maritime transport of New Zealand's imports and exports. *Energy Policy, 39*(3), 1521–1531. doi:10.1016/j.enpol.2010.12.026

Foote, J. (1999). *An overview of audio information retrieval. In Multimedia Systems*. Springer.

Forbes. (2013). *Collaborating in the Cloud*. Retrieved February 28, 2015 from https://www.cisco.com/c/dam/en/us/solutions/collateral/collaboration/hosted-collaboration-solution/forbes_cisco_cloud_collaboration_business.pdf

Forsgren, M., & Johanson, J. (2014). *Managing networks in international business*. Routledge.

Frakes, W. B. (1992). Introduction to information storage and retrieval systems. *Space, 14*, 10.

Freeman, C., & Soete, L. (1997). *The economics of industrial innovation*. Cambridge, MA: MIT Press.

Freire, J. (2014, April 22). *Nomad* [Web log post]. Retrieved from http://nomada.blogs.com/

Freire, J. (n.d.). *Cultures of innovation and design thinking, new paradigms of management*. Universidade da Coruña and EOI School of Industrial Organization. Retrieved from http://laboratoriodetendencias.com/wp-content/uploads/2011/09/DesignThinking.pdf

Furst, S., Blackburn, R., & Rosen, B. (1999). Virtual team effectiveness: A proposed research agenda. *Information Systems Journal, 9*(4), 249–269. doi:10.1046/j.1365-2575.1999.00064.x

Furumo, K. (2009). The impact of conflict and conflict management style on deadbeats and deserters in virtual teams. *Journal of Computer Information Systems, 49*, 66–73.

Furumo, K., & Pearson, M. (2007). Gender based communication styles, trust and satisfaction in virtual teams. *Journal of Information, Information Technology, and Organizations, 2*, 47–60.

Galbraith, J. (1972). Organization design: An information processing view. In J. Lorsch & P. Lawrence (Eds.), *Organization planning: Cases and concepts* (pp. 49–74). Homewood, IL: Richard D. Irwin, Inc.

Galegher, J., & Kraut, R. E. (1994). Computer-mediated communication for intellectual teamwork: An experiment in group writing. *Information Systems Research*, *5*(2), 110–138. doi:10.1287/isre.5.2.110

Garden & Bent. (2004). *Working together, apart the web as project infrastructure*. Retrieved from www.intranetjournal.com/feathures/idmo398.html

Geiger, S., & Turley, D. (2005). Personal selling as knowledge-based activity: Communities of practice in the sales force. *Irish Journal of Management*, *26*(1), 61–71.

Geisler, B. (2014). *Virtual Teams*. Retrieved from http://www.newfoundations.com/OrgTheory/Geisler721.html

Gera, S. (2013). Virtual teams versus face to face teams: A review of literature. *Journal of Business and Management*, *11*(2). Retrieved from http://www.academia.edu/4858172/Virtual_teams_versus_face_to_face_teams_A_review_of_literature

Ghaffari, M., Sheikhahmadi, F., & Safakish, G. (2014). Modeling and risk analysis of virtual project team through project life cycle with fuzzy approach. *Computers & Industrial Engineering*, *72*, 98–105. doi:10.1016/j.cie.2014.02.011

Gibbs, J., & Boyraz, M. (2015). International HRM's role in managing global teams. In D. G. Collings, G. Wood, & P. Caligiuri (Eds.), *The Routledge companion to international human resource management* (pp. 532–551). New York, NY: Routledge.

Gibbs, J., Eisenberg, J., Rozaidi, N. A., & Gryaznova, A. (2015). The "megapozitiv" role of enterprise social media in enabling cross-boundary communication in a distributed russian organization. *The American Behavioral Scientist*, *59*(1), 75–102. doi:10.1177/0002764214540511

Gibbs, J., Rozaidi, N. A., & Eisenberg, J. (2013). Overcoming the "ideology of openness": Probing the affordances of social media for organizational knowledge sharing. *Journal of Computer-Mediated Communication*, *19*(1), 102–120. doi:10.1111/jcc4.12034

Gibson, C. B., & Gibbs, J. L. (2006). Unpacking the concept of virtuality: The effects of geographic dispersion, electronic dependence, dynamic structure, and national diversity on team innovation. *Administrative Science Quarterly*, *51*(3), 451–495.

Gibson, C., & Cohen, G. (2003). *Virtual Teams that work: Creating conditions for virtual team effectiveness.* San Francisco: Jossey Bass.

Gibson, C., & Cohen, S. (2003). *Virtual teams that work: Creating conditions for virtual team effectiveness.* San Francisco, CA: Jossey Bass.

Gibson, C., & Manuel, J. (2003). Building trust. In C. B. Gibson & S. Cohen (Eds.), *Virtual teams that work: Creating conditions for virtual team effectiveness* (pp. 59–86). San Francisco, CA: John Wiley & Sons Inc.

Gilson, L., Maynard, M. T., Jones Young, N. C., Vartiainen, M., & Hakonen, M. (2015). Virtual teams research: 10 years, 10 themes, and 10 opportunities. *Journal of Management*, *41*(5), 1313–1337. doi:10.1177/0149206314559946

Giuri, P., Rullani, F., & Torrisi, S. (2008). Explaining leadership in virtual teams: The case of open source software. *Information Economics and Policy*, *20*(4), 305–315. doi:10.1016/j.infoecopol.2008.06.002

Glaser, B. G., & Strauss, A. L. (1967). *The discovery of grounded theory: Strategies for qualitative research*. New York, NY: Aldine.

Glikson, E., & Erez, M. (2015). Emotion display norms in virtual teams. *Journal of Personnel Psychology*.

Gluckler, J., & Schrott, G. (2007). Leadership and performance in virtual teams: Exploring brokerage in electronic communication. *International Journal of e-Collaboration*, *3*(3), 31–52. doi:10.4018/jec.2007070103

Goffin, K., & Koners, U. (2011). Tacit Knowledge, Lessons Learnt, and New Product Development. *Journal of Product Innovation Management*, *28*(2), 300–318. doi:10.1111/j.1540-5885.2010.00798.x

Gohmann, S. F., Guan, J., Barker, R. M., & Faulds, D. J. (2005). Perceptions of sales force automation: Differences between sales force and management. *Industrial Marketing Management*, *34*(4), 337–343. doi:10.1016/j.indmarman.2004.09.014

Goldman, S., Nagel, R., & Preiss, K. (1995). Agile Competitors and virtual organizations. New York, NY: van Nostrand Reinhold.

Goleman, D. (1995). *Emotional intelligence*. Bantam Books.

Gong, W. B. (1992). Rational representation for performance functions of queuing systems. In Princeton Conf. System & Information Science, (pp. 204-210).

Gong, W. B., & Nananukul, S. (1996). *Rational interpolation for rare event probabilities. In Stochastic Networks: Stability and Rare Events*. Springer – Verlag.

Gong, W. B., & Yang, H. (1995). Rational approximants for same performance analysis problems. *IEEE Transactions on Computers*, *44*(12), 1394–1404. doi:10.1109/12.477245

Gonzales, V. M. G., Nardi, B., & Mark, G. (2009). Ensembles: Understanding the instantiation of activities. *Information Technology & People*, *22*(2), 109–131.

Goodrum, A. A. (2000). Image Information Retrieval: An Overview of Current Research. *Informing Science*, *3*(2).

Goranson, H. T. (1999). *The agile virtual enterprise: Cases, metrics, tools*. Westport, CT: Quorum Books and Greenwood Publishing Group, Inc.

Gordon, R. A. (1994). The Integrals of Lebesgue, Denjoy, Perron and Henstock. Grad. Stud. In Math., 4, 234-245.

Goulielmos, A. M., & Plomaritou, E. (2014). The Shipping Marketing Strategies within the Framework of Complexity Theory. *British Journal of Economics, Management & Trade*, *4*(7), 1128–1142.

Govindarajan, V., & Gupta, A. K. (2001). Building an effective global business team. *MIT Sloan Management Review*, *42*(4), 63–71.

Graham, C., & Nikolova, M. (2013). Does access to information technology make people happier? Insights from well-being surveys from around the world. *Journal of Socio-Economics*, *44*, 126–139. doi:10.1016/j.socec.2013.02.025

Grandon, E., & Pearson, J. M. (2004). Electronic commerce adoption: An empirical study of small and medium US business. *Information & Management*, *42*(1), 197–216. doi:10.1016/j.im.2003.12.010

Grant, R. M. (1996). Toward a knowledge-based theory of the firm. *Strategic Management Journal*, *17*(S2), 109–122. doi:10.1002/smj.4250171110

Gratton, L., & Erickson, T. J. (2007). *Eight ways to build collaborative teams*. Retrieved from https://hbr.org/2007/11/eight-ways-to-build-collaborative-teams/

Gratton, L., & Erickson, T. J. (2007). Eight ways to build collaborative teams. *Harvard Business Review*, *85*(11), 100. PMID:18159790

Green, D., & Roberts, G. (2010). Personnel implications of public sector virtual organizations. *Public Personnel Management*, *39*(1), 47–57.

Greengard, S. (2011). Following the Crowd. *Communications of the ACM, 54*(2), 20–22. doi:10.1145/1897816.1897824

Greer, T. W., & Payne, S. C. (2014). Overcoming telework challenges: Outcomes of successful telework strategies. *The Psychologist Manager Journal, 17*(2), 87–111. doi:10.1037/mgr0000014

Griffith, T. L., Sawyer, J. E., & Neale, M. A. (2003). Virtualness and knowledge in teams: Managing the love triangle of organizations, individuals, and information technology. *Management Information Systems Quarterly, 27*(2), 265–287.

Griffith, T., Mannix, E., & Neale, T. (2003). *Conflict and virtual teams in Virtual teams that work: creating conditions for virtual team effectiveness* (S. Cohen, Ed.). San Francisco: Jossey Bass.

Gross, C. (2002). Managing communication within virtual intercultural teams. *Business Communication Quarterly, 65*(4), 22–38. doi:10.1177/108056990206500404

Gudykunst, W. B. (1997). Cultural variability in communication. *Communication Research, 24*(4), 327–348. doi:10.1177/009365097024004001

Gupta-Sunderji, M. (2013). *Long-Distance Leadership: Managing Virtual Teams*. Retrieved May 14, 2015 from http://www.hrvoice.org/long-distance-leadership-managing-virtual-teams/

Gupta, V., MacMillan, I. C., & Surie, G. (2004). Entrepreneurial leadership: Developing and measuring a cross-cultural construct. *Journal of Business Venturing, 19*(2), 241–260. doi:10.1016/S0883-9026(03)00040-5

Gurtner, A., Kolbe, M., & Boos, M. (2007). Satisfaction in virtual teams and in organizations. The *Electronic Journal for Virtual Organizations and Networks*.[Special Issue]. *The Limits of Virtual Work, 9*, 9–29.

Gutierrez-Rubi, A., & Freire, J. (2013). *Manifesto crowd. The company and the intelligence of the crowds*. Retrieved from https://books.google.com.co/books?id=7XIkJWE7qeoC&printsec=frontcover&vq=Crowd+wisdom&hl=es&source=gbs_ge_summary_r&cad=0#v=onepage&q=Crowd%20wisdom&f=false

Hackman, J. R. (1983). *A normative model of work team effectiveness*. New Haven, CT: Yale School of Organization and Management, Research Program on Groups Effectiveness.

Hackman, J. R. (1987). The design of work teams. In J. Lorsch (Ed.), *Handbook of Organizational Behavior*. Englewood Cliffs, NJ: Prentice-Hall.

Haines, R. (2014). Group development in virtual teams: An experimental reexamination. *Computers in Human Behavior, 39*, 213–222. doi:10.1016/j.chb.2014.07.019

Halavais, R. A. (2004). *Load bearing structure for composite ecological shipping pallet*. Google Patents.

Hale, J. R., & Fields, D. L. (2007). Exploring servant leadership across cultures: A study of followers in Ghana and the USA. *Leadership, 3*(4), 397–417. doi:10.1177/1742715007082964

Hallam, G. (1997). Seven common beliefs about teams: Are they true? *Leadership in Action, 17*(3), 1–4.

Hambley, L. A., O'Neill, T. A., & Kline, T. J. B. (2007). Virtual team leadership: The effects of leadership style and communication medium on team interaction styles and outcomes. *Organizational Behavior and Human Decision Processes, 103*(1), 1–20. doi:10.1016/j.obhdp.2006.09.004

Hambley, L., & Kline, T. (2007). Virtual team leadership: Perspectives from the field. *International Journal of e-Collaboration, 3*(1), 40–64. doi:10.4018/jec.2007010103

Hanna, V., & Walsh, K. (2000). Alliances: The small firm perspective. In *Proceedings of 4th International Conference on Managing Innovative Manufacturing* (pp. 333-340), Aston University.

Harrigan, K. R. (1984). Formulating vertical integration strategies. *Academy of Management Review, 9,* 638–652.

Harrigan, K. R. (1985). Vertical integration and corporate strategy. *Academy of Management Journal, 28*(2), 397–425. doi:10.2307/256208

Harris, J. (2009). *Tönnies, Community and Civil Society.* Cambridge, UK: Cambridge Univerity Press.

Harris, R. (2009). Improving tacit knowledge transfer within SMEs through e-collaboration. *Journal of European Industrial Training, 33*(3), 215–231. doi:10.1108/03090590910950587

Hart, P. J. (2014). *Investigating Issues Influencing Knowledge Sharing in a Research Organization Using AIM.* (Thesis). University of Portsmouth.

Harwood, S. (2003). *ERP: The implementation cycle.* Burlington: Butterworth-Heinemann.

Hasteer, N., Bansal, A., & Murthy, B. K. (2013). Pragmatic assessment of research intensive areas in cloud: A systematic review. *Software Engineering Notes, 38*(3), 1. doi:10.1145/2464526.2464533

Hastings, R. (2008). *Set Ground Rules for Virtual Team Communications.* SHRM Online. Retrieved from http://www.shrm.org

Hastings, R. (2010). *Fostering Virtual Working Relationships Isn't Easy.* SHRM Online. Retrieved from http://www.shrm.org

Hauser, K., Sigurdsson, H. S., & Chudoba, K. M. (2010). EDSOA: An event-driven service-oriented architecture model for enterprise applications. *International Journal of Management & Information Systems, 14*(3), 37–47.

Health Insurance Exchanges: An Update from the Administration. (2013, November 6). United States Senate Committee on Finance. Retrieved November 9, 2013, from http://www.finance.senate.gov/hearings/hearing/?id=3dd91089-5056-a032-5290-f158359b9247

Healthcare.gov Website Rollout. (2014, March 31). In *Ballotpedia.* Retrieved August 8, 2015, from http://ballotpedia.org/Healthcare.gov_website_rollout

Heckman, R., Crowston, K., & Misiolek, N. (2007). A structurational perspective on leadership in virtual teams. In K. Crowston, & S. Seiber (Eds.), *Proceedings of the IFIP Working Group 8.2/9.5 Working Conference on Virtuality and Virtualization* (pp. 151–168). Portland, OR: Springer. doi:10.1007/978-0-387-73025-7_12

Heim, M. H. (1993). *The Metaphysics of Virtual Reality* (1st ed.). New York, NY: Oxford University Press, Inc.

Henderson, L. (2008). The impact of project managers ' communication competencies. *Project Management Journal, 39*(June), 48–59. doi:10.1002/pmj.20044

Henricks, T. S. (2010). Callois´ *Man, Play, and Games.* An appreciation and evaluation. *American Journal of Play,* 157–185.

Herbert, D. E. J. (2011). Theorizing religion and media in contemporary societies: An account of religious 'publicization'. *European Journal of Cultural Studies, 14*(6), 626-648.

Herbsleb, J. D., & Mockus, A. (2003). An empirical study of speed and communication in globally-distributed software development. *IWWW Transactions on Software Engineering, 29*(6), 1–134.

Herrington, T. (2004). Where in the world is the Global Classroom Project? In J. Di Leo & W. Jacobs (Eds.), *If classrooms matter: Progressive visions of educational environments* (pp. 197–210). New York: Routledge.

Hertel, G., Geister, S., & Konradt, U. (2005). Managing virtual teams: A review of current empirical research. *Human Resource Management Review, 15*(1), 69–95. doi:10.1016/j.hrmr.2005.01.002

Hertel, G., Konradt, U., & Orlikowski, B. (2004). Managing distance by interdependence: Goal setting, task interdependence, and team-based rewards in virtual teams. *European Journal of Work and Organizational Psychology, 13*(1), 1–28. doi:10.1080/13594320344000228

He, X. (2004). The ERP challenge in China: A resource-based perspective. *Information Systems Journal, 14*(2), 153–167. doi:10.1111/j.1365-2575.2004.00168.x

Hicks, D. A., & Stecke, K. E. (1995). The ERP maze: Enterprise resource planning and other production and inventory control software. *IIE Solutions, 27*, 12–16.

Hiller, N. J., DeChurch, L., Murase, T., & Doty, D. (2011). Searching for outcomes of leadership: A 25-year review. *Journal of Management, 37*(4), 1137–1177. doi:10.1177/0149206310393520

Hill, N. S. (2005). Leading together, working together: The role of team shared leadership in building collaborative capital in virtual teams. *Advances in Interdisciplinary Studies of Work Teams, 11*(05), 183–209. doi:10.1016/S1572-0977(05)11007-3

Hinds, P. J., & Bailey, D. E. (2003). Out of sight, out of sync: Understanding conflict in distributed teams. *Organization Science, 14*(6), 615–632. doi:10.1287/orsc.14.6.615.24872

Hinds, P., & Kiesler, S. (1995). Communication across boundaries: Work, structure, and use of communication technologies in a large organization. *Organization Science, 6*(4), 373–393. doi:10.1287/orsc.6.4.373

Hitt, L., Wu, D., & Zhou, X. (2002). Investment in enterprise resource planning: Business impact and productivity measures. *Journal of Management Information Systems, 19*(1), 71–98.

Hoch, J. E., & Dulebohn, J. H. (2013). Shared leadership in enterprise resource planning and human resource management system implementation. *Human Resource Management Review, 23*(1), 114–125. doi:10.1016/j.hrmr.2012.06.007

Hoch, J. E., & Kozlowski, S. W. J. (2014). Leading virtual teams: Hierarchical leadership, structural supports, and shared team leadership. *The Journal of Applied Psychology, 99*(3), 390–403. doi:10.1037/a0030264 PMID:23205494

Hoffman, W. J. (2007). Strategies for managing a portfolio of alliances. *Strategic Management Journal, 28*(8), 827–856. doi:10.1002/smj.607

Hofmann, P. (2008). ERP is dead, long live ERP. *IEEE Internet Computing, 12*(4), 84–88. doi:10.1109/MIC.2008.78

Hofstede, G. (1980). *Culture's consequences: international differences in work-related values*. Beverly Hills, CA: Sage.

Hofstede, G. (2001). *Cultures consequences: comparing values, behaviors, institutions, and organizations across nations* (2nd ed.). Thousand Oaks, CA: SAGE Publications.

Hollenstein, H. (2004). Determinants of the adoption of information and communication technologies. *Structural Change and Economic Dynamics, 15*(3), 315–342. doi:10.1016/j.strueco.2004.01.003

Hollingshead, A., Mcgrath, J., & O'Connor, K. (1993). *Group task performance and communication technology: A longitudinal study of computer-mediated versus face-to-face work groups*. Thousand Oaks, CA: Sage.

Hope, J., & Hope, T. (1997). *Competing in the third wave: The ten management issues of the information age*. Boston, MA: Harvard Business School Press.

Howe, J. (2006). The rise of crowdsourcing. *Wired Magazine*. Retrieved November, 2014, from http://www.wired.com/wired/archive/14.06/crowds_pr.html

Howe, J. (2008). *Crowdsourcing*. New York: Crown Business.

Hoyle, B. S. (1989). The port-city interface: Trends, problems, and examples. *Geoforum, 20*(4), 429–435. doi:10.1016/0016-7185(89)90026-2

Huang, R., Kahai, S., & Jestice, R. (2010). The contingent effect of leadership on team collaboration in virtual teams. *ScienceDirect: Computers in Human Behavior, 26*, 1098–1110.

Huang, R., Kahai, S., & Jestice, R. (2010). The contingent effects of leadership on team collaboration in virtual teams. *Computers in Human Behavior, 26*(5), 1098–1110. doi:10.1016/j.chb.2010.03.014

Huettner, B., Brown, M. K., & James-Tanny, C. (2006). *Managing Virtual Teams: Getting the Most from Wikis, Blogs, and Other Collaborative Tools*. Plano, TX: Wordware Publishing Inc.

Huizinga, J. (1949). Homo ludens: A study of the play-element in culture. London: Routledge & Kegan Paul.

Hunton, J. (2002). Blending information and communication technology with accounting research. *Accounting Horizons, 16*(1), 55–67. doi:10.2308/acch.2002.16.1.55

Hunton, J. E., Lippincott, B., & Reck, J. L. (2003). Enterprise resource planning (ERP) systems: Comparing firm performance of adopters and non-adopters. *International Journal of Accounting Information Systems, 4*(3), 165–184. doi:10.1016/S1467-0895(03)00008-3

Hu, Z., & Wu, W. (2014). A Game Theoretic Model of Software Crowdsourcing. In *Proceedings of 8th International Symposium on Service Oriented System Engineering (SOSE)*. Oxford, UK: IEEE. doi:10.1109/SOSE.2014.79

Hyvonen, T., Jarvinen, J., & Pellinen, J. (2008). A virtual integration – the management control system in a multinational enterprise. *Management Accounting Research, 19*(1), 45–61. doi:10.1016/j.mar.2007.08.001

I'm a Writer. (2014). *What is crowdcreation?* [Web log post]. Retrieved from http://www.soyescritor.com/?q=crowdcreation

Immen, W. (2009, August 23). Power has its own price. *The Globe and Mail*. Retrieved from http://www.theglobeandmail.com/report-on-business/power-has-its-own-price/article4215325/

Im, Y., & Lee, O. (2004). Pedagogical implications of online discussion for preservice teacher training. *Journal of Research on Technology in Education, 36*(2), 155–170. doi:10.1080/15391523.2003.10782410

Indiramma, M. M., & Anandakumar, K. R. (2009). Behavioral analysis of team members in virtual organization based on trust dimension and learning. *Proceedings of World Academy of Science: Engineering & Technology, 39*(3), 269–274.

Ingram, J. (2011). A food systems approach to researching food security and its interactions with global environmental change. *Food Security, 3*(4), 417–431. doi:10.1007/s12571-011-0149-9

Ipeirotis, P. G. (2010). Analyzing the Amazon Mechanical Turk Marketplace. *ACM XRDS, 17*(2), 16–21. doi:10.1145/1869086.1869094

Ismail, N. A., & King, M. (2014). Factors influencing the alignment of accounting information systems in small and medium sized Malaysian manufacturing firms. *Journal of Information Systems and Small Business, 1*(1-2), 1–20.

Itoh, H., Tiwari, P., & Doi, M. (2002). An analysis of cargo transportation behaviour in Kita Kanto (Japan). *International Journal of Transport Economics, 29*, 319–335.

Iuisi. (2012). *The use of the collective wisdom and operational means of intelligence analysis.* Retrieved from http://www.iuisi.es/15_boletines/15_ISIe/doc_ISIe_13_2012.pdf

Iverson, J. H., Mathiassen, L., & Nielsen, P. A. (n.d.). Managing Risk in Software Process Improvement: An Action Research Approach. *Management Information Systems Quarterly, 28*(3), 395–433.

Jackson, S., Joshi, A., & Erhardt, N. (2003). Recent research on team and organizational diversity: Swot analysis and implications. *Journal of Management, 29*(6), 801–830. doi:10.1016/S0149-2063(03)00080-1

Jacobs, F. R. (2007). Enterprise resource planning (ERP) – a brief history. *Journal of Operations Management, 25*(2), 357–363. doi:10.1016/j.jom.2006.11.005

Jacobs, J. (2005). Exploring defect causes in product developed by virtual team. *Information and Software Technology,* 47–60.

Jacobs, W., Ducruet, C., & De Langen, P. W. (2010). Integrating world cities into production networks: The case of port cities. *Global Networks, 10*(1), 92–113. doi:10.1111/j.1471-0374.2010.00276.x

Jacobs, W., Koster, H. R. A., & Hall, P. V. (2011). The Location and Global Network Structure of Maritime Advanced Producer Services. *Urban Studies (Edinburgh, Scotland), 48*(13), 2749–2769. doi:10.1177/0042098010391294

Jagdev, H. S., & Browne, J. (1998). The extended enterprise – a context for manufacturing. *Production Planning and Conitrol, 9*(3), 216–229. doi:10.1080/095372898234190

Jagdev, H. S., & Thoben, K. D. (2001). Anatomy of enterprise collaboration. *Production Planning and Control, 12*(5), 437–451. doi:10.1080/09537280110042675

Jagdev, H., Vasiliu, L., Browne, J., & Zaremba, M. (2008). A semantic web service environment for B2B and B2C auction applications within extended and virtual enterprises. *Computers in Industry, 59*(8), 786–797. doi:10.1016/j.compind.2008.04.001

Jardine, N., & van Rijsbergen, C. J. (1971). The use of hierarchic clustering in information retrieval. *Information Storage and Retrieval, 7*(5), 217-240.

Jarmon, L., Traphagan, T., Mayrath, M., & Trivedi, A. (2009). Virtual world teaching, experiential learning, and assessment: An interdisciplinary communication course in Second Life. *Computers & Education, 53*(1), 169–182. doi:10.1016/j.compedu.2009.01.010

Jarvenpaa, S. L., & Leidner, D. E. (1998). Communication and trust in global virtual teams. *Journal of Computer-Mediated Communication, 3*(4).

Jarvenpaa, S. L., & Leidner, D. E. (1998). Communication and trust in Global virtual teams. *Journal of Computer-Mediated Communication, 3*(4). Available at http://hyperion.math.upatras.gr/commorg/jarvenpaa/

Jarvenpaa, S. L., & Leidner, D. E. (1999). Is anybody out there? Antecedents of trust in global virtual teams. *Journal of Management Information Systems, 14*(4), 29–64.

Jarvenpaa, S. L., Shaw, T., & Staples, S. (2004). The Role of Trust in Global Virtual Teams. *Information Systems Research, 15*(3), 250–267. doi:10.1287/isre.1040.0028

Jarvenpaa, S., & Leidner, D. (1998). Communication and trust in global virtual teams. *Journal of Computer-Mediated Communication, 3*(4), 1–53.

Jarvenpaa, S., & Leidner, D. E. (1999). Communication and trust in global virtual teams. *Organization Science, 10*(6), 791–815. doi:10.1287/orsc.10.6.791

Johns, T., & Gratton, L. (2013). *The third wave of virtual work*. Retrieved from https://hbr.org/2013/01/the-third-wave-of-virtual-work

Johnson, S. D., Suriya, C., Won Yoon, S., Berrett, J. V., & La Fleur, J. (2002). Team development and group processes of virtual learning teams. *Computers & Education, 39*(4), 379–393. doi:10.1016/S0360-1315(02)00074-X

Johnson, T., Lorents, A. C., Morgan, J., & Ozmun, J. (2004). A customized ERP/SAP model for business curriculum integration. *Journal of Information Systems Education, 15*(3), 245–253.

Jones, T. L. (2009). *Virtual team communication and collaboration in army and corporate applications*. (Master's thesis). Fort Leavenworth, Kansas. Retrieved from file:///C:/Users/m/Downloads/ADA502091.pdf

Jones. (1997). *Virtual culture: Identity &communication in cyber society*. Sag Press.

Jones, G. (2008). *Organizational theory. Design and organizational change*. Mexico City, Mexico: Prentice Hall.

Jones, R., Oyung, R., & Pace, L. (2005). *Working virtually: Challenges of virtual teams*. Hershey, PA: Cybertech Publishing. doi:10.4018/978-1-59140-585-6

Jonti, D. (2013). *How does creativity work in a highly virtual environment?* Retrieved from https://www.linkedin.com/groups/How-does-creativity-work-in-3980060.S.260012070

Jorgenson, D. W., & Vu, K. (2007). Information technology and the world growth resurgence. *German Economic Review, 8*(2), 125–145. doi:10.1111/j.1468-0475.2007.00401.x

Joshi, A., & Lazarova, M. (2005). Do "global" teams need "global" leaders? Identifying leadership competencies in multinational teams. In *Managing Multinational Teams*. Global Perspectives.

Joshi, A., Lazarova, M. B., & Liao, H. (2009). Getting everyone on board: The role of inspirational leadership in geographically dispersed teams. *Organization Science, 20*(1), 240–252. doi:10.1287/orsc.1080.0383

Joskow, P.L. (2003). *Vertical integration, Handbook of New Institutional Economics*. Boston, MA: Kluwer.

Joyanes, L. (1997). *Cibersociedad*. Madrid, Spain: McGraw-Hill.

Kaihara, T., & Fujii, S. (2002). IT based virtual enterprise coalition strategy for agile manufacturing environment. In *Proc. of the 35th CIRP Int. Seminar on Manufacturing Systems*, (pp. 32-37).

Kam, H.-J., & Katerattanakul, P. (2014). Structural model of team-based learning using Web 2.0 collaborative software. *Computers & Education, 76*, 1–12. doi:10.1016/j.compedu.2014.03.003

Kankanhalli, A., Teo, H. H., Tan, B. C. Y., & Wei, K. K. (2003). An integrative study of information systems security effectiveness. *International Journal of Information Management, 23*(2), 139–154. doi:10.1016/S0268-4012(02)00105-6

Kantabutra, S. (2014). Visionary leadership at a Thai apparel manufacturer: Surprising evidence? *International Journal of Business Excellence, 7*(2), 168–187. doi:10.1504/IJBEX.2014.059547

Katzy, B. R., & Dissel, M. (2001). A toolset for building the virtual enterprise. *Journal of Intelligent Manufacturing, 12*(2), 121–131. doi:10.1023/A:1011248409830

Kayworth, T., & Leidner, D. (2002). Leadership Effectiveness in Global Virtual Teams. *Journal of Management Information Systems, 18*(3), 7-40.

Kayworth, T. R., & Leidner, D. E. (2002). Leadership effectiveness in global virtual teams. *Journal of Management Information Systems, 18*(3), 7–40.

Kayworth, T., & Leidner, D. (2000). The global virtual manager: A prescription for success. *European Management Journal, 18*(2), 183–194.

Keller, M. R. (2014). *Effective global virtual teams: The impact of culture, communication, and trust.* University of Maryland University College.

Kennedy, J., & Eberhart, R. (1995). Particle Swarm Optimization. In *Proceedings of the International Conference on Neural Networks.* doi:10.1109/ICNN.1995.488968

Kennedy, J., & Eberhart, R. (2009). *Swarm Intelligence.* San Francisco, CA: Kauffman publishers.

Kilgour, F. G. (1992). Entrepreneurial leadership. *Library Trends, 40*(3), 457–474.

Kim, D. H. (1998). The link between individual and organizational learning. *The Strategic Management of Intellectual Capital,* 41–62.

Kimball, L. (1999). *The virtual team: Strategies to optimize performance.* Retrieved from http://www.groupjazz.com/pdf/opt-perf.pdf

Kimble, C., Li, F., & Barlow, A. (2000) Effective virtual teams trough communities of Practice. Unpublished manuscript, Strathclyde Business School, University of Strathclyde, Glasglow, Scotland.

Kim, J. Y. (2000). Social interaction in computer-mediated communication. *Bulletin of the American Society for Information Science and Technology, 26*(3), 15–17. doi:10.1002/bult.153

Kirk, D. J. (2014). The "Singapore of the middle east": The role and attractiveness of the Singapore model and TIMSS on education policy and borrowing in the Kingdom of Bahrain. In *Education for a Knowledge Society in Arabian Gulf Countries.* Emerald Group Publishing Limited.

Kirkman, B. L., Rosen, B., Tesluk, P., & Gibson, C. (2004). The impact of team empowerment on virtual team performance: The moderating role of face-to-face interaction. *Academy of Management Journal, 47*(2), 175–192. doi:10.2307/20159571

Kirkman, B., & Rosen, B. (1999). Beyond self-management: Antecedents and consequences of team empowerment. *Academy of Management Journal, 42*(1), 58–74. doi:10.2307/256874

Kisielnicki, J. (2008). *Virtual technologies: Concepts, methodologies, tools, and applications.* Hershey, PA: Information Science Reference / IGI Global. doi:10.4018/978-1-59904-955-7

Kissler & Gray. (2000). E-leadership. *Organizational Dynamics, 30*(2).

Kittur, A. (2010). Crowdsourcing, collaboration and creativity. *ACM XRDS, 17*(2), 22–26. doi:10.1145/1869086.1869096

Klaus, H., Rosemann, M., & Gable, G. G. (2000). What is ERP? *Information Systems Frontiers, 2*(2), 141–162. doi:10.1023/A:1026543906354

Kleemann, F., & Voß, G. G. (2008). Un(der)paid Innovators: The Commercial Utilization of Consumer Work through Crowdsourcing. *Science. Technology & Innovation Studies, 4*(1), 6–26.

Klein, H., & Myers, M. (1999). A set of principles for conducting and evaluating interpretive field studies in information systems. *Management Information Systems Quarterly, 23*(1), 67–93. doi:10.2307/249410

Klein, M. (1991). Supporting conflict resolution in cooperative design systems. *IEEE Transactions on Systems, Man, and Cybernetics, 21*(6), 1379–1390. doi:10.1109/21.135683

Kleis, L., Chwelos, P., Ramirez, R. V., & Cockburn, I. (2012). Information technology and intangible output: The impact of IT investment on innovation productivity. *Information Systems Research, 23*(1), 42–59. doi:10.1287/isre.1100.0338

Klitmøller, A., & Lauring, J. (2013). When global virtual teams share knowledge: Media richness, cultural difference and language commonality. *Journal of World Business*, *48*(3), 398–406. doi:10.1016/j.jwb.2012.07.023

Klotz-Young, H. (2012). *The Virtual' Marketing Team*. Security Distributing & Marketing.

Knockaert, M., Ucbasaran, D., Wright, M., & Clarysse, B. (2011). The Relationship Between Knowledge Transfer, Top Management Team Composition, and Performance: The Case of Science-Based Entrepreneurial Firms. *Entrepreneurship Theory and Practice*, *35*(4), 777–803. doi:10.1111/j.1540-6520.2010.00405.x

Kock, N. (2005). What is e-collaboration? *International Journal of e-Collaboration*, *1*(1), i–vii.

Kock, N. (2011). *E-Collaboration Technologies and Organizational Performance: Current and Future*. New York: Information Science Reference. doi:10.4018/978-1-60960-466-0

Kock, N., & Lynn, G. S. (2012). Electronic media variety and virtual team performance: The mediating role of task complexity coping mechanisms. *IEEE Transactions on Professional Communication*, *55*(4), 325–344. doi:10.1109/TPC.2012.2208393

Kock, N., & Nosek, J. (2005). Expanding the boundaries of e-collaboration. *IEEE Transactions on Professional Communication*, *48*(1), 1–9.

Ko, D., & Fink, D. (2010). Information technology governance: An evaluation of the theory-practice gap. *Corporate Governance*, *10*(5), 662–674. doi:10.1108/14720701011085616

Koivunen, M., Hätonen, H., & Välimäki, M. (2008). Barriers to facilitators influencing the implementation of an interactive internet-portal application for patient education in psychiatric hospital. *Patient Education and Counseling*, *70*(3), 412–419. doi:10.1016/j.pec.2007.11.002 PMID:18079085

Kolowich, L. (2014, October 22). *Will Telecommuting Replace the Office? How Technology Is Shaping the Workplace*. Retrieved February 28, 2015 from http://blog.hubspot.com/marketing/technologyremoteworkstatsinfographic

Kossaï, M., & Piget, P. (2014). Adoption of information and communication technology and firm profitability: Empirical evidence from Tunisian SMEs. *The Journal of High Technology Management Research*, *25*(1), 9–20. doi:10.1016/j.hitech.2013.12.003

Kostner, J. (1994). *Virtual leadership: Secrets from the Round Table for the multi-site manager*. New York, NY: Warner Books.

Kratzer, J., Leenders, R., & Van Engelen, J. (2006). Managing creative team performance in virtual environments: An empirical study in 44 R&D teams. *Technovation*, *26*(1), 42–49. doi:10.1016/j.technovation.2004.07.016

Kraut, R., Maher, M., Lou, U. S. N., Olson, J., & Thomas, J. C. (2010). Scientific Foundations: A Case For Technology Mediated Social Participation Theory. *IEEE Computer*, *43*(11), 22-28.

Krishna, S., Sahay, S., & Walsham, G. (2004). Cross-cultural issues in global software outsourcing. *Communications of the ACM*, *47*(4), 62–66. doi:10.1145/975817.975818

Kücklich, J. (2014). *Play and playability as key concepts in new media studies*. Retrieved from http://es.scribd.com/doc/205038181/Play-and-Playability-as-Key-Concepts-in-New-Media-Studies#scribd

Kudyba, S., & Vitaliano, D. (2003). Information technology and corporate profitability: A focus on operating efficiency. *Information Resources Management Journal*, *16*(1), 1–13. doi:10.4018/irmj.2003010101

Kumar, K., & van Hillegersberg, J. (2000). ERP experiences and evolution. *Communications of the ACM*, *43*(4), 23–26.

Kuo, F.-y., & Yu, C. (2009). An Exploratory Study of Trust Dynamics in Work-Oriented Virtual Teams. *Journal of Computer-Mediated Communication, 14*(4), 823–854. doi:10.1111/j.1083-6101.2009.01472.x

Kuratko, D. F., & Hornsby, J. S. (1999). Corporate entrepreneurial leadership for the 21st Century. *Journal of Leadership & Organizational Studies, 5*(2), 27–39. doi:10.1177/107179199900500204

Kuruppuarachci, P. (2009). Virtual team concepts in projects: A case study. *Project Management Journal, 40*(2), 19–53. doi:10.1002/pmj.20110

Kvapil, L. A. (2012). *The Agricultural Terraces of Korphos-Kalamianos: A Case Study of the Dynamic Relationship Between Land Use and Socio-Political Organization in Prehistoric Greece.* University of Cincinnati.

La Monte, D. P., & Woytowitz, P. J. (2015). *Landscape controller with feature module.* Google Patents.

Ladimeji, K. (2014). *Why you should hire candidates for EQ rather than IQ.* Retrieved from https://www.recruiter.com/i/why-you-should-hire-candidates-for-eq-rather-than-iq/

Laftit, S., Proth, J. M., & Xie, X. L. (1992). Optimization of invariant criteria for event graphs. *IEEE Transactions on Automatic Control, 37*(12), 547–555. doi:10.1109/9.135488

Lahm, R. J. Jr. (2013). Obamacare and small business: Delays and "glitches" exacerbate uncertainty and economic consequences. *Journal of Management and Marketing Research, 16*, 1–16.

Lai, E. (2008). *Size matters: Yahoo claims 2-petabyte database is world's biggest, busiest.* Retrieved from http://www.computerworld.com/article/2535825/business-intelligence/size-matters--yahoo-claims-2-petabyte-database-is-world-s-biggest--busiest.html

Lambotte, F. (2013). *Managing & Working in Virtual Teams.* Retrieved February 28, 2015 from http://www.academia.edu/4415499/How_to_manage_a_virtual_team_How_to_communicate_in_virtual_teams

Larsline. (2006). *Virtual engineering teams: Strategy and implementation.* Retrieved from www.itcon,org/1007/3/paper.html

Larsson, A., Larsson, T., Bylund, N., & Isaksson, O. (2007). Rethinking virtual teams for streamlined development. In S. P. MacGregor & T. Torres-Coronas (Eds.), *Higher creativity for virtual teams: Developing platforms for co-creation* (pp. 138–156). Hershey, PA: Information Science Reference. doi:10.4018/978-1-59904-129-2.ch007

Latoza, T. D., Ben Towne, W., Adriano, C. M., & Van Der Hoek, A. (2014). Microtask Programming: Building Software with a Crowd. In *Proceedings of User Interface Software and Technology Symposium* (UIST). Honolulu, HI: ACM. doi:10.1145/2642918.2647349

Lau & Sarker. (2004). On managing virtual teams. *HCSM, 2nd quarter.*

Laudon, K., & Laudon, J. (2004). *Management Information Systems.* Mexico City, Mexico: Prentice Hall.

Laurey & Raisinghani. (2001). An impirical study of best practices in virtual team. *Information & Management,* 523–544.

Law, R., & Jogaratnam, G. (2005). A study of hotel information technology applications. *International Journal of Contemporary Hospitality Management, 17*(2–3), 170–180. doi:10.1108/09596110510582369

Leiss, W. (2013). *Social communication in advertising: Consumption in the mediated marketplace.* Routledge.

Lekushoff, A. (2012). Lifestyle-driven virtual teams: A new paradigm for professional services firms. *Ivey Business Journal, 76*(5). Retrieved February 28, 2015 from http://iveybusinessjournal.com/ publication/lifestyle-driven-virtual-teams-a-new-paradigm-for-professional-services-firms/

Lemmex, S. (2005). *Successfully Managing Remote Teams*. Retrieved February 28, 2015 from ftp://ftp.software.ibm.com/software/emea/dk/frontlines/SuccesfullyManagingRemoteTeams-S.pdf

Leonard, B. (2011). Managing virtual teams. *HRMagazine*, *56*(6), 38.

Leonard, E., & Trusty, K. (2015). *Supervision: Concepts and practices of management*. Cengage Learning.

Lepsinger, R. (2014). Virtual team failure: Six common reason why virtual teams do not succeed. *Business Know-How*. Retrieved from http://www.businessknowhow.com/manage/virtualteam.htm

Lerner, J., & Tirole, J. (2002). Some simple economics of open source. *The Journal of Industrial Economics*, *L*(2), 197–234.

Letaief, R., Favier, M., & Le Coat, F. (2006). *Creativity and the creation process in global virtual teams: Case study of the intercultural virtual project*. DBLP.

Levenson, A., & Cohen, S. (2003). Meeting the performance challenge: Calculating return of investment for virtual teams. In C. Gibson & S. Cohen (Eds.), *Virtual teams that work: Creating conditions for virtual team effectiveness* (pp. 145–174). San Francisco, CA: Jossey–Bass.

Levin, G., & Rad, P. (2007). *Key people skills for virtual project managers*. Retrieved from www.aapm.com

Levina, N., & Vaast, E. (2008). Innovating or doing as told? Status differences and overlapping boundaries in offshore collaboration. *Management Information Systems Quarterly*, *32*(2), 307–332.

Levin, B., He, Y., & Robbins, H. (2004) Comparative study of synchronous and asynchronous online case discussions. In C. Crawford et al. (Eds.), *Proceedings of society for InformationTechnology and Teacher Education International Conference* (pp. 551–558). Chesapeake, VA: AACE.

Levy, M., Loebbecke, C., & Powell, P. (2003). SMEs, co-opetition and knowledge sharing: The role of information systems. *European Journal of Information Systems*, *12*(1), 3–17. doi:10.1057/palgrave.ejis.3000439

Li, C. (1999). ERP packages: What's next?[electronic version]. *Information Systems Management*, *16*(3), 31–36. doi:10.1201/1078/43197.16.3.19990601/31313.5

Li, E. (1995). Marketing information systems in US companies: A longitudinal analysis. *Information & Management*, *28*(1), 13–31. doi:10.1016/0378-7206(94)00030-M

Li, F., & Williams, H. (1999). Interfirm collaboration through interfirm networks. *Information Systems Journal*, *9*(2), 103–115. doi:10.1046/j.1365-2575.1999.00053.x

Limbu, Y. B., Jayachandran, C., & Babin, B. J. (2014). Does information and communication technology improve job satisfaction? The moderating role of sales technology orientation. *Industrial Marketing Management*, *43*(7), 1236–1245. doi:10.1016/j.indmarman.2014.06.013

Lin, C. P. (2011). Modeling job effectiveness and its antecedents from a social capital perspective: A survey of virtual teams within business organizations. *Computers in Human Behavior*, *27*(2), 915–923. doi:10.1016/j.chb.2010.11.017

Lines, R. (2007). Using power to install strategy: The relationships between expert power, position power, influence tactics and implementation success. *Journal of Change Management*, *7*(2), 143–170. doi:10.1080/14697010701531657

Linnes, C. (2014). College Student Perception of Electronic Textbook Usage. *American Journal of Information Technology*, *4*(2), 21–39.

Lipnack, J., & Stamps, J. (1997). *"Virtual teams", reaching across space, time, and organizations with technology*. New York, NY: Wiley.

Lipnack, J., & Stamps, J. (1997). *Virtual teams: Reaching across space, time and organizations wit technology.* New York, NY: John Wiley & Sons.

Lipnack, J., & Stamps, J. (1997). *Virtual teams: Reaching across space, time, and organizations with technology.* Jeffrey Stamps.

Lipnack, J., & Stamps, J. (2000). *Virtual teams: People working across boundaries with technology.* New York: John Wiley & Co.

Lister, K., & Harnish, T. (2011, June). *The State of Telework in the U.S.* Retrieved February 24, 2015, from http://www. workshifting.com/downloads/downloads/Telework-Trends-US.pdf

Lister, K., & Harnish, T. (2011). *The State of Telework in the US.* Telework Research Network.

Litan, R. E., & Rivlin, A. M. (2001). Projecting the economic impact of the internet. *The American Economic Review, 91*(2), 313–317. doi:10.1257/aer.91.2.313

Lockwood, N. (2010). *Successfully Transitioning to a Virtual Organization: Challenges, Impact and Technology.* Alexandria, VA: SHRM Research Quarterly.

Luo, X., Slotegraaf, R. J., & Pan, X. (2006). Cross-functional "coopetition": The simultaneous role of cooperation and competition within firms. *Journal of Marketing, 70*(2), 67–80. doi:10.1509/jmkg.70.2.67

Luse, A., McElroy, J. C., Townsend, A. M., & DeMarie, S. (2013). Personality and cognitive style as predictors of preference for working in virtual teams. *Computers in Human Behavior, 29*(4), 1825–1832. doi:10.1016/j.chb.2013.02.007

Lyman, K. B., Caswell, N., & Biem, A. (2009). Business value network concepts for the extended enterprise. In P. H. M. Vervest, D. W. Liere, & L. Zheng (Eds.), *Proc. of the Network Experience.* Berlin: Springer. doi:10.1007/978-3-540-85582-8_9

Lynch, R. (2003). *Corporate strategy* (3rd ed.). Harlow: Prentice-Hall Financial Times.

MacBeth, D. K. (2002). Emergent strategy in managing cooperative supply chain change. *International Journal of Operations & Production Management, 22*(7), 728–740. doi:10.1108/01443570210433517

MacMahon. (2004). *Virtual Project Management for software.* McGraw-Hill.

Madu, C. N., & Kuei, C. (2004). *ERP and supply chain management.* Fairfield, CT: Chi Publishers.

Mahajan, N. (2014). *The importance of creativity in business.* Retrieved from http://knowledge.ckgsb.edu.cn/2014/08/13/marketing/the-importance-of-creativity-in-business/

Mahoney, J. T. (1992). The choice of organisational form: Vertical financial ownership versus other methods of vertical integration. *Strategic Management Journal, 13*(8), 559–584. doi:10.1002/smj.4250130802

Maiga, A. S., Nilsson, A., & Jacobs, F. A. (2014). Assessing the interaction effect of cost control systems and information technology integration on manufacturing plant financial performance. *The British Accounting Review, 46*(1), 77–90. doi:10.1016/j.bar.2013.10.001

Majchrzak, A., Rice, R. E., King, N., Malhotra, A., & Sulin, B. (2000). Computer-mediated inter-organizational knowledge-sharing: Insights from a virtual team innovating using a collaborative tool. *Information Resources Management Journal, 13*(2), 44–53. doi:10.4018/irmj.2000010104

Malhotra, A. & Majchrzak, A. (2005). Virtual Workplace Technologies. *MIT Sloan Management Review, 46*(2), 11-16.

Malhotra, A. (2010). *Managing an A-Team of Far-flung Experts Requires Special Leadership Tactics*. Retrieved from http://www.kenan-flagler.unc.edu/~/media/files/documents/Malhotra-leadership-tactics

Malhotra, A., Majchrzak, A., Carman, R., & Lott, V. (2001). Radical innovation without collacation: A case study at Boein-Rocketdyne. *Management Information Systems Quarterly*, *25*(2), 229–249. doi:10.2307/3250930

Malhotra, A., Majchrzak, A., & Rosen, B. (2007). Leading virtual teams. *The Academy of Management Perspectives*, *21*(February), 60–71. doi:10.5465/AMP.2007.24286164

Maliniak, D. (2001). Design Teams Collaborate Using Internet Fast Track. *Electronic Design*, *49*(11), 69.

Management Study Guide. (n.d). *Advantages and Disadvantages of Virtual Teams*. Retrieved on February 24, 2015 from http://www.managementstudyguide.com/virtual-teams-advantages-and-disadvantages.htm

Manning, C. D., Raghavan, P., & Schütze, H. (2008). *Introduction to information retrieval* (Vol. 1). Cambridge, UK: Cambridge University Press. doi:10.1017/CBO9780511809071

Mansor, N. N. B., Mirahsani, S., & Saidi, M. I. (2012). Investigating possible contributors towards "organizational trust" in effective "virtual team" collaboration context. *Procedia: Social and Behavioral Sciences*, *57*, 283–289. doi:10.1016/j.sbspro.2012.09.1187

Manz, C. C. (1986). Self-leadership: Toward an expanded theory of self-influence processes in organizations. *Academy of Management Review*, *11*(3), 585–600.

Marantz Henig, R. (2008, February 17). *The New York Times*. Retrieved from http://www.nytimes.com/2008/02/17/magazine/17play.html?pagewanted=all&_r=0

Marketwired. (2011). *Canadian study reveals lack of team collaboration despite its positive business impact*. Retrieved from http://www.marketwired.com/press-release/canadian-study-reveals-lack-team-collaboration-despite-its-positive-business-impact-lse-inf-1581172.htm

Marks, M. A., Sabella, M. J., Burke, C. S., & Zaccaro, S. J. (2002). The impact of cross-training on team effectiveness. *The Journal of Applied Psychology*, *87*(1), 3.

Markus, M. L., Manville, B., & Agres, C. (2000). What makes a virtual organization work? *Sloan Management Review*, (Fall): 13–26.

Markus, M. L., Manville, B., & Agres, C. E. (2014). What makes a virtual organization work: Lessons from the open-source world. *Image*.

Markus, M. L., & Tanis, C. (2000). The enterprise system experience – from adoption to success. In R. W. Zmud (Ed.), *Framing the domains of IT management: Projecting the future through the past* (pp. 173–207). Cincinnatti, OH: Pinnaflex Educational Resources, Inc.

Maron, M. E. (2008). An historical note on the origins of probabilistic indexing. *Information Processing & Management*, *44*(2), 971–972. doi:10.1016/j.ipm.2007.02.012

Marquina, J. (2013, April 18). *Collective intelligence: crowdsourcing* [Message in a blog]. Retrieved from http://www.julianmarquina.es/tag/web-social/

Martinez, M. T., Fouletier, P., Park, K. H., & Faurel, J. (2001). Virtual enterprise: Organization, evolution and control. *International Journal of Production Economics*, *74*(1-3), 225–238. doi:10.1016/S0925-5273(01)00129-3

Martin, L., & Matlay, H. (2001). Blanket' approaches to promoting ICT in small firms: Some lessons from the DTI ladder adoption model in the UK. *Internet Research: Electronic Networking Applications and Policy, 11*(5), 399–410. doi:10.1108/EUM0000000006118

Martins, L. L., & Shalley, C. E. (2009). *Creativity in virtual work: Effects of demographic differences.* Retrieved from http://icos.umich.edu/sites/icos6.cms.si.umich.edu/files/lectures/MartinsShalley.pdf

Martins, L. L., Shalley, C. E., & Gilson, L. L. (2009). Virtual teams and creative performance. *Proceedings of the (42th) Annual Hawaii International Conference on Systems Science.*

Martins. (2004). Virtual team: What do we know and where do we go from here. *Journal of Management,* 30-41.

Martins, L. L., Gilson, L. L., & Maynard, M. T. (2004). Virtual teams: What do we know and where do we go from here? *Journal of Management, 30*(6), 805–835. doi:10.1016/j.jm.2004.05.002

Mason, W. (2014). *How videoconferencing in the cloud can transform business collaboration.* Retrieved from http://www.techradar.com/us/news/world-of-tech/how-video-conferencing-in-the-cloud-can-transform-business-collaboration-1250490

Massey, A., Montoya-Weiss, M., & Hung, Y. (2003). Because Time Matters: Temporal Coordination in Global Virtual Project Teams. *Journal of Management Information Systems, 19*(4), 129–155.

Maurizio, A., Girolami, L., & Jones, P. (2007). EAI and SOA: Factors and methods influencing the integration of multiple ERP systems (in an SAP environment) to comply with the Sarbanes-Oxley Act. *Journal of Enterprise Information Management, 20*(1), 14–31. doi:10.1108/17410390710717110

Maynard, H., & Mehrtens, S. (1993). *The fourth wave: Business in the 21ˢᵗ century.* San Francisco, CA: Berrett-Koehler Publishers.

Maynard, M. T., & Gilson, L. (2013). The role of shared mental model development in understanding virtual team effectiveness. *Group & Organization Management, 39*(1), 3–32. doi:10.1177/1059601113475361

Maznevski, M. L., & Chudoba, K. M. (2000). Bridging space over time: Global virtual team dynamics and effectiveness. *Organization Science, 11*(5), 473–492. doi:10.1287/orsc.11.5.473.15200

McConnell, M. (2002). Capacity building for a sustainable shipping industry: a key ingredient in improving coastal and ocean and management. *Ocean & Coastal Management, 45*(10), 617-632. doi:10.1016/S0964-5691(02)00089-3

McDonough & Kenneth. (2001). An investigation of the use global virtual and collocated new product development teams. *Journal of Product Innovation Management.*

McDonough, E. F., Kahnb, K. B., & Barczaka, G. (2001). An investigation of the use of global, virtual, and colocated new product development teams. *Journal of Product Innovation Management, 18*(2), 110–120. doi:10.1016/S0737-6782(00)00073-4

McEvily, B., Perrone, V., & Zaheer, A. (2003). Trust as an organizing principle. *Organization Science, 14*(1), 91–103. doi:10.1287/orsc.14.1.91.12814

McGannon, B. (2014, June 23). *Managing Remote Employees.* Retrieved February 28, 2015 from http://www.lynda.com/Business-Skills-tutorials/Managing-remote-employees/156090/179147-4.html

McGrath, J. E., & Hollingshead, A. B. (1994). *Groups interacting with technology.* Thousand Oaks, CA: Sage Publications.

McKnight, D. H., Cummings, L. L., & Chervany, N. L. (1998). Initial trust formation in new organizational relationships. *Academy of Management Review, 23*(3), 473–490.

McLean, J. (2007). Managing global virtual teams. *British Journal of Administrative Management, 59*(2), 16–17.

Medellin, C. (2013). *Build innovation.* Mexico City, Mexico: Fese.

Meier, P. (2015). Digital Humanitarians: How Big Data is changing the face of humanitarian response. Taylor & Francis Press. Spring 2015.

Meier, P., & Munro, R. (2010). The unprecedented role of SMS in disaster response: Learning from Haiti. *SAIS Review of International Affairs, 30*(2), 91–103.

Mell, P., & Grance, T. (2011). *The NIST Definition of Cloud Computing Recommendations of the National Institute of Standards and Technology.* NIST Special publication 800-145.

Mental Health Roundtable. (2011). Retrieved from http://www.mentalhealthroundtable.ca/

Meyer, P., & Swatman, P. (2009). *Virtual worlds: The role of rooms and avatars in virtual teamwork.* Paper presented at the 2009 Americas Conference on Information Systems (AMCIS), San Francisco, CA.

Michalewicz, Z., & Michalewicz, M. (1997). Evolutionary computation techniques and their applications. *IEEE International Conference on Intelligent Processing Systems, ICPIS.*

Michel, R. (2000). *The road to extended ERP.* Retrieved May 8, 2009, from www.manufacturingsystems.com/extend-edenterprise

Micromecenazgo. (2015, January 6). In *Wikipedia, the free encyclopedia* [Electronic version]. Retrieved from http://es.wikipedia.org/w/index.php?title=Micromecenazgo&oldid=79243220

Midha, V., & Nandedkar, A. (2012). Impact of similarity between avatar and their users on their perceived identifiability: Evidence from virtual teams in Second Life platform. *Computers in Human Behavior, 28*(3), 929–932. doi:10.1016/j.chb.2011.12.013

Mihalic, J. (2013, October). Leading the way with thought leadership in the project management community. *PMI Today,* 3-5.

Mihhailova, G. (2007). *From ordinary to virtual teams: A model for measuring the virtuality of a teamwork.* Retrieved at: http://managementstudyguide.com/degree-of-virtuality-in-teams.htm

Miller, D. (1988). Relating porter's business strategies to environment and structure: Analysis and performance implications. *Academy of Management Journal, 31*(2), 280–308. doi:10.2307/256549

Millward, L., & Kyriakidou, O. (2004). Effective virtual teamwork. In S. H. Godar & S. P. Ferris (Eds.), *Leading from Afar: Strategies for Effectively Leading Virtual Teams* (pp. 20–34). Hershey, PA: Idea Group Publishing.

Minton-Eversole, T. (2012). *Virtual Teams Used Most by Global Organizations, Survey Says.* Accessed on December 10, 2014 from http://www.shrm.org/hrdisciplines/orgempdev/articles/pages/virtualteamsusedmostbyglobalorganizations,surveysays.aspx

Miola, A., Marra, M., & Ciuffo, B. (2011). Designing a climate change policy for the international maritime transport sector: Market-based measures and technological options for global and regional policy actions. *Energy Policy, 39*(9), 5490–5498. doi:10.1016/j.enpol.2011.05.013

Mocanu, M. D. (2014). Virtual Teams–An Opportunity in the Context of Globalization. *Business Excellence and Management, 4*(1), 47–53.

Mohamed, N., & Singh, J. K. G. (2012). A conceptual framework for information technology governance effectiveness in private organizations. *Information Management & Computer Security, 20*(2), 88–106. doi:10.1108/09685221211235616

Moller, C. (2005). ERPII: A conceptual framework for next-generation enterprise systems? *Journal of Enterprise Information Management, 18*(4), 483–497. doi:10.1108/17410390510609626

Monk, E. F., & Wagner, B. J. (2009). *Concepts in enterprise resource planning* (3rd ed.). Cambridge, MA: Course Technology, Cengage Learning.

Monteverde, K., & Teece, D. J. (1982). Supplier switching costs and vertical integration in the automobile industry. *The Bell Journal of Economics, 13*(1), 206–213. doi:10.2307/3003441

Montoya, M., Massey, A., & Lockwood, N. (2011). 3D Collaborative Virtual Environments: Exploring the Link between Collaborative Behaviors and Team Performance. *Decision Sciences, 42*(2), 451–476. doi:10.1111/j.1540-5915.2011.00318.x

Montoya-Weiss, M., Massey, A., & Song, M. (2001). Getting it together: Temporal oordination and conflict management in global virtual teams. *Academy of Management Journal, 44*(6), 1251–1262. doi:10.2307/3069399

Morcillo, P. (2011). *Natural innovating. The pass says it all.* Retrieved from https://books.google.com.co/books?id=9F j3pIPqmzgC&pg=PA73&dq=Crowdcreating&hl=es&sa=X&ei=xyCgVMa8IZHUgwTGoILgAQ&ved=0CDIQuwU wAw#v=onepage&q=Crowdcreating&f=false, 73.

Muethel, M., & Hoegl, M. (2010). Cultural and societal influences on shared leadership in globally dispersed teams. *Journal of International Management, 16*(3), 234–246. doi:10.1016/j.intman.2010.06.003

Muniapan, B. (2014). The Bhagavad-Gita and Business Ethics: A Leadership Perspective. *Asian Business and Management Practices: Trends and Global Considerations: Trends and Global Considerations*, 232.

Muscatello, J. R., Small, M. H., & Chen, I. J. (2003). Implementing enterprise resource planning (ERP) systems in small and midsize manufacturing firms. *International Journal of Operations & Production Management, 23*(8), 850–871. doi:10.1108/01443570310486329

Mustafa, G., & Lines, R. (2012). The triple role of values in culturally adapted leadership styles. *International Journal of Cross Cultural Management*, 1470595812452636.

Mutual Support. (2014, December 11). In *Wikipedia, the free encyclopedia* [Electronic version]. Retrieved from http:// es.wikipedia.org/w/index.php?title=Apoyo_mutuo&oldid=78703967

Nag, S., Heffan, I., Alvar, S.-O., & Lydon, M. (2012). SPHERES Zero Robotics Software Development: Lessons on Crowdsourcing and Collaborative Competition. In *Proceedings of Aerospace Conference*. IEEE. doi:10.1109/ AERO.2012.6187452

Nananukul, S., & Gong, W. B. (1999). Rational interpolation for stochastic DES's coverage issues. *IEEE Transactions on Automatic Control, 44*(5), 1070–1078. doi:10.1109/9.763231

Nandhakumar & Baskervil. (2001). Trusting online:nurturing trust in virtual teams. *The 9th European Conference on Information Systems.* Slovenia.

Naur, P., & Randell, B. (1968). Software Engineering: Report of a Conference Sponsored by the NATO Science Committee. Garmisch, Germany: NATO.

Nemiro, J. E. (2001). Assessing the climate for creativity in virtual teams. In M. Beyerlein, D. Johnson, & S. Beyerlein (Eds.), *Virtual teams: Advances in interdisciplinary studies of work teams* (pp. 59–84). Bingley, UK: Emerald Group Publishing. doi:10.1016/S1572-0977(01)08019-0

Nemiro, J. E. (2004). *Creativity in virtual teams: Key components for success.* New York: John Wiley & Co.

Neufeld, D. J., & Fang, Y. (2005). Individual, social and situational determinants of telecommuter productivity. *Information & Management, 42*(7), 1037–1049.

Neufeld, D., Wan, Z., & Fang, Y. (2008). Remote leadership, communication effectiveness and leader performance. *Journal of Group Decision and Negotiation, 19*(3), 227–246. doi:10.1007/s10726-008-9142-x

Niman, N. B. (2014). *The Gamification of Higher Education: Developing a Game-based Business Strategy in a Disrupted Marketplace.* Palgrave Macmillan. doi:10.1057/9781137331465

Noll & Scachi. (1999). Supporting software development in virtual enterprise. *Journal of Digital Information, 1*(4).

Nonaka, I. (1991). The knowledge-creating company. *Harvard Business Review, 69*(6), 96–104.

Nunamaker, J. F., Jr., & Chen, M. (1990, January). Systems development in information systems research. In System Sciences, 1990. In *Proceedings of the Twenty-Third Annual Hawaii International Conference* (Vol. 3, pp. 631-640). IEEE.

Nydegger, R., & Nydegger, L. (2010, March). Challenges in Managing Virtual Teams. *Journal of Business & Economics Research, 8*(3), 69-82. Retrieved on February 24, 2015 from http://www.cluteinstitute.com/ojs/index.php/JBER/article/view/690

O'Hara-Devereaux, M., & Johansen, R. (1994). *Globalwork: Bridging distance, culture, and time.* San Francisco, CA: Jossey-Bass.

O'Leary, M., & Mortensen, M. (2005). *Subgroups with attitude: Imbalance and isolation in geographically dispersed teams.* Presented at the Academy of Management Conference, Honolulu, HI.

O'Leary, M. B., & Mortensen, M. (2008). A surprising truth about geographically distributed teams. *MIT Sloan Management Review, 49*(4), 5–6.

O'Leary, M. B., Wilson, J. M., & Metiu, A. (2014). Beyond being there: The symbolic role of communication and identification in perceptions of proximity to geographically dispersed colleagues. *Management Information Systems Quarterly, 38*(4), 1219–1243.

O'Leary, M., & Cummings, J. (2007). The spatial, temporal, and configurational characteristics of geographic dispersion in teams. *Management Information Systems Quarterly, 31*(3), 433–452.

O'Neill, H., & Sackett, P. (1994). The extended manufacturing enterprise paradigm. *Management Decision, 32*(8), 42–49. doi:10.1108/00251749410069453

O'Neil, W. A. (2003). The human element in shipping. *WMU Journal of Maritime Affairs, 2*(2), 95–97. doi:10.1007/BF03195037

O'sallivan. (2003). Dispersed collaboration in a multi-firm, multi-team, product-development projects. *Journal Technology Management*, 93-116.

Obra, A., Camara, S., & Melendez, A. (2002). Internet usage and competitive advantage: The impact of the Internet on an old economy industry in Spain. *Benchmarking: An International Journal, 12*(5), 391–401.

Obstfeld, D. (2005). Social networks, the tertius iungens orientation, and involvement in innovation. *Administrative Science Quarterly, 50*(1), 100–130.

Okolloh, O. (2009). Ushahidi, or 'testimony': Web 2.0 tools for crowdsourcing crisis information. *Participatory Learning and Action, 59*(1), 65-70.

Okumus, F. (2013). Facilitating knowledge management through information technology in hospitality organizations. *Journal of Hospitality and Tourism Technology, 4*(1), 64–80. doi:10.1108/17579881311302356

Olariu, C., & Aldea, C. C. (2014). Managing processes for virtual teams: A BPM approach. *Procedia: Social and Behavioral Sciences, 109*, 380–384. doi:10.1016/j.sbspro.2013.12.476

Oliner, S. D., & Sichel, D. E. (2003). Information technology and productivity: Where are we now and where are we going? *Journal of Policy Modeling, 25*(5), 477–503. doi:10.1016/S0161-8938(03)00042-5

Olson, D. (2014, October 2). *Manage Projects with Wunderlist*. Retrieved February 28, 2015 from http://moultriejournal.net/2014/10/02/manage-projects-with-wunderlist/

Olson, G. M., & Olson, J. S. (2000). Distance matters. *Human-Computer Interaction, 15*(2–3), 139–179. doi:10.1207/S15327051HCI1523_4

Olteanu, A., Castillo, C., Diaz, F., & Vieweg, S. (2014, May). CrisisLex: A lexicon for collecting and filtering micro-blogged communications in crises. In *Proceedings of the 8th International AAAI Conference on Weblogs and Social Media (ICWSM'14)*.

Online Etymology Dictionary. (2014). Retrieved from http://www.etymonline.com/

Onpoint Consulting. (2010). *Six lessons for effective virtual teams: A recipe for success*. Retrieved from http://www.onpointconsultingllc.com/2010/05/six-lessons-for-effective-virtual-teams-a-recipe-for-success/

Ortiz de Guinea, A., Webster, J., & Staples, D. S. (2012). A meta-analysis of the consequences of virtualness on team functioning. *Information & Management, 49*(6), 301–308. doi:10.1016/j.im.2012.08.003

Owen, L., Goldwasser, C., Choate, K., & Blitz, A. (2008). Collaborative innovation throughout the extended enterprise. *Strategy and Leadership, 36*(1), 39–45. doi:10.1108/10878570810840689

Oxford Dictionary Online. (2015). Retrieved from http://www.oxforddictionaries.com/definition/english/team

Padar, K., Pataki, B., & Sebestyen, Z. (2011). A comparative analysis of stakeholder and role theories in project management and change management. *International Journal of Management Cases, 13*(4), 252–260. doi:10.5848/APBJ.2011.00134

Palen, L., Anderson, K. M., Mark, G., Martin, J., Sicker, D., Palmer, M., & Grunwald, D. (2010, April). A vision for technology-mediated support for public participation & assistance in mass emergencies & disasters. In *Proceedings of the 2010 ACM-BCS visions of computer science conference* (p. 8). British Computer Society.

Palen, L., Hiltz, S. R., & Liu, S. B. (2007). Online forums supporting grassroots participation in emergency preparedness and response. *Communications of the ACM, 50*(3), 54–58. doi:10.1145/1226736.1226766

Pal, N., & Pantaleo, D. C. (2005). *The agile enterprise: Reinventing your organization for success in an on-demand world*. New York, NY: Springer SciencetBusiness Media, Inc.

Pankaj, J. (2010). *Software Engineering: A Precise Approach*. Wiley India.

Papa, P. (2013). US and EU strategies for maritime transport security: A comparative perspective. *Transport Policy, 28*(0), 75–85. doi:10.1016/j.tranpol.2012.08.008

Parkinson & Hudson. (2001). Extending the learning exprence using the web &a knowledge –based virtual environment. *Computer &Education*, 95-101.

Park, K., & Kusiak, A. (2005). Enterprise resource planning (ERP) operations support system for maintaining process integration. *International Journal of Production Research, 43*(19), 3959–3982. doi:10.1080/00207540500140799

Paulus, T., & Phipps, G. (2008). Approaches to case analyses in synchronous and asynchronous environments. *Journal of Computer-Mediated Communication, 13*(2), 459–484. doi:10.1111/j.1083-6101.2008.00405.x

Pavic, S., Koh, S. C. L., Simpson, M., & Padmore, J. (2007). Could e-business create a competitive advantage in UK SMEs? *Benchmarking: An International Journal, 14*(3), 320–351. doi:10.1108/14635770710753112

Pawar, B., & Eastman, K. K. (1997). The nature and implications contextual influences on transformational leadership: A conceptual examination. *Academy of Management Review, 22*(1), 80–109.

Pearce, C. L., & Ensley, M. D. (2004). *A reciprocal and longitudinal investigation of the innovation process: The central role of shared vision in product and process innovation teams (ppits).* Academic Press.

Pearce, C. L. (2004). The future of leadership: Combining vertical and shared leadership to transform knowledge work. *The Academy of Management Executive, 18*(1), 47–57. doi:10.5465/AME.2004.12690298

Pearce, C. L., & Conger, J. a. (2003). *Shared leadership: Reframing the hows and whys of leadership.* Thousand Oaks, CA: Sage Publications.

Pearce, C. L., & Sims, H. P. (2001). Shared leadership: Toward a multi-level theory of leadership. *Advances in Interdisciplinary Studies of Work Teams, 7*, 115–139. doi:10.1016/S1572-0977(00)07008-4

Pearce, C. L., & Sims, H. P. (2002). Vertical versus shared leadership as predictors of the effectiveness of change management teams: An examination of aversive, directive, transactional, transformational, and empowering leader behaviors. *Group Dynamics, 6*(2), 172–197. doi:10.1037/1089-2699.6.2.172

Pearce, C. L., Yoo, Y., & Alavi, M. (2004). Leadership, social work, and virtual teams: The relative influence of vertical vs. shared leadership in the nonprofit section. In R. Reggio & S. Smith Orr (Eds.), *Improving leadership in nonprofit organizations.* San Francisco: Jossey-Bass.

Peñarroja, V., Orengo, V., Zornoza, A., Sánchez, J., & Ripoll, P. (2015). How team feedback and team trust influence information processing and learning in virtual teams: A moderated mediation model. *Computers in Human Behavior, 48*, 9–16. doi:10.1016/j.chb.2015.01.034

Pena-Shaef, J., Martin, W., & Gray, G. (2001). An epistemological framework for analyzing student interactions in computer-mediated communication environments. *Journal of Interactive Learning Research, 12*(1), 41–68.

Pendharkar, P. C. (2013). Genetic learning of virtual team member preferences. *Computers in Human Behavior, 29*(4), 1787–1798. doi:10.1016/j.chb.2013.02.015

Peslak, A. R. (2012). An analysis of critical information technology issues facing organizations. *Industrial Management & Data Systems, 112*(5), 808–827. doi:10.1108/02635571211232389

Pesquera, M. (2013, September 3). *The programmable university in scale-free networks.* [Slideshare]. Retrieved from http://es.slideshare.net/mapesquera/la-universidad-programable-en-redes-libres-de-escala-map-2103?next_slideshow=1

Peter, K., Laura, D., Sara, K., & Haakon, F. (2013). Co-worker transparency in microtask marketplace. In *CSCW'13 Proceedings of 2013 conference on computer supported cooperative work.* San Antonio, TX: ACM.

Peter, S. (1995). *Fifth Discipline.* Barcelona, Spain: Granica SA.

Phuong, T. (2008). Internet use, customer relationships and loyalty in the Vietnamese travel industry. *Asia Pacific Journal of Marketing and Logistics, 20*(2), 190–210. doi:10.1108/13555850810864551

Piccoli, G., Powell, A., & Ives, B. (2004). Virtual teams: Team control structure, work processes, and team effectiveness. *Information Technology & People, 15*(4), 389–406.

Piotr Kropotkin. (2014, December 30). In *Wikipedia, the free encyclopedia* [Electronic version]. Retrieved from http://es.wikipedia.org/w/index.php?title=Piotr_Kropotkin&oldid=79108122

Politis, J., & Politis, D. (2011). The Big Five Personality Traits and the art of Virtual Leadership. In *Proceedings of the European Conference on Management. Academic Conferences, Ltd.*

Poltrock, S. E., & Engelbeck, G. (1999). Requirements for a virtual collocation environment. *Information and Software Technology, 41*(6), 331–339. doi:10.1016/S0950-5849(98)00066-4

Ponis, S. T., & Spanos, A. C. (2009). ERPII systems to support dynamic, reconfigurable and agile virtual enterprises. *International Journal of Applied Systemic Studies, 2*(3), 265–283. doi:10.1504/IJASS.2009.027664

Poston, R., & Grabski, S. (2001). Financial impact of enterprise resource planning implementations. *International Journal of Accounting Information Systems, 2*(4), 271–294. doi:10.1016/S1467-0895(01)00024-0

Powell, A., Piccoli, G., & Ives, B. (2004). Virtual teams: A review of current literature and directions for future research. *The Data Base for Advances in Information Systems, 35*(1), 6–36. doi:10.1145/968464.968467

Prabhakaran, T., Lathabai, H. H., & Changat, M. (2015). Detection of paradigm shifts and emerging fields using scientific network: A case study of Information Technology for Engineering. *Technological Forecasting & Social Change: An International Journal, 91*, 124–145. doi:10.1016/j.techfore.2014.02.003

Pressman, R. S. (2014). *Software Engineering A Practitioner's Approach (7th ed.)*. Mc Graw Hill Education.

Protalinski, E. (2014). *Facebook passes 1.23 billion monthly active users, 945 million mobile users, and 757 million daily users.* Retrieved February 28, 2015 from http://thenextweb.com/facebook/2014/01/29/facebook-passes-1-23-billion-monthly-active-users-945-million-mobile-users-757-million-daily-users/

Proth, J. M., & Xie, X. L. (1994). Cycle time of stochastic event graphs: Evaluation and marking optimization. *IEEE Transactions on Automatic Control, 39*(7), 1482–1486. doi:10.1109/9.299640

Psychometrics Canada. (2009). *Warring egos, toxic individuals, feeble leadership. A study of conflict in the Canadian workplace.* Retrieved from http://www.psychometrics.com/docs/conflict%20study%20news%20release%20feb%202009.pdf

Pui, S. (2010). Maintaining packaged software: A revelatory study. *Journal of Information Technology, 25*(1), 65–90. doi:10.1057/jit.2009.8

Purdue University. (2008). *Virtual teams are here to stay: to succeed, we must learn how to work with them.* Retrieved June 26, 2015, from http://www.purdue.edu/uns/x/2008b/080730O-BeyerleinTeams.html

Purohit, H., Castillo, C., Diaz, F., Sheth, A., & Meier, P. (2013). Emergency-relief coordination on social media: Automatically matching resource requests and offers. *First Monday, 19*(1). doi:10.5210/fm.v19i1.4848

Purohit, H., Hampton, A., Bhatt, S., Shalin, V. L., Sheth, A. P., & Flach, J. M. (2014). Identifying Seekers and Suppliers in Social Media Communities to Support Crisis Coordination. *Computer Supported Cooperative Work, 23*(4-6), 513–545. doi:10.1007/s10606-014-9209-y

Purvanova, R. K. (2014). Face-to-face versus virtual teams: What have we really learned? *The Psychologist Manager Journal, 17*(1), 2–29. doi:10.1037/mgr0000009

Purvanova, R., & Bono, J. (2009). Transformational leadership in context: Face-to-face and virtual teams. *The Leadership Quarterly, 20*(3), 343–357. doi:10.1016/j.leaqua.2009.03.004

Qmarkets. (2014). *Learn from the wisdom of the crowd* [Web log post]. Retrieved from http://www.qmarkets.net/additional-products/crowd-voting

Quarantelli, E. L. (1988). Disaster Crisis Management: A Summary of Research Findings. *Journal of Management Studies*, 25(4), 373–385. doi:10.1111/j.1467-6486.1988.tb00043.x

Rafaeli, S., & Ravid, G. (2003). Information sharing as enabler for the virtual team: An experimental approach to assessing the role of electronic mail in disintermediation. *Information Systems Journal*, 13(2), 191–206. doi:10.1046/j.1365-2575.2003.00149.x

Rahim, M. A. (1992). *Managing Conflict in Organizations* (2nd ed.). New York: Praeger.

Rains, S. A. (2007). The impact of anonymity on perceptions of source credibility and influence in computer-mediated group communication: A test of two competing hypotheses. *Communication Research*, 34(1), 100–125. doi:10.1177/0093650206296084

Rao, A. (2014). *Constructing a team identity*. Retrieved from http://www.selfgrowth.com/articles/constructing-a-team-identity

Rappa, M. A. (2004). The utility business model and the future of computing services. *IBM Systems Journal*, 43(1), 32–42. doi:10.1147/sj.431.0032

Rashid, M. A., Hossain, L., & Patrick, J. D. (2002). The evolution of ERP systems: A historical perspective. In L. Hossain, J. D. Patrick, & M. A. Rashid (Eds.), *Enterprise Resource Planning: Global opportunities and challenges* (pp. 1–16). Hershey, PA: Idea Group Publishing. doi:10.4018/978-1-931777-06-3.ch001

Rawson, R. (2010). *The 8 Best Collaboration Tools for Virtual Teams*. Time Doctor. Retrieved May 4, 2015, from http://blog.timedoctor.com/2010/12/03/the-8-best-collaboration-tools-for-virtual-teams

Ray, B. (2010). Apple wipes smile off FaceTime in the Middle East. *The Register*. Retrieved November 20, 2014, from: http://www.theregister.co.uk/2010/10/19/facetime/

Raymond, L., & St-Pierre, J. (2005). Antecedents and performance outcomes of advanced manufacturing systems sophistication in SMEs. *International Journal of Operations & Production Management*, 25(6), 514–533. doi:10.1108/01443570510599692

Rayport, J. F., & Sviokla, J. J. (1995). Exploiting the virtual value chain. *The McKinsey Quarterly*, 1, 21–36.

Reason, J. (1997). Managing the Risks of Organizational Accidents. Ashgate, UK: Aldershot.

Recalde, L., s.a. (1999). Modeling and analysis of sequential processes that cooperate through buffers, *IEEE Trans. on Rob. and Autom.*, 14(2), 267-277.

Reed, A., & Knight, L. (2010). Project risk differences between virtual and co-located teams. *Journal of Computer Information Systems*, (Fall), 19–30.

Reich. (2000). Knowledge traps in IT projects. *CACM*, 36, 63-77.

Reinig, B. A., & Mejias, R. J. (2004). The effects of national culture and anonymity on flaming and criticalness in GSS supported discussions. *Small Group Research*, 35(6), 698–723. doi:10.1177/1046496404266773

Reips, U. D., & Matzat, U. (2014). Mining "Big Data" using Big Data Services. *International Journal of Internet Science*, 9(1), 1–8.

Reis, R. A. D., & Freitas, M. D. C. D. (2014). Critical factors on information technology acceptance and use: An analysis on small and medium Brazilian clothing industries. *Procedia Computer Science*, 31, 105–114. doi:10.1016/j.procs.2014.05.250

Renko, M., El Tarabishy, A., Carsrud, A. L., & Brännback, M. (2015). Understanding and measuring entrepreneurial leadership style. *Journal of Small Business Management*, *53*(1), 54–74. doi:10.1111/jsbm.12086

Reuter, C., Heger, O., & Pipek, V. (2013). Combining real and virtual volunteers through social media. In *Proceedings of the 10th International ISCRAM Conference*. KIT.

Reynolds, B. (2014, October 1). *Survey: People Who Want Flexible Jobs and Why*. Retrieved February 25, 2015, from http://www.flexjobs.com/blog/post/survey-people-who-want-flexible-jobs-and-why/

Richardson, J. (1996). Vertical integration and rapid response in fashion apparel. *Organization Science*, *7*(4), 400–412. doi:10.1287/orsc.7.4.400

Right Management. (2006). *Virtual teaming study*. Retrieved from http://www.right.com/global/includes/pdfs/Virtual_Teaming_Study.pdf

Robert, L., & You, S. (2013). Are you satisfied yet? shared leadership, trust and individual satisfaction in virtual teams. In iConference 2013 (pp. 461–466).

Robertson Associates. (2013). *Which generation are you? X/Y/Z? Lost?* Retrieved April 10, 2014 from: http://www.robertson-associates.eu/blog/2013/11/29/which-generation-are-you-xyz-lost

Rockart, J. F., Ear, M. J., & Ross, J. W. (1996). Eight imperatives for the new IT organization. *MIT Sloan Management Review*, *38*(1), 43–56.

Rodriguez, O. F., Fernandez, F., & Torres, R. S. (2011). Impact of information technology certifications in Puerto Rico. *Management Research: The Journal of the Iberoamerican Academy of Management*, *9*(2), 137–153.

Roller, L. H., & Waverman, L. (2001). Telecommunications infrastructure and economic development: A simultaneous approach. *The American Economic Review*, *91*(4), 909–923. doi:10.1257/aer.91.4.909

Romano, N. C. Jr, Pick, J. B., & Roztocki, N. (2010). A motivational model for technology-supported cross-organizational and cross-border collaboration. *European Journal of Information Systems*, *19*(2), 117–133. doi:10.1057/ejis.2010.17

Romero, D., & Molina, A. (2011). Collaborative networked organisations and customer communities: value co-creation and co-innovation in the networking era. *Production Planning & Control: The Management of Operations*, *22*(5-6), 447-472.

Root-Bernstein, R. S., & Root-Bernstein, M. M. (2013). *Sparks of genius: The thirteen thinking tools of the world's most creative people*. Houghton Mifflin Harcourt.

Rosen, B., Furst, S., & Blackburn, R. (2006). Training for virtual teams: An investigation of current practices and future needs. *Human Resource Management*, *45*(2), 229–247. doi:10.1002/hrm.20106

Rosen, B., Furst, S., & Blackburn, R. (2007). Overcoming barriers to knowledge sharing in virtual teams. *Organizational Dynamics*, *36*(3), 259–273. doi:10.1016/j.orgdyn.2007.04.007

Rothaermel, F. T., Hitt, M. A., & Jobe, L. A. (2006). Balancing vertical integration and strategic outsourcing: Effects on product portfolio, product success, and firm performance. *Strategic Management Journal*, *27*(11), 1033–1056. doi:10.1002/smj.559

Rourke, L., & Anderson, T. (2002). Exploring social presence in computer onferencing. *Journal of Interactive Learning Research*, *13*(3), 259–275.

Rousseau, V., Aube, C., & Savoie, A. (2006). Teamwork behaviors: A review and an integration of frameworks. *Small Group Research*, *37*(5), 540–570. doi:10.1177/1046496406293125

Royce, W. W. (1970). Managing the development of large software systems. In *Proceedings of International Conference on Software Enginerring (ICSE-9)*. IEEE.

Rushton, A., Croucher, P., & Baker, P. (2010). The Handbook of Logistics & Distribution Management (4th ed.). Kogan Page.

Saafein, O., & Shaykhian, G. A. (2014). Factors affecting virtual team performance in telecommunication support environment. *Telematics and Informatics*, *31*(3), 459–462. doi:10.1016/j.tele.2013.10.004

Saeed, K. A., Malhotra, M. K., & Grover, V. (2011). Interorganizational system characteristics and supply chain integration: An empirical research. *Decision Sciences*, *42*(1), 7–42. doi:10.1111/j.1540-5915.2010.00300.x

Salminen-Karlsson, M. (2013). Swedish and Indian Teams: Consensus Culture Meets Hierarchy Culture in Offshoring. In *Proceedings of the European Conference on Information Management & Evaluation. Academic Conferences & Publishing International Ltd*.

Salton, G., Wong, A., & Yang, C. S. (1975). A vector space model for automatic indexing. *Communications of the ACM*, *18*(11).

Sarker, S., Ahuja, M., Sarker, S., & Kirkeby, S. (2011). The role of communication and trust in global virtual teams: A social network perspective. *Journal of Management Information Systems*, *28*(1), 273–310. doi:10.2753/MIS0742-1222280109

Sarker, S., & Sahay, S. (2004). Implications of space and time for distributed work: An interpretive study of US-Norwegian system development teams. *European Journal of Information Systems*, *13*(1), 3–20. doi:10.1057/palgrave.ejis.3000485

Sarker, S., & Valacich, J. (2010). An Alternative To Methodological Individualism: A Non-Reductionist Approach To Studying technology Adoption By Group. *Management Information Systems Quarterly*, *34*(4), 779–808.

Schaper, L. K., & Pervan, G. P. (2007). ICTs & OTs: A model of information and communications technology acceptance and utilisation by occupational therapists. *International Journal of Medical Informatics*, *76*(1), S212–S221. doi:10.1016/j.ijmedinf.2006.05.028 PMID:16828335

Schiller, S. Z., Mennecke, B. E., Nah, F. F. H., & Luse, A. (2014). Institutional boundaries and trust of virtual teams in collaborative design: An experimental study in a virtual world environment. *Computers in Human Behavior*, *35*, 565–577. doi:10.1016/j.chb.2014.02.051

Schimmer, T. (2011). *Enough with the Late Penalties!* Retrieved February 1, 2015 from http://tomschimmer.com/2011/02/21/enough-with-the-late-penalties/

Schmidt, J., Montoya-Weiss, M., & Massey, A. (2001). New Product Development Decision-Making Effectiveness: Comparing Individuals, Face-to-Face Teams, and Virtual Teams. *Decision Sciences*, *32*(4), 575–600. doi:10.1111/j.1540-5915.2001.tb00973.x

Schultze, U., & Orlikowski, W. J. (2010). Research commentary – Virtual worlds: A performative perspective on globally distributed, immersive work. *Information Systems Research*, *21*(4), 810–821. doi:10.1287/isre.1100.0321

Schwaber, K. (2009). *Scrum Guide*. ScrumAlliance.

Scott, J. E., & Vessey, I. (2000). Implementing enterprise resource planning systems: The role of learning from failure. *Information Systems Frontiers*, *2*(2), 213–232. doi:10.1023/A:1026504325010

Sedlaczek, K. & Eberhard, P. (2007). Augmented Lagrangian Particle Swarm Optimization in Mechanism Design. *Journal of System Design and Dynamics, 1*(3).

Seeger, M. W., Sellnow, T. L., & Ulmer, R. R. (1998). Communication, organization, and crisis. *Communication Yearbook*, *21*, 231–275.

Seely, A. (2014). /var/log/manager: Rock Stars and Shift Schedules. login:, 39(3).

Sein, M. K., Henfridsson, O., Puraro, S., Rossi, M., & Lindgren, R. (2011). Action Design Research. *Management Information Systems Quarterly*, *35*(1), 37–56.

Senge, P. (2006). *The fifth discipline: The art and practice of the learning organization* (2nd ed.). Melbourne, Australia: Random House.

Sharif, A. M. (2010). It's written in the cloud: The hype and promise of cloud computing. *Journal of Enterprise Information Management*, *23*(2), 131–134. doi:10.1108/17410391011019732

Sharif, A. M., Irani, Z., & Love, P. E. D. (2005). Integrating ERP with EAI: A model for post-hoc evaluation. *European Journal of Information Systems*, *14*(2), 162–174. doi:10.1057/palgrave.ejis.3000533

Sharma, K. L. (2014). *Entrepreneurial performance in role perspective*. Abhinav Publications.

Sharma, G., & Baoku, L. (2013). Customer satisfaction in Web 2.0 and information technology development. *Information Technology & People*, *26*(4), 347–367. doi:10.1108/ITP-12-2012-0157

Sharp, J. M., Irani, Z., & Desai, S. (1999). Working towards agile manufacturing in the UK industry. *International Journal of Production Economics*, *62*(1-2), 155–169. doi:10.1016/S0925-5273(98)00228-X

Shehab, E., Sharp, M., Supramaniam, L., & Spedding, T. (2004). Enterprise resource planning: An integrative review. *Business Process Management Journal*, *10*(4), 359–386. doi:10.1108/14637150410548056

Shih, W., & Allen, M. (2007). Working with Generation-D: Adopting and adapting to cultural learning and change. *Library Management*, *28*(1/2), 89–100. doi:10.1108/01435120710723572

Shuffler, M., Wiese, C., Salas, E., & Burke, S. (2010). Leading one another across time and space: Exploring shared leadership functions in virtual teams. *Revista de Psicología del Trabajo y de las Organizaciones*, *26*(1), 3–17. doi:10.5093/tr2010v26n1a1

Siebdrat, F., Hoegl, M., & Ernst, H. (2009). *How to manage virtual teams*. Retrieved from http://sloanreview.mit.edu/article/how-to-manage-virtual-teams/

Singhal, A. (2001). Modern information retrieval: A brief overview. *IEEE Data Eng. Bull.*, *24*(4), 35–43.

Singh, J. (1979). *Operations Research*. Middlesex, UK: Penguin Books.

Sivunen, A. (2006). Strengthening identification with the team in virtual teams: The leaders' perspective. *Group Decision and Negotiation*, *15*(4), 345–366. doi:10.1007/s10726-006-9046-6

Smith, C. (2013). *Working in a Virtual Team Using Technology to Communicate Collaborate*. Retrieved February 28, 2015 from www.mindtools.com/pages/article/working-virtual-team.htm

Smith, H., McKeen, J., & Singh, S. (2007). Tacit Knowledge transfer: Making it Happen. *Journal of Information Science and Technology*, *4*(2), 23–44.

Snyderman, N. (2009). *Article*. Retrieved from http://www.huffingtonpost.com/living/the-blog/featured-posts/2009/11/11/

Society for Human Resource Management. (2012). *Virtual teams*. Retrieved from http://www.shrm.org/research/survey-findings/articles/pages/virtualteams.aspx

Songini, M. L. (2002). J.D. Edwards pushes CRM, ERP integration. *Computerworld, 36*(25), 4.

Sosik, J., Avolio, B., Kahai, S., & Jung, D. I. (1998). Computer-supported work group potency and effectiveness: The role of transformational leadership, anonymity, and task interdependence. *Computers in Human Behavior, 14*(3), 491–511. doi:10.1016/S0747-5632(98)00019-3

Specialist, H. R. (2011). *9 things employees want from their managers (and 5 things they don't).* Retrieved February 1, 2015 from http://www.thehrspecialist.com/print.aspx?id=32033

Sprague, D., & Greenwell, R. (1992). Project management: Are employees trained to work in project teams? *Project Management Journal, 23*(1), 22–26.

Sproull, L., & Keisler, S. (1986). Reducing social context cues: Electronic mail in organizational communication. *Management Science, 32*(11), 1492–1512. doi:10.1287/mnsc.32.11.1492

Stahl, G., & Hesse, F. (2007). Welcome to the future. *International Journal of Computer-Supported Collaborative Learning, 2*(1). Retrieved from http://ijcscl.org/?go=contents&article=24

Stahl, H. (1989). On the convergence of generalized Padé approximants. *Constructive Approximation, 5*(1), 221–240. doi:10.1007/BF01889608

Stalk, G., Evans, P., & Shulman, L. E. (1992). Competing on capabilities: The new rules of corporate strategy. *Harvard Business Review*, (March-April), 57–69. PMID:10117369

Stansfield, M. H. (1997), *The Effect Of Computer Based Technology In Attempting To Enhance A Subjective Method Of Knowledge Elicitation.* (Thesis) University of Paisley.

Staples, D. D., & Zhao, L. (2006). The effects of cultural diversity in virtual teams versus face-to-face teams. *Group Decision and Negotiation, 15*(4), 389–406. doi:10.1007/s10726-006-9042-x

Starbird, K. (2011). Digital volunteerism during disaster: Crowdsourcing information processing. In *Conference on Human Factors in Computing Systems.*

Starke-Meyerring, D., & Andrews, D. (2006). Building a shared virtual learning culture. *Business Communication Quarterly, 69*(1), 25–49. doi:10.1177/1080569905285543

Starvish, M. (2012). *Why leaders need to rethink teamwork.* Retrieved from http://www.forbes.com/sites/hbsworking-knowledge/2012/12/28/why-leaders-need-to-rethink-teamwork/

Statista. (2010). *Number of global social network users 2010-2018.* Retrieved February 28, 2015 from: http://www.statista.com/statistics/278414/number-of-worldwide-social-network-users/

Stepper, J. (2013). *The best office design for collaboration is also the cheapest.* Retrieved from http://johnstepper.com/2013/02/23/the-best-office-design-for-collaboration-is-also-the-cheapest/

Stevens, C. P. (2003). Enterprise resource planning: A trio of resources. *Information Systems Management, 20*(3), 61–71. doi:10.1201/1078/43205.20.3.20030601/43074.7

Stoel, M. D., & Muhanna, W. A. (2009). IT capabilities and firm performance: A contingency analysis of the role of industry and IT capability type. *Information & Management, 46*(4), 181–189. doi:10.1016/j.im.2008.10.002

Stol, K., & Fitzgerald, B. (2014). Two's Company, Three's a Crowd: A Case Study of Crowdsourcing Software Development. In *Proceedings of ICSE 2014.* Hyderabad, India: IEEE. doi:10.1145/2568225.2568249

Storey, M.-A., Singer, L., Cleary, B., Figueira Filho, F., & Zagalsky, A. (2014). The (R) Evolution of social media in software engineering. In *Proceedings of the Future of Software Engineering - FOSE 2014*. Hyderabad, India: IEEE.

Stough, S., Eom, S., & Buckenmyer, J. (2000). Virtual teaming: A strategy for moving your organization into the new millennium. *Industrial Management & Data Systems*, *100*(8), 370–378. doi:10.1108/02635570010353857

Stowell, F. A. (1989). *Change, organizational power and the metaphor 'commodity'*. (Unpublished PhD Thesis). Department of Systems, University of Lancaster.

Stowell, F. A. (2012). The Appreciative Inquiry Method – A Suitable Framework for Action Research? *Systems Research and Behavioral Science*, *30*(1), 15–30. doi:10.1002/sres.2117

Stowell, F. A. (2014). Organizational Power and the Metaphor Commodity. *International Journal of Systems and Society*, *1*(1), 12–20. doi:10.4018/ijss.2014010102

Stowell, F. A., & Welch, C. (2012). *The Managers Guide to Systems Practice, Making Sense of Complex Problems*. Chichester: Wiley. doi:10.1002/9781119208327

Strauss, A. (1987). *Qualitative Analysis for Social Scientists*. Cambridge: Cambridge University Press. doi:10.1017/CBO9780511557842

Strauss, A., & Corbin, J. (1990). *Basics of qualitative research: Grounded theory procedures and techniques*. Newbury Park, CA: Sage.

Sulonen & Alho. (2000). Supporting virtual software project on the web. *Computer in Industry*.

Susman, G., & Evered, R. (1978). An Assessment of the merits of scientific action research. *Administrative Science Quarterly*, *23*(December), 583–603.

Sutton, S. G. (2006). Extended-enterprise systems' impact on enterprise risk management. *Journal of Enterprise Information Management*, *19*(1), 97–114. doi:10.1108/17410390610636904

Svendsen, A. S. (1981). *The role of the entrepreneur in the shipping industry*. Academic Press.

Symons, J., & Stenzel, C. (2007). Virtually borderless: An examination of culture in virtual teaming. *Journal of General Management*, *32*(3), 1–17.

Szulanski, G. (2000). The process of knowledge transfer: A diachronic analysis of stickiness. *Organizational Behavior and Human Decision Processes*, *82*(1), 9–27. doi:10.1006/obhd.2000.2884

Talley, W. (2013). *Maritime safety, security and piracy*. CRC Press.

Talley, W. K., & Ng, M. W. (2013). Maritime transport chain choice by carriers, ports and shippers. *International Journal of Production Economics*, *142*(2), 311–316. doi:10.1016/j.ijpe.2012.11.013

Tapscott, D., Ticoll, D., & Lowy, A. (2000). *Digital capital*. Boston, MA: Harvard Business School Press.

Tarling, N. (2013). *Status and Security in Southeast Asian States*. Routledge.

Tastoglou & Milious. (2005). Virtual culture: Work and play on the internet. *Social Computer Review*.

Taylor, P. J. (2004). *World City Network: A Global Urban Analysis*. London: Routledge.

Tencati, A., & Zsolnai, L. (2009). The collaborative enterprise. *Journal of Business Ethics*, *85*(3), 367–376. doi:10.1007/s10551-008-9775-3

Thai, V. V. (2007). Impacts of Security Improvements on Service Quality in Maritime Transport: An Empirical Study of Vietnam. *Maritime Econ Logistics, 9*(4), 335-356.

The Virtual Leader. (2013). *Using telework to lean out the enterprise.* Retrieved from https://thevirtualleader.wordpress.com/2013/07/02/using-telework-to-lean-out-the-enterprise/

Themistocleous, M., Irani, Z., & O'Keefe, R. (2001). ERP and application integration: Exploratory survey. *Business Process Management Journal, 7*(3), 195–204. doi:10.1108/14637150110392656

TheTop10BestOnlineBackup.com. (2014). *Review of the best cloud storage services.* Retrieved November 20, 2014, from: http://www.thetop10bestonlinebackup.com/cloud-storage

Thomas, D., Bostrom, R., & Gouge, M. (2007). Making knowledge work in virtual teams. *Communications of the ACM, 50*(1), 85–90. doi:10.1145/1297797.1297802

Thomas, K. W., & Kilmann, R. H. (1974). *Thomas-Kilmann Conflict Mode Instrument.* Mountain View, CA: Xicom.

Thompson, J. (2011, September 30). *Is Nonverbal Communication a Numbers Game?* Retrieved June 26, 2015, from http://www.psychologytoday.com/blog/beyond-words/201109/is-nonverbal-communication-numbers-game

Thompson. (2006). *Leading virtual team.* Retrieved from www.qualitydiyst.com/septoo/teams.html

Thompson, L., & Couvert, M. (2003). Teamwork online: The effects of computer conferencing on perceived confusion, satisfaction and post discussion accuracy. *Group Dynamics, 7*(2), 135–151. doi:10.1037/1089-2699.7.2.135

Thun, J. H. (2010). Angles of integration: An empirical analysis of the alignment of internet-based information technology and global supply chain integration. *Journal of Supply Chain Management, 46*(2), 30–44. doi:10.1111/j.1745-493X.2010.03188.x

TopCoder. (n.d.). Retrieved November, 2014 from https://www.topcoder.com

Torbacki, W. (2008). SaaS – direction of technology development in ERP/MRP systems. *Archives of Materials Science and Engineering, 31*(1), 57–60.

Townsend, A. M., DeMarie, S. M., & Hendrickson, A. R. (1998). Virtual teams: Technology and the workplace of the future. *The Academy of Management Executive, 12*(3), 17–29.

Triantafillakis, A., Kanellis, P., & Martakos, D. (2004). Data warehousing interoperability for the extended enterprise. *Journal of Database Management, 15*(3), 73–82. doi:10.4018/jdm.2004070105

Truzzi, M. (1971). *Sociology: The Classic Statements.* New York: Oxford University Press.

Tsai, W. (2002). Social structure of "coopetition" within a multiunit organization: Coordination, competition, and intra-organizational knowledge sharing. *Organization Science, 13*(2), 179–190. doi:10.1287/orsc.13.2.179.536

Tsai, W.-T., Wu, W., & Huhns, M. N. (2014). Cloud-Based Software Crowdsourcing. *Internet Computing, 18*(3), 78–83. doi:10.1109/MIC.2014.46

Tugend, A. (2014, March 7). It's Unclearly Defined, but Telecommuting Is Fast on the Rise. *The New York Times.* Retrieved February 28, 2015 from http://www.nytimes.com/2014/03/08/your-money/when-working-in-your-pajamas-is-more-productive.html

Tuli, A., Hasteer, N., Sharma, M., & Bansal, A. (2014). Empirical Investigation of agile software development: A Cloud Perspective. *Software Engineering Notes, 39*(4), 1–6. doi:10.1145/2632434.2632447

Ulrich. (2002). *Managing virtual web organization in the, 20th century: Issues & challenges.* Ipea Great Press.

UNCTAD. (2014). *Review of Maritime Transport*. New York, Geneva.

University of Phoenix. (2011). *University of Phoenix survey reveals nearly seven-in-ten workers have been part of dysfunctional teams*. Retrieved from http://www.phoenix.edu/news/releases/2013/01/university-of-phoenix-survey-reveals-nearly-seven-in-ten-workers-have-been-part-of-dysfunctional-teams.html

Valacich, J., Dennis, A. R., & Nunamaker, A. F. Jr. (1992). Group size and anonymity effects on computer-mediated idea generation. *Small Group Research*, *23*(1), 49–73. doi:10.1177/1046496492231004

Vallespir, B., & Kleinhans, S. (2001). Positioning a company in enterprise collaborations: Vertical integration and make-or-buy decisions. *Production Planning and Control*, *12*(5), 478–487. doi:10.1080/09537280110042701

Van De Ven, A. H., Delbecq, A., & Koenig, R. (1976). Determinants of coordination modes within organizations. *American Sociological Review*, *41*(2), 322–338. doi:10.2307/2094477

Van den Bulte, C., & Moenaert, R. (1998). The effects of R&D team co-location on communication patterns among R&D, marketing, and manufacturing. *Management Science*, *44*(11), 1–18. doi:10.1287/mnsc.44.11.S1

Vathanophas, V. (2007). Business process approach towards an inter-organizational enterprise system. *Business Process Management Journal*, *13*(3), 433–450. doi:10.1108/14637150710752335

Vazquez-Bustelo, D., & Avella, L. (2006). Agile manufacturing: Industrial case studies in spain. *Technovation*, *26*(10), 1147–1161. doi:10.1016/j.technovation.2005.11.006

Velasco, J. (2013, February 7). *Think Big* [Web log post]. Retrieved from http://blogthinkbig.com/crowdsourcing-colaboracion-motor-ideas

Venkatesh, V., Hoehle, H., & Aljafari, R. (2014). A usability evaluation of the Obamacare website. *Government Information Quarterly*, *31*(4), 669–680.

Vickers, G. (1983). *The Art of Judgement*. London: Harper and Rowe.

Virtual College. (2014). *Most global firms to use gamification in e-learning by 2014*. Retrieved from http://www.virtual-college.co.uk/news/Most-global-firms-to-use-gamification-in-elearning-by-2014-newsitems-801528054.aspx

Vukovic, M. (2009). Crowdsourcing for Enterprises Maja Vukovi. In *Congress on Services-I*. IEEE. doi:10.1109/SERVICES-I.2009.56

Wadsworth, M. B., & Blanchard, A. L. (2015). Influence tactics in virtual teams. *Computers in Human Behavior*, *44*, 386–393. doi:10.1016/j.chb.2014.11.026

Wakefield, R. L., Leidner, D. E., & Garrison, G. (2008). A model of conflict, leadership, and performance in virtual teams. *Information Systems Research*, *19*(4), 434–455. doi:10.1287/isre.1070.0149

Walker, A., & Shuangye, C. (2007). Leader authenticity in intercultural school contexts. *Educational Management Administration & Leadership, 35*(2), 185-204.

Waller, M., Conte, J., Gibson, C., & Carpenter, M. (2001). The effect of individual perceptions of deadlines on team performance. *Academy of Management Review*, *26*(4), 586–600.

Walters, D. (2004). New economy – new business models – new approaches. *International Journal of Physical Distribution & Logistics Management*, *34*(3/4), 219–229. doi:10.1108/09600030410533556

Walumbwa, F. O., Orwa, B., Wang, P., & Lawler, J. J. (2005). Transformational leadership, organizational commitment, and job satisfaction: A comparative study of Kenyan and US financial firms. *Human Resource Development Quarterly, 16*(2), 235–256. doi:10.1002/hrdq.1135

Wan, Y., & Clegg, B. T. (2010). Enterprise management and ERP development: Case study of Zoomlion using Dynamic Enterprise Reference Grid. In *CENTERIS 2010 – Conference on ENTERprise Information Systems* (pp. 191-198). Springer Verlag.

Wang, G., Oh, I. S., Courtright, S. H., & Colbert, E. (2011). Transformational leadership and performance across criteria and levels: A meta-analytic review of 25 years of research. *Group & Organization Management, 36*(2), 223–270. doi:10.1177/1059601111401017

Watkins, M. (2013). *Making virtual teams work: Ten basic principles*. Retrieved from https://hbr.org/2013/06/making-virtual-teams-work-ten/

Watkins, K., & Marsick, V. (1993). *Sculpting the learning organization*. San Francisco, CA: Jossey-Bass.

Watkins, M. (2003). *The first 90 days: Critical success strategies for new leaders at all levels*. Boston: Harvard Business School Press.

Watson-Manheim, M. B., Chudoba, K. M., & Crowston, K. (2002). Discontinuities and continuities: A new way to understand virtual work. *Information Technology & People, 15*(3), 191–209. doi:10.1108/09593840210444746

Watson-Manheim, M., & Belanger, F. (2002). Support for communication-based work processes in virtual work. *e-Service Journal, 1*(3), 61–82. doi:10.2979/ESJ.2002.1.3.61

West & Stansfield. (1999). Systems maps for interpretive inquiry: Some comments and experiences. *Computing and Information Systems, 6*, 64–82.

West & Thomas, L. (2005). Looking for the Bigger Picture: An Application of the Appreciative Inquiry Method in RCUS. In *Conference proceedings 'Information Systems Unplugged'*. Northumbria University.

West, D. (1995). The Appreciative Inquiry Method: A Systemic Approach To information Systems Requirements Analysis. In F. A. Stowell (Ed.), *Information systems Provision: The Contribution of Soft Systems Methodology* (pp. 140–158). Maidenhead, UK: McGraw-Hill.

Wight, O. (1984). *Manufacturing Resource Planning: MRPII. Williston: Oliver Wight Ltd*. Publications.

Wikinomics. (2014, August 31). In *Wikipedia, the free encyclopedia* [Electronic version]. Retrieved from http://es.wikipedia.org/wiki/Wikinom%C3%ADa

Wikipedia. (2015). *Serious Games*. Retrieved from wikipedia.org/wiki/Serious_game#Definition_and_scope

Wild, J., Wild, K. L., & Han, J. C. Y. (2014). *International business*. Pearson Education Limited.

Wilkes, L., & Veryard, R. (2004, April). Service-oriented architecture: Considerations for agile systems. *Microsoft Architect Journal*. Retrieved May 16, 2010, from www.msdn2.microsoft.com

Williams, M. D., & Williams, J. (2007). A change management approach to evaluating ICT investment initiatives. *Journal of Enterprise Information Management, 20*(1), 32–50. doi:10.1108/17410390710717129

Wilson, J. M., Boyer O'Leary, M., Metiu, A., & Jett, Q. R. (2008). Perceived proximity in virtual work: Explaining the paradox of far-but-close. *Organization Studies, 29*(7), 979–1002. doi:10.1177/0170840607083105

Wisdom of Crowds. (2014, August 5). In *Wikipedia, the free encyclopedia* [Electronic version]. Retrieved from http://es.wikipedia.org/wiki/Sabidur%C3%ADa_de_los_grupos

Wisdom of the Crowd. (2015, January 3). In *Wikipedia, The Free Encyclopedia* [Electronic version]. Retrieved from http://en.wikipedia.org/wiki/Wisdom_of_the_crowd

Wong, W. Y. C. (2012). *Cognitive metaphor in the West and the East: A comparison of metaphors in the speeches of Barack Obama and Wen Jiabao*. Academic Press.

Wood, & Associates. (2003). *Why emotional intelligence matters on the job*. Retrieved from http://www.woodassociates.net/toi/Search/PDF/Why%20Emotional%20Intelligence%20Matters%20on%20the%20Job.pdf

Wood, B. (2010). *ERP vs ERPII vs ERPIII future enterprise applications*. Retrieved October 3, 2010, from www.r3now.com/erp-vs-erp-ii-vs-erp-iii-future-enterprise-applications

Wu, W., Tsai, W., & Li, W. (2013). Creative software crowdsourcing: From components and algorithm development to project concept formations. *International Journal of Creative Computing*, *1*(1), 57–91. doi:10.1504/IJCRC.2013.056925

Xue, L., Ray, G., & Sambamurthy, V. (2012). Efficiency or innovation: How do industry environments moderate the effects of firms' IT assets portfolios. *Management Information Systems Quarterly*, *36*(2), 509–528.

Xu, W., Wei, Y., & Fan, Y. (2002). Virtual enterprise and its intelligence management. *Computers & Industrial Engineering*, *42*(2-4), 199–205. doi:10.1016/S0360-8352(02)00053-0

Yang, H. A. (1995). Efficient calculation of cell loss in ATM multiplexers. In *Proc. GLOBECOM*. doi:10.1109/GLOCOM.1995.502598

Yang, K. H., Lee, S. M., & Lee, S. (2007). Adoption of information and communication technology. *Industrial Management & Data Systems*, *107*(9), 1257–1275. doi:10.1108/02635570710833956

Ye, F., & Wang, Z. (2013). Effects of information technology alignment and information sharing on supply chain operational performance. *Computers & Industrial Engineering*, *65*(3), 370–377. doi:10.1016/j.cie.2013.03.012

Yeh, C. H., Lee, G. G., & Pai, J. C. (2012). How information system capability affects e-business information technology strategy implementation: An empirical study in Taiwan. *Business Process Management Journal*, *18*(2), 197–218. doi:10.1108/14637151211225171

Yen, J., Fan, X., Sun, S., Want, R., Chen, C., Kamali, K., & Volz, R. A. (2003). *Implementing Shared Mental Models for Collaborative Teamwork*. Retrieved from http://agentlab.psu.edu/lab/publications/cast_iat03wkshp.pdf

Yılmaz, F. G. K., Yılmaz, R., Ozturk, H. T., Sezer, B., & Karademir, T. (2015). Cyberloafing as a barrier to the successful integration of information and communication technologies into teaching and learning environments. *Computers in Human Behavior*, *45*, 290–298. doi:10.1016/j.chb.2014.12.023

Yoo, Y., & Kanawattanachai, P. (2001). Developments of transactive memory systems and collective mind in virtual teams. *The International Journal of Organizational Analysis*, *9*(2), 187–208. doi:10.1108/eb028933

Yukl, G. (2012). Effective leadership behavior: What we know and what questions need more attention. *The Academy of Management Perspectives*, *26*(November), 66–85. doi:10.5465/amp.2012.0088

Yukl, G., & Tracey, J. B. (1992). Consequences of influence tactics used with subordinates, peers, and the boss. *The Journal of Applied Psychology*, *77*(4), 525–535. doi:10.1037/0021-9010.77.4.525

Yu, P. (2003). Virtual instrument parameter calibration with particle swarm optimization. *IEEE Swarm Intelligence Symposium*.

Zaccaro, S. J., & Burke, C. (1998). *Team versus crew leadership: Differences and similarities.* Academic Press.

Zaccaro, S. J., & Banks, D. J. (2001). Leadership, vision, and organizational effectiveness. In *The nature of organizational leadership: Understanding the performance imperatives confronting today's leaders* (pp. 181–218). Uratko.

Zafiropoulos, C., Vrana, V., & Paschaloudis, D. (2006). Research in brief the Internet practices analysis from Greece. *International Journal of Contemporary Hospitality Management, 18*(2), 156–163. doi:10.1108/09596110610646709

Zander, L., Zettinig, P., & Makela, K. (2013). Leading global virtual teams to success. *Organizational Dynamics, 42*(3), 228–237. doi:10.1016/j.orgdyn.2013.06.008

Zenun, M. M. N., Loureiro, G., & Araujo, C. S. (2007). The Effects of Teams' Colocation on Project Performance. In G. Loureiso & R. Curran (Eds.), *Complex Systems Concurrent Engineering Collaboration, Technology Innovation and ustainability.* London: Springer. doi:10.1007/978-1-84628-976-7_79

Zhu, Y., & Li, H. (1993). The Mac Laurin expansion for a GI/G/1 queue with Markov-modulated arrivals and services. *Queueing Systems, 14*(1-2), 125–134. doi:10.1007/BF01153530

Zibetti, E., Chevalier, A., & Eyraud, R. (2012). What type of information displayed on digital scheduling software facilitates reflective planning tasks for students? Contributions to the design of a school task management tool. *Computers in Human Behavior, 28*(2), 591–607. doi:10.1016/j.chb.2011.11.005

Zigaras, I. (2003). leadership in virtual team:oxymoron or opportunity. *Organizational Dynamics, 31*(4).

Zigurs, I. (2003). Leadership in virtual teams: Oxymoron or opportunity? *Organizational Dynamics, 31*(4), 339–351. doi:10.1016/S0090-2616(02)00132-8

Zigurs, I., Poole, M. S., & DeSanctis, G. (1988). A study of influence in computer-mediated group decision making. *Management Information Systems Quarterly, 12*(December), 625–644. doi:10.2307/249136

Zineldin, M. (2004). Co-opetition: The organization of the future. *Marketing Intelligence & Planning, 22*(7), 780–789. doi:10.1108/02634500410568600

Zivick, J. (2012). Mapping global virtual team leadership actions to organizational roles. *Business Review (Federal Reserve Bank of Philadelphia), 19*(2), 18–25.

Zrimsek, B. (2003). *ERPII vision.* Paper presented at US Symposium/ITxpo, Gartner Research (25C, SPG5, 3/03), San Diego, CA.

Zurawski, R., & Zhon, M. C. (1994). Petri nets and industrial applications: A tutorial. *IEEE Transactions on Industrial Electronics, 41*(6), 567–583. doi:10.1109/41.334574

About the Contributors

Christian Graham, prior to coming to UMaine, was the Chair of the Business Administration and Computer Technology programs at Andover College located in Portland and Lewiston ME. He also taught web design and computer application courses at the University of Southern Maine's Lewiston and Auburn campus and for Central Maine Community College. Prior to his teaching career, he worked in Insurance, business to business credit analysis, Internet technology, and owned his own business. The business was a Bodega which focused the sales of Spanish and Jamaican groceries. When not teaching or studying, he enjoys collecting comic books. He is the proud owner of a # 1 first edition X-Men comic book published in 1963. Christian has a Ph.D. (Information Systems) Nova Southeastern University, Ms.Ed. (Business Education) Southern NH University, BS (Business Administration) Husson University. He is the recipient of UMaine's 2011 Faculty Technology Stipend for developing an in-class SMS text messaging system to increase student engagement in large classes. His research interests include information security, computer privacy, online learning environments, open-source web technologies, and workplace e-learning. In 2011, he was the program coordinator for International Educator Group's conference titled: Education in 2025.

* * *

Surendran Balachandran is a senior digital campaigner with over 5 years of experience in online advocacy and campaign design. He writes about Technology & Social Change, Privacy, Public/Open Data, and Data Visualization.

Abhay Bansal is a Head and Professor at Department of Computer Science & Engineering, Amity School of Engineering & Technology and Director of DICET, Amity University Uttar Pradesh, Noida, India. He holds a PhD degree in the area of data mining. His research interest includes Web Technologies and Software Design and Development. He has published more than 50 papers in various journals and conferences of repute.

Shavindrie Cooray is an Assistant Professor in Management Information Systems at Curry College, Boston. Dr. Cooray has presented at international conferences and published in the areas of soft systems thinking and participatory design. Her present research focuses on navigating the gap in understanding between non-technical end users and technical developers using soft systems thinking methods. Dr. Cooray holds a PhD in MIS and a MSc in Internet Systems Development from the University of

Portsmouth, and a BSc (Hons) from the Manchester Metropolitan University. Prior to joining Curry, she worked as a Lecturer at Boston University and University of Massachusetts, Boston and as an Analyst in the corporate world.

Mamta Dalal works in the IT industry and is an enthusiast of emerging technology trends. Beyond a passion for supporting social causes, she loves writing, and her works have been published in various national and international venues.

William Dario Avila Diaz, Postdoctoral in Economy, Society and the Construction of Knowledge in the Modern World of the National University of Cordoba, Argentina. Postdoctoral in Communication, Education and Culture of the University Santo Tomas of Bogota, Colombia and National University of Cordoba, Argentina. PhD in Business Administration of the Newport International University, CA, United States. Magister in Systems Engineering and Computing of the University of the Andes, Bogotá, Colombia. Specialist in Administrative Management of the Central University of Colombia. Systems Engineering and distinguished graduate of the Catholic University of Colombia.

Julia Eisenberg (PhD, Rutgers University) is an Assistant Professor in the Department of Management and Management Science at Pace University's Lubin School of Business. Her research interests include leadership, innovation, and collaborative processes among geographically distributed colleagues. Prior to joining academia, she worked for a decade in the financial industry with roles spanning product development and management. Her work has been published in the Journal of Computer-Mediated Communication and the American Behavioral Scientist.

Niclas Erhardt (PhD from Rutgers University) is an Associate Professor in Management at Maine Business School and teaches undergraduate and graduate courses in Human Resource Management and Organizational Behavior. He has a wide research interest at multiple levels, including team-based knowledge work, creativity, virtual teamwork, workplace diversity, human resource management, and organizational design and culture. His work has been published in journals including Journal of Management, Management Learning, Human Resource Management Review, and Corporate Governance: An International Review.

Howard Bennett Esbin is the creator of Prelude, which is informed by his professional and academic background. He has three decades of senior management experience in the private sector, international development, and philanthropy. This includes merchandising a chain of 300 retail outlets, training members of a Kenyan carvers cooperative, and managing a multimillion-dollar fundraising non-profit. He has been an advisor to organizations such as the Ottawa Community Foundation and the National Life Work Centre Real Games Program. He also served on the board of Oxfam Canada as the managing director of its fair trade organization. The International Labour Organization, Education Canada, and UNESCO have published his work. This includes: 'The Challenge Of Vocational Training, Guidance Counsellng, and Basic Education For The 21st Century' and 'Lifespan Career Education: Getting Serious Play'. Howard received his doctoral degree in education in second languages from McGill University in 1998. His thesis "Carving Lives From Stone: Visual Literacy In An African Cottage Industry" has been described as "an excellent, sophisticated, very creative piece of work and a significant pedagogical contribution to educational research."

Jennifer L. Gibbs (Ph.D., University of Southern California) is an Associate Professor of Communication and Director of the Masters of Communication and Information Studies (MCIS) program at Rutgers University. Her research focuses on collaboration in global virtual teams and other distributed work arrangements as well as affordances of new communication technologies for strategic communication processes. Her work has been published in leading journals from a variety of disciplines including Administrative Science Quarterly, American Behavioral Scientist, Communication Research, Communication Yearbook, Computers in Human Behavior, Human Relations, The Information Society, Journal of Computer-Mediated Communication, Journal of Social & Personal Relationships, and Organization Science.

Nitasha Hasteer is a faculty at Amity School of Engineering & Technology, Amity University Uttar Pradesh, Noida, India. She holds a graduation and masters in engineering and is presently heading the Department of Information Technology at the School of Engineering. Her research interest includes Software Engineering, Software Process Modelling and Project Management. She has fifteen years of experience in industry and academics. She is a member of IEEE, ACM, IET(UK), IACSIT and Computer Society of India along with other professional bodies. She has presented and published many papers at National and International Conferences.

Nory Jones received her PhD from the University of Missouri in 2001 with a focus on knowledge management and diffusion of innovations. She is a Professor at the University of Maine Business School where she teaches classes in business information systems, e-business, knowledge management, and marketing. Her research focuses on tacit knowledge transfer as well as distance technologies and e-business.

Tarek Taha Mohamed Kandil, PhD Banking and Finance, the University of Plymouth – Plymouth- United Kingdom, is an assistant Professor of global finance and banking in multi-cultural higher educational institution in England, Egypt and recently in the United Arab Emirate–Dubai. He has multi disciplines of researches in different fields of international banking, Islamic Multi Cultural methods of finance, cross culture leadership styles and virtual teamwork of multi-national organisations of shipping and Maritime industries in United Arab Emirates UAE. In 2011 the Author of A Tale of Two Different Cultures & Financial Performance Relationships - The Art of Cross-border Mergers and Acquisitions in the Global and 2014 published a book in Emerald about the Islamic Cross-border Merger and Acquisitions and Global Financial Crisis.

Kijpokin Kasemsap received his BEng degree in Mechanical Engineering from King Mongkut's University of Technology Thonburi, his MBA degree from Ramkhamhaeng University, and his DBA degree in Human Resource Management from Suan Sunandha Rajabhat University. He is a Special Lecturer at Faculty of Management Sciences, Suan Sunandha Rajabhat University based in Bangkok, Thailand. He is a Member of International Association of Engineers (IAENG), International Association of Engineers and Scientists (IAEST), International Economics Development and Research Center (IEDRC), International Association of Computer Science and Information Technology (IACSIT), International Foundation for Research and Development (IFRD), and International Innovative Scientific and Research Organization (IISRO). He also serves on the International Advisory Committee (IAC) for International

Association of Academicians and Researchers (INAAR). He has numerous original research articles in top international journals, conference proceedings, and book chapters on business management, human resource management, and knowledge management published internationally.

Raheel Khursheed leads News, Politics & Government verticals in India at Twitter Inc., with over 6 years of experience as a cross platform journalist & storyteller for outlets like CNN-IBN, and BBC-PRI. He passionately supports crisis responses.

Vidya Krishnan is a Delhi based Health journalist. She specializes in mapping Indian government's public health policy, and is also the International Reporting Project Fellow for 2015. She actively supports volunteering initiatives.

Harsh Kushwah currently works as a consultant in a non-profit organization, while employing his skills from Masters in Communication and Computer Engineering. He is passionate about volunteering for social causes.

Cathrine Linnes is an Associate Professor in Information Systems in the College of Business at Hawaii Pacific University where she teach in the Master of Science in Information Systems program. Her focus in in software engineering and business analytics. She received her Ph.D. from Nova Southeastern University.

Vaibhav Madhok is a quantum physicist based in Vancouver. He studies computer science, physics and evolutionary biology.

Amir Manzoor holds a bachelor's degree in engineering from NED University, Karachi, an MBA from Lahore University of Management Sciences (LUMS), and an MBA from Bangor University, United Kingdom. He has many years of diverse professional and teaching experience working at many renowned national and internal organizations and higher education institutions. His research interests include electronic commerce and technology applications in business. He is a member of Chartered Banker Institute of UK and Project Management Institute, USA.

Vijaya Moorthy is a professional in the socio-political space with expertise in policy, political engagement campaigns and development projects. Volunteering and coordinating crisis response on social media during disasters has been her parallel pursuit for over a decade. She holds a management degree and has professionally associated with corporates, non-profits and political organizations.

B. K. Murthy is an Executive Director of Centre for Development of Advanced Computing at Noida, India and was a senior director at Department of Electronics and Information Technology (DeitY), Ministry of Communications & IT, Government of India. He is responsible for the National Knowledge Network, Human Resources Development, and Productivity Enhancement & Employment Generation Divisions of the department. He holds a Ph.D. degree from IIT Delhi. His research interests include Artificial Intelligence, Natural Language Processing, Knowledge Engineering, Object Oriented Design, Development and Semantic Web. He has published and presented more than 40 papers in various journals and conferences.

Shereen Nassar is an Assistant Professor in Operations and Supply Chain Management. Dr. Nassar has had her doctorate degree in Supply Chain and Logistics Management in 2012 from School of Management (a top-ranking UK management school with a world prominent reputation for management studies) at Bath University, which is one of the leading and prestigious UK universities. Dr. Nassar was invited for an academic visit for six months at the Logistics Research Center, School of Management and Languages at Heriot-Watt University. During her visit, she collaborated in research and teaching activities with her colleagues in the area of Operations and Supply Chain Visibility. For nearly two years after completing her PhD, she worked at one of a highly regarded Egyptian academic institution in Managerial Sciences as an Assistant Professor in Operations Management. During these two years, she worked closely with the top management team through her participation as a member of the Learning and Teaching Development Committee. Ten years before she started her PhD, she got her MBA degree in 2001. She worked as an academic instructor in Operations Management. Dr. Shereen Nassar main research interest is Supply Chain visibility driven by the advancement of tracking and tracing technology specifically RFID. She has got a number of research papers that have been presented at European Operations Management Associations Conference (which is the leading Operations Management conference in Europe) and other international academic events. Her current research interest includes risk management in association with the implementation of innovative digital supply chain applications. She has extended her research interest to investigate how responsible supply chain can minimize product recalls. Dr. Shereen Nassar collaborates in her research with a number of leading UK scholars in the field of Operations and Supply Chain Management.

Bhavana Nissima is a communication consultant, writer, and long-term social activist. She has served as a volunteer in multiple disasters internationally, both on ground as well as on social media.

Hemant Purohit is an assistant professor of Information Sciences and Technology at George Mason University, with Kno.e.sis Center as alma mater. His passion of computing for Social Good has led him to research on improving crisis coordination.

Aashish Rajgaria is a chartered accountant professional and actively supports volunteering efforts for crises.

Navin Rustagi is a Big Data Scientist, working on Genomic Big Data and interested in emerging technologies.

Andrew Seely has over 20 years of experience delivering solutions in the Information Technology domain, spanning systems administration, software development, entrepreneurship, education, professional development, and leadership in commercial, government, and military settings in the United States, Asia, and Europe. Andrew is currently the Engineering Division Manager and Chief Engineer for Science Applications International Corporation (SAIC)'s Enterprise Support Contract in Tampa, Florida. Previous to working with SAIC, Andrew worked in systems, engineering, and leadership roles on large defense contracts for Computer Sciences Corporation, Northrop Grumman, and General Dynamics, as well as "dot-comming" as a self-employed contractor and for San Mateo, California start-up There.com. Andy served in the U.S. Air Force for seven years as a computer systems operator at duty stations in Mississippi, Korea, and Germany. He teaches Computer and Information Science courses for the distance education program at the University of Maryland University College, where he has taught programming and operating systems courses for over a decade, and he is an occasional visiting professor for the Systems Administration graduate program in the Technology, Art, and Design department at Oslo and Akershus University College of Applied Sciences in Oslo, Norway. He currently teaches Information Systems courses for the Information and Technology Management department at the University of Tampa in Tampa, Florida. Andrew is a regular columnist for the USENIX Association publication ";login:", where he writes about managing technical teams. He was a Program Committee member for the 2012 Large Installation System Administration (LISA) conference, and he has been the annual chair of LISA's Government and Military systems administration workshop since 2008. He was the chairman for the Global Command and Control Systems Administration and Engineering international conference in 2005 and 2010. Andrew holds a bachelor's degree in Computer and Information Science from the University of Maryland University College in Heidelberg, Germany, and a master's degree in Computer Science from Nova Southeastern University in Ft. Lauderdale, Florida.

Parminder Singh is the managing director of Southeast Asia, India and MENA region at Twitter Inc. with over 14 years of cross-functional experience in strategy and brand marketing. He actively supports social causes, and philanthropy.

Frank Stowell is Emeritus Professor of Systems and Information Systems at the University of Portsmouth. He has a PhD in Organizational Change and his research centers around methods of participative design. He has supervised a number of research projects from modeling complex decision-making in mental health care, knowledge management, through to methods for client-led information systems development. He has been co-chair of a number of research council funded projects notably the Systems Practice for Managing Complexity project, designed to help managers address complex issues, which has developed into a self sustaining network. His latest publication The Managers Guide to Systems Practice (2012, Wiley Chichester) is a text written expressly with the kind of managers in mind who have attended the workshops over the past decade. He is past President of the UK Academy of Information Systems and the UK Systems Society (http://www.ukss.org.uk/). He presently occupies the chair of the Council of Information Systems Professors and has recently joined the Board of the World Organization of Systems and Cybernetics. He has published papers and texts in the field and presented papers at a number of international conferences in Europe and the United States. Prior to his academic career he was employed by central government as a consultant within the Management Systems Development Group and has experience of defining and developing IT supported management information systems.

Lisa Toler is a Project Management Professional and Manager at Brookhaven National Laboratory. She currently leads the Nonproliferation Policy and Safeguards Implementation Team. Dr. Toler has been working at Brookhaven Lab for 31 years. Her research interests and Ph.D. are in organization and management. Dr. Toler also teaches Organizational Behavior at the graduate level and Organizational Change at the undergraduate level as an adjunct Associate Professor at Ashford University, Forbes School of Business. In addition, Dr. Toler has several publications including an article on virtual teams that is located at the Project Management Institute's Virtual Library, and has a featured story in "Their Journey to the PhD: Stories of Personal Perseverance and Academic Achievement", published by Hawthorne Press. Finally, Dr. Toler has several publications pending in addition to this book chapter including one on the effects of workplace bullying on employee performance.

Arun Vemuri heads Brand Measurement Commercialization for Google in APAC, with over two decades of experience in brand management, and analytics. He supports multiple causes owing to his passion about social responsibility and crisis response.

Yi Wan is a doctoral researcher at Aston Business School. He has a Bachelor from Birmingham City University and a Masters from Warwick Business School. His research, governing multi-organizational collaboration using 3rd generation enterprise resource planning systems, focuses on information systems strategy and enterprises integration perspectives. He specializes in the areas of enterprise resource planning development and management, inter-company management, supply chain management, strategic operations management practice, and grounded theory research; and has published several papers in the international referred journals, international conferences and books.

Index

Printed in the United States
By Bookmasters